Advances in Neurobiology

Volume 13

Series Editor
Arne Schousboe

More information about this series at http://www.springer.com/series/8787

Arne Schousboe • Ursula Sonnewald
Editors

The Glutamate/ GABA-Glutamine Cycle

Amino Acid Neurotransmitter Homeostasis

Springer

Editors
Arne Schousboe
Department of Drug Design
and Pharmacology
Faculty of Health and Medical Sciences
University of Copenhagen
Copenhagen, Denmark

Ursula Sonnewald
Department of Neuroscience
Norwegian University of Science
and Technology (NTNU)
Trondheim, Norway

Department of Drug Design
and Pharmacology
Faculty of Health and Medical Sciences
University of Copenhagen
Copenhagen, Denmark

ISSN 2190-5215 ISSN 2190-5223 (electronic)
Advances in Neurobiology
ISBN 978-3-319-45094-0 ISBN 978-3-319-45096-4 (eBook)
DOI 10.1007/978-3-319-45096-4

Library of Congress Control Number: 2016956107

Printed on acid-free paper

This Springer imprint is published by Springer Nature
The registered company is Springer International Publishing AG
The registered company address is: Gewerbestrasse 11, 6330 Cham, Switzerland

Contents

Contributors

Jan Albrecht Department of Neurotoxicology, Mossakowski Medical Research Centre, Polish Academy of Sciences, Warsaw, Poland

Ana I. Amaral Anne McLaren Laboratory, Wellcome Trust Medical Research Council Cambridge Stem Cell Institute and Department of Clinical Neurosciences, University of Cambridge, Cambridge, UK

Michael Aschner Department of Molecular Pharmacology, Albert Einstein College of Medicine, Bronx, NY, USA

Eva Brekke Department of Pediatrics, Nordland Hospital Trust, Bodo, Norway

José A. Campos-Sandoval Departamento de Biología Molecular y Bioquímica, Canceromics Lab, Facultad de Ciencias, Universidad de Málaga, Málaga, Spain

Instituto de Investigación Biomédica de Málaga (IBIMA), Málaga, Spain

M.E. Conway Department of Applied Sciences, University of the West of England, Bristol, UK

Megan Culbreth Department of Molecular Pharmacology, Albert Einstein College of Medicine, Bronx, NY, USA

Niels Christian Danbolt, M.D., Ph.D. Department of Molecular Medicine, Institute for Basic Medical Sciences, University of Oslo, Oslo, Norway

Roni Dhaher, Ph.D. Department of Laboratory Medicine, Yale School of Medicine, New Haven, CT, USA

Tore Eid, M.D., Ph.D. Department of Laboratory Medicine, Yale School of Medicine, New Haven, CT, USA

Gustavo C. Ferreira Department of Pediatrics, University of Maryland School of Medicine, Baltimore, MD, USA

Institute of Medical Biochemistry Leopoldo de Meis, Federal University of Rio de Janeiro, Rio de Janeiro, RJ, Brazil

Shaun E. Gruenbaum, M.D. Department of Anesthesiology, Yale School of Medicine, New Haven, CT, USA

Leif Hertz Laboratory of Brain Metabolic Diseases, Institute of Metabolic Disease Research and Drug Development, China Medical University, Shenyang, China

S.M. Hutson Department of Human Nutrition, Foods and Exercise, Integrated Life Sciences Building (0913), Blacksburg, VA, USA

Arumugam R. Jayakumar Laboratory of Neuropathology, Veterans Affairs Medical Center, Miami, FL, USA

Mark R.N. Kotter Anne McLaren Laboratory, Wellcome Trust Medical Research Council Cambridge Stem Cell Institute and Department of Clinical Neurosciences, University of Cambridge, Cambridge, UK

Tih-Shih W. Lee, M.D., Ph.D. Department of Psychiatry, Yale School of Medicine, New Haven, CT, USA

Renata Leke Department of Clinical Neuroscience and Rehabilitation, Institute of Neuroscience and Physiology Sahlgrenska Academy, University of Gothenburg, Göteborg, Sweden

Javier Márquez Departamento de Biología Molecular y Bioquímica, Canceromics Lab, Facultad de Ciencias, Universidad de Málaga, Málaga, Spain

Instituto de Investigación Biomédica de Málaga (IBIMA), Málaga, Spain

José M. Matés Departamento de Biología Molecular y Bioquímica, Canceromics Lab, Facultad de Ciencias, Universidad de Málaga, Málaga, Spain

Instituto de Investigación Biomédica de Málaga (IBIMA), Málaga, Spain

Mary C. McKenna, Ph.D. Department of Pediatrics, University of Maryland School of Medicine, Baltimore, MD, USA

Program in Neuroscience, University of Maryland School of Medicine, Baltimore, MD, USA

Tora Sund Morken Department of Ophthalmology, Trondheim University Hospital, Trondheim, Norway

Department of Laboratory Medicine, Children's and Women's Health, Norwegian University of Science and Technology (NTNU), Trondheim, Norway

Michael D. Norenberg Laboratory of Neuropathology, Veterans Affairs Medical Center, Miami, FL, USA

Department of Pathology, University of Miami School of Medicine, Miami, FL, USA

Departments of Biochemistry and Molecular Biology, University of Miami School of Medicine, Miami, FL, USA

Vladimir Parpura Department of Neurobiology, Center for Glial Biology in Medicine, Atomic Force Microscopy and Nanotechnology Laboratories, Civitan International Research Center, Evelyn F. McKnight Brain Institute, University of Alabama, Birmingham, AL, USA

Liang Peng Laboratory of Metabolic Brain Diseases, Institute of Metabolic Disease Research and Drug Development, China Medical University, Shenyang, P. R. China

Douglas L. Rothman Magnetic Resonance Research Center, Radiology and Biomedical Engineering, Yale University, New Haven, CT, USA

Arne Schousboe Department of Drug Design and Pharmacology, Faculty of Health and Medical Sciences, University of Copenhagen, Copenhagen, Denmark

Ursula Sonnewald Department of Neuroscience, Norwegian University of Science and Technology (NTNU), Trondheim, Norway

Department of Drug Design and Pharmacology, Faculty of Health and Medical Sciences, University of Copenhagen, Copenhagen, Denmark

Luca Steardo Department of Psychiatry, University of Naples SUN, Largo Madonna delle Grazie, Naples, Italy

Monika Szeliga Department of Neurotoxicology, Mossakowski Medical Research Centre, Polish Academy of Sciences, Warsaw, Poland

Joana M. Tavares Anne McLaren Laboratory, Wellcome Trust Medical Research Council Cambridge Stem Cell Institute and Department of Clinical Neurosciences, University of Cambridge, Cambridge, UK

Tetsufumi Ueda Molecular and Behavioral Neuroscience Institute, The University of Michigan, Ann Arbor, MI, USA

Department of Pharmacology, University of Michigan Medical School, Ann Arbor, MI, USA

Department of Psychiatry, Medical School, The University of Michigan, Ann Arbor, MI, USA

Alexei Verkhratsky Faculty of Life Sciences, The University of Manchester, Manchester, UK

Achucarro Center for Neuroscience, IKERBASQUE, Basque Foundation for Science, Bilbao, Spain

Department of Neurosciences, University of the Basque Country UPV/EHU, Leioa, Spain

Helle Waagepetersen Department of Drug Design and Pharmacology, Faculty of Health and Medical Sciences, University of Copenhagen, Copenhagen, Denmark

Anne B. Walls Department of Drug Design and Pharmacology, Faculty of Health and Medical Sciences, University of Copenhagen, Copenhagen, Denmark

Yun Zhou, Ph.D. Department of Molecular Medicine, Institute for Basic Medical Sciences, University of Oslo, Oslo, Norway

Chapter 1
Introduction to the Glutamate–Glutamine Cycle

Ursula Sonnewald and Arne Schousboe

Abstract The term 'glutamate–glutamine cycle' was coined several decades ago based on the observation that using certain ^{14}C-labeled precursors for studies of brain metabolism the specific radioactivity of glutamine generated from glutamate was higher than that of glutamate, its immediate precursor. This is metabolically impossible unless it is assumed that at least two distinct pools of these amino acids exist. This combined with the finding that the enzyme synthesizing glutamine from glutamate was expressed in astrocytes but not in neurons formed the basis of the notion that a cycle must exist in which glutamate released from neurons is transported into astrocytes, converted to glutamine which is subsequently returned to neurons and converted to glutamate by an enzyme the activity of which is much higher in neurons than in astrocytes. Originally this cycle was supposed to function in a stoichiometric fashion but more recent research has seriously questioned this.

This volume of Advances in Neurobiology is intended to provide a detailed discussion of recent developments in research aimed at delineating the functional roles of the cycle taking into account that in order for this system to work there must be a tight coupling between metabolism of glutamate in astrocytes, transfer of glutamine to neurons and de novo synthesis of glutamine in astrocytes. To understand this, knowledge about the activity and regulation of the enzymes and transporters involved in these processes is required and as can be seen from the table of contents these issues will be dealt with in detail in the individual chapters of the book.

Keywords Glutamate • Glutamine • Transport • Synthesis metabolism

U. Sonnewald (✉)
Department of Neuroscience, Norwegian University of Science and Technology (NTNU), Trondheim 7489, Norway

Department of Drug Design and Pharmacology, Faculty of Health and Medical Sciences, University of Copenhagen, Copenhagen 2100, Denmark
e-mail: Ursula.sonnewald@ntnu.no

A. Schousboe
Department of Drug Design and Pharmacology, Faculty of Health and Medical Sciences, University of Copenhagen, Copenhagen 2100, Denmark
e-mail: arne.schousboe@sund.ku.dk

A. Schousboe, U. Sonnewald (eds.), *The Glutamate/GABA-Glutamine Cycle*, Advances in Neurobiology 13, DOI 10.1007/978-3-319-45096-4_1

1.1 Introduction

The concept of a mechanism in the brain which could handle the large amounts of glutamate being cycled as a neurotransmitter was fostered by biochemical investigations of glutamate and glutamine metabolism in different brain tissue preparations using radioactively labelled ([^{14}C]) precursors. Depending on the precursor used, the specific radioactivity of these two amino acids was found to follow the predicted substrate–product relationship, i.e. the precursor (glutamate) exhibiting the highest specific radioactivity or the opposite, i.e. the highest specific radioactivity was seen in the product, glutamine (for references, see Berl and Clarke (1969) and Schousboe (2012)). The latter result could not be easily explained unless the assumption was made that at least two compartmentalized pools of the precursor glutamate must exist (Garfinkel 1966; Berl et al. 1968; van den Berg and Garfinkel 1971). This combined with the seminal observation that the enzyme responsible for glutamine synthesis, glutamine synthetase (GS) is selectively expressed in astrocytes not neurons (Norenberg and Martinez-Hernandez 1979) led to the concept of the glutamate–glutamine cycle in which glutamate from glutamatergic neurons is taken up into astrocytes and subsequently returned to the neurons in the form of glutamine.

1.2 Release and Uptake of Glutamate

Due to its high concentration in the brain it was not easy to accept that glutamate could function as an excitatory neurotransmitter. Even though it was demonstrated that glutamate exerts a depolarizing action on spinal cord neurons, its legitimate status as a neurotransmitter was not immediately accepted since the depolarizing action could be mimicked by other amino acids as well (Curtis et al. 1959). However, the discovery of specific agonists and antagonists of glutamate receptors finally allowed the firm conclusion that glutamate is a *bona fide* neurotransmitter (Watkins and Evans 1981). Moreover, clearly the release of glutamate from synaptic terminals and glutamatergic neurons in culture fulfils the criteria for a neurotransmitter (Nicholls 1989; Ehrhart-Bornstein et al. 1991). In addition, in keeping with the status as a neurotransmitter there is a specific mechanism operating for inactivation of the receptor-mediated excitatory action of glutamate since extremely efficient high affinity transporters for glutamate have been identified (Schousboe 1981; Danbolt 2001), transporters which have been shown to be preferentially expressed in astrocytes (Gegelashvili and Schousboe 1997; Danbolt 2001).

The purification of a glutamate transporter from a rat brain synaptosomal preparation (Danbolt et al. 1990) paved the way for cloning of altogether five different high affinity glutamate transporters, GLAST/EAAT-1, GLT1/EAAT-2, EAAC-1/EAAT-3, EAAT-4 and EAAT-5 (Storck et al. 1992; Pines et al. 1992; Kanai and Hediger 1992; Fairman et al. 1995; Arriza et al. 1997) of which the former two, that were found to be localized primarily in astrocytes, account for the majority of the glutamate transport capacity in the brain (Danbolt 2001). As these transporters clearly are of fundamental functional importance not only as direct partners in the regulation of excitatory neurotransmission but also in prevention of the excitotoxic actions of elevated extracellular concentrations of glutamate in the brain (Choi 1988), it is not surprising that their expression level is tightly regulated. This was first demonstrated by the observation that neuronally secreted factors were able to increase the expression of the high affinity glutamate transporter activity in cultured astrocytes (Drejer et al. 1983). This finding was subsequently strengthened by the demonstration that exposure of cultured astrocytes to either conditioned medium from neuronal cultures or by direct contact with neurons resulted in a significant increase in the expression of GLT-1, a transporter not normally expressed to any significant extent in cultured astrocytes (Gegelashvili et al. 1997; Swanson et al. 1997; Schlag et al. 1998). It may also be of interest that other factors such as arachidonic acid, pH and the redox state seem to affect the expression level and activity of astrocytic glutamate transporters (Levy 2002).

The functional activity of the glutamate transporters depends on an intact cellular sodium gradient and hence, the operation of the Na^+–K^+–ATPase (Danbolt 2001). It is therefore not surprising that the transporters have been shown to be tightly coupled to the ATPase in the plasma membrane (Cholet et al. 2002; Rose et al. 2009; Bauer et al. 2012; Matos et al. 2013; Roberts et al. 2014; Illarionova et al. 2014). As the sodium-dependent transport obviously requires chemical energy in the form of ATP it is of considerable interest that the astrocytic glutamate transporters have been shown to be closely associated with both glycolytic enzymes and mitochondria (Genda et al. 2011; Bauer et al. 2012; Jackson et al. 2015). It seems of particular importance that there is a correlation between glutamate transporter activity and the mobility of mitochondria in the fine astrocytic processes (Jackson et al. 2014). Thus, an increase in neuronal activity and glutamate uptake leads to arrest of mitochondrial mobility anchoring the mitochondria in close proximity to the transporters (Jackson et al. 2014).

1.3 Metabolism of Glutamate in Astrocytes

As described earlier, extracellular glutamate is very efficiently taken up into astrocytes via transporters and the glutamate content inside the astrocytes increases with increasing extracellular glutamate concentrations (McKenna et al. 1996). It is well established that astrocytes metabolize glutamate taken up from the extracellular milieu to glutamine (Waniewski and Martin 1986; Zielke et al. 1990; Farinelli and Nicklas 1992) but additionally glutamate can be oxidatively metabolized (Yu et al. 1982; McKenna et al. 1996). This conversion of extracellular glutamate to glutamine is a part of the so-called glutamate–glutamine cycle (Hertz 1979).

There are conflicting results regarding the relative amount of glutamate metabolized via the TCA cycle versus the amount directly converted to glutamine (Yu et al. 1982; Yudkoff et al. 1986; Waniewski and Martin 1986; Zielke et al. 1990;

Sonnewald et al. 1993). McKenna et al. (1996) showed an increase in the % of glutamate metabolized via the TCA cycle as compared to the amount converted to glutamine directly. At extracellular glutamate concentrations of 0.01, 0.05, 0.25 and 0.5 mM it has been shown by several investigators (Waniewski and Martin 1986; Yu et al. 1982; Yudkoff et al. 1986; Sonnewald et al. 1993; McKenna et al. 1996) that a significant amount of glutamate is converted to α-ketoglutarate but at concentrations of 0.01 mM in the presence of 10 % fetal calf serum (Zielke et al. 1990) or 0.05 mM in the presence of 1 mM pyruvate (Farinelli and Nicklas 1992), glutamate appears to be almost exclusively converted to glutamine.

In order for exogenous glutamate to be metabolized in the TCA cycle it must be converted to α-ketoglutarate a process which can occur as a transamination or a deamination. Studies of glutamate metabolism in astrocytes (Farinelli and Nicklas 1992; Sonnewald et al. 1993; Waniewski and Martin 1986; Yu et al. 1982; Yudkoff et al. 1986; Zielke et al. 1990) have yielded somewhat conflicting results regarding the significance and mechanism of oxidation through the TCA cycle. The significance of transamination can be probed by using aminooxyacetic acid (AOAA), a non-specific inhibitor of pyridoxal phosphate-dependent enzymes (Yu et al. 1982). AOAA inhibits the major transaminases involved in glutamate metabolism: aspartate aminotransferase (AAT) and alanine aminotransferase (ALAAT), both in the cytosol and in mitochondria (Kauppinen et al. 1987; Cooper 1988). To establish the significance of transamination for glutamate metabolism in astrocytes, cells were incubated with [U-^{13}C]glutamate in the presence or absence of AOAA. The ^{13}C labelling pattern in compounds present in extracts of the cells and the incubation media were subsequently determined using ^{13}C magnetic resonance spectroscopy. Since glutamate was uniformly labelled, detection of incorporation of label into metabolites is unambiguously due to ^{13}C-^{13}C-spin coupling patterns. Biochemical methods were used to quantify ammonia, lactate, amino acids and citrate. After blocking transamination of glutamate with AOAA, entry into the TCA cycle of the carbon units from glutamate was still taking place since lactate and citrate labelling with ^{13}C were observed. Glutamate and glutamine synthesis from α-ketoglutarate could, however, not be detected under this condition showing the importance of transamination for glutamate formation from α-ketoglutarate.

As described earlier, a considerable proportion of the glutamate taken up by astrocytes is metabolized via the TCA cycle (Yu et al. 1982; Waniewski and Martin 1986; Schousboe et al. 1993a, b; Bachelard et al. 1994; McKenna et al. 1996) and converted to TCA cycle-related products, including lactate. Sonnewald et al. (1993) have demonstrated using ^{13}C magnetic resonance spectroscopy that the carbon skeleton of extracellular [U-^{13}C]glutamate is converted to lactate to a large extent. In order to accomplish conversion of glutamate to lactate, glutamate must obviously enter the TCA cycle for conversion to metabolites like malate, oxaloacetate, citrate, α-ketoglutarate and aspartate, which can subsequently leave the mitochondria. Citrate leaving the mitochondria may give rise to oxaloacetate in the cytoplasm and subsequently to pyruvate *via* the activity of malic enzyme or by the concerted action of phosphoenolpyruvate carboxykinase and pyruvate kinase. Subsequently, pyruvate may be metabolized to lactate, alanine or acetyl-CoA. The release of lactate by astrocytes has been demonstrated by several groups (Walz and Mukerji 1988; Larrabee 1992; Sonnewald et al. 1993) and considerable evidence exists that lactate is an effective substrate for maintaining energy metabolism in synaptic terminals (Schurr et al. 1988; McKenna et al. 1993, 1994).

As is apparent from the results presented earlier, glutamate metabolism is highly regulated and in addition, there is evidence of the existence of several glutamate pools in astrocytes: one pool, with low capacity, in which glutamine formation and some transamination is taking place and another pool with large capacity, in which detoxification of large amounts of glutamate taken up from the extracellular space might take place. This latter pool allows the use of glutamate as an energy substrate via the TCA cycle. This pool has a low requirement for α-keto acids, since deamination does not require them, a clear advantage under conditions where these substances may not be present in adequate amounts (Yudkoff et al. 1994a, b). The above-mentioned results confirm earlier reports of compartmentation of glutamate metabolism in which it was shown that extracellular, radioactively labelled glutamine gave rise to a higher specific labelling of intracellular glutamate than glutamine and that aspartate formation was different when extracellular glutamate or glutamine was used as precursors (Schousboe et al. 1993a, b).

References

Arriza JL, Eliasof S, Kavanaugh MP, Amara SG (1997) Excitatory amino acid transporter 5, a retinal glutamate transporter coupled to chloride conductance. Proc Natl Acad Sci U S A 94:4155–4160

Bachelard H, Badar-Goffer R, Morris P, Thatcher N (1994) Magnetic resonance spectroscopy studies on Ca^{2+}, Zn^{2+} and energy metabolism in superfused brain slices. Biochem Soc Trans 22:988–991

Bauer DE, Jackson JG, Genda EN, Montoya MM, Yudkoff M, Robinson MB (2012) The glutamate transporter, GLAST, participates in a macromolecular complex that supports glutamate metabolism. Neurochem Int 61:566–574

Berl S, Clarke DD (1969) Compartmentation of amino acid metabolism. In: Lajtha A (ed) Handbook of neurochemistry. Plenum Press, New York, pp 447–472

Berl S, Nicklas WJ, Clarke DD (1968) Compartmentation of glutamic acid metabolism in brain slices. J Neurochem 15:131–140

Choi DW (1988) Glutamate neurotoxicity and diseases of the nervous system. Neuron 1:623–634

Cholet N, Pellerin L, Magistretti PJ, Hamel E (2002) Similar perisynaptic glial localization for the Na+, K+-ATPase α2 subunit and the glutamate transporters GLAST and GLT-1 in the rat somatosensory cortex. Cereb Cortex 12:515–525

Cooper AJL (1988) L-Glutamate (2-oxoglutarate) aminotransferases. In: Kvamme E (ed) Glutamine and glutamate in mammals, vol 1. CRC, Boca Raton, FL, pp 123–152

Curtis DR, Phillis JW, Watkins JC (1959) Chemical excitation of spinal neurones. Nature 183:611–612

Danbolt NC (2001) Glutamate uptake. Prog Neurobiol 65:1–105

Danbolt NC, Pines G, Kanner BI (1990) Purification and reconstitution of the sodium- and potassium-coupled glutamate transport glycoprotein from rat brain. Biochemistry 29:6734–6740

Drejer J, Meier E, Schousboe A (1983) Novel neuron-related regulatory mechanisms for astrocytic glutamate and GABA high affinity uptake. Neurosci Lett 37:301–306

Ehrhart-Bornstein M, Treiman M, Hansen GH, Schousboe A, Thorn NA, Frandsen A (1991) Parallel expression of synaptophysin and evoked neurotransmitter release during development of cultured neurons. Int J Dev Neurosci 9:463–471

Fairman WA, Vandenberg RJ, Arriza JL, Kavanaugh MP (1995) An excitatory amino-acid transporter with properties of a ligand-gated chloride channel. Nature 375:599–603

Farinelli SE, Nicklas WJ (1992) Glutamate metabolism in rat cortical astrocyte cultures. J Neurochem 58:1905–1915

Garfinkel D (1966) A simulation study of the metabolism and compartmentation in brain of glutamate, aspartate, Krebs cycle and related metabolites. J Biol Chem 241:3918–3929

Gegelashvili G, Schousboe A (1997) High affinity glutamate transporters: regulation of expression and activity. Mol Pharmacol 52:6–15

Gegelashvili G, Danbolt NC, Schousboe A (1997) Neuronal soluble factors differentially regulate the expression of the GLT1 and GLAST glutamate transporters in cultured astroglia. J Neurochem 69:2612–2615

Genda EN, Jackson JG, Sheldon AL, Locke SF, Greco TM, O'Donnell JC, Spruce LA, Xiao R, Guo W, Putt M, Seeholzer S, Ischiropoulos H, Robinson MB (2011) Co-compartmentalization of the astroglial glutamate transporter, GLT-1, with glycolytic enzymes and mitochondria. J Neurosci 31:18275–18288

Hertz L (1979) Functional interactions between neurons and astrocytes I. Turnover and metabolism of putative amino acid transmitters. Prog Neurobiol 13:277–323

Illarionova NB, Brismar H, Aperia A, Gunnarson E (2014) Role of Na, K-ATPase alpha1 and alpha2 isoforms in the support of astrocyte glutamate uptake. PLoS One 9:e98469

Jackson JG, O'Donnell JC, Takano H, Coulter DA, Robinson MB (2014) Neuronal activity and glutamate uptake decrease mitochondrial mobility in astrocytes and position mitochondria near glutamate transporters. J Neurosci 34:1613–1624

Jackson JG, O'Donnell JC, Krizman E, Robinson MB (2015) Displacing hexokinase from mitochondrial voltage-dependent anion channel impairs GLT-1 mediated glutamate uptake but does not disrupt interactions between GLT-1 and mitochondrial proteins. J Neurosci Res 93:999–1083

Kanai Y, Hediger MA (1992) Primary structure and functional characterization of a high-affinity glutamate transporter. Nature 360:467–471

Kauppinen RA, Sihra TS, Nicholls DG (1987) Aminooxyacetic acid inhibits the malate-aspartate shuttle in isolated nerve terminals and prevents the mitochondria from utilizing glycolytic substrates. Biochim Biophys Acta 930:173–178

Larrabee MG (1992) Extracellular intermediates of glucose metabolism: fluxes of endogenous lactate and alanine through extracellular pools in embryonic sympathetic ganglia. J Neurochem 59:1041–1052

Levy LM (2002) Structure, function and regulation of glutamate transporters. In: Egebjerg J, Schousboe A, Krogsgaard-Larsen P (eds) Glutamate and GABA receptors and transporters: structure, function and pharmacology. Taylor and Francis, London, pp 307–336

Matos M, Augusto E, Agostinho P, Cunha RA, Chen JF (2013) Antagonistic interaction between adenosine A2A receptors and Na^+/K^+-ATPase-alpha2 controlling glutamate uptake in astrocytes. J Neurosci 33:18492–18502

McKenna MC, Tildon JT, Stevenson JH, Boatright R, Huang S (1993) Regulation of energy metabolism in synaptic terminals and cultured rat brain astrocytes: differences revealed using aminooxyacetate. Dev Neurosci 15:320–329

McKenna MC, Tildon JT, Stevenson JH, Hopkins IB (1994) Energy metabolism in cortical synaptic terminals from weanling and mature rat brain: evidence for multiple compartments of tricarboxylic acid cycle activity. Dev Neurosci 16:291–300

McKenna MC, Sonnewald U, Huang X, Stevenson J, Zielke HR (1996) Exogenous glutamate concentration regulates the metabolic fate of glutamate in astrocytes. J Neurochem 66:386–393

Nicholls DG (1989) Release of glutamate, aspartate and gamma-aminobutyric acid from isolated terminals. J Neurochem 52:331–341

Norenberg MD, Martinez-Hernandez A (1979) Fine structural localization of glutamine synthetase in astrocytes of rat brain. Brain Res 161:303–310

Pines G, Danbolt NC, Bjørås M, Zhang Y, Bendahan A, Eide L, Koepsell H, Storm-Mathisen J, Seeberg E, Kanner BI (1992) Cloning and expression of a rat brain L-glutamate transporter. Nature 360:464–467

Roberts RC, Roche JK, McCullumsmith RE (2014) Localization of excitatory amino acid transporters EAAT1 and EAAT2 in human postmortem cortex: a light and electron microscopic study. Neuroscience 77:522–540

Rose EM, Koo JC, Antflick JE, Ahmed SM, Angers S, Hampson DR (2009) Glutamate transporter coupling to Na, K-ATPase. J Neurosci 29:8143–8155

Schlag BD, Vondrasek JR, Munir M, Kalandaze A, Zelenaia OA, Rothstein JD, Robinson MB (1998) Regulation of the glial Na^+-dependent glutamate transporters by cyclic AMP analogs and neurons. Mol Pharmacol 53:355–369

Schousboe A (1981) Transport and metabolism of glutamate and GABA in neurons and glial cells. Int Rev Neurobiol 22:1–45

Schousboe A (2012) Studies of brain metabolism: a historical perspective. Adv Neurobiol 4:909–920

Schousboe A, Westergaard N, Hertz L (1993a) Neuronal-astrocytic interactions in glutamate metabolism. Biochem Soc Trans 21:49–53

Schousboe A, Westergaard N, Sonnewald U, Petersen SB, Huang R, Peng L, Hertz L (1993b) Glutamate and glutamine metabolism and compartmentation in astrocytes. Dev Neurosci 15:359–366

Schurr A, West CA, Rigor BM (1988) Lactate-supported synaptic function in the rat hippocampal slice preparation. Science 240:1326–1328

Sonnewald U, Westergaard N, Petersen SB, Unsgard G, Schousboe A (1993) Metabolism of [U-^{13}C]glutamate in astrocytes studied by ^{13}C NMR spectroscopy: incorporation of more label into lactate than into glutamine demonstrates the importance of the tricarboxylic acid cycle. J Neurochem 61:1179–1182

Storck T, Schulte S, Hofman K, Stoffel W (1992) Structure, expression, and functional analysis of a Na^+-dependent glutamate/aspartate transporter from rat brain. Proc Natl Acad Sci U S A 89:10955–10959

Swanson RA, Miller JW, Rothstein JD, Farrell K, Stein BA, Longuemare MC (1997) Neuronal regulation of glutamate transporter subtype expression in astrocytes. J Neurosci 17:932–940

van den Berg CJ, Garfinkel D (1971) A simulation study of brain compartments. Metabolism of glutamate and related substances in mouse brain. Biochem J 123:211–218

Walz W, Mukerji S (1988) Lactate release from cultured astrocytes and neurons: a comparison. Glia 1:366–370

Waniewski RA, Martin DL (1986) Exogenous glutamate is metabolized to glutamine and exported by rat primary astrocyte cultures. J Neurochem 47:304–313

Watkins JC, Evans RH (1981) Excitatory amino acid transmitters. Ann Rev Pharmacol Toxicol 21:165–204

Yu AC, Schousboe A, Hertz L (1982) Metabolic fate of 14C-labeled glutamate in astrocytes in primary cultures. J Neurochem 39:954–960

Yudkoff M, Nissim I, Hummeler K, Medow M, Pleasure D (1986) Utilization of [^{15}N]glutamate by cultured astrocytes. Biochem J 234:185–192

Yudkoff M, Daikhin Y, Lin ZP, Nissim I, Stern J, Pleasure D (1994a) Interrelationships of leucine and glutamate metabolism in cultured astrocytes. J Neurochem 62:1192–1202

Yudkoff M, Nelson D, Daikhin Y, Erecinska M (1994b) Tricarboxylic acid cycle in rat brain synaptosomes. Fluxes and interactions with aspartate aminotransferase and malate/aspartate shuttle. J Biol Chem 269:27414–27420

Zielke HR, Tildon JT, Landry ME, Max SR (1990) Effect of 8-bromo-cAMP and dexamethasone on glutamate metabolism in rat astrocytes. Neurochem Res 15:1115–1122

Chapter 2
Glucose, Lactate, β-Hydroxybutyrate, Acetate, GABA, and Succinate as Substrates for Synthesis of Glutamate and GABA in the Glutamine–Glutamate/GABA Cycle

Leif Hertz and Douglas L. Rothman

Abstract The glutamine–glutamate/GABA cycle is an astrocytic-neuronal pathway transferring precursors for transmitter glutamate and GABA from astrocytes to neurons. In addition, the cycle carries released transmitter back to astrocytes, where a minor fraction (~25 %) is degraded (requiring a similar amount of resynthesis) and the remainder returned to the neurons for reuse. The flux in the cycle is intense, amounting to the same value as neuronal glucose utilization rate or 75–80 % of total cortical glucose consumption. This glucose:glutamate ratio is reduced when high amounts of β-hydroxybutyrate are present, but β-hydroxybutyrate can at most replace 60 % of glucose during awake brain function. The cycle is initiated by α-ketoglutarate production in astrocytes and its conversion via glutamate to glutamine which is released. A crucial reaction in the cycle is metabolism of glutamine after its accumulation in neurons. In glutamatergic neurons all generated glutamate enters the mitochondria and its exit to the cytosol occurs in a process resembling the malate-aspartate shuttle and therefore requiring concomitant pyruvate metabolism. In GABAergic neurons one half enters the mitochondria, whereas the other one half is released directly from the cytosol. A revised concept is proposed for the synthesis and metabolism of vesicular and nonvesicular GABA. It includes the well-established neuronal GABA reuptake, its metabolism, and use for resynthesis of vesicular GABA. In contrast, mitochondrial glutamate is by transamination to α-ketoglutarate and subsequent retransamination to releasable glutamate essential for the transaminations occurring during metabolism of accumulated GABA and subsequent resynthesis of vesicular GABA.

L. Hertz
Laboratory of Brain Metabolic Diseases, Institute of Metabolic Disease Research and Drug Development, China Medical University, Shenyang, China
e-mail: lhertz538@gmail.com

D.L. Rothman (✉)
Magnetic Resonance Research Center, Radiology and Biomedical Engineering, Yale University, New Haven, CT, USA
e-mail: douglas.rothman@yale.edu

A. Schousboe, U. Sonnewald (eds.), *The Glutamate/GABA-Glutamine Cycle*, Advances in Neurobiology 13, DOI 10.1007/978-3-319-45096-4_2

Keywords Astrocyte • β-Hydroxybutyrate • Branched chain amino acids • GABAergic neuron • Glucose • Glutamatergic neuron • Glutamate–glutamine cycle • Neuronal GABA uptake • Succinate

Abbreviations

^{13}C MRS	^{13}C magnetic resonance spectroscopy
α-KG	α-ketoglutarate
AAT	Aspartate aminotransferase
Ac.CoA	Acetyl coenzyme A
ADP	Adenosine diphosphate
AGC	Glutamate/aspartate exchanger (in brain AGC1 or aralar)
ANLS	Astrocyte neuron lactate shuttle
AOAA	Alpha-aminooxyacetic acid
ATP	Adenosine triphosphate
AV	Arteriovenous
BCAA	Branched chain amino acid
BCAT	Branched chain amino acid transaminase
BCATc	Cytosolic branched chain amino acid transferase
BCATm	Mitochondrial branched chain amino acid transferase
DNA, RNA	Deoxyribonucleic acid, ribonucleic acid
GABA	γ-aminobutyric acid
GABA-T	GABA transaminase
GAD65, GAD67	Glutamate amino decarboxylase 65 and 67 KD isoforms
GAT1, GAT2, GAT3	GABA transporters 1, 2, 3
GDH	Glutamate dehydrogenase
GLAST	Glutamate aspartate transporter
GLT-1	Glutamate type 1 transporter
GS	Glutamine synthase
KIC	Alpha-ketoisocaproic acid
MAS	Malate-aspartate shuttle
MCT1	Monocarboxylic acid transporter 1
MCT2, MCT3	Monocarboxylic acid transporters 2, 3
MCTs	Monocarboxylic acid transporters
NADH/NAD$^+$	Nicotinamide adenine dinucleotide
NADPH	Nicotinamide adenine dinucleotide phosphate
OAA	Oxaloacetate
PAG	Phosphate activated glutaminase
PC	Pyruvate carboxylase
PDH	Pyruvate dehydrogenase complex
SAT1, SAT2	System A glutamine transporters 1 and 2
SN1	System N glutamine transporter 1
SSA	Succinic semialdehyde
TCA cycle	Tricarboxylic acid cycle

2.1 Introduction

The brain is almost unique among organs in having an obligate requirement of glucose for function. However, the brain can use a variety of other substrates including lactate, β-hydroxybutyrate, acetate, and medium and long chain fatty acids. Although the brain's relative preference for substrates was well known from the pioneering AV difference studies of Kety and colleagues (Siesjö 1978), the reason for requirement for glucose has remained unexplained. In this paper, we examine evidence ranging from isolated enzymes, to cell culture, to in vivo studies to address this question. Our overall hypothesis is that the obligate requirement for glucose is due to the special needs for supporting the large glutamate and γ-aminobutyric acid (GABA) neurotransmitter cycling fluxes in the awake brain. We describe a model for glutamate–glutamine cycling that can explain the observed glucose dependence, as well as a revised model that explains the reduced dependence on glucose in GABAergic neurons. These models are consistent with the present results over a wide range of systems. Although the definitive in vivo studies remain to be done to test these models they will hopefully be useful in guiding experimentation.

2.2 Brain Metabolism of Glucose, Lactate, and β-Hydroxybutyrate

2.2.1 Glucose

The adult mammalian brain consumes about 20 % of total bodily oxygen and thus calorie utilization (Raichle and Gusnard 2002), and it normally uses glucose as its only or main substrate (Sokoloff et al. 1977). Studies in the rodent and human brain in vivo using ^{13}C-based magnetic resonance spectroscopy (MRS) and labeled glucose or acetate, which is oxidized in astrocytes but not in neurons (Muir et al. 1986; Waniewski and Martin 1998) showed that in the awake human brain astrocytes contribute ~25 % of the oxidative metabolism in brain gray matter (Gruetter et al. 2001; Lebon et al. 2002; Mason et al. 2007). Most of this glucose is metabolized via formation of two molecules of pyruvate in the cytosol from one molecule of glucose (glycolysis) followed by mitochondrial formation of acetyl coenzyme A (ac.CoA) from pyruvate (Fig. 2.1). The latter reaction is catalyzed by the pyruvate dehydrogenase complex (PDH), which is present in all brain cells. In the tricarboxylic acid (TCA) cycle the acetyl part of acetyl CoA (containing two carbon atoms) condenses with a four-carbon TCA intermediate, oxaloacetate (OAA), to form a six-carbon TCA cycle constituent (citrate), which via two decarboxylations regenerates OAA (4 carbon atoms), ready to condense with another molecule of acetyl CoA.

TCA cycle activity followed by oxidative phosphorylation in the respiratory chain creates a large amount of energy in the form of adenosine triphosphate (ATP) in both astrocytes and neurons. Since OAA is consumed at the beginning of the TCA cycle and regenerated at the end of the cycle, this process continues as long as

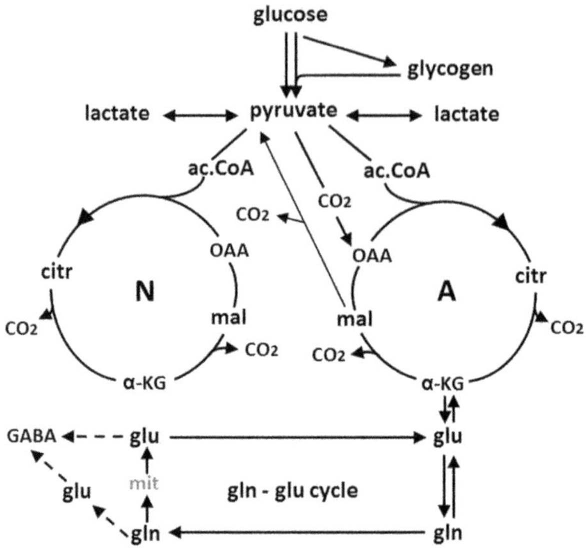

Fig. 2.1 Cartoon of glucose metabolism via pyruvate in neurons (*left*—N) and astrocytes (*right*—A) and of glutamine–glutamate (GABA) cycling. One molecule glucose is metabolized by glycolysis in the cytosol to two molecules of pyruvate in a complex and strictly regulated pathway, where one oxidative process requires transfer of reducing equivalents to the mitochondria (see, e.g., Hertz and Dienel 2002). In both neurons and astrocytes pyruvate metabolism via acetyl Coenzyme A (ac.CoA) leads to formation of citrate by condensation with preexisting oxaloacetate (OAA) in the tricarboxylic acid (TCA), an end result of the previous turn of the cycle. Citrate oxidation in the TCA cycle includes two decarboxylations, leading to reformation of oxaloacetate, ready for another turn of the cycle, and to reduction of NADH, creating large amounts of energy (ATP) via reoxidation in the electron transport chain. Pyruvate carboxylation, which is active in astrocytes, but absent in neurons (reviewed in Hertz 2013), creates a new molecule of oxaloacetate, which after condensation with acetyl Coenzyme A, derived from a second molecule of pyruvate, forms a new molecule of citrate. α-Ketoglutarate (α-KG), one of the intermediates of the TCA cycle can leave the cycle to form glutamate (glu), catalyzed by either aspartate aminotransferase or glutamate dehydrogenase. Further metabolism by the cytosolic and astrocyte-specific enzyme glutamine synthetase leads to the formation of glutamine (gln). In glutamatergic neurons all glutamate formed by deamidation of glutamine enters the mitochondria (mit) and is returned to the cytosol in a complex process, which requires simultaneous glucose metabolism. In GABAergic neurons this is only the case for some of the glutamate, whereas the remainder enters the cytosol directly. This mechanism, which is further described in Fig. 2.6, its legend, and associated text may make GABA production less sensitive to replacement of glucose as the substrate with β-hydroxybutyrate in patients receiving very high amounts of ketone bodies to prevent seizures. Released glutamate is almost quantitatively reaccumulated in astrocytes, together with at least part of the released GABA [upper line of the glutamine–glutamate/GABA cycle (glu–gln cycle)] and reaccumulated in the astrocytic cytosol. Here, about 75 % is converted to glutamine and reenters the glutamine–glutamate/GABA cycle. The remaining ~25 % is oxidatively degraded, via one of two partly different pathways. In both α-ketoglutarate is reconverted to malate. In one malate exits to the cytosol, is decarboxylated by cytosolic malic enzyme to pyruvate, which is oxidized in the TCA cycle via acetyl Coenzyme A. In the other malate does not exit the TCA cycle but may be further metabolized to α-ketoglutarate after condensation with acetyl Coenzyme A, allowing resynthesis of another molecule of glutamate from only one molecule pyruvate. In either case the degraded glutamate must in the long term be replaced by a quantitatively similar production of glutamate from glucose, in the first case by complete de novo synthesis from one molecule glucose, in the second from one half of a glucose molecule. However, temporary fluctuations in the content of glutamate occur. The initial part of GABA metabolism is different, as all GABA is metabolized via succinic semialdehyde, succinate, and α-ketoglutarate to glutamate. Modified from Hertz (2013)

ac.CoA is available and energy is consumed, but it cannot give rise to a new molecule of any TCA cycle constituent. Nevertheless, it does lead to labeling of glutamate and its derivatives, a labeling that forms the basis of ^{13}C-NMR spectroscopy. However, this labeling is due to a transamination-mediated bidirectional exchange between α-KG and glutamate (Fitzpatrick et al. 1990).

Recently, experiments by Patel et al. (2005, 2015) and Duarte and Gruetter (2013) have shown a considerable flow of neurotransmitter GABA to astrocytes via the GABA/glutamine cycle. GABA after its release from inhibitory neurons is metabolized in astrocytes via succinic semialdehyde and succinate. It will later be discussed that a similar reaction may occur in GABAergic neurons. Succinate is a TCA cycle intermediate between α-ketoglutarate (α-KG) and malate (Fig. 2.1) and in the TCA cycle it is metabolized via OAA to citrate, requiring uptake of one molecule ac.CoA. Citrate is further metabolized to α-KG and transaminated to glutamate. Subsequently, the generated glutamate can be returned to neurons via astrocytes in the conventional glutamine–glutamate/GABA cycle via glutamine (Fig. 2.1). Duarte and Gruetter (2013) showed using metabolic modeling that the ac.CoA flow needed in the astrocyte to accommodate GABA metabolism constitutes a considerable fraction of total astrocyte ac.CoA production. The formation of ac.CoA needed for metabolism to citrate of OAA derived from GABA is included in the 25 % of total brain glucose consumption assigned to astrocytes; however, from the standpoint of ATP production the flow from succinate also needs to be considered.

Although pyruvate formation from glucose takes place in the cytosol one oxidative process, formation of diphosphoglycerate from glyceraldehyde 3-phosphate is involved. As an oxidation this process is coupled to formation of nicotinamide dinucleotide NADH from NAD^+. Since NADH is unable to cross the mitochondrial membrane for reoxidation in the mitochondria a reducing equivalent must be carried across the mitochondrial membrane for oxidation with return of the corresponding oxidized equivalent. In the brain this transport is mediated by the malate-aspartate shuttle (MAS). In the MAS (Fig. 2.2), cytosolic NADH is reoxidized to NAD^+ by reduction of cytosolic OAA to malate. Malate traverses the mitochondrial membrane but not in direct exchange for OAA (which cannot leave the mitochondria), but in exchange for α-KG, using the malate/α-ketoglutarate carrier (OGC-Slc25a11), and it is reoxidized to OAA in the TCA cycle. In the mitochondria OAA is transaminated by aspartate transaminase (AAT) to the corresponding amino acid, aspartate, which then exits the mitochondrial membrane in exchange for glutamate, using the glutamate/aspartate exchanger (AGC), in brain AGC1 or aralar (Slc25a12). This exchanger is abundantly expressed not only in neurons but also in astrocytes (Lovatt et al. 2007; Li et al. 2012). The glutamate that enters the mitochondria is simultaneously transaminated to α-KG, its corresponding keto acid. α KG exits the mitochondria in exchange with incoming malate in the process catalyzed by malate/α-ketoglutarate carrier. This carrier is also operating in brain (Passarella et al. 1987), including astrocytes (Pardo et al. 2011). In the cytosol α-KG is transaminated to glutamate with simultaneous conversion of aspartate to OAA, closing the cycle (Fig. 2.2). AAT activity is high in most cells and in retina, a cerebral tissue, it is expressed both in the cytosol and in mitochondria (Ross and Godfrey 1987).

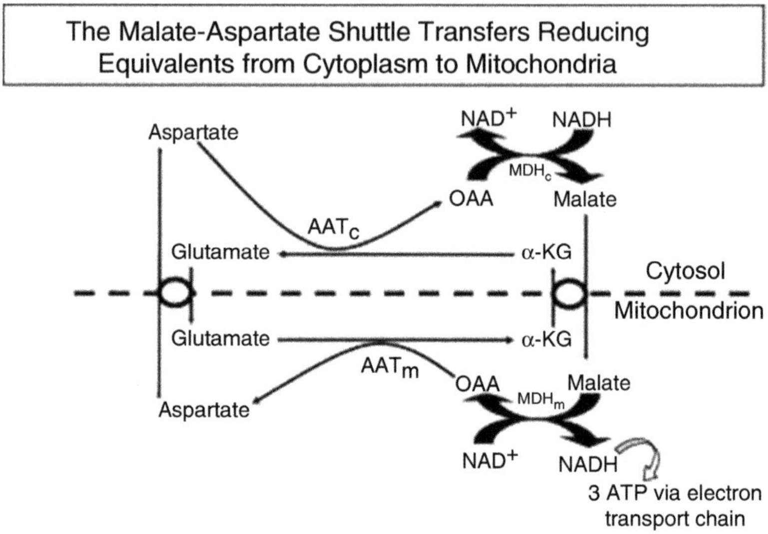

Fig. 2.2 The malate-aspartate shuttle (MAS). In the malate-aspartate shuttle (MAS), which serves to transfer reducing equivalents across the mitochondrial membrane, cytosolic malate dehydrogenase (MDHc) oxidizes NADH and converts oxaloacetate (OAA) to malate (top right of figure), which enters the mitochondria in exchange with α-ketoglutarate (α-KG). The mitochondrial malate dehydrogenase (MDHm) reoxidizes malate to OAA, which is transaminated to aspartate by the mitochondrial aspartate aminotransferase (AATm). Aspartate leaves the mitochondria in exchange with glutamate, requiring aspartate-glutamate exchanger (aralar in brain). In the mitochondria glutamate conversion to α-KG is essential for AATm activity forming aspartate from OAA and delivering α-KG for mitochondrial export. The glutamate imported into the mitochondria had been formed by cytosolic aspartate aminotransferase (AATc) from α-KG after its entry into the cytosol. From Hertz and Dienel (2002)

Due to the dependence of oxidative glycolysis on MAS activity, complete and direct oxidation of glucose requires not only that pyruvate enters the mitochondria as ac.CoA, but also that not only malate and aspartate, but also glutamate and α-KG are transported across the mitochondrial membrane. Note also that the dependency between these processes is mutual: the MAS would not be able to operate without NAD^+ reduction.

In addition to entering the TCA cycle via ac.CoA, pyruvate can also be carboxylated to OAA, an anaplerotic process, which expands the amount of TCA cycle intermediates and is catalyzed by the astrocyte-specific pyruvate carboxylase (PC). This enzyme is virtually absent in neurons both in the brain in vivo (Shank et al. 1984; Hutson et al. 2008) and in cultured cells (Yu et al. 1983). Condensation of one molecule OAA with one molecule of acetyl-CoA forms a new molecule of citrate. This molecule as well as all molecules derived from it, including α-KG and glutamate will be labeled from [^{13}C]glucose in different positions than corresponding molecules generated by PDH activity alone, which is the key principle used for determination of rate of pyruvate carboxylation in the brain by ^{13}C-MRS (Gruetter et al. 2001; Sibson et al. 2001; Mason et al. 2007). Although pyruvate carboxylation is necessary to replace lost TCA intermediates, by far its main role in the brain is to produce glutamate which is then transferred to neurons for use as transmitter glutamate and, after decarboxylation, transmitter GABA. A minor part of synthesized glutamate is also used by the astrocytes themselves as gliotransmitter, for synthesis of glutathione or for detoxification of ammonia to glutamine. Since carboxylation represents ~10 % of total pyruvate metabolism and it must be combined with a similar PDH-mediated metabolism, a large fraction of the 25 % of total pyruvate metabolism occurring in astrocytes must serve to produce glutamate. This does not mean that no energy is produced since combined production and oxidative degradation of glutamate, which also mainly occurs in astrocytes, produces almost as much ATP as direct oxidation of glucose (Gruetter et al. 2001; Hertz et al. 2007).

Besides the pathway shown in Fig. 2.1 a small amount of glucose (2–5 % of total glucose utilization in brain) is metabolized via the pentose phosphate pathway (Gaitonde et al. 1987; Hostetler and Landau 1967; Ben-Yoseph et al. 1995). This pathway generates pentoses, necessary for synthesis of DNA and RNA. It also produces NADPH, a reducing equivalent not replaceable by NADH, used for synthesis of lipids and in a few metabolic pathways, including reduction of oxidized glutathione (Vogel et al. 1999). Although this pathway is relatively small it has been recently suggested as responsible for part of the 5–10 % mismatch[1] between total brain glucose uptake and oxidation (Raichle 2015).

2.2.2 *Lactate*

Pyruvate is rapidly reduced to lactate and vice versa and most normal cells contain considerably higher concentrations of lactate than of pyruvate. Both can also cross the cell membrane in either direction catalyzed by equilibrating monocarboxylate transporters (MCTs). Lactate uptake is followed by its oxidation to pyruvate with concomitant reduction of NAD^+ to NADH. This has the consequences that lactate oxidation competes with glucose oxidation and that metabolism of lactate, like that of glucose activates MAS. MCT transporters have different kinetics in astrocytes (displaying MCT1 and 3) and neurons (displaying MCT2). It is important that these transporters, like the glucose transporters, are equilibrating, not concentrative, because that means that they are unable to accumulate (or release) the unaltered compound to a concentration beyond that from which it is accumulated or released (reviewed by Hertz and Dienel 2005). However, once the native compound is removed by metabolism, uptake or release can continue. It has been suggested that

[1] Considerable work has been done studying this mismatch and the larger mismatches that occur during functional activation. There are several theories to explain apparent uncoupling of glucose uptake and oxidation including the astrocyte neuron lactate shuttle (ANLS), discussed again later, which has received a great deal of attention. As this chapter focuses on oxidative metabolism we refer the reader to two recent reviews on opposite sides of the ANLS controversy (Pellerin and Magistretti 2012; Dienel 2012).

astrocytically released lactate is taken up and metabolized largely by neurons after transfer by the astrocyte neuron lactate shuttle ANLS (Pellerin and Magistretti 2012) and this may be critical for complex functions such as memory formation (Steinman et al. 2016), but considerable evidence is available *against* this hypothesis (e.g., Dienel 2012; Patel et al. 2014). Furthermore, astrocytes have a high ability to transport lactate between themselves through gap junctions which may substantially reduce transport to neurons (Gandhi et al. 2009). Nevertheless a small amount of glycogenderived lactate seems to be transferred from astrocytes to neurons during learning (Steinman et al. 2016). If this lactate is accumulated into the minute dendritic spines which have no mitochondria it may create ATP by oxidation to pyruvate without interfering with glucose metabolism (Hertz and Chen 2016). The established importance of glial cells for memory formation depends upon glycogen but is most likely independent of a major lactate transfer between cell types (Hertz and Chen 2016).

The unlikelihood of astrocytes supplying neurons with glucose-derived lactate does not mean that lactate metabolism is unable to cover a substantial fraction of the brain's energy demand. Thus, Van Hall et al. (2009) showed that lactate at a plasma level of 6.9 mM contributed 27 % of energy demand in human brain; high exercise intensity can decrease brain glucose uptake by almost 40 % (Kemppainen et al. 2005), and a similar decrease occurs in the anesthetized rat when plasma lactate concentration is increased to 5.8 % (Wyss et al. 2011). Boumezbeur et al. (2010) used ^{13}C magnetic resonance spectroscopy (^{13}C MRS) to estimate the relationship between plasma lactate level and cortical lactate metabolism. They found under resting awake conditions that plasma lactate at physiological levels (low mM range) accounted for ~10 % of consumption by the TCA cycle. However, based on the transport kinetics determined it was concluded that it could account for up to 60 % under maximally elevated conditions, a value comparable with those cited earlier. The labeling patterns observed were the same as with glucose, suggesting a similar percentage contribution to neuronal and glial metabolism. In mature cultured cortical astrocytes, glucose utilization can similarly be reduced by 60–70 % in the presence of lactate (Swanson and Benington 1996; Rodrigues et al. 2009), but in very young cultures of cortical neurons lactate can cover even more of total energy consumption (Bouzier-Sore et al. 2003, 2006). In this context it should be remembered that lactate oxidation to pyruvate competes with the oxidation of glyceraldehyde 3-phosphate for cytosolic NADH and thereby may inhibit glucose utilization. Nevertheless, a small increase in lactate is under certain conditions capable of increasing glucose uptake into brain at low glucose concentrations without major effect on lactate contribution to metabolism (De Feyter et al. 2013; Herzog et al. 2013). This effect appears to be more marked in neurons than in astrocytes. It is unknown whether it might be related to the signaling effect of physiological concentrations of lactate on noradrenergic neurons (Tang et al. 2014).

As a substrate lactate is able to sustain all the functions of glucose shown in Fig. 2.1 except glycolysis. It can support both PDH and PC activity and lactate and pyruvate can even give rise to glycogen synthesis (Dringen et al. 1993; Huang et al. 1994; Dringen et al. 2005). Lactate interacts with MAS. It cannot be used as a substrate for the pentose phosphate shunt; however, cytosolic NADPH can be produced from lactate conversion by pyruvate carboxylase into OAA and by malic enzyme into pyruvate with the conversion of an $NADP^+$ to an NADPH.

2.2.3 Acetate, Fatty Acids, and Ketone Bodies

Most dietary lipids are triglycerides. After lipolysis glycerol is converted via glyceraldehyde 3-phosphate to pyruvate (Zabłocki and Bryła 1988), and fatty acids are β-oxidized leading to the formation of ac.CoA (Lynen 1953). The formed ac.CoA can either be metabolized as such by condensation with OAA (e.g., in the heart), or two molecules of ac.CoA can be converted to the ketone bodies acetoacetate and its reduced metabolite β-hydroxybutyrate, a reaction occurring in the liver (McPherson and McEneny 2012). After release from liver cells circulating ketone bodies are transported by MCTs (Vannucci and Simpson 2003) and utilized as metabolic substrates in several organs, but normally not to an major degree in adult brain. This is in contrast to brain energy metabolism in young animals where MCT activity in the blood–brain barrier is high (Cremer et al. 1976) and ketone bodies plentiful in blood. However, ketone bodies, mainly in the reduced form of β-hydroxybutyrate, accumulate when glucose metabolism is failing, e.g., in diabetes. Ketone bodies also accumulate during fasting (due to metabolism of fatty acids), and Owen et al. (1967) showed (1) a considerable arteriovenous difference for ketone bodies across the brain of obese patients treated with starvation and (2) that ketone bodies accounted for ~60 % of brain oxygen consumption in these patients. An increased ketone body concentration in the systemic circulation and upregulation of blood–brain barrier permeability may have contributed to the increase (Robinson and Williamson 1980), but an elevation in D-β-hydroxybutyrate dehydrogenase activity has also been described in the brains of fasting rats (Smith et al. 1969).

That ketone body oxidation to approximately 50–60 % can replace glucose oxidation in intact brain has repeatedly been confirmed (Zhang et al. 2013). Chowdhury et al. (2014) compared rate of β-hydroxybutyrate metabolism in awake, mildly anaesthetized (halothane), and in deeply anaesthetized (pentobarbital) rats with isoelectric brain activity. In the awake animals ketone body oxidation increased with its plasma concentration and at a saturating plasma β-hydroxybutyrate concentration of 17 mM reached a plateau of ~60 % of total substrate oxidation in brain cortex, a similar value as that observed by Owen et al. (1967). However, in the deeply anaesthetized rats the much less active neuronal TCA cycle was fully supported by ketone bodies at a β-hydroxybutyrate plasma level of ~12 mM, in spite of the fact that the animals were euglycemic. Accordingly, ketone bodies can provide complete energetic support of basal (nonsignaling) processes in brain when available in sufficient concentration, but they are unable to fully support signaling processes even at the highest concentrations. By subtracting the nonsignaling dependent component Chowdhury et al. (2014) concluded that in the awake rat cerebral cortex approximately 50 % of neuronal energetics required for signaling cannot be replaced by β-hydroxybutyrate and has to derive from glucose oxidation (Fig. 2.4).

The functional capabilities of ketone bodies compared to glucose are likely to be related to the different metabolic pathways followed in their oxidative degradation. β-Hydroxybutyrate is oxidized to acetoacetate intramitochondrially, and acetoacetate is converted to two molecules of ac.CoA which then condense with OAA to form citrate. Since the oxidation of β-hydroxybutyrate is intramitochondrial it does not compete with oxidation of glyceraldehyde 3-phosphate. This is in contrast to lactate, and metabolism of β-hydroxybutyrate does also not share the ability of glucose and lactate to interact with MAS. In contrast to lactate, it can only support PDH-mediated activity, not pyruvate carboxylation. It supports not only neuronal (Amaral 2013; Chowdhury et al. 2014) but also astrocytic PDH activity (Melø et al. 2006; Gibbs et al. 2009; Jiang et al. 2011; McKenna 2012). Similarly, Pan et al. (2002) showed in awake human subjects that the labeling pattern from β-hydroxybutyrate was similar to that of glucose and calculated a similar relative percentage contribution to neuronal and glial oxidative metabolism as has been found with glucose. This is in contrast to acetate, which in cell culture is almost exclusively metabolized (although only via ac.CoA) by astrocytes (Waniewski and Martin 1998), confirming previous findings in brain and retina (Muir et al. 1986). In ^{13}C MRS studies on humans and rodents the labeling kinetics have also shown glial localization of acetate metabolism (Lebon et al. 2002; Jiang et al. 2013). Although acetate is normally found in low levels in the blood stream it can be substantially elevated by alcohol consumption to levels over 1 mM. At these levels it has been measured using ^{13}C MRS to account for as much as 50 % of glial oxidative ATP needs, particularly in subject groups exposed to hypoglycemia such as intensively insulin-treated type 1 diabetes patients (Gulanski et al. 2013). Acetate has even been reported to have protective effect against hypoglycemic seizures (Urion et al. 1979), which would suggest a major contribution by astrocytes to the seizures. Consumers of large amounts of alcohol also show increased acetate metabolism, reflecting that alcohol is metabolized to acetate (Jiang et al. 2013; Volkow et al. 2013, 2015).

2.3 The Glutamine–Glutamate/GABA Cycle

2.3.1 Synthesis and Degradation of Glutamate and GABA

In the adult brain perfusion techniques have shown very limited transport of glutamate and its precursor glutamine from blood to the brain, and changes in plasma concentration of glutamine have only little influence on brain glutamate content (Smith 2000; Patel et al. 2015). Accordingly, neurotransmitter glutamate and its decarboxylation product GABA must be synthesized within the brain. As discussed in Sect. 2.2.2 formation of glutamate requires synthesis of a new molecule of a TCA cycle intermediate and both PDH and PC activities. It therefore occurs in astrocytes but not in neurons. Transport to neurons is mediated by the glutamine–glutamate/GABA cycle where formation of neuronal glutamate accounts for about 75–80 % of the flux (measured from neurons toward astrocytes) and the remaining 20–25 % is made up by GABA (Patel et al. 2005, 2015; Duarte and Gruetter 2013). A recent study of glutamate amino decarboxylase 67 KD isoform GAD67 knock out mice suggested lower contributions by GABA, but the possibility of an underestimate was mentioned (Walls et al. 2015).

The cycle matures slowly, i.e. after 1 month in rats (reviewed by Chowdhury et al. 2007; Hertz 2013; Brekke and Morken 2015) and in the adult brain it has a flux equal to rate of glucose utilization in neurons (Sibson et al. 1998; Chowdhury et al. 2007; Hyder et al. 2013a, b) or ~75 % of the total rate of glucose utilization in brain. However, only a minor part (25 % or less) of the glutamate flux from astrocytes to neurons represents newly synthesized glutamate, whereas the remainder is made up of glutamate that has already been released as transmitter but subsequently is accumulated by astrocytes and returned to neurons in the cycle (Rothman et al. 2011). Almost all released transmitter glutamate is treated in this manner, whereas only little is reaccumulated directly into glutamatergic neurons (Danbolt 2001; Zhou and Danbolt 2013). Some neuronal reuptake has been described in cultured hippocampal neurons (Coco et al. 1997) but might be due to the young age of the cultures (see Hertz 2013).

In the long term the glutamate concentration in adult healthy brain must stay unaltered (Lebon et al. 2002; Sonnewald 2014), meaning that a fraction similar to that synthesized de novo (~25 %) must be oxidatively degraded in astrocytes, or to a much lesser extent lost from the brain as glutamine, a process mainly occurring during ammonia detoxification (Zielińska et al. 2014). For oxidative degradation glutamate is converted to α-KG, which after circling in the TCA cycle to malate can exit into the cytosol and be converted by the astrocyte-enriched (Kurz et al. 1993) cytosolic malic enzyme to pyruvate. Pyruvate can then enter the TCA cycle and be oxidatively degraded (Fig. 2.1). However, Sonnewald (2014) hypothesized that a relatively low degree of pyruvate recycling (turnover of pyruvate or one of its metabolites in the TCA cycle to regenerate pyruvate) suggested that instead pyruvate was released from the cells as lactate, which would constitute a similarly effective cataplerosis. This hypothesis is supported by her previous observation that lactate generated from TCA cycle intermediates is found in the medium (Sonnewald et al. 1993), but these experiments were performed in cultured cells known to release very large amounts of lactate. On the other hand, labeled ac.CoA was derived from TCA cycle intermediates in brain mitochondria, demonstrating malic enzyme activity and pyruvate recycling by Bakken et al. (1997) from the same group. Also, release from brain of specifically glutamate-derived lactate would cause a very considerable decrease in the amount of ATP produced by astrocytes. A complication in determining pyruvate recycling using labeling methods is also that the astrocyte-derived pattern of labeling it produces in lactate will be substantially diluted by the large amount of lactate from neuronal metabolism. Furthermore, similar scrambling occurs in glucose released by the liver which can make determination ambiguous.

A third possibility is that some of the ~25 % of returning glutamate which is converted to α-KG is only partly degraded. This would happen if malate did not leave the astrocytic TCA cycle but was converted to OAA, condensed with a new molecule of ac.CoA and further metabolized via α-KG to glutamate. Such a process would allow formation of another molecule of glutamate and glutamine using only one pyruvate molecule and requiring no pyruvate carboxylation. Accordingly it would not require glucose or lactate utilization but could also use β-hydroxybutyrate as the source. Maciejewski and Rothman (2008) described several alternate pathways of glutamate neurotransmitter cycling that involved glutamate oxidation but did not require pyruvate carboxylation, but at present there is no evidence of their importance for the glutamine–glutamate/GABA cycle in the brain in vivo, at least under normal conditions (Rothman et al. 2012).

In the short term glutamate content in brain can transiently increase, e.g., during learning (Gibbs et al. 2007), in some cases of epileptic seizures and during visual stimulation (Bjørnsen et al. 2007; Peca et al. 2010; Mangia et al. 2010, 2012; Perez et al. 2012). The involvement of astrocytic metabolism during anaplerosis/cataplerosis provides huge possibilities for astrocytic regulation. However, even astrocytic involvement only during return of previously released transmitter without conversion of glutamate to α-KG enables some astrocytic regulation, because glutamine can move within the astrocytic syncytium (Cruz et al. 2007). Perhaps glutamate released from neurons might even be redirected to GABA-ergic neurons and vice versa. Formation of a new molecule of glutamate by conversion of glutamate-derived malate to α-KG and glutamate in the TCA cycle and incorporation of ac. CoA during this process would increase the astrocytic influence.

2.3.2 *Individual Steps of Glutamate Formation and Degradation*

An important, debated question is whether the initial conversion of α-KG to glutamate in astrocytes is catalyzed by glutamate dehydrogenase (GDH) or by aspartate aminotransferase (AAT). Westergaard et al. (1996) showed that in cultured astrocytes conversion of a-KG (Greek alpha) to glutamate is catalyzed by AAT wheres glutamate oxidation is mediated by GDH. The equilibrium constant for AAT is close to unity (Krebs 1953), making the reaction easily reversible. In contrast, in the GDH reaction the reductive amination to glutamate is thermodynamically favored (Engel and Dalziel 1967). However, a high K_m for ammonia together with a high NAD^+:NADH ratio in brain may enable oxidative deamination of glutamate (Zaganas et al. 2001, 2009). GDH is highly regulated by allosteric activators and inhibitors (Schousboe et al. 2013), whereas AAT is dependent upon the presence of suitable transactivation partners, for example, conversion of aspartate to OAA concomitant with glutamate formation from α-KG.

One of us (L.H) bears some responsibility for the concept of a large involvement of GDH in glutamate oxidation, since this process was first suggested to depend on GDH in his laboratory, based on findings in cultured astrocytes (Yu et al. 1982). This finding has repeatedly been confirmed (e.g., McKenna et al. 1996), and GDH is also the enzyme involved in glutamate degradation in astrocytes obtained from brain tissue (Whitelaw and Robinson 2013). However, in both of these situations

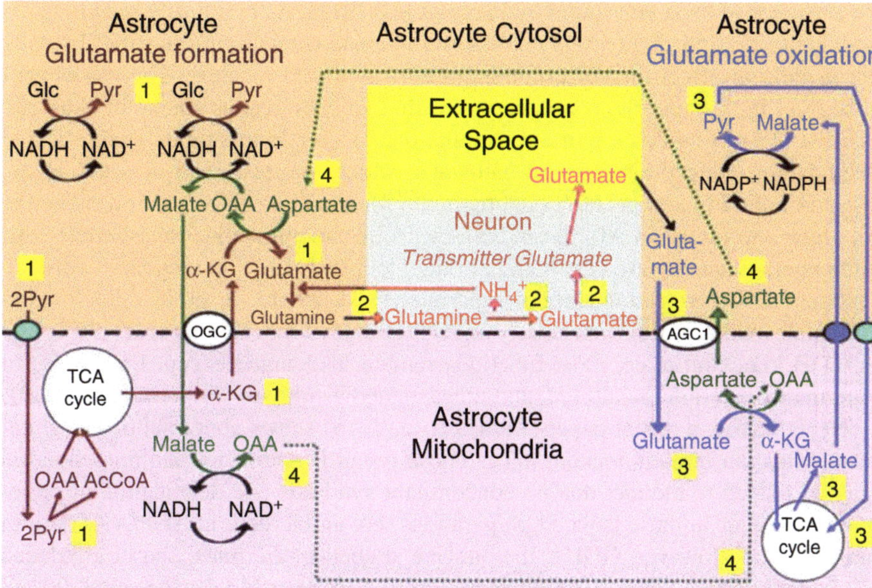

Fig. 2.3 Proposed pathway for coupled production and metabolism of transmitter glutamate using aspartate transamination for exchange between α-ketoglutarate and glutamate. Joint pyruvate carboxylase and pyruvate dehydrogenase activation generates a "new" molecule of citrate (*lower left corner*) as detailed in Fig. 2.1. Citrate-derived α-ketoglutarate exiting the mitochondrial membrane leaves the astrocytic TCA cycle and is transaminated with aspartate to form glutamate, with concomitant oxaloacetate (OAA) formation from aspartate. The mitochondrial exit of α-ketoglutarate occurs via the ketoglutarate/malate exchanger, generally acknowledged to be expressed in astrocytes, and the cytosolic malate with which is exchanged, is generated via NADH-supported reduction of oxaloacetate formed from aspartate (Pardo et al. 2011). Glutamate is amidated to glutamine (pathway 1), which is transferred to glutamatergic neurons (without indication of any extracellular space in the figure) and extracellular release as transmitter glutamate (pathway 2), and subsequent reuptake of glutamate and oxidative metabolism in astrocytes (pathway 3). During oxidative metabolism transamination of glutamate to α-ketoglutarate generates aspartate that can be used in the transamination of α-ketoglutarate to glutamate in pathway 1 after transfer via pathway 4. Metabolism of α-ketoglutarate via malate is only shown to pyruvate. Biosynthesis of glutamine is shown in brown and metabolic degradation of glutamate in blue. Redox shuttling and astrocytic release of glutamine and uptake of glutamate are shown in *black*, and neuronal uptake of glutamine, hydrolysis to glutamate, and its release is shown in *red*. Reactions involving or resulting from transamination between aspartate and oxaloacetate (OAA) are shown in green. *Small blue circle* shows pyruvate carrier into mitochondria and *small purple circle* malate carrier out from mitochondria. *AGC1* aspartate/glutamate exchanger, aralar; *α-KG* α-ketoglutarate; *Glc* glucose; *Pyr* pyruvate; *OGC* malate/α-ketoglutarate exchanger (From Hertz 2013)

the astrocytes are isolated without the possibility of functional interactions with neurons and a complete glutamine–glutamate/GABA cycle. A suggested, but not proven, functional interaction between glutamate anaplerosis and cataplerosis (Hertz 2011a, 2013) is therefore not able to operate. The suggestion of such a correlation between glutamate formation and degradation (Fig. 2.3) was triggered by a large stimulation of glutamate/glutamine formation in astrocytes in intact brain in the presence of aspartate found by Pardo et al. (2011). As previously mentioned, a high concentration of aspartate facilitates glutamate formation from α-KG by AAT but it will not affect the corresponding GDH-mediated reaction. The model proposed in Fig. 2.3 represents an attempt to find an endogenous source of aspartate capable of exerting this important stimulation. As seen in this figure synthesis and degradation along the suggested pathways, which are described in detail in the legend of this Fig and in Hertz (2011a, b, 2013), would be able to supply aspartate for glutamate synthesis. Moreover, both cytosolic and mitochondrial aspartate aminotransferase activity are very high in brain (Cooper 2013), allowing rapid nitrogen exchange between glutamate and aspartate. Furthermore, in brain mitochondria glutamate oxidation depends mainly on AAT (Balazs 1965; Dennis et al. 1977), and in GDH knockout mice, most functions remain unchanged except for a reduced glutamate oxidation in cultured, and thus isolated astrocytes (Frigerio et al. 2012).

Nevertheless, a recent paper (Karaca et al. 2015) shows abnormalities in glutamate oxidation in the knockout mice. Those found in astrocytes and mitochondria may be related to the fact that no concomitant synthesis and degradation of glutamate can occur in these isolated preparations, but an increase in ADP/ATP ratio in brain cannot. However, GDH1, the enzyme studied in the mice, shows a reduced K0.5 for ammonia when ADP is lowered from 1 to 0.1 mM (Zaganas et al. 2013), which may enhance glutamate oxidation when GDH is operating. Elimination of this mechanism in the knockout animals might possibly explain the increased ADP/ATP ratio. However, the findings do suggest that also in the brain in vivo GDH does contribute to glutamate formation/oxidation although the contribution might be minor. This would be consistent with a ten times higher activity of AAT than of GDH in mitochondria (compare Zaganas et al. 2001 with McKenna et al. 2006), and it would not disagree with the observations by Balazs (1965) and Dennis et al. (1977). They also confirm that glutamate is an important metabolic substrate (McKenna 2012, 2013) perhaps more because of its direct access to the TCA cycle than of the amount oxidized which must be much smaller than that of glucose, since brain glutamate exclusively originates from its astrocytic production from glucose and astrocytes only account for about one quarter of brain volume and metabolism (reviewed by Hertz 2011b). Finally, although the model suggested in Fig. 2.3 requires that glutamate formation from α-KG and glutamate degradation to α-KG are catalyzed by the same enzyme that enzyme may not catalyze 100 % of the interconversion.

An alternative mechanism for anaplerotic resynthesis of glutamate from α-KG was originally proposed by Yudkoff et al. (1996), who found a rapid uptake of the branched chain amino acid (BCAA) leucine into cultured astrocytes (V_{max} 54 nmol/mg of protein/min and K_m 450 μM). Astrocytic accumulation of leucine was three times greater in the presence of alpha-aminooxyacetic acid (AOAA), a transamination inhibitor, suggesting that these isolated astrocytes rapidly transaminate leucine to alpha-ketoisocaproic acid (KIC), which they then release into the extracellular fluid. This finding was linked to glutamate oxidation by Hutson and coworkers (Hutson et al. 1998, 2001), who suggested that mitochondrial branched chain amino acid transferase (BCATm) donates an amino nitrogen group to α-KG, forming glutamate which is then converted by glutamine synthase to glutamine. In their original model the amino nitrogen was fixed in the neuron by GDH and transferred by cytosolic neuronal BCAT (BCATn) to a branched chain keto acid, which then shuttled from the neuron to the astrocyte as a BCAA. However, it is now believed that this is more likely an internal glial reaction, based on lack of evidence that neuronal GDH can mediate glutamate synthesis. Rothman et al. (2012) determined the flux via BCAT in brain tissue vivo to ~0.1 μmol/nmol/mg, which would just be enough to catalyze transamination of the 20 % of glutamate synthesis that represents anaplerosis. Although it is unusual that a transaminase activity is similar to the net rate of the actual process (usually being several to many times greater) BCAT activity might therefore catalyze anaplerotic synthesis of glutamate. An advantage of using BCAT is that BCAAs are continuously supplied to the brain from the blood, and the nitrogen all four BCAAs together bring is almost stoichiometrically equivalent to the amount that leaves the brain as glutamine (Rothman et al. 2012 and see later). This means that the reaction described by Yudkoff et al. (1996) would suffice to generate glutamate from α-KG. The rate of α-KG synthesis (20–25 % of ~0.6 nmol/mg of protein/min) is at least as high, and in the pathways suggested in Fig. 2.3 the amount of aspartate supplied equals the glutamate uptake and metabolism in astrocytes (Hertz et al. 1978; McKenna 2012), which is much more intense than those of the BCAAs. Important evidence against participation of BCAT in glutamine–glutamate/GABA cycle glutamate synthesis is that the 50 % stimulation of glutamate synthesis caused by administration of aspartate could not be replicated by administration of the branched chain amino acid leucine (Pardo et al. 2011). Thus, use of BCAT for glutamate synthesis related to neurotransmitter cycling, as opposed to nitrogen balance, in intact tissue under normal conditions is not likely. However, in the situations where synthesis of glutamate exceeds its metabolic degradation (and thus rate of aspartate formation) BCATs could play a role in glutamate synthesis by converting BCAAs to branched chain keto acids. The latter could be reconverted to BCAAs when cataplerosis catches up with the increased anaplerosis. GDH maight also play a role. Glutamate is converted to glutamine by the astrocyte-specific (Martinez-Hernandez et al. 1977; Norenberg and Martinez-Hernandez 1979; Hutson et al. 2008; Anlauf and Derouiche 2013) enzyme glutamine synthetase (GS), which is localized in the cytosol. Inhibition of this enzyme in retinal Müller cells (an astrocyte-like cell) almost immediately interrupts glutamatergic activity in the retina (Barnett et al. 2000). The inhibition also increases retinal cell death (Gorovits et al. 1997; Bringmann et al. 2013), whereas a hormonally induced increase in Müller cell glutamine synthase GS protects against neuronal injury (Ola et al. 2011). Similarly acute administration of MSO to inhibit GS in vivo leads to rapid elevation of extracellular glutamate and can result in seizure and chronic administration will lead to the development of epilepsy (Eid et al. 2008).

Enhanced metabolism of BCAAs also counteract ammonia toxicity (Watanabe et al. 1986; Muto et al. 2005), and a nonnegligible oxidative metabolism (following transamination) and incorporation of BCAAs into protein in cultured astrocytes is strongly inhibited by a toxic concentration of ammonia (Murthy and Hertz 1987). At one time ammonia detoxification was believed to be the major pathway of glutamine synthesis but it was shown to be extremely slow compared to that in the glutamate–glutamine cycle, even during hepatic encephalopathy (Shen et al. 1998; Sibson et al. 2001; Keiding et al. 2006; Cudalbu et al. 2012). Under hyperammonemic conditions the glutamate formed for detoxification is probably formed by glutamate dehydrogenase GDH working in the glutamate synthesis direction due to the elevated NH_4^+ concentration (Shen et al. 1998; Cooper 2012; Cudalbu et al. 2012; Kanamori and Ross 1995), but the ammonia interactions with BCAAs may suggest that they are also involved.

Glutamine exit from astrocytes and entry intro neurons are equally important as glutamine synthesis for regulation of de novo synthesis of glutamate and GABA and it is essential even during the return of released transmitter glutamate to neurons after its initial uptake in astrocytes. The amino acid transporter SN1 mediates electroneutral and bidirectional glutamine transport in astrocytes (Nissen-Meyer and Chaudhry 2013). Its activity is regulated by many factors, e.g., by extracellular pH, because protons compete with Na^+, which is required for its transport activity. There are also consistent observations that SN1 is down-regulated by protein kinase C phosphorylation, probably by internalization (Nissen-Meyer and Chaudhry 2013). After its release to the extracellular space glutamine is accumulated into glutamatergic neurons by a different transporter which is coupled to Na^+, SAT1, or at some places SAT2. It is subsequently converted to glutamate by phosphate-activated glutaminase (PAG).

As discussed in more detail by Hertz (2013) and Schousboe et al. (2013), the conversion of glutamine to glutamate in glutamatergic neurons is remarkably complex. After PAG-mediated deamidation within the mitochondrial membrane glutamate enters the mitochondrial matrix and is returned from here to the cytoplasm in a "pseudomalate-aspartate shuttle." This has been demonstrated both in cultured astrocytes (Palaiologos et al. 1988, 1989) and in mitochondria (Ziemińska et al. 2004; Bak et al. 2008). The "pseudomalate-aspartate shuttle" is identical to the 'real' MAS (Fig. 2.2) with the exception that glutamate on the cytosolic side is generated by deamidation of glutamine by PAG, abolishing the need for reaccumulation into mitochondria of cytosolic glutamate formed from α-KG by transamination. This glutamate thus becomes freely available in the cytosol for use as transmitter glutamate. Since cytosolic malate formation is an indispensable part of MAS (Fig. 2.2) and thus also of pseudo-MAS, formation of transmitter glutamate must also coincide with cytosolic reduction of NAD^+ to NADH. This will in general mean formation of one molecule of pyruvate from glucose, i.e., utilization of half a molecule of glucose. The energy obtained from this process and subsequent oxidation of glucose does not need to be used in the glutamine–glutamate cycle, which only requires little ATP to function (for glutamate and glutamine uptake and glutamine synthesis). However, it contributes substantially (1/2 molecule glucose per molecule glutamate formed) to the ratio between glucose (i.e., 2 pyruvate) utilization and glutamate flux, which as shown in Fig. 2.4 equals approximately 85 % of cerebral-cortical neuronal glucose oxidation in the resting awake state and 60–70 % of total cerebralcortical glucose oxidation (with the remainder due to glia and GABAergic neurons primarily) (Sibson et al. 1998; Rothman et al. 2011). Since the "pseudo-MAS" only operates in glutamate formation which accounts for 80 % of the total flux the contribution must be about 40 %. Anaplerotic astrocytic glutamate formation from glucose which accounts for about one quarter of total flux must account for another 20 %.

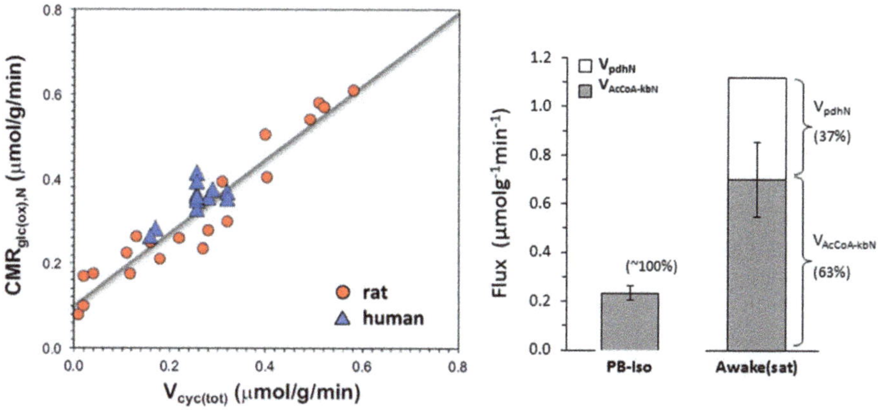

Fig. 2.4 Rate of the glutamate/glutamine cycle versus glucose oxidation and ability of ketone oxidation to displace glucose. *Left*: Measured relation between the rate of the glutamate/glutamine cycle in rat cerebral cortex versus neuronal glucose oxidation based on 12 published studies in rodents and 9 in humans (adapted from Hyder et al. 2013b). The relation is close to linear throughout the range with a slope of close to 1:1. *Right*: Saturating ketones dark part of columns can displace all of the glucose oxidation under isoelectric conditions but only approximately 50 % of the glucose oxidation needed to support neuronal signaling in glutamatergic neurons (adapted from Chowdhury et al. 2014). This percentage is in agreement with the prediction of the pseudo malate asparate shuttle model which is discussed in the text (and see Figs. 2.1, 2.2, and 2.3) VAcCoA-kbN, rate of neuronal acetyl-CoA utilization from ketone bodies; VpdhN, neuronal pyruvate dehydrogenase flux (i.e., rate of acetyl-CoA utilization from glucose)

After glutamate has been released from neurons as a transmitter most glutamate is accumulated in astrocytes by the powerful glutamate transporters Glut-1 and GLAST (Danbolt 2001; Zhou and Danbolt 2013). Although some glutamate may also be taken up by axonal terminals in glutamatergic neurons this is not sufficient to maintain glutamate homeostasis (Bjørnsen et al. 2014). In astrocytes a part (20–25 %) is oxidized as already discussed, whereas the remaining 75–80 % is again converted to glutamine and returned to neurons for reuse as transmitter (Rothman et al. 2011). However, total flux in the glutamine–glutamate/GABA cycle is so large and the association between glutamate uptake and metabolism so close that the fraction of glutamate which is oxidized in astrocytes is more than high enough to support its own energy-consuming uptake (McKenna 2012). Metabolic utilization of glutamate is strongly supported by a close association between astrocytic glutamate transporters and metabolic enzymes demonstrated by the Robinson group (Genda et al. 2011; Bauer et al. 2012; Whitelaw and Robinson 2013; Jackson et al. 2014). A close functional interaction between formation and degradation of glutamate (Fig. 2.3) is also consistent with the quantitative similarity between anaplerosis and cataplerosis and the evidence that the latter also mainly occurs in astrocytes. However, the conclusion that glutamate is a metabolizable substrate is in distinct contrast to a view expressed in a recent "Neuron" review by Magistretti and Allaman (2015), despite the studies discussed earlier and findings that human, cat, and guinea pig brain slices show at least as high rates of oxygen uptake with glutamate alone as the substrate as with glucose alone (McIlwain 1953; Kratzing 1953; Takagaki and Tsukada 1957).

The presumed pathway for oxidative metabolism in intact tissue (Fig. 2.3) is relatively simple: Glutamate enters the mitochondria in exchange with aspartate, formed from OAA generated during glutamate synthesis. It is subsequently transaminated to α-KG in the same transamination process that formed aspartate from OAA, and α-KG enters the TCA cycle. One problem with this hypothesis is that glutamate may not necessarily be synthesized and degraded in the same astrocyte. One potential solution to this problem may be that lactate, which is rapidly transported between astrocytes (Gandhi et al. 2009), may be metabolized and facilitate glutamate synthesis when return of aspartate is deficient as a result of reduced glutamate oxidation (Hertz et al. 2014). For example, lactate might support production of OAA and thus of aspartate in pathway 4 in Fig. 2.3 if insufficient glutamate is being oxidized in a specific astrocyte. As already mentioned BCAA transamination or GDH activity may also alleviate this problem.

2.3.3 *Individual Steps of GABA Formation and Degradation*

It has been mentioned that GABA-derived glutamate accounts for ~20 % of the total glutamine–glutamate/GABA cycle flux from neurons toward astrocytes in the brain in vivo (Patel et al. 2005), and an approximately similar magnitude was found by Duarte and Gruetter (2013) and Patel et al. (2015). This means that in spite of a considerable rate of GABA reuptake into GABA-ergic neurons (the functional importance of which will be discussed later) GABA synthesis ultimately requires operation of the glutamine–glutamate/GABA cycle. It is likely that glutamine production and release in astrocytes for GABA production proceeds along similar pathways as those for glutamate production. It is therefore of interest to establish whether astrocytic formation and degradation of GABA might be interconnected in a similar manner as suggested for glutamate. Figure 2.5 shows that this might well be the case, although the needed conversion of GABA to glutamate before it can be returned to astrocytes in the glutamine–glutamate/GABA cycle makes the pathway for astrocytic return of GABA as glutamate more complex than that of transmitter glutamate.

GABA uptake has repeatedly been demonstrated in cultured astrocytes (Schousboe et al. 1977b; Hertz et al. 1978), and high GABA transporter currents have been measured in striatal astrocytes by patch-clamp intracellular recording (Goubard et al. 2011). Initially released GABA must after its cytosolic uptake be transferred to the mitochondria, since the conversion of GABA to succinic semialdehyde (SSA) is initiated by GABA transaminase (GABA-T) (Wong et al. 1974), which is a mitochondrial enzyme (Schousboe et al. 1977a). Figure 2.5 suggests a glutamate/GABA antiporter at the mitochondrial membrane, although such a transporter has presently not been demonstrated in mitochondria. It has, however, been shown in the bacterial membrane (Sa et al. 2015).

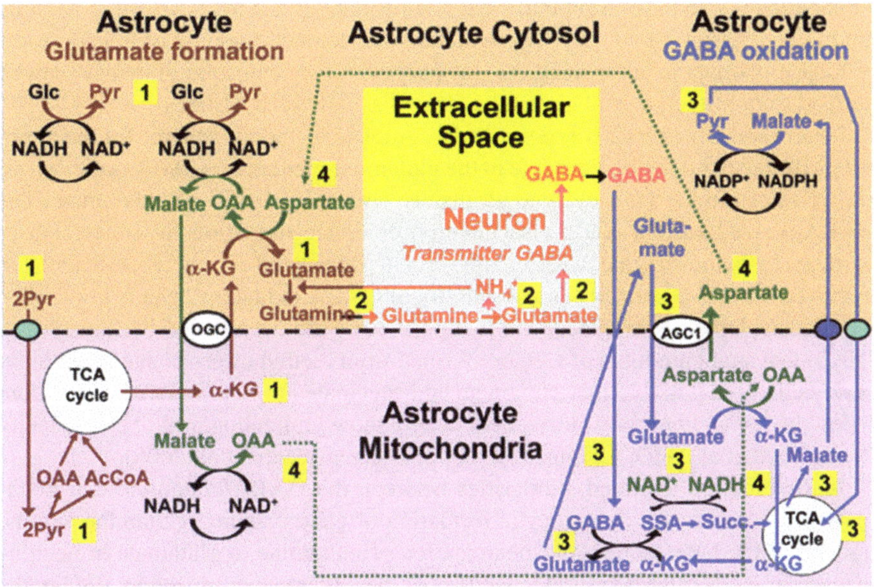

Fig. 2.5 Proposed pathway for coupled production and metabolism of transmitter GABA using aspartate transamination for exchange between α-ketoglutarate and glutamate. Pathways, color coding, and most abbreviations are as in Fig. 2.3. Most differences are in the lower right corner of the figure. The suggested glutamate-GABA exchange at the mitochondrial membrane has not been described in mammalian brain, but metabolism of GABA via succinic semialdehyde (SSA) to succinate (Succ.), a TCA cycle constituent is a well-established pathway. Metabolism of succinate via malate is only shown to pyruvate. As in Fig. 2.3 glutamate metabolism provides aspartate that can be used in the transamination of α-ketoglutarate to glutamate in pathway 1. It also generates α-ketoglutarate needed for the transamination of GABA to SSA

SSA is further oxidized to succinate, which enters the mitochondria. Figure 2.5 also suggests that glutamate reenters the mitochondria through the glutamate-aspartate exchanger (requiring aralar) and that the simultaneous exit of aspartate from mitochondria to cytosol as in Fig. 2.3 may secure the AAT substrate needed for transamination-mediated glutamate production at the initial step of GABA production. As in the metabolism of accumulated glutamate this aspartate is formed from OAA generated during the initial formation of precursor glutamate (Pathway 4) in the transamination process converting glutamate to α-KG. Instead of being oxidized α-KG retransamination to glutamate enables concomitant GABA transamination to SSA and at the same time creates the glutamate suggested to exit the mitochondria in exchange with GABA. McKenna and Sonnewald (2005) have provided some experimental evidence for these suggestions. In a study of metabolism in cultured astrocytes they found increased aspartate content and larger formation of [U-^{13}C]aspartate from added [U-^{13}C]glutamate when the cells were incubated with nonlabeled GABA. This was tentatively interpreted as caused by increased entry of [U-^{13}C] glutamate into the TCA cycle to allow for the transamination of GABA. This explanation is strikingly similar to what is shown in Fig. 2.5 for GABA degradation, but in intact tissue, where GABA production and degradation may be linked, extracellular glutamate is not needed for this purpose, since cytosolic glutamate can be derived from mitochondrial glutamate formed as described earlier.

Duarte and Gruetter (2013) described a complete metabolic scheme for the return of GABA taken up by the astrocyte in the glutamine–glutamate/GABA cycle, where malate stayed in the TCA cycle to ultimately give rise to α-KG and glutamate. The GABA-derived carbon could also undergo complete oxidation by conversion to pyruvate by malic enzyme, which may be reintroduced and completely metabolized in the TCA cycle (Fig. 2.5). The possibility of GABA oxidation, which would then require anaplerosis for replacement is supported by the observation by Zhang et al. (1995) that small amounts of CO_2 are formed from methyl esters of succinate in rat brain cells. It is possible that glutamate and succinate are accumulated in different types of mitochondria, since astrocytes show mitochondrial heterogeneity (Sonnewald et al. 1993; Collins et al. 2002; Waagepetersen et al. 2006).

In spite of the assumed similarities between the GABA/glutamine and glutamate/glutamine cycles in astrocytic formation of glutamate and glutamine and the transfer of the latter to neurons, deamidation of glutamine to glutamate in neurons may occur in a very different manner. Part of the glutamate produced by PAG in the intermembrane space of the mitochondria is released directly to the cytosol (the direct pathway) whereas another fraction is transaminated to α-KG in the mitochondria (the indirect pathway) (Waagepetersen et al. 2001; Leke et al. 2011). The contribution of each pathway to GABA synthesis is about the same. Both pathways were concluded to contribute to vesicular GABA release and to be involved in determining GABAergic tone resulting from exit of cytosolic GABA via reversal of uptake (Waagepetersen et al. 2001; Walls et al. 2011; Leke et al. 2011), for example, by exposure to elevated extracellular K^+ (Romei et al. 2014). The glutamate decarboxylase GAD 67 is functioning in both pools, whereas GAD 65 operates only in the vesicular pool (Waagepetersen et al. 2001) and is crucial for biosynthesis of synaptic GABA (Kaufman et al. 1991; Tian et al. 1999; Walls et al. 2011) and is the primary isoform in vivo responsible for activity-dependent GABA metabolism (Patel et al. 2006).

Waagepetersen et al. (1999, 2001) and Schousboe et al. (2013) have described that the α-KG entering the mitochondria undergoes a full turn of TCA cycle flux, regenerating α-KG. This conclusion was based on the finding of incorporation of isotope from 0.5 mM [U-^{13}C]glucose into GABA during a 4 h incubation of GABAergic cerebrocortical neurons. An important consequence of this complete turn in the TCA cycle is introduction of ac.CoA between OAA and citrate. If this ac.CoA is derived from glucose or lactate it can enable exit of α-KG from the mitochondria to the cytosol and its conversion to glutamate similar to what was described for astrocytes (pathway 1 in Figs. 2.3 and 2.5). This will, however, not be the case if the ac.CoA is derived from β-hydroxybutyrate, which enters the mitochondria directly. Cytosolic glutamate can then be decarboxylated by glutamate decarboxylase (GAD), a cytosolic enzyme (Balazs et al. 1966) to GABA. Thus, cycling of α-KG within the mitochondria during GABA synthesis via the indirect pathway may enable exit of glutamate from mitochondria, but it requires that the ac.CoA required to metabolize OAA to citrate is formed from glucose or lactate.

In contrast to the low rate of reaccumulation of glutamate in glutamatergic neurons, GABA is reaccumulated into GABA-ergic neurons in brain slices (Iversen and Neal 1968). Cell culture experiments have indicated that neuronal uptake exceeds that into astrocytes (Larsson et al. 1983; Yu et al. 1984; Schousboe et al. 2013). Müller cells in intact retina also show a high rate of GABA uptake, and in addition GABA-T is activated and NAD(P)H fluorescence increased at the mitochondrial location after GABA uptake (Biedermann et al. 2002). By aid of immunocytochemistry it has been shown that the transporter 3 (GAT 3) is expressed in astrocytes and in some animals also oligodendrocytes, whereas GAT 1 is expressed in neurons with very little astrocytic expression (Scimemi 2014a). In the mouse GABA from vesicular sources is accumulated by GAT1, while that from nonvesicular sources is taken up by the astrocytic GAT3 and 4 (Song et al. 2013). Blockade of uptake by the GAT1-specific inhibitor tiagabine during continued release of GABA should therefore be expected to reduce the neuronal content of GABA. Nevertheless, tiagabine has no effect on tissue amino acid levels and on ^{13}C enrichments from [2-^{13}C]acetate in lightly anesthetized rats (Patel et al. 2015). This is the case although similar concentrations of tiagabine cause a large increase in extracellular GABA (Fink-Jensen et al. 1992; Ipponi et al. 1999). Rather than enhancing neurotransmitter cycling tiagabine tended to reduce rates of neurotransmitter cycling in both glutamatergic and GABAergic neurons slightly without any change in the ratio between the two (Patel et al. 2015). These observations suggest that GABA taken up into GABAergic neurons from the extracellular space is either metabolized or released. In this context it is relevant that Romei et al. (2015) concluded that high-affinity uptake of GABA into nerve terminals seems to have functions other than recapture of the transmitter. The small impact of tiagabine on the astrocytic GABA/glutamine cycle may also reflect different K_m's of the transporters with much higher transient levels of GABA anticipated from vesicular release than back transport of GABA during depolarization. The high affinity of the neuronal GABA transporter is consistent with this possibility.

Romei et al. (2015) used synaptosomes from mouse cerebellum prelabeled with [^3H]GABA and superfused with GABA. Influx of GABA through GAT1 transporters stimulated efflux of [^3H]GABA. The authors suggested that this was partly by homoexchange, although such a process had not been found for GABA in cultured neurons by Hertz et al. (1978). Similarly Scimemi (2014a, b) presented evidence that not only GABA biosynthesis but also GABA uptake is essential to sustain GABAergic synaptic transmission. Based on these results, those by Duarte and Gruetter (2013) and Chowdhury et al. (2014) and the observation by Biedermann et al. (2002) that Müller cells show enhanced GABA-T expression and stimulation of mitochondrial function after GABA uptake we hypothesize that metabolism of accumulated GABA in neurons may facilitate the transport between mitochondrial and cytosolic compartments needed for production of GABA via the indirect pathway. This suggestion is not in disagreement with the experimental findings by Waagepetersen et al. (1999, 2001), since the 4-h incubation period used by these authors provides ample time for GABA release, reuptake, and metabolism.

In accordance with the observations by Waagepetersen et al. (2001) the suggested pathway assumes that for every two glutamine molecule one is converted in the cytosol to glutamate (light blue pathway in Fig. 2.6) from which GABA is formed by decarboxylation and released by arrow 1. The other glutamine molecule enters the mitochondria where it is deamidated to glutamate and converted by transamination to α-KG, contingent upon the presence of OAA for simultaneous conversion to aspartate (sea-blue pathway). This OAA becomes available when released GABA, which at the beginning originates exclusively from the cytosol, is reaccumulated (gray arrow and arrow 2) and metabolized. As in astrocytes, GABA is assumed to enter the mitochondria in exchange with glutamate. Here it is transaminated to SSA (in contrast to the pathways shown in Figs. 2.3 and 2.5 this process has to be a transamination), with concomitant transamination of the α-KG, formed from the glutamine that initially entered the mitochondria, to glutamate. This glutamate exits to the cytosol via a glutamate carrier (sea-blue pathway) and from there the extracellular space, from where it is accumulated by astrocytes for return in the glutamine–glutamate/GABA cycle or for metabolic degradation. Accordingly, the glutamine accumulated into the mitochondria does not give rise to any formation of GABA, but it is essential for the transamination of incoming GABA to SSA, and it also delivers aspartate for the subsequent conversion of succinate-derived alpha-KG to glutamate and glutamate for return to astrocytes. SSA is oxidized to succinate which is metabolized via OAA, condensing with ac.CoA to alpha-KG which is transaminated to glutamate, and conversion of aspartate to OAA. This glutamate molecule exits to the cytosol (in exchange with GABA) and is again decarboxylated to a molecule of GABA which is released as synaptic GABA (arrow 3). Once the vesicular release is established, it is accordingly this GABA which is reaccumulated (arrow 4 and the pathway shown in red in Fig. 2.6), since Song et al. (2013) have shown that it is vesicularly released GABA which is accumulated by the neuronal GAT1. Thus, although the released and reaccumulated GABA does not give rise to any increase in GABA concentration it is used as the source for production of another glutamate molecule which can enter the cytosol and form releasable GABA. The whole process (direct and indirect pathways together) produces two molecules of GABA from two molecules of glutamine, but they are both formed from the glutamine molecule that entered the cytosol by the direct pathway. The indirect pathway is not used for direct production of GABA, but it facilitates

Fig. 2.6 (continued) from there to the extracellular fluid and probably accumulated into astrocytes (*Italics*) in the glutamine–glutamate/GABA cycle. Exchange of OAA, α-KG, and Asp between the two pathways is indicated in the figure, and there is stoichiometric balance between synthesis and utilization. Thus for each 2 Gln molecules the neurons receive from the astrocytes they return 1 Glu and 1 GABA molecule in the glutamine/glutamate/GABA cycle. These pathways are more complex than that originally suggested by Waagepetersen et al. (1999, 2001) but fully compatible with the experimental data obtained by these authors since the 4-h. incubation period used allows time for GABA uptake and metabolism and release via *arrow 3*. In vivo data by other authors supporting the proposed pathways are cited in the text. To prove or disprove this pathway it will be necessary to study the effect of extracellular GABA on GABA content in GABAergic neurons in the presence of extracellular glutamine

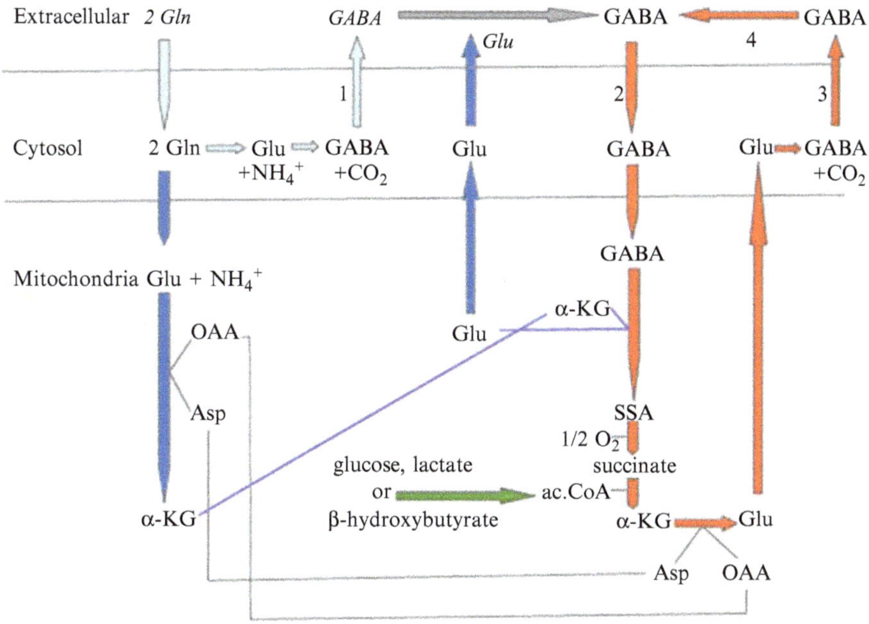

Fig. 2.6 Proposed pathways for production of cytosolically and vesicularly released GABA in GABAergic neurons from glutamine (Gln) released from astrocytes and accumulated into the astrocytic cytosol. Since initially one half of accumulated Gln deamidated to glutamate by glutaminase (PAG) located in the intramembranaceous space enters directly in the cytosol (*light blue*), whereas another one half enters the mitochondria in a similar manner as in glutamatergic neurons (Waagepetersen et al. 1999, 2001) the figure depicts the metabolic fate of two molecules of glutamine (*Italics*), whereas each of the two pathways follows the metabolism of one molecule Glu. Glu released directly to the cytosol is decarboxylated to GABA and released to the extracellular space (*light blue and arrow 1*). When GABAergic signaling is initiated it is reaccumulated into the cytosol of the GABAergic neuron (*red and arrow 2*) and from there into the mitochondria, probably in exchange with glutamate. Here it is metabolized via succinic semialdehyde (SSA) to succinate which is accumulated into the TCA cycle, where it is metabolized to α-ketoglutarate (α-KG), probably explaining the incorporation of radioactivity from the TCA cycle in released GABA shown by Waagepetersen et al. (1999, 2001). This cycling requires condensation of oxaloacetate (OAA) with acetyl coenzyme A (co.A), which may be derived from either glucose, lactate, or β-hydroxybutyrate (green arrow). α-KG is transaminated to Glu, which exits from the mitochondria (probably in exchange with incoming GABA), is decarboxylated to GABA, which is released to the extracellular space (*red and arrow 3*). Results by Song et al. (2013) show that during established GABAergic signaling in vivo it is probably mainly or exclusively the GABA formed by this mechanism (released via *arrow 3*) that becomes reaccumulated into the GABAergic neuron (*arrow 4*). Thus during established GABAergic signaling formation of vesicularly released GABA sustains itself by the neuronal uptake of GABA and its metabolism to "new" releasable GABA, whereas neuronal reuptake of GABA formed from glutamate released directly to the cytosol (unlabelled gray arrow) ceases. This allows uptake of this GABA (*Italics*) into astrocytes in the glutamine–glutamate/GABA cycle. The Gln initially released into the mitochondria is converted to α-KG (as also concluded by Waagepetersen et al. 1999, 2001), and this α-KG is reconverted to Glu in the transamination of GABA to SSA. This has the consequence that the cycle of vesicular release and neuronal reuptake and production of "new" GABA only is possible when another molecule of Gln simultaneously is converted to mitochondrial Glu and α-KG. Glu is released to the cytosol and

resynthesis of GABA from accumulated and metabolized GABA. According to the proposed mechanism it is thus not the mitochondrial α-KG formed from astrocytically delivered glutamine which is circled a full turn in the TCA cycle but succinate (which is formed in the TCA cycle by decarboxylation of α-KG) which is converted via almost a complete turn of the cycle to α-KG. An important difference from the original concept by Waagepetersen et al. (1999, 2001) is that the exogenous ac.CoA which is needed to convert mitochondrial OAA to citrate can equally well be supplied by glucose and β-hydroxybutyrate. This would make GABA synthesis much less sensitive to replacement of most glucose with a ketogenic diet than glutamate synthesis. This conclusion is in agreement with a better metabolic efficiency of β-hydroxybutyrate of rats on ketogenic diet found by Roy et al. (2015). The same authors found a 44 % increase in brain content of the BCAAs leucine and isoleucine associated with the presence of free leucine and isoleucine in the liver that was never seen in control rats. Along similar lines Chowdhury et al. (2014, see description earlier and Fig. 2.4) had found a very drastic reduction of glucose:glutamate ratio in the animals treated with high doses of β-hydroxybutyrate (from 1:1 in controls to 0.4:1). Together, the findings by these two groups suggest more drastic changes in glutamine–glutamate/GABA cycling than only replacement of glucose with β-hydroxybutyrate in the processes indicated in this review.

The proposed pathway might explain how the neuron disposes of excess TCA cycle constituents although it shows no pyruvate recycling, indicating lack of formation of pyruvate from TCA cycle constituents (Waagepetersen et al. 2002) and express little if any malic enzyme (Kurz et al. 1993). Of the two glutamine molecules followed in Fig. 2.6 one is returned to astrocytes as released glutamate, the other by the GABA released by arrow 1 after it is no longer reaccumulated by the neurons (cessation of the process shown by the gray arrow) after the reaccumulation of synaptically released GABA has been established. This GABA is converted to glutamate as described by Duarte and Gruetter (2013), but if the pathway scheduled in Fig. 2.6 is followed only one half of this conversion (one GABA molecule) occurs in astrocytes, the other (the GABA molecule metabolized in the pathway shown in red) in the GABAergic neurons. The mechanism seems to be consistent with the conclusion by Van den Berg et al. (1974) 'that there are two metabolic spaces involved in the degradation of GABA.... One of the GABA degradation metabolic spaces leads to a high labeling of glutamine, the other not.' The space leading to the high labeling of glutamine is astrocytic. The other neuronal, where it can be seen that no glutamine is formed in the pathway shown in red. It can probably also be concluded that GAD 67, which participates in the formation of both cytosolic and vesicular GABA catalyzes the formation of the GABA released by nonvesicular means (arrow 1). GAD 65, which selectively leads to formation of vesicular GABA the release of which is shown in arrow 3.

GABA production and degradation also contribute to the 1:1 ratio between glutamate flux and neuronal glucose utilization in individuals on a normal diet. It was concluded earlier that the requirement for pyruvate formation during conversion of glutamine to glutamate in glutamatergic neurons and pyruvate utilization for anaplerosis together could account for 60 % of the glucose consumption. De novo GABA production depends also on glucose metabolism because newly synthesized astrocytic glutamate normally will be derived from glucose, but this is already included in the 60 %. However, an additional 10–12 % (one half of the 20–25 % of the glutamine–glutamate/GABA cycle used for GABA synthesis) can be explained by the complex processes occurring both in astrocytes (Duarte and Gruetter 2013) and according to Fig. 2.6 also in GABAergic neurons, when GABA is converted to glutamate for recycling (see green arrow in Fig. 2.6). In addition, the rates of GABAergic neuronal glucose oxidation and glutamate/glutamine cycling increase proportionate to glutamatergic neuron glucose oxidation and glutamate/glutamine cycling indicating that the proportionality will be maintained across changes in activity. Thus, under normal conditions with glucose as the only important substrate the combined turnover of glutamate and GABA can account for 70 % of the 1:1 ratio between glutamate cycling and neuronal glucose metabolism.

2.4 Concluding Remarks

The glutamine–glutamate/GABA cycle is an astrocytic-neuronal pathway which is essential for production of transmitter glutamate and GABA and thus for normal brain function. Glucose or lactate is essential precursor for some processes involved in the cycle, whereas they can be replaced by ketone bodies, a BCAA, or acetate at other steps. Formation of glutamate from glutamine is complex in both glutamatergic and GABAergic neurons. It is completely dependent upon concomitant glucose metabolism in glutamatergic neurons, whereas β-hydroxybutyrate can replace glucose in GABAergic neurons. Accordingly, formation of GABA may be less sensitive than that of glutamate to inhibition by diets highly enriched in fats. Energetically glutamate and GABA production via the glutamine–glutamate/GABA cycle is much more expensive than neuronal production, reuptake, and oxidation of glutamate (and GABA) would have been. However, a higher neuronal reuptake capacity would mean that the levels of glutamate and GABA in the extracellular space would be influenced to a greater extent by membrane depolarization as opposed to vesicular release, which could interfere with neuronal communication especially in the case of glutamate. Such a process seems to occur in the very immature brain and the development of the glutamine–glutamate/GABA cycle may be a key factor explaining the difference between the functional capability of the immature and mature brain. It also provides astrocytes with the possibility to coregulate glutamatergic and GABAergic transmission. The modified pathway for synthesis of GABA shown in Fig. 2.6 seems to be able to explain several previous findings of association of specific GAD isozymes and transporters with cytosolic and vesicular released GABA and it may also be relevant for tonic and phasic GABA signaling.

Acknowledgements D.L.R. acknowledges support from the National Institute of Health grant R01NS087568A.

Conflicts of interest The authors declare no conflicts of interest.

References

Amaral AI (2013) Effects of hypoglycaemia on neuronal metabolism in the adult brain: role of alternative substrates to glucose. J Inherit Metab Dis 36(4):621–634

Anlauf E, Derouiche A (2013) Glutamine synthetase as an astrocytic marker: its cell type and vesicle localization. Front Endocrinol (Lausanne) 4:144 (Hertz L, Rodrigues TB (Eds). eCollection)

Bak LK, Ziemńska E, Waagepetersen HS, Schousboe A, Albrecht J (2008) Metabolism of [U-13C] glutamine and [U-13C]glutamate in isolated rat brain mitochondria suggests functional phosphate-activated glutaminase activity in matrix. Neurochem Res 233(2):273–278

Bakken IJ, Sonnewald U, Clark JB, Bates TE (1997) [U-13C]glutamate metabolism in rat brain mitochondria reveals malic enzyme activity. Neuroreport 8(7):1567–1570

Balazs R (1965) Control of glutamate oxidation in brain and liver mitochondrial systems. Biochem J 95:497–508

Balazs R, Dahl D, Harwood JR (1966) Subcellular distribution of enzymes of glutamate metabolism in rat brain. J Neurochem 13:897–905

Barnett NL, Pow DV, Robinson SR (2000) Inhibition of Müller cell glutamine synthetase rapidly impairs the retinal response to light. Glia 30(1):64–73

Bauer DE, Jackson JG, Genda EN, Montoya MM, Yudkoff M, Robinson MB (2012) The glutamate transporter, GLAST, participates in a macromolecular complex that supports glutamate metabolism. Neurochem Int 61(4):566–574

Ben-Yoseph O, Camp DM, Robinson TE, Ross BD (1995) Dynamic measurements of cerebral pentose phosphate pathway activity in vivo using [1,6-13C2,6,6-2H2]glucose and microdialysis. J Neurochem 64(3):1336–1342

Biedermann B, Bringmann A, Reichenbach A (2002) High-affinity GABA uptake in retinal glial (Müller) cells of the guinea pig: electrophysiological characterization, immunohistochemical localization, and modeling of efficiency. Glia 39(3):217–228

Bjørnsen LP, Eid T, Holmseth S, Danbolt NC, Spencer DD, de Lanerolle NC (2007) Changes in glial glutamate transporters in human epileptogenic hippocampus: inadequate explanation for high extracellular glutamate during seizures. Neurobiol Dis 25(2):319–330

Bjørnsen LP, Hadera MG, Zhou Y, Danbolt NC, Sonnewald U (2014) The GLT-1 (EAAT2; slc1a2) glutamate transporter is essential for glutamate homeostasis in the neocortex of the mouse. J Neurochem 128(5):641–649

Boumezbeur F, Petersen KF, Cline GW, Mason GF, Behar KL, Shulman GI, Rothman DL (2010) The contribution of blood lactate to brain energy metabolism in humans measured by dynamic ^{13}C nuclear magnetic resonance spectroscopy. J Neurosci 30(42):13983–13991

Bouzier-Sore AK, Voisin P, Canioni P, Magistretti PJ, Pellerin L (2003) Lactate is a preferential oxidative energy substrate over glucose for neurons in culture. J Cereb Blood Flow Metab 23(11):1298–12306

Bouzier-Sore AK, Voisin P, Bouchaud V, Bezancon E, Franconi JM, Pellerin L (2006) Competition between glucose and lactate as oxidative energy substrates in both neurons and astrocytes: a comparative NMR study. Eur J Neurosci 24(6):1687–1694

Brekke E, Morken TS (2015) Sonnewald U (2015) Glucose metabolism and astrocyte-neuron interactions in the neonatal brain. Neurochem Int 82:33–41

Bringmann A, Grosche A, Pannicke T, Reichenbach A (2013) GABA and glutamate uptake and metabolism in retinal glial (Müller) cells. Front Endocrinol (Lausanne) 4:48 (Hertz L, Rodrigues TB (Eds). eCollection)

Chowdhury GM, Patel AB, Mason GF, Rothman DL, Behar KL (2007) Glutamatergic and GABAergic neurotransmitter cycling and energy metabolism in rat cerebral cortex during postnatal development. J Cereb Blood Flow Metab 27(12):1895–1907

Chowdhury GM, Jiang L, Rothman DL, Behar KL (2014) The contribution of ketone bodies to basal and activity-dependent neuronal oxidation in vivo. J Cereb Blood Flow Metab 34(7):1233–1242

Coco S, Verderio C, Trotti D, Rothstein JD, Volterra A, Matteoli M (1997) Non-synaptic localization of the glutamate transporter EAAC1 in cultured hippocampal neurons. Eur J Neurosci 9:1902–1910

Collins TJ, Berridge MJ, Lipp P, Bootman MD (2002) Mitochondria are morphologically and functionally heterogeneous within cells. EMBO J 21(7):1616–1627

Cooper AJL (2012) The role of glutamine synthetase and glutamate dehydrogenase in cerebral ammonia homeostasis. Neurochem Res 37(11):2439–2455

Cooper AJ (2013) Quantitative analysis of neurotransmitter pathways under steady state conditions—a perspective. Front Endocrinol (Lausanne) 4:179 (Hertz L, Rodrigues TB (Eds). eCollection)

Cremer JE, Braun LD, Oldendorf WH (1976) Changes during development in transport processes of the blood-brain barrier. Biochim Biophys Acta 448(4):633–637

Cruz NF, Ball KK, Dienel GA (2007) Functional imaging of focal brain activation in conscious rats: impact of [(14)C]glucose metabolite spreading and release. J Neurosci Res 85(15):3254–3266

Cudalbu C, Lanz B, Duarte JM, Morgenthaler FD, Pilloud Y, Mlynárik V, Gruetter R (2012) Cerebral glutamine metabolism under hyperammonemia determined in vivo by localized (1)H and (15)N NMR spectroscopy. J Cereb Blood Flow Metab 32(4):696–708

Danbolt NC (2001) Glutamate uptake. Prog Neurobiol 65(1):1–105

De Feyter HM, Mason GF, Shulman GI, Rothman DL, Petersen KF (2013) Increased brain lactate concentrations without increased lactate oxidation during hypoglycemia in type 1 diabetic individuals. Diabetes 62(9):3075–3080

Dennis SC, Lai JC, Clark JB (1977) Comparative studies on glutamate metabolism in synaptic and non-synaptic rat brain mitochondria. Biochem J 164(3):727–736

Dienel GA (2012) Brain lactate metabolism: the discoveries and the controversies. J Cereb Blood Flow Metab 32(7):1107–1138

Dringen R, Schmoll D, Cesar M, Hamprecht B (1993) Incorporation of radioactivity for gluconeogenesis in brain cells. Biol Chem Hoppe Seyler

Dringen R, Schmoll D, Cesar M, Hamprecht B (2005) Incorporation of radioactivity from [14C] lactate into the glycogen of cultured mouse astroglial cells. Evidence for functional roles in brain cells. J Neurosci Res 79(1–2):11–18

Duarte JM, Gruetter R (2013) Glutamatergic and GABAergic energy metabolism measured in the rat brain by (13) C NMR spectroscopy at 14.1 T. J Neurochem 126(5):579–590

Eid T, Ghosh A, Wang Y, Beckström H, Zaveri HP, Lee TS, Lai JC, Malthankar-Phatak GH, de Lanerolle NC (2008) Recurrent seizures and brain pathology after inhibition of glutamine synthetase in the hippocampus in rats. Brain 131(Pt 8):2061–2070

Engel PC, Dalziel K (1967) The equilibrium constants of the glutamate dehydrogenase systems. Biochem J 105(2):691–695

Fink-Jensen A, Suzdak PD, Swedberg MD, Judge ME, Hansen L, Nielsen PG (1992) The gamma-aminobutyric acid (GABA) uptake inhibitor, tiagabine, increases extracellular brain levels of GABA in awake rats. Eur J Pharmacol 220(2–3):197–201

Fitzpatrick SM, Hetherington HP, Behar KL, Shulman RG (1990) The flux from glucose to glutamate in the rat brain in vivo as determined by 1H-observed, 13C-edited NMR spectroscopy. J Cereb Blood Flow Metab 10(2):170–179

Frigerio F, Karaca M, De Roo M, Mlynárik V, Skytt DM, Carobbio S, Pajęcka K, Waagepetersen HS, Gruetter R, Muller D, Maechler P (2012) Deletion of glutamate dehydrogenase 1 (Glud1) in the central nervous system affects glutamate handling without altering synaptic transmission. J Neurochem 123(3):342–348

Gaitonde MK, Jones J, Evans G (1987) Metabolism of glucose into glutamate via the hexose monophosphate shunt and its inhibition by 6-aminonicotinamide in rat brain in vivo. Proc R Soc Lond B Biol Sci 231(1262):71–90

Gandhi GK, Cruz NF, Ball KK, Dienel GA (2009) Astrocytes are poised for lactate trafficking and release from activated brain and for supply of glucose to neurons. J Neurochem 111(2):522–536

Genda EN, Jackson JG, Sheldon AL, Locke SF, Greco TM, O'Donnell JC, Spruce LA, Xiao R, Guo W, Putt M, Seeholzer S, Ischiropoulos H, Robinson MB (2011) Co-compartmentalization of the astroglial glutamate transporter, GLT-1, with glycolytic enzymes and mitochondria. J Neurosci 31(50):18275–18288

Gibbs ME, Lloyd HG, Santa T, Hertz L (2007) Glycogen is a preferred glutamate precursor during learning in 1-day-old chick: biochemical and behavioral evidence. J Neurosci Res 85(15): 3326–3333

Gibbs ME, Gibbs Z, Hertz L (2009) Rescue of Abeta(1-42)-induced memory impairment in day-old chick by facilitation of astrocytic oxidative metabolism: implications for Alzheimer's disease. J Neurochem 109(Suppl 1):230–236

Gorovits R, Avidan N, Avisar N, Shaked I, Vardimon L (1997) Glutamine synthetase protects against neuronal degeneration in injured retinal tissue. Proc Natl Acad Sci U S A 94(13): 7024–7029

Goubard V, Fino E, Venance L (2011) Contribution of astrocytic glutamate and GABA uptake to corticostriatal information processing. J Physiol 589(Pt9):2301–2319

Gruetter R, Seaquist ER, Ugurbil K (2001) A mathematical model of compartmentalized neurotransmitter metabolism in the human brain. Am J Physiol Endocrinol Metab 281(1): E100–E112

Gulanski BI, De Feyter HM, Page KA, Belfort-DeAguiar R, Mason GF, Rothman DL, Sherwin RS (2013) Increased brain transport and metabolism of acetate in hypoglycemia. J Clin Endocrinol Metab 98(9):3811–3820

Hertz L (2011a) Brain glutamine synthesis requires neuronal aspartate: a commentary. J Cereb Blood Flow Metab 31(1):384–387

Hertz L (2011b) Astrocytic energy metabolism and glutamate formation—relevance for 13C-NMR spectroscopy and importance of cytosolic/mitochondrial trafficking. Magn Reson Imaging 29(10):1319–1329

Hertz L (2013) The glutamate-glutamine (GABA) cycle: importance of late postnatal development and potential reciprocal interactions between biosynthesis and degradation. Front Endocrinol (Lausanne) 4:59, Hertz L, Rodrigues TB (Eds). eCollection

Hertz L, Dienel GA (2002) Energy metabolism in the brain. Int Rev Neurobiol 51:1–102

Hertz L, Dienel GA (2005) Lactate transport and transporters: general principles and functional roles in brain cells. J Neurosci Res 79(1–2):11–18

Hertz L, Wu PH, Schousboe A (1978) Evidence for net uptake of GABA into mouse astrocytes in primary cultures—its sodium dependence and potassium independence. Neurochem Res 3(3):313–323

Hertz L, Peng L, Dienel GA (2007) Energy metabolism in astrocytes: high rate of oxidative metabolism and spatiotemporal dependence on glycolysis/glycogenolysis. J Cereb Blood Flow Metab 27(2):219–249

Hertz L, Gibbs ME, Dienel GA (2014) Fluxes of lactate into, from, and among gap junction-coupled astrocytes and their interaction with noradrenaline. Front Neurosci 8:261

Hertz L, Chen Y. (2016) Editorial: all 3 types of glial cells are important for memory formation. Front Integr Neurosci (in press)

Herzog RI, Jiang L, Herman P, Zhao C, Sanganahalli BG, Mason GF, Hyder F, Rothman DL, Sherwin RS, Behar KL (2013) Lactate preserves neuronal metabolism and function following antecedent recurrent hypoglycemia. J Clin Invest 123(5):1988–1998

Hostetler KY, Landau BR (1967) Estimation of the pentose cycle contribution to glucose metabolism in tissue in vivo. Biochemistry 6(10):2961–2964

Huang R, Peng L, Chen Y, Hajek I, Zhao Z, Hertz L (1994) Signaling effect of elevated potassium concentrations and monoamines on brain energy metabolism at the cellular level. Dev Neurosci 16(5–6):337–351

Hutson SM, Berkich D, Drown P, Xu B, Aschner M, LaNoue KF (1998) Role of branched-chain aminotransferase isoenzymes and gabapentin in neurotransmitter metabolism. J Neurochem 71(2):863–874

Hutson SM, Lieth E, LaNoue KF (2001) Function of leucine in excitatory neurotransmitter metabolism in the central nervous system. J Nutr 131(3):846S–850S

Hutson SM, Cole JT, Sweatt AJ, LaNoue KF (2008) Is the anaplerotic enzyme pyruvate carboxylase (PC) only expressed in astrocytes? J Neurochem 104(Suppl 1):58–59

Hyder F, Fulbright RK, Shulman RG, Rothman DL (2013a) Glutamatergic function in the resting awake human brain is supported by uniformly high oxidative energy. J Cereb Blood Flow Metab 33(3):339–347

Hyder F, Rothman DL, Bennett MR (2013b) Cortical energy demands of signaling and nonsignaling components in brain are conserved across mammalian species and activity levels. Proc Natl Acad Sci U S A 110(9):3549–3554

Ipponi A, Lamberti C, Medica A, Bartolini A, Malmberg-Aiello P (1999) Tiagabine antinociception in rodents depends on GABA(B) receptor activation: parallel antinociception testing and medial thalamus GABA microdialysis. Eur J Pharmacol 368(2–3):205–211

Iversen LL, Neal MJ (1968) The uptake of [3H]GABA by slices of rat cerebral cortex. J Neurochem 15:1141–1910

Jackson JG, O'Donnell JC, Takano H, Coulter DA, Robinson MB (2014) Neuronal activity and glutamate uptake decrease mitochondrial mobility in astrocytes and position mitochondria near glutamate transporters. J Neurosci 34(5):1613–1624

Jiang L, Mason GF, Rothman DL, de Graaf RA, Behar KL (2011) Cortical substrate oxidation during hyperketonemia in the fasted anesthetized rat in vivo. J Cereb Blood Flow Metab 31(12):2313–2323

Jiang L, Gulanski BI, De Feyter HM, Weinzimer SA, Pittman B, Guidone E, Koretski J, Harman S, Petrakis IL, Krystal JH, Mason GF (2013) Increased brain uptake and oxidation of acetate in heavy drinkers. J Clin Invest 123(4):1605–1614

Kanamori K, Ross BD (1995) Steady-state in vivo glutamate dehydrogenase activity in rat brain measured by 15N NMR. J Biol Chem 270(42):24805–24809

Karaca M, Frigerio F, Migrenne S, Martin-Levilain J, Skytt DM, Pajecka K, Martin-del-Rio R, Gruetter R, Tamrit-Rodriguez J, Waagepetersen HS, Magnan C, Maechler P (2015) GDH-dependent glutamate oxidation in the brain dictates peripheral energy substrate distribution. Cell Rep 13:365–375

Kaufman DL, Houser CR, Tobin AJ (1991) Two forms of the gamma-aminobutyric acid synthetic enzyme glutamate decarboxylase have distinct intraneuronal distributions and cofactor interactions. J Neurochem 56(2):720–723

Keiding S, Sørensen M, Bender D, Munk OL, Ott P, Vilstrup H (2006) Brain metabolism of 13N-ammonia during acute hepatic encephalopathy in cirrhosis measured by positron emission tomography. Hepatology 43(1):42–50

Kemppainen J, Aalto S, Fujimoto T, Kalliokoski KK, Långsjö J, Oikonen V, Rinne J, Nuutila P, Knuuti J (2005) High intensity exercise decreases global brain glucose uptake in humans. J Physiol 568(Pt 1):323–332

Kratzing CC (1953) The ability of some carboxylic acids to maintain phosphate level and support electrical stimulation in cerebral tissues. Biochem J 54(2):312–317

Krebs HA (1953) Equilibria in transamination systems. Biochem J 54(1):82–86

Kurz GM, Wiesinger H, Hamprecht B (1993) Purification of cytosolic malic enzyme from bovine brain, generation of monoclonal antibodies, and immunocytochemical localization of the enzyme in glial cells of neural primary cultures. J Neurochem 60(4):1467–1474

Larsson OM, Johnston GA, Shousboe A (1983) Differences in uptake kinetics of cis-3-aminocyclohexane carboxylic acid into neurons and astrocytes in primary cultures. Brain Res 260(2):279–285

Lebon V, Petersen KF, Cline GW, Shen J, Mason GF, Dufour S, Behar KL, Shulman GI, Rothman DL (2002) Astroglial contribution to brain energy metabolism in humans revealed by 13C nuclear magnetic resonance spectroscopy: elucidation of the dominant pathway for neurotransmitter glutamate repletion and measurement of astrocytic oxidative metabolism. J Neurosci 22(5):1523–1531

Leke R, Bak LK, Iversen P, Sørensen M, Keiding S, Vilstrup H, Ott P, Portela LV, Schousboe A, Waagepetersen HS (2011) Synthesis of neurotransmitter GABA via the neuronal tricarboxylic

acid cycle is elevated in rats with liver cirrhosis consistent with a high GABAergic tone in chronic hepatic encephalopathy. J Neurochem 117(5):824–832
Li B, Hertz L, Peng L (2012) Aralar mRNA and protein levels in neurons and astrocytes freshly isolated from young and adult mouse brain and in maturing cultured astrocytes. Neurochem Int 61(8):1325–1332
Lovatt D, Sonnewald U, Waagepetersen HS, Schousboe A, He W, Lin JH, Han X, Takano T, Wang S, Sim FJ, Goldman SA, Nedergaard M (2007) The transcriptome and metabolic gene signature of protoplasmic astrocytes in the adult murine cortex. J Neurosci 27(45):12255–12266
Lynen F (1953) Mechanism of beta oxidation of fatty acids. Bull Soc Chim Biol (Paris) 35(10):1061–1083
Maciejewski PK, Rothman DL (2008) Proposed cycles for functional glutamate trafficking in synaptic neurotransmission. Neurochem Int 52(4–5):809–825
Magistretti PJ, Allaman I (2015) A cellular perspective on brain energy metabolism and functional imaging. Neuron 86(4):883–901
Mangia S, Garreffa G, Maraviglia B, Giove F (2010) Metabolic correlatives of brain activity in a FOS epilepsy patient. NMR Biomed 23(2):170–178
Mangia S, Giove F, Dinuzzo M (2012) Metabolic pathways and activity-dependent modulation of glutamate concentration in the human brain. Neurochem Res 37(11):2554–2561
Martinez-Hernandez A, Bell KP, Norenberg MD (1977) Glutamine synthetase: glial localization in brain. Science 195:1356–1358
Mason GF, Petersen KF, de Graaf RA, Shulman GI, Rothman D (2007) Measurements of the anaplerotic rate in the human cerebral cortex using 13C magnetic resonance spectroscopy and [1-13C] and [2-13C] glucose. J Neurochem 100:73–86
McIlwain H (1953) Substances which support respiration and metabolic response to electrical impulses in human cerebral tissues. J Neurol Neurosurg Psychiatry 16(4):257–266
McKenna MC (2012) Substrate competition studies demonstrate oxidative metabolism of glucose, glutamate, glutamine, lactate and 3-hydroxybutyrate in cortical astrocytes from rat brain. Neurochem Res 37(11):2613–2626
McKenna MC (2013) Glutamate pays its own way in astrocytes. Front Endocrinol (Lausanne) 16(4):191
McKenna MC, Sonnewald U (2005) GABA alters the metabolic fate of [U-13C]glutamate in cultured cortical astrocytes. J Neurosci Res 79(1–2):81–87
McKenna MC, Sonnewald U, Huang X, Stevenson J, Zielke HR (1996) Exogenous glutamate concentration regulates the metabolic fate of glutamate in astrocytes. J Neurochem 66(1):386–393
McKenna MC, Hopkins IB, Lindauer SL, Bamford P (2006) Aspartate aminotransferase in synaptic and nonsynaptic mitochondria: differential effect of compounds that influence transient hetero-enzyme complex (metabolon) formation. Neurochem Int 48(6–7):629–636
McPherson PA, McEneny J (2012) The biochemistry of ketogenesis and its role in weight management, neurological disease and oxidative stress. J Physiol Biochem 68(1):141–151
Melø TM, Nehlig A, Sonnewald U (2006) Neuronal-glial interactions in rats fed a ketogenic diet. Neurochem Int 48(6–7):498–507
Muir D, Berl S, Clarke DD (1986) Acetate and fluoroacetate as possible markers for glial metabolism in vivo. Brain Res 380:336–340
Murthy CR, Hertz L (1987) Comparison between acute and chronic effects of ammonia on branched-chain amino acid oxidation and incorporation into protein in primary cultures of astrocytes and of neurons. J Neurosci Res 17(3):271–276
Muto Y, Sato S, Watanabe A, Moriwaki H, Suzuki K, Kato A, Kato M, Nakamura T, Higuchi K, Nishiguchi S, Kumada H (2005) Long-Term Survival Study Group. Effects of oral branched-chain amino acid granules on event-free survival in patients with liver cirrhosis. Clin Gastroenterol Hepatol 3(7):705–713
Nissen-Meyer LS, Chaudhry FA (2013) Protein kinase C phosphorylates the system N glutamine transporter SN1 (Slc38a3) and regulates its membrane trafficking and degradation. Front Endocrinol (Lausanne) 4:138

Norenberg MD, Martinez-Hernandez A (1979) Fine structural localization of glutamine synthetase in astrocytes of rat brain. Brain Res 161(2):303–310

Ola MS, Hosoya K, LaNoue KF (2011) Regulation of glutamate metabolism by hydrocortisone and branched chain keto acids in cultured rat retinal Müller cells (TR-MUL). Neurochem Int 59(5):656–663

Owen OE, Morgan AP, Kemp HG, Sullivan JM, Herrera MG, Cahill GF Jr (1967) Brain metabolism during fasting. J Clin Invest 46(10):1589–1595

Palaiologos G, Hertz L, Schousboe A (1988) Evidence that aspartate aminotransferase activity and ketodicarboxylate carrier function are essential for biosynthesis of transmitter glutamate. J Neurochem 51(1):317–320

Palaiologos G, Hertz L, Schousboe A (1989) Role of aspartate aminotransferase and mitochondrial dicarboxylate transport for release of endogenously and exogenously supplied neurotransmitter in glutamatergic neurons. Neurochem Res 14(4):359–366

Pan JW, de Graaf RA, Petersen KF, Shulman GI, Hetherington HP, Rothman DL (2002) [2,4-13 C2]-beta-Hydroxybutyrate metabolism in human brain. J Cereb Blood Flow Metab 22(7):890–898

Pardo B, Rodrigues TB, Contreras L, Garzón M, Llorente-Folch I, Kobayashi K, Saheki T, Cerdan S, Satrústegui J (2011) Brain glutamine synthesis requires neuronal-born aspartate as amino donor for glial glutamate formation. J Cereb Blood Flow Metab 31(1):90–101

Passarella S, Atlante A, Barile M, Quagliariello E (1987) Anion transport in rat brain mitochondria: fumarate uptake via the dicarboxylate carrier. Neurochem Res 12(3):255–264

Patel AB, de Graaf RA, Mason GF, Rothman DL, Shulman RG, Behar KL (2005) The contribution of GABA to glutamate/glutamine cycling and energy metabolism in the rat cortex in vivo. Proc Natl Acad Sci U S A 102(15):5588–5593

Patel AB, de Graaf RA, Martin DL, Battaglioli G, Behar K (2006) Evidence that GAD65 mediates increased GABA synthesis during intense neuronal activity in vivo. J Neurochem 97(2):385–396

Patel AB, Lai JC, Chowdhury GM, Hyder F, Rothman DL, Shulman RG, Behar KL (2014) Direct evidence for activity-dependent glucose phosphorylation in neurons with implications for the astrocyte-to-neuron lactate shuttle. Proc Natl Acad Sci U S A 111(14):5385–5390

Patel AB, de Graaf RA, Rothman DL, Behar KL (2015) Effects of GAT1 inhibition by tiagabine on brain glutamate and GABA metabolism in the anesthetized rat in vivo. J Neurosci Res 93(7):1101–1108

Peca S, Carnì M, Di Bonaventura C, Aprile T, Hagberg GE, Giallonardo AT, Manfredi M, Mangia S, Garreffa G, Maraviglia B, Giove F (2010) Metabolic correlatives of brain activity in a FOS epilepsy patient. NMR Biomed 23(2):170–178

Pellerin L, Magistretti PJ (2012) Sweet sixteen for ANLS. J Cereb Blood Flow Metab 32(7):1152–1166

Perez EL, Lauritzen F, Wang Y, Lee TS, Kang D, Zaveri HP, Chaudhry FA, Ottersen OP, Bergersen LH, Eid T (2012) Evidence for astrocytes as a potential source of the glutamate excess in temporal lobe epilepsy. Neurobiol Dis 47(3):331–337

Raichle ME (2015) The restless brain: how intrinsic activity organizes brain function. Philos Trans R Soc Lond B Biol Sci 19(370):20140172

Raichle ME, Gusnard DA (2002) Appraising the brain's energy budget. Proc Natl Acad Sci U S A 99(16):10237–10239

Robinson AM, Williamson DH (1980) Physiological role of ketone bodies as substrates and signals in mammalian tissues. Physiol Rev 60(1):143–187

Rodrigues TB, López-Larrubia P, Cerdán S (2009) Redox dependence and compartmentation of [13C]pyruvate in the brain of deuterated rats bearing implanted C6 gliomas. J Neurochem 109(Suppl 1):237–245

Romei C, Sabolla C, Raiteri L (2014) GABA release provoked by disturbed Na(+), K(+) and Ca(2+) homeostasis in cerebellar nerve endings: roles of Ca(2+) channels, Na(+)/Ca(2+) exchangers and GAT1 transporter reversal. Neurochem Int 72:1–9

Romei C, Sabolla C, Raiteri L (2015) High-affinity GABA uptake by neuronal GAT1 transporters provokes release of [(3)H]GABA by homoexchange and through GAT1-independent Ca(2+)-mediated mechanisms. Neuropharmacology 88:164–170

Ross CD, Godfrey DA (1987) Distribution of activities of aspartate aminotransferase isoenzymes and malate dehydrogenase in guinea pig retinal layers. J Histochem Cytochem 35(6): 669–674
Rothman DL, De Feyter HM, de Graaf RA, Mason GF, Behar KL (2011) ^{13}C MRS studies of neuroenergetics and neurotransmitter cycling in humans. NMR Biomed 24(8):943–957
Rothman DL, De Feyter HM, Maciejewski PK, Behar KL (2012) Is there in vivo evidence for amino acid shuttles carrying ammonia from neurons to astrocytes? Neurochem Res 37(11):2597–25612
Roy M, Beauvieux MC, Naulin J, El Hamrani D, Gallis JL, Cunnane SC, Bouzier-Sore AK (2015) Rapid adaptation of rat brain and liver metabolism to a ketogenic diet: an integrated study using (1)H- and (13)C-NMR spectroscopy. J Cereb Blood Flow Metab 25(7):1153–1162
Sa HD, Park JY, Jeong SJ, Lee KW, Kim JH (2015) Characterization of glutamate decarboxylase (GAD) from Lactobacillus sakei A156 isolated from Jeot-gal. J Microbiol Biotechnol 25(5):696–703
Schousboe I, Bro B, Schousboe A (1977a) Intramitochondrial localization of the 4-aminobutyrate-2-oxoglutarate transaminase from ox brain. Biochem J 162(2):303–307
Schousboe A, Hertz L, Svenneby G (1977b) Uptake and metabolism of GABA in astrocytes cultured from dissociated mouse brain hemispheres. Neurochem Res 2(2):217–229
Schousboe A, Bak LK, Waagepetersen HS (2013) Astrocytic control of biosynthesis and turnover of the neurotransmitters glutamate and GABA. Front Endocrinol (Lausanne) 4:102
Scimemi A (2014a) Plasticity of GABA transporters: an unconventional route to shape inhibitory synaptic transmission. Front Cell Neurosci 8:128, eCollection
Scimemi A (2014b) Structure, function, and plasticity of GABA transporters. Front Cell Neurosci 8:161. doi:10.3389/fncel.2014.00161, eCollection
Shank RP, Bennett GS, Freytag SO, Campbell GL (1984) Pyruvate carboxylase: an astrocyte-specific enzyme implicated in the replenishment of amino acid neurotransmitter pools. Brain Res 29(1–2):364–367
Shen J, Sibson NR, Cline G, Behar KL, Rothman DL, Shulman RG (1998) ^{15}N-NMR spectroscopy studies of ammonia transport and glutamine synthesis in the hyperammonemic rat brain. Dev Neurosci 20(4–5):434–443
Sibson NR, Dhankhar A, Mason GF, Rothman DL, Behar KL, Shulman RG (1998) Stoichiometric coupling of brain glucose metabolism and glutamatergic neuronal activity. Proc Natl Acad Sci U S A 95(1):316–321
Sibson NR, Mason GF, Shen J, Cline GW, Herskovits AZ, Wall JE, Behar KL, Rothman DL, Shulman RG (2001) In vivo ^{13}C NMR measurement of neurotransmitter glutamate cycling, anaplerosis and TCA cycle flux in rat brain during [2-^{13}C]glucose infusion. J Neurochem 76(4):975–989
Siesjö BK (1978) Brain energy metabolism. Wiley, New York
Smith QR (2000) Transport of glutamate and other amino acids at the blood-brain barrier. J Nutr 130(4S Suppl):1016S–1022S
Smith AL, Satterthwaite HS, Sokoloff L (1969) Induction of brain D(--)-beta-hydroxybytrate dehydrogenase activity by fasting. Science 163(3862):79–81
Sokoloff L, Reivich M, Kennedy C, Des Rosiers MH, Patlak CS, Pettigrew KD, Sakurada O, Shinohara M (1977) The [14C]deoxyglucose method for the measurement of local cerebral glucose utilization: theory, procedure, and normal values in the conscious and anesthetized albino rat. J Neurochem 28(5):897–916
Song I, Volynski K, Brenner T, Ushkaryov Y, Walker M, Semyanov A (2013) Different transporter systems regulate extracellular GABA from vesicular and non-vesicular sources. Front Cell Neurosci 7:23
Sonnewald U (2014) Glutamate synthesis has to be matched by its degradation—where do all the carbons go? J Neurochem 131(4):399–406
Sonnewald U, Westergaard N, Hassel B, Müller TB, Unsgård G, Fonnum F, Hertz L, Schousboe A, Petersen SB (1993) NMR spectroscopic studies of 13C acetate and 13C glucose metabolism

in neocortical astrocytes: evidence for mitochondrial heterogeneity. Dev Neurosci 15(3–5):351–358

Steinman MQ, Gao V, Alberini CM (2016) The role of lactate-mediated metabolic coupling between astrocytes and neurons in long term memory formation. Front Intergr Neurosci 3(10):10

Swanson RA, Benington JH (1996) Astrocyte glucose metabolism under normal and pathological conditions in vitro. Dev Neurosci 18(5–6):515–521

Takagaki G, Tsukada Y (1957) The effect of some inorganic ions on brain slices metabolizing glucose or pyruvate. J Neurochem 2(1):21–29

Tang F, Lane S, Korsak A, Paton JF, Gourine AV, Kasparov S, Teschemacher AG (2014) Lactate-mediated glia-neuronal signalling in the mammalian brain. Nat Commun 5:3284

Tian N, Petersen C, Kash S, Baekkeskov S, Copenhagen D, Nicoll R (1999) The role of the synthetic enzyme GAD65 in the control of neuronal gamma-aminobutyric acid release. Proc Natl Acad Sci U S A 96(22):12911–12916

Urion D, Vreman HJ, Weiner MW (1979) Effect of acetate on hypoglycemic seizures in mice. Diabetes 28(11):1022–1026

Van den Berg CJ, Matheson DF, Ronda G (1974) A model of glutamate metabolism in brain. A biochemical analysis of a heterogeneous structure. In: Berl S, Clarke DD, Schneider D (eds) Metabolic compartmentation and neurotransmission. Plenum Press, New York, pp 515–540

Van Hall G, Strømstad M, Rasmussen P, Jans O, Zaar M, Gam C, Quistorff B, Secher NH, Nielsen HB (2009) Blood lactate is an important energy source for the human brain. J Cereb Blood Flow Metab 29(6):1121–1129

Vannucci SJ, Simpson IA (2003) Developmental switch in brain nutrient transporter expression in the rat. Am J Physiol Endocrinol Metab 285(5):E1127–E1134

Vining EP, Freeman JM, Ballaban-Gil K, Camfield CS, Camfield PR, Holmes GL, Shinnar S, Shuman R, Trevathan E, Wheless JW (1998) A multicenter study of the efficacy of the ketogenic diet. Arch Neurol 55(11):1433–1437

Vogel R, Wiesinger H, Hamprecht B, Dringen R (1999) The regeneration of reduced glutathione in rat forebrain mitochondria identifies metabolic pathways providing the NADPH required. Neurosci Lett 275(2):97–100

Volkow ND, Kim SW, Wang GJ, Alexoff D, Logan J, Muench L, Shea C, Telang F, Fowler JS, Wong C, Benveniste H, Tomasi D (2013) Acute alcohol intoxication decreases glucose metabolism but increases acetate uptake in the human brain. Neuroimage 64:277–283

Volkow ND, Wang GJ, Shokri Kojori E, Fowler JS, Benveniste H, Tomasi D (2015) Alcohol decreases baseline brain glucose metabolism more in heavy drinkers than controls but has no effect on stimulation-induced metabolic increases. J Neurosci 35(7):3248–3255

Waagepetersen HS, Sonnewald U, Larsson OM, Schousboe A (1999) Synthesis of vesicular GABA from glutamine involves TCA cycle metabolism in neocortical neurons. J Neurosci Res 57(3):342–349

Waagepetersen HS, Sonnewald U, Gegelashvili G, Larsson OM, Schousboe A (2001) Metabolic distinction between vesicular and cytosolic GABA in cultured GABAergic neurons using 13C magnetic resonance spectroscopy. J Neurosci Res 63(4):347–355

Waagepetersen HS, Qu H, Hertz L, Sonnewald U, Schousboe A (2002) Demonstration of pyruvate recycling in primary cultures of neocortical astrocytes but not in neurons. Neurochem Res 27(11):1431–1437

Waagepetersen HS, Hansen GH, Fenger K, Lindsay JG, Gibson G, Schousboe A (2006) Cellular mitochondrial heterogeneity in cultured astrocytes as demonstrated by immunogold labeling of alpha-ketoglutarate dehydrogenase. Glia 53(2):225–231

Walls AB, Nilsen LH, Eyjolfsson EM, Vestergaard HT, Hansen SL, Schousboe A, Sonnewald U, Waagepetersen H (2011) Knockout of GAD65 has major impact on synaptic GABA synthesized from astrocyte-derived glutamine. J Cereb Blood Flow Metab 31(2):494–503

Walls AB, Waagepetersen HS, Bak LK, Schousboe A, Sonnewald U (2015) The glutamine/glutamate/GABA cycle: function, regional differences in glutamate and GABA production and effects of interference with GABA metabolism. Neurochem Res 40(2):402–409

Waniewski RA, Martin DL (1998) Preferential utilization of acetate by astrocytes is attributable to transport. J Neurosci 18(14):5225–5233
Watanabe A, Shiota T, Takei N, Fujiwara M, Nagashima H (1986) Ammonia detoxification by accelerated oxidation of branched chain amino acids in brains of acute hepatic failure rats. Biochem Med Metab Biol 3:367–375
Westergaard N, Drejer J, Scjhouseboe A, Sonnewald U (1996) Evaluation of the importance of transamination versus deamination in astrocytic metabolism of [U-13C]glutamate. Glia 17(2):160–168
Whitelaw BS, Robinson MB (2013) Inhibitors of glutamate dehydrogenase block sodium-dependent glutamate uptake in rat brain membranes. Front Endocrinol (Lausanne) 4:123 (Hertz L, Rodrigues TB (Eds). eCollection)
Wong E, Schousboe A, Saito K, Wu JY, Roberts E (1974) Immunochemical studies of brain glutamate decarboxylase and GABA-transaminase of six inbred strains of mice. Brain Res 68(1):133–142
Wyss MT, Jolivet R, Buck A, Magistretti PJ, Weber B (2011) In vivo evidence for lactate as a neuronal energy source. J Neurosci 31(20):7477–7485
Yu AC, Schousboe A, Hertz L (1982) Metabolic fate of 14C-labeled glutamate in astrocytes in primary cultures. J Neurochem 4:954–960
Yu AC, Drejer J, Hertz L, Schousboe A (1983) Pyruvate carboxylase activity in primary cultures of astrocytes and neurons. J Neurochem 41(5):1484–1487
Yu AC, Hertz E, Hertz L (1984) Alterations in uptake and release rates for GABA, glutamate, and glutamine during biochemical maturation of highly purified cultures of cerebral cortical neurons, a GABAergic preparation. J Neurochem 42(4):951–960
Yudkoff M, Daikhin Y, Grunstein L, Nissim I, Stern J, Pleasure D, Nissim I (1996) Astrocyte leucine metabolism: significance of branched-chain amino acid transamination. J Neurochem 66(1):378–385
Zabłocki K, Bryła J (1988) Effect of glycerol on gluconeogenesis in isolated rabbit kidney cortex tubules. Biochim Biophys Acta 970(3):231–240
Zaganas I, Waagepetersen HS, Georgopoulos P, Sonnewald U, Plaitakis A, Schousboe A (2001) Differential expression of glutamate dehydrogenase in cultured neurons and astrocytes from mouse cerebellum and cerebral cortex. J Neurosci Res 66(5):909–913
Zaganas I, Kanavouras K, Mastorodemos V, Latsoudis H, Spanaki C, Plaitakis A (2009) The human GLUD2 glutamate dehydrogenase: localization and functional aspects. Neurochem Int 55(1–3):52–63
Zaganas I, Pajęcka K, Wendel Nielsen C, Schousboe A, Waagepetersen HS, Plaitakis A (2013) The effect of pH and ADP on ammonia affinity for human glutamate dehydrogenases. Metab Brain Dis 28(2):127–131
Zhang TM, Rasschaert J, Malaisse WJ (1995) Metabolism of succinic acid methyl esters in neural cells. Biochem Mol Med 54(2):112–116
Zhang Y, Kuang Y, Xu K, Harris D, Lee Z, LaManna J, Puchowicz MA (2013) Ketosis proportionately spares glucose utilization in brain. J Cereb Blood Flow Metab 33(8):1307–1311
Zhou Y, Danbolt NC (2013) GABA and glutamate transporters in brain. Front Endocrinol (Lausanne) 4:165 (Hertz L, Rodrigues TB (Eds). eCollection)
Zielińska M, Popek M, Albrecht J (2014) Roles of changes in active glutamine transport in brain edema development during hepatic encephalopathy: an emerging concept. Neurochem Res 39(3):599–604
Ziemińska E, Hilgier W, Waagepetersen HS, Hertz L, Sonnewald U, Schousboe A, Albrecht J (2004) Analysis of glutamine accumulation in rat brain mitochondria in the presence of a glutamine uptake inhibitor, histidine, reveals glutamine pools with a distinct access to deamidation. Neurochem Res 29(11):2121–2123

Chapter 3
Anaplerosis for Glutamate Synthesis in the Neonate and in Adulthood

Eva Brekke, Tora Sund Morken, Anne B. Walls, Helle Waagepetersen, Arne Schousboe, and Ursula Sonnewald

Abstract A central task of the tricarboxylic acid (TCA, Krebs, citric acid) cycle in brain is to provide precursors for biosynthesis of glutamate, GABA, aspartate and glutamine. Three of these amino acids are the partners in the intricate interaction between astrocytes and neurons and form the so-called glutamine–glutamate (GABA) cycle. The ketoacids α-ketoglutarate and oxaloacetate are removed from the cycle for this process. When something is removed from the TCA cycle it must be replaced to permit the continued function of this essential pathway, a process termed anaplerosis. This anaplerotic process in the brain is mainly carried out by pyruvate carboxylation performed by pyruvate carboxylase. The present book chapter gives an introduction and overview into this carboxylation and additionally anaplerosis mediated by propionyl-CoA carboxylase under physiological conditions in the adult and in the developing rodent brain. Furthermore, examples are given about pathological conditions in which anaplerosis is disturbed.

Keywords Pyruvate carboxylation • Neonate • Adult • Anaplerosis • Pyruvate • Glutamate

---

E. Brekke
Department of Pediatrics, Nordland Hospital Trust, Bodo, Norway

T.S. Morken
Department of Ophthalmology, Trondheim University Hospital, Trondheim 7006, Norway

Department of Laboratory Medicine, Children's and Women's Health, Norwegian University of Science and Technology (NTNU), Trondheim 7489, Norway

A.B. Walls • H. Waagepetersen • A. Schousboe
Department of Drug Design and Pharmacology, Faculty of Health and Medical Sciences, University of Copenhagen, Copenhagen 2100, Denmark

U. Sonnewald (✉)
Department of Neuroscience, Norwegian University of Science and Technology (NTNU), Postboks 8905, Trondheim 7489, Norway

Department of Drug Design and Pharmacology, Faculty of Health and Medical Sciences, University of Copenhagen, Copenhagen 2100, Denmark
e-mail: ursula.sonnewald@ntnu.no

A. Schousboe, U. Sonnewald (eds.), *The Glutamate/GABA-Glutamine Cycle*, Advances in Neurobiology 13, DOI 10.1007/978-3-319-45096-4_3

Abbreviations

AD	Alzheimer's disease
AMPA	α-Amino-3-hydroxy-5-methyl-4-isoxazolepropionic acid
AMPK	AMP activated protein kinase
CBZ	Carbamazepine
GS	Glutamine synthetase
ME	Malic enzyme
MRS	Magnetic resonance spectroscopy
P	Postnatal day
PAG	Phosphate activated glutaminase
PC	Pyruvate carboxylase
PCC	Propionyl-CoA carboxylase
PDH	Pyruvate dehydrogenase
PEPCK	Phosphoenolpyruvate carboxykinase
PPP	Pentose phosphate pathway
PTZ	Pentylenetetrazole
TCA	Tricarboxylic acid

3.1 Introduction

The major fuel for the brain is glucose and relevant metabolic pathways are shown in Fig. 3.1. Glucose metabolism is necessary for glutamate synthesis which is essential for brain function since the blood–brain barrier prevents glutamate from entering the brain. Synthesis of this excitatory amino acid neurotransmitter in neurons is particularly important since 90 % of synapses in the forebrain are glutamatergic (Attwell and Laughlin 2001) and glutamate is the precursor of the other major amino acid neurotransmitter, GABA. The synthesis of the carbon skeleton of these neurotransmitters is dependent on the production of the anaplerotic substrate oxaloacetate from glucose (Fig. 3.2). Sir Hans Kornberg coined the expression: *anaplerotic sequences* to describe a series of enzymatic reactions or pathways that replenish the pools of metabolic intermediates in the tricarboxylic acid (TCA) cycle. Since both glutamate and GABA are derived from the TCA cycle intermediate α-ketoglutarate the anaplerotic reactions securing the synthesis of α-ketoglutarate in brain are important. However, because of the coupling of this process with cataplerosis, i.e. the exit of intermediates from the cycle, both pathways are equally important for the regulation of amino acid, glucose and fatty acid homeostasis (Sonnewald 2014).

One role of this cycle in all mammalian cells is the oxidation of acetyl-CoA to carbon dioxide and energy production. However, in brain synthesis of precursors for the neurotransmitters glutamate, GABA and aspartate are also an essential part of its function. Pyruvate carboxylase (PC) is the enzyme with major responsibility for oxaloacetate production from glucose and thus anaplerosis in

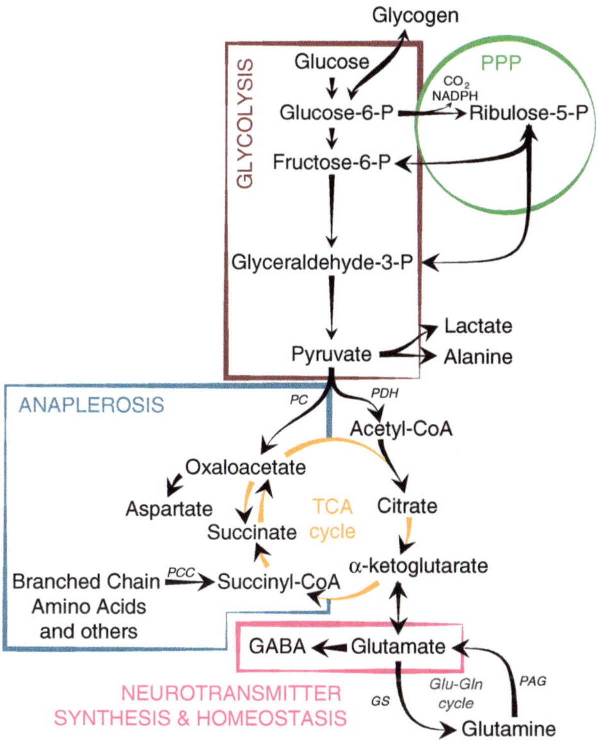

Fig. 3.1 Pathways for glucose metabolism. *Gln* glutamine, *Glu* glutamate, *KG* ketoglutarate, *ME* malic enzyme, *GS* glutamine synthetase, *NADPH* nicotinamide adenine dinucleotide phosphate, *PAG* phosphate-activated glutaminase, *PC* pyruvate carboxylase, *PCC* propionyl-CoA carboxylase, *PDH* pyruvate dehydrogenase, *PPP* pentose phosphate pathway, *TCA* tricarboxylic acid

brain (Mohler et al. 1974; Patel 1974b; Kanamatsu and Tsukada 1999). However, pyruvate carboxylation and neurotransmitter synthesis and release do not occur in the same compartment (Fig. 3.3). In the brain only astrocytes express PC (Yu et al. 1983; Shank et al. 1985; Cesar and Hamprecht 1995) but the two amino acid neurotransmitters are produced in the neurons. This dependence of neurons on astrocytes has been investigated extensively in the adult brain of animals and humans (McKenna et al. 2012) and is defined by the details of the glutamate–glutamine cycle (Fig. 3.3) which can also encompass GABA (Berl and Clarke 1983). Using ^{13}C magnetic resonance spectroscopy (MRS) and ^{13}C-labelled glucose it has been shown that pyruvate carboxylation occurs in humans, rats and mice in vivo as well as in vitro in neural tissue and cell preparations (Badar-Goffer et al. 1990; Shank et al. 1993; Hassel and Sonnewald 1995; Griffin et al. 1999; Waagepetersen et al. 2001a; Rae et al. 2005; Mason et al. 2007). An excellent review article gives an overview over the different MRS studies performed to evaluate the relationship between pyruvate carboxylation and dehydrogenation (Hertz 2011). In order to perform de novo synthesis of glutamate, an

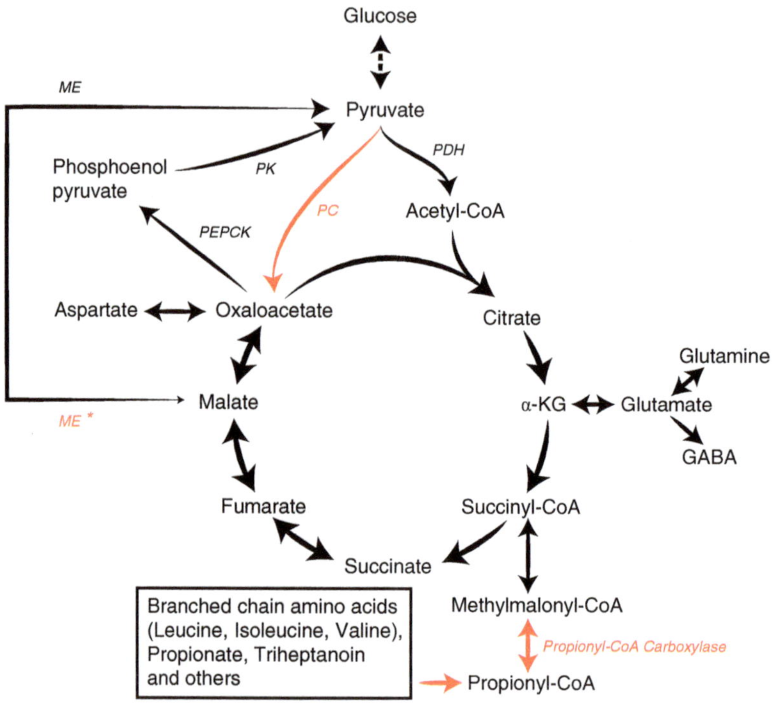

Fig. 3.2 Schematic presentation of reactions pertinent to pyruvate carboxylation. *In the brain ME appear to operate primarily in the decarboxylation direction, see Section 3.2. *KG* ketoglutarate, *ME* malic enzyme, *PC* pyruvate carboxylase, *PDH* pyruvate dehydrogenase, *PEPCK* phosphoenolpyruvate carboxykinase, *PK* pyruvate kinase

anaplerotic substrate must be transferred from astrocytes to neurons, since the latter cell type does not contain PC. In contrast to glutamate, glutamine can safely be released to the extracellular milieu without a disturbing interaction with receptors for the two mentioned amino acid neurotransmitters. When glutamine is released from astrocytes into the synapse it is taken up by specific transporters into neurons (Varoqui et al. 2000) where it can be converted to glutamate by phosphate activated glutaminase (PAG) (Hogstad et al. 1988). As can be seen in Fig. 3.3, after release in the neurotransmission process, glutamate is taken up via specialized, high affinity transporters located primarily in astrocytes (Danbolt et al. 1992; Danbolt 2001). In astrocytes, glutamate is either rapidly converted to glutamine via the enzyme glutamine synthetase (GS), localized only in astrocytes (Norenberg and Martinez-Hernandez 1979), or converted to α-ketoglutarate, thereby entering the TCA cycle (McKenna et al. 1996; Schousboe et al. 2014). This sequence of processes is termed the glutamate–glutamine cycle (Fig. 3.3) (references in (McKenna et al. 2012)), which is essential in order to ensure precise neural signalling, preserve carbon atoms for neuronal glutamate synthesis and limit excitotoxicity which may result from excessive glutamate receptor stimulation (Danbolt 2001). Only limited information is

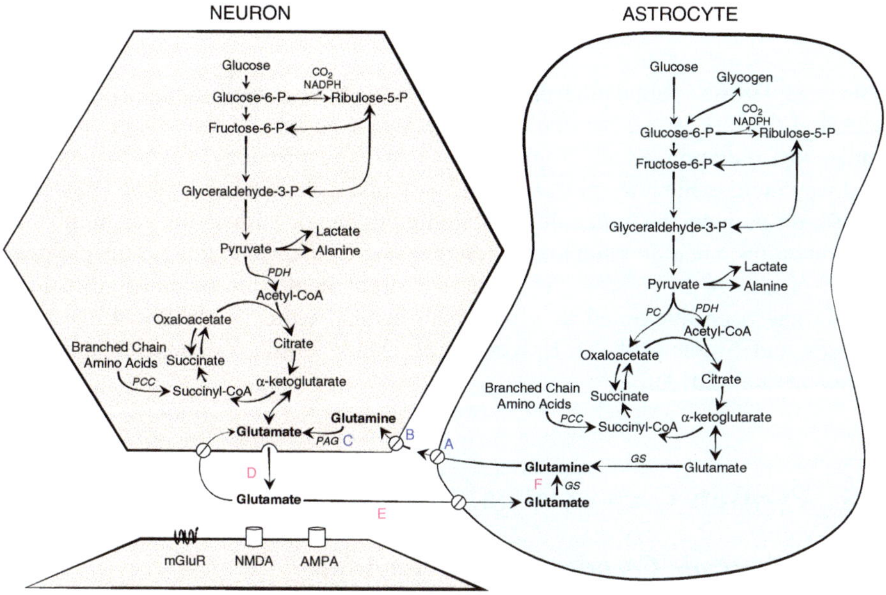

Fig. 3.3 The glutamine–glutamate cycle. *AMPA* α-amino-3-hydroxy-5-methyl-4-isoxazolepropionic acid receptor, *A,B* glutamine transport, *C,D* glutamate transport, *E* conversion of glutamate to glutamine, *GS* glutamine synthetase, *NADPH* nicotinamide adenine dinucleotide phosphate, *NMDA N*-methyl-D-aspartate receptor, *PAG* phosphate activated glutaminase, *PC* pyruvate carboxylase, *PCC* propionyl-CoA carboxylase, *PDH* pyruvate dehydrogenase, *mGluR* metabotropic glutamate receptors

available on the above-mentioned processes in the neonatal brain. However, there are indications that the glutamate–glutamine cycle at this stage does not operate as a cycle, but rather as a delivery system of glutamine from astrocytes to neurons, since transport of glutamate from neurons to astrocytes is negligible while that of glutamine from astrocytes to neurons is higher than in the adult brain (Morken et al. 2013; Brekke et al. 2014).

3.2 Carboxylation in the Brain

There are several enzymes that theoretically would have the potential for catalysing carboxylation processes in the brain (Fig. 3.2) such as pyruvate carboxylase (PC), propionyl-CoA carboxylase, phosphoenolpyruvate carboxykinase (PEPCK) and malic enzyme (ME). The latter three enzymes are present both in astrocytes and neurons (Kurz et al. 1993; Alves et al. 1995; McKenna et al. 1995, 2000; Vogel et al. 1998a, b). However, in the brain PEPCK and ME appear to operate primarily if not exclusively in the decarboxylation direction (Kurz et al. 1993; Alves et al. 1995; McKenna et al. 1995, 2000; Vogel et al. 1998a, b).

3.3 Propionyl CoA Carboxylation

The anaplerotic pathway from propionyl-CoA via methylmalonyl-CoA to succinyl-CoA (Fig. 3.2) is known to operate in peripheral tissues. Anaplerotic molecules metabolized to propionyl-CoA and the propionyl-CoA carboxylase pathway include the branched chain amino acids, isoleucine and valine (Bak et al. 2009), propionate and molecules containing fatty acids with an uneven number of carbon atoms, such as triheptanoin (Borges and Sonnewald 2012). This pathway has been shown to be active in brain and supplementation of diet with triheptanoin has been attempted as a treatment against epilepsy (Willis et al. 2010; Borges and Sonnewald 2012; Kim et al. 2013; Marin-Valencia et al. 2013; Smeland et al. 2013).

3.4 Pyruvate Carboxylation

3.4.1 Pyruvate Carboxylation in the Adult

Pyruvate carboxylase is an ATP-dependent mitochondrial enzyme catalysing the carboxylation of pyruvate to form oxaloacetate. It is a nuclear encoded homotetramer found in most tissue from eukaryotes and in many prokaryotes and is a member of the family of biotin-dependent carboxylases (Wallace et al. 1998). In the brain it is involved in various anabolic pathways such as lipogenesis and the synthesis of neurotransmitter substances (Wallace et al. 1998; Schousboe et al. 2015). A defect in the expression or biotinylation of pyruvate is a rare autosomal recessively inherited disorder in humans resulting in developmental delays and failure to thrive starting in the neonatal or early infantile period, early death, or severe debilitating psychomotor retardation (Wallace et al. 1998; Schiff et al. 2006).

While the cellular location of PC in astrocytes but not neurons (Yu et al. 1983) is not disputed, pyruvate carboxylation has been suggested to take place also in neurons (Hassel 2001; Merle et al. 2002). However, a whole range of experiments, using different preparations, in different laboratories have essentially unequivocally confined this carboxylation to glia. Primary mono-cultures of neurons and astrocytes were used to obtain information about the cellular location and magnitude of pyruvate carboxylation using ^{13}C labelled precursors. However, the interpretation of results obtained in these studies may be ambiguous since oxaloacetate can be converted not only to citrate but also via several steps to the symmetrical molecule fumarate (back cycling) and subsequently from oxaloacetate and citrate (cycling). The latter pathway will lead to scrambling of label resulting in misinterpretation of some labelling results. Merle et al. (2002) suggested incomplete back cycling for astrocytes and complete back cycling for cerebellar neurons. If back cycling in neurons is indeed complete, the study mentioned in the previous sentence, utilizing [1-^{13}C]glucose, was not capable of showing carboxylation in neurons. In order to clarify this, Waagepetersen et al. (2001a) have performed a study using [U-^{13}C]glucose or [U-^{13}C]lactate and 3-nitropropionic acid to specifically block the TCA cycle at the succinate dehydrogenase step (Alston et al. 1977). In this way multiple cycling of TCA cycle constituents was avoided and carboxylation could be clearly detected in labelling of the C-3 position, regardless of back cycling. Analysing the ^{13}C labelling patterns in amino acids, pyruvate carboxylation was detected in astrocytes but not in neurons (Waagepetersen et al. 2001a). Furthermore, also in studies of brain tissue slices and cell cultures carboxylation was detected in astrocytes (Griffin et al. 1998, 2003; Waagepetersen et al. 2001a). In a very elegant study using unrestrained awake rats which were continuously infused with a combination of $H^{14}CO_3^-$ and [1-^{13}C] glucose in over 50 infusions ranging from 5 to 60 min it was shown that synthesis of glutamine from $H^{14}CO_3^-$ was substantial, amounting to 32 % of the glutamate/glutamine cycle (Xu et al. 2004).

3.4.2 *Pyruvate Carboxylation in the Neonate*

Also in the neonatal brain pyruvate carboxylation is predominantly localized in glia since the level of [2,3-^{13}C]glutamine derived from [1,2-^{13}C]glucose is higher than that of [2,3-^{13}C]glutamate (Morken et al. 2013; Brekke et al. 2014). As pointed out earlier, synthesis of the neurotransmitters glutamate and GABA is coupled to pyruvate carboxylation and this is particularly important in development when neurotransmitter levels increase. GABA levels are only slightly lower at postnatal day 7 (P7) compared to later in life, but glutamate has a sharp increase early in brain development (Chowdhury et al. 2007; Morken et al. 2013). In humans, glutamate levels double from gestational week 32 until term and continue to increase during the first year of life (Kreis et al. 2002). In rats, glutamate levels double between P7 and adulthood. This increase in de novo synthesis of glutamate must happen via anaplerosis catalysed by PC.

PC activity in the rat brain is very low at birth, but increases sharply during the first weeks of life (Larsson et al. 1985), coinciding with the major period of gliogenesis (Schousboe 1972; Bandeira et al. 2009). During the first days of life, the activity of the TCA cycle-related enzymes is also low (Wilbur and Patel 1974; Larsson et al. 1985). Thus, at the time of birth, astrocytes may not have the same capacity as in the adult brain for metabolic support of neurons but the relative contribution of PC to pyruvate metabolism might still be significant. Indeed, a larger part of the available glucose is channelled into this pathway in the P7 rat than in the adult (Morken et al. 2013; Brekke et al. 2014). In fact, for every glucose molecule metabolized through glycolysis to become pyruvate, half is metabolized via pyruvate dehydrogenase, and the other half is used for anaplerosis in the P7 rat (Morken et al. 2013; Brekke et al. 2014).

Since pyruvate carboxylation occurs in astrocytes, it is possible to calculate the relative glucose utilization in astrocytes compared to neurons at the level of pyruvate using results from experiments with [1,2-^{13}C]glucose injection. This was calculated to be 34 ± 3 % of glucose metabolism at the level of pyruvate in the P7 rat brain (Morken et al. 2013; Brekke et al. 2014). This is in surprisingly good accordance with results from the adult brain based on a corresponding equation using [U-^{13}C]glucose. Qu et al. (2000) found that astrocytes metabolize maximally 34 % of glucose at the level of pyruvate in the adult brain, and similar results have been reported by Hassel et al. (1995). This could indicate that in rats both astrocytes and neurons have lower glucose metabolism at P7 compared to adult animals. However, the distribution between nonneuronal cells and neurons changes considerably from P7 until adult age, since the number of astrocytes increases dramatically in this period to outnumber neuronal cells at adult age (Baburamani et al. 2013; Bandeira et al. 2009). This may suggest that each astrocyte has a higher metabolic rate relative to each neuron in the neonatal brain compared to later in life.

The exchange of glutamate and glutamine between neurons and astrocytes in rats is radically different in the neonatal brain compared to the adult brain (Morken et al. 2013; Brekke et al. 2014). The necessary transporters for glutamine to enter neurons are present from late gestation (Weiss et al. 2003) (Fig. 3.3B) and the number of glutamine transporters on astrocytes actually reaches higher levels than the adult brain on postnatal day 14 (Boulland et al. 2003) (Fig. 3.3A). However, the expression of astrocytic glutamate transporters (GLT-1 and GLAST) is low early in brain development (Danbolt 2001) (Fig. 3.3E), indicating that the ability of astrocytes to take up glutamate from the synapse may be limited in the neonatal brain. Indeed, the transport of newly synthesized glutamine from astrocytes to glutamatergic neurons is higher in the P7 rat brain than the adult rat brain (Morken et al. 2013, 2014) while the reciprocal transfer of glutamate from neurons to astrocytes was very low at P7 compared to adults. The latter may also be due to low release of glutamate from neurons (Fig. 3.3D) at this stage and/or low rate of conversion into glutamine in the astrocytes (Fig. 3.3F). In the neonatal rat brain the action potentials are smaller in amplitude and the firing rate is lower (McCormick and Prince 1987) and the EEG does not develop an adult pattern until P12 (Snead and Stephens 1983). Moreover, glutamine synthetase activity, just like PC activity, is very low at birth, but increases sharply during the first weeks of life (Wilbur and Patel 1974; Hertz et al. 1978; Juurlink et al. 1981; Larsson et al. 1985). Thus, several factors may lead to this difference regarding astrocyte–neuron interactions in the neonatal and the adult brain.

3.5 What Factors Affect Pyruvate Carboxylation?

3.5.1 Increased Pyruvate Carboxylation

Cerebral hyperammonemia is believed to play a pivotal role in the development of hepatic encephalopathy (HE), a debilitating condition arising due to acute or chronic liver disease (Butterworth 2003). In the brain, ammonia is thought to be detoxified via the activity of the astrocytic enzyme glutamine synthetase. However, glutamine cannot be synthesized without increased activity of PC as pointed out earlier. Zwingmann

(2007) showed that an elevated ammonia concentration increased pyruvate carboxylation both in astrocytes in culture and in animal models. This carboxylation might be coupled to the increased glutamine formation which is a hallmark of ammonia detoxification (Zwingmann 2007). Glutamine formation is the only efficient method of ammonia fixation in brain and also in other cell culture studies increased production of glutamine was detected (Leke et al. 2011). It may be of interest that upon inhibition of the glutamine synthesizing enzyme glutamine synthetase ammonia fixation is brought about by the concerted action of glutamate dehydrogenase and alanine aminotransferase leading to net synthesis of alanine, a process stimulating glycolysis to provide pyruvate for alanine production (Dadsetan et al. 2011, 2013).

The ketogenic diet is a high-fat, adequate-protein, low-carbohydrate diet that is used primarily to treat refractory epilepsy in children (Yudkoff et al. 2004). The diet causes an increase in the metabolism of fatty acids rather than carbohydrates. The effect of ketone bodies on pyruvate carboxylation by rat brain mitochondria has been investigated (Patel 1974a) and it was shown that pyruvate carboxylation increased subsequent to feeding rats a ketogenic diet (Melø et al. 2006). This increase in pyruvate carboxylation was coupled to an elevated pyruvate recycling (Melø et al. 2006), i.e. conversion of oxaloacetate to pyruvate which re-enters the TCA cycle via pyruvate dehydrogenase, as expected since synthesis of a metabolite has to be equalled by its degradation.

3.5.1.1 Neurotransmission

Activation of the glutamatergic α-amino-3-hydroxy-5-methyl-4-isoxazolepropionic acid (AMPA) receptor induces large post-synaptic action potentials (Rao and Finkbeiner 2007) and should therefore stimulate release of neurotransmitter glutamate. In experiments with brain tissue slices metabolizing [1-^{13}C]glucose it was possible to show evidence of pyruvate carboxylation in aspartate which is obtained by transamination of oxaloacetate, the product of pyruvate carboxylation by pyruvate carboxylase (Rae et al. 2009). Increased pyruvate carboxylation was detected when the slices were incubated with AMPA (Rae et al. 2009) suggesting that increased neuronal activity increases carboxylation of pyruvate, a finding compatible with the notion that glutamate biosynthesis requires anaplerosis coupled to pyruvate carboxylation. Furthermore, it was shown that pyruvate carboxylation in the awake rat brain was several fold higher than under deep pentobarbital anaesthesia (Oz et al. 2004).

3.5.2 *Decreased Pyruvate Carboxylation*

3.5.2.1 Anxiety and Epilepsy

Decreased pyruvate carboxylation has been reported after pentylenetetrazole (PTZ) injection in rats (Eloqayli et al. 2004). PTZ administration is a generally accepted model for epileptic seizures but when given in low doses PTZ can cause anxiety

(Eloqayli et al. 2004). Interestingly, the antiepileptic drug carbamazepine (CBZ) decreased the biotin concentration in both humans and rats and decreased pyruvate carboxylase activity in rat brain (Rathman et al. 2003). These results support the use of biotin supplementation as a concurrent strategy during CBZ administration to help maintain pyruvate carboxylation.

3.5.2.2 Ischaemia

Not surprisingly, ischaemia has a profound effect on pyruvate carboxylation, a pathway dependent on adenosine triphosphate availability. In adult rats subjected to 120 min of middle cerebral artery occlusion followed by 120 min of reperfusion it could be shown that the utilization of precursors in astrocytic metabolism originating from the pyruvate carboxylase pathway was markedly reduced compared to that of precursors originating from the pyruvate dehydrogenase pathway. Hence, glutamate synthesis was reduced and that of GABA completely stopped in the ischaemic core area (Håberg et al. 2006). Furthermore, in the re-perfused penumbra, glutamatergic and GABAergic neurons used relatively more astrocytic metabolites derived from the pyruvate carboxylase pathway than in the ischaemic core (Håberg et al. 2006). In the canine cardiac arrest model of global ischaemia, pyruvate carboxylation was decreased in the hippocampus in animals that had undergone hyperoxic resuscitation (Scafidi et al. 2009).

Interestingly, in the Rice–Vannucci model of neonatal hypoxia–ischaemia in postnatal day 7 rats, PC activity is relatively preserved compared to that of the pyruvate dehydrogenase pathway. Labelling via pyruvate dehydrogenase was reduced by 44 % while labelling via pyruvate carboxylase was reduced by 30 % in the hypoxic–ischaemic hemisphere in the Rice–Vannucci model (Brekke et al. 2014) while, as pointed out, in adult models of middle cerebral artery occlusion, PC activity is more profoundly reduced than pyruvate dehydrogenase activity (Haberg, Qu et al. 1998, 2006). This illustrates how the neonatal and adult brain can respond differently to similar insults and may need different treatments.

The consequences of relatively preserved PC activity in the neonatal brain following hypoxia–ischaemia may be ambiguous. As mentioned earlier there is negligible transport of glutamate from neurons to astrocytes during normal conditions in the neonatal brain (Morken et al. 2014). During hypoxia–ischaemia, there is a large release of glutamate that may originate from neurons as well as astrocytes through reversal of glutamate transporters (Puka-Sundvall, Sandberg et al. 1997; Kusaka et al. 2004). However, no increase in transfer of glutamate from neurons to astrocytes was found following neonatal hypoxia–ischaemia at postnatal day 7 in rats. Both glutamate uptake and glutamine synthesis are energy demanding processes, which may explain the lacking augmentation in transfer following hypoxia–ischaemia. Contradicting this, transfer of glutamine from astrocytes to neurons as well as glutamine synthesis and PC activity, which are ATP demanding processes were preserved at this time point following injury suggesting that astrocytes have at least partly maintained or re-established their energy balance (Morken et al. 2014). It is conceivable that preservation of anaplerosis and transfer of newly synthesized glutamine to neurons in the immediate recovery phase following hypoxia–ischaemia may have harmful effects, but this remains to be investigated. On the other hand, following the massive release of glutamate into the extracellular space during hypoxia–ischaemia, neurons might be depleted of TCA cycle intermediates (such as α-ketoglutarate). Under these conditions anaplerosis might be beneficial since de novo synthesized glutamine transferred from astrocytes may be necessary for neurons to re-establish oxidative metabolism.

Surprisingly, at 6 h after hypoxia–ischaemia the transfer of glutamate from neurons to astrocytes was transiently decreased, even though at this point, the energy levels are almost back to normal (Yager et al. 1992). The exact mechanism of this has yet to be explored.

3.5.2.3 Glutamate Exposure

Qu et al. (2001) showed that exogenous glutamate had an inhibitory effect on pyruvate carboxylation in astrocytes, presumably by formation of oxaloacetate from α-ketoglutarate derived from glutamate metabolism. Inhibition of pyruvate carboxylation was also observed when glutamine was added to astrocytes (Hassel and Sonnewald 2002). When metabolism of ^{13}C-labelled glucose was investigated in cultured mouse cerebellar astrocytes, it could be shown that carboxylation of pyruvate was more important for biosynthesis of releasable glutamine and citrate, compared with their intracellular pools (Waagepetersen et al. 2001b). Activation of the cellular energy sensor, AMP-activated protein kinase (AMPK), has been shown to reduce the metabolism of glutamate in the TCA cycle of astrocytes. Specifically, the synthesis of citrate from glutamate involving malic enzyme and pyruvate carboxylase was reduced. Since AMPK is known to upregulate catabolic processes, this may suggest that glutamate has an anabolic function in astrocytes (Voss et al. 2015).

3.5.2.4 Alzheimer's Disease

Regional hypometabolism of glucose in the brain is a hallmark of Alzheimer's disease (AD) (Gibson et al. 1998, 1999; Huang et al. 2003; Klivenyi et al. 2004). However, little is known about the specific alterations of neuronal and astrocytic metabolism involved in homeostasis of glutamate and GABA in AD. In a recent study, the effects of amyloid β pathology on energy and neurotransmitter metabolism were investigated in the transgenic McGill-R-Thy1-APP rat model of AD, at age 15 months (Nilsen et al. 2014). Rats were injected with [1-^{13}C]glucose and [1,2-^{13}C]acetate, and brain extracts of the hippocampal formation as well as several cortical regions were analysed using ^{1}H- and ^{13}C nuclear magnetic resonance spectroscopy and high performance liquid chromatography. A decrease of pyruvate carboxylation in the hippocampal formation and retrosplenial/cingulate cortex was observed in the McGill-R-Thy1-APP rat and this decrease might have affected the level of glutamine in the hippocampal formation and levels of glutamate, glutamine, GABA and aspartate in the retrosplenial/cingulate cortex (Nilsen et al. 2014). Future treatment of AD patients might be targeted at increasing anaplerosis in the brain.

References

Alston TA, Mela L, Bright HJ (1977) 3-Nitropropionate, the toxic substance of Indigofera, is a suicide inactivator of succinate dehydrogenase. Proc Natl Acad Sci U S A 74:3767–3771

Alves PM, McKenna MC, Sonnewald U (1995) Lactate metabolism in mouse brain astrocytes studied by [^{13}C]NMR spectroscopy. Neuroreport 6:2201–2204

Attwell D, Laughlin SB (2001) An energy budget for signaling in the grey matter of the brain. J Cerebr Blood Flow Metab 21:1133–1145

Baburamani AA, Lo C, Castillo-Melendez M, Walker DW (2013) Morphological evaluation of the cerebral blood vessels in the late gestation fetal sheep following hypoxia in utero. Microvasc Res 85:1–9

Badar-Goffer RS, Bachelard HS, Morris PG (1990) Cerebral metabolism of acetate and glucose studied by ^{13}C-n.m.r. spectroscopy. A technique for investigating metabolic compartmentation in the brain. Biochem J 266:133–139

Bak LK, Iversen P, Sorensen M, Keiding S, Vilstrup H, Ott P, Waagepetersen HS, Schousboe A (2009) Metabolic fate of isoleucine in a rat model of hepatic encephalopathy and in cultured neural cells exposed to ammonia. Metab Brain Dis 24:135–145

Bandeira F, Lent R, Herculano-Houzel S (2009) Changing numbers of neuronal and non-neuronal cells underlie postnatal brain growth in the rat. Proc Natl Acad Sci U S A 106:14108–14113

Berl S, Clarke DD (1983) The metabolic compartmentation concept. In: Hertz L et al (eds) Glutamine, glutamate and GABA in the central nervous system. Alan R. Liss, Inc., New York, pp 205–217

Borges K, Sonnewald U (2012) Triheptanoin—a medium chain triglyceride with odd chain fatty acids: a new anaplerotic anticonvulsant treatment? Epilepsy Res 100:239–244

Boulland J-L, Rafiki A, Levy LM, Storm-Mathisen J, Chaudhry FA (2003) Highly differential expression of SN1, a bidirectional glutamine transporter, in astroglia and endothelium in the developing rat brain. Glia 41:260–275

Brekke EM, Morken TS, Wideroe M, Haberg AK, Brubakk AM, Sonnewald U (2014) The pentose phosphate pathway and pyruvate carboxylation after neonatal hypoxic-ischemic brain injury. J Cereb Blood Flow Metab 34:724–734

Butterworth RF (2003) Hepatic encephalopathy. Alcohol Res Health 27:240–246

Cesar M, Hamprecht B (1995) Immunocytochemical examination of neural rat and mouse primary cultures using monoclonal antibodies raised against pyruvate carboxylase. J Neurochem 64:2312–2318

Chowdhury GM, Patel AB, Mason GF, Rothman DL, Behar KL (2007) Glutamatergic and GABAergic neurotransmitter cycling and energy metabolism in rat cerebral cortex during postnatal development. J Cereb Blood Flow Metab 27:1895–1907

Dadsetan S, Bak LK, Sorensen M, Keiding S, Vilstrup H, Ott P, Leke R, Schousboe A, Waagepetersen HS (2011) Inhibition of glutamine synthesis induces glutamate dehydrogenase-dependent ammonia fixation into alanine in co-cultures of astrocytes and neurons. Neurochem Int 58:482–488

Dadsetan S, Kukolj E, Bak LK, Sorensen M, Ott P, Vilstrup H, Schousboe A, Keiding S, Waagepetersen HS (2013) Brain alanine formation as an ammonia-scavenging pathway during hyperammonemia: effects of glutamine synthetase inhibition in rats and astrocyte-neuron co-cultures. J Cereb Blood Flow Metab 33:1235–1241

Danbolt NC (2001) Glutamate uptake. Prog Neurobiol 65:1–105

Danbolt NC, Storm-Mathisen J, Kanner BI (1992) An [Na+ + K+]coupled L-glutamate transporter purified from rat brain is located in glial cell processes. Neuroscience 51:295–310

Eloqayli H, Dahl CB, Gotestam KG, Unsgard G, Sonnewald U (2004) Changes of glial-neuronal interaction and metabolism after a subconvulsive dose of pentylenetetrazole. Neurochem Int 45:739–745

Gibson GE, Sheu KF, Blass JP (1998) Abnormalities of mitochondrial enzymes in Alzheimer disease. J Neural Transm 105:855–870

Gibson GE, Park LC, Zhang H, Sorbi S, Calingasan NY (1999) Oxidative stress and a key metabolic enzyme in Alzheimer brains, cultured cells, and an animal model of chronic oxidative deficits. Ann N Y Acad Sci 893:79–94

Griffin JL, Rae C, Dixon RM, Radda GK, Matthews PM (1998) Excitatory amino acid synthesis in hypoxic brain slices: does alanine act as a substrate for glutamate production in hypoxia? J Neurochem 71:2477–2486

Griffin JL, Rae C, Radda GK, Matthews PM (1999) Lactate-induced inhibition of glucose catabolism in guinea pig cortical brain slices. Neurochem Int 35:405–409

Griffin JL, Keun H, Richter C, Moskau D, Rae C, Nicholson JK (2003) Compartmentation of metabolism probed by [2-13C]alanine: improved 13C NMR sensitivity using a CryoProbe detects evidence of a glial metabolon. Neurochem Int 42:93–99

Håberg A, Qu H, Haraldseth O, Unsgard G, Sonnewald U (1998) In vivo injection of [1-13C]glucose and [1,2-13C]acetate combined with ex vivo 13C nuclear magnetic resonance spectroscopy: a novel approach to the study of middle cerebral artery occlusion in the rat. J Cereb Blood Flow Metab 18(11):1223–1232

Håberg A, Qu H, Sonnewald U (2006) Glutamate and GABA metabolism in transient and permanent middle cerebral artery occlusion in rat: importance of astrocytes for neuronal survival. Neurochem Int 48:531–540

Hassel B (2001) Pyruvate carboxylation in neurons. J Neurosci Res 66:755–762

Hassel B, Sonnewald U (1995) Glial formation of pyruvate and lactate from TCA cycle intermediates: implications for the inactivation of transmitter amino acids? J Neurochem 65:2227–2234

Hassel B, Sonnewald U (2002) Effects of potassium and glutamine on metabolism of glucose in astrocytes. Neurochem Res 27:167–171

Hassel B, Sonnewald U, Fonnum F (1995) Glial-neuronal interactions as studied by cerebral metabolism of [2-^{13}C]acetate and [1-^{13}C]glucose: an ex vivo 13C NMR spectroscopic study. J Neurochem 64:2773–2782

Hertz L (2011) Astrocytic energy metabolism and glutamate formation—relevance for ^{13}C-NMR spectroscopy and importance of cytosolic/mitochondrial trafficking. Magn Reson Imaging 29:1319–1329

Hertz L, Bock E, Schousboe A (1978) GFA content, glutamate uptake and activity of glutamate metabolizing enzymes in differentiating mouse astrocytes in primary cultures. Dev Neurosci 1:226–238

Hogstad S, Svenneby G, Torgner IA, Kvamme E, Hertz L, Schousboe A (1988) Glutaminase in neurons and astrocytes cultured from mouse brain: kinetic properties and effects of phosphate, glutamate, and ammonia. Neurochem Res 13:383–388

Huang HM, Zhang H, Xu H, Gibson GE (2003) Inhibition of the alpha-ketoglutarate dehydrogenase complex alters mitochondrial function and cellular calcium regulation. Biochim Biophys Acta 1637:119–126

Juurlink BH, Schousboe A, Jorgensen OS, Hertz L (1981) Induction by hydrocortisone of glutamine synthetase in mouse primary astrocyte cultures. J Neurochem 36:136–142

Kanamatsu T, Tsukada Y (1999) Effects of ammonia on the anaplerotic pathway and amino acid metabolism in the brain: an ex vivo 13C NMR spectroscopic study of rats after administering [2-^{13}C]] glucose with or without ammonium acetate. Brain Res 841:11–19

Kim TH, Borges K, Petrou S, Reid CA (2013) Triheptanoin reduces seizure susceptibility in a syndrome-specific mouse model of generalized epilepsy. Epilepsy Res 103:101–105

Klivenyi P, Starkov AA, Calingasan NY, Gardian G, Browne SE, Yang L, Bubber P, Gibson GE, Patel MS, Beal MF (2004) Mice deficient in dihydrolipoamide dehydrogenase show increased vulnerability to MPTP, malonate and 3-nitropropionic acid neurotoxicity. J Neurochem 88:1352–1360

Kreis R, Hofmann L, Kuhlmann B, Boesch C, Bossi E, Huppi PS (2002) Brain metabolite composition during early human brain development as measured by quantitative in vivo 1H magnetic resonance spectroscopy. Magn Reson Med 48:949–958

Kurz GM, Wiesinger H, Hamprecht B (1993) Purification of cytosolic malic enzyme from bovine brain, generation of monoclonal antibodies, and immunocytochemical localization of the enzyme in glial cells of neural primary cultures. J Neurochem 60:1467–1474

Kusaka T, Matsuura S, Fujikawa Y, Okubo K, Kawada K, Namba M, Okada H, Imai T, Isobe K, Itoh S (2004). Relationship between cerebral interstitial levels of amino acids and phosphorylation potential during secondary energy failure in hypoxic-ischemic newborn piglets. Pediatr Res 55(2):273–279

Larsson OM, Drejer J, Kvamme E, Svenneby G, Hertz L, Schousboe A (1985) Ontogenetic development of glutamate and GABA metabolizing enzymes in cultured cerebral cortex interneurons and in cerebral cortex in vivo. Int J Dev Neurosci 3:177–185

Leke R, Bak LK, Anker M, Melo TM, Sorensen M, Keiding S, Vilstrup H, Ott P, Portela LV, Sonnewald U, Schousboe A, Waagepetersen HS (2011) Detoxification of ammonia in mouse cortical GABAergic cell cultures increases neuronal oxidative metabolism and reveals an emerging role for release of glucose-derived alanine. Neurotox Res 19:496–510

Marin-Valencia I, Good LB, Ma Q, Malloy CR, Pascual JM (2013) Heptanoate as a neural fuel: energetic and neurotransmitter precursors in normal and glucose transporter I-deficient (G1D) brain. J Cereb Blood Flow Metab 33:175–182

Mason GF, Petersen KF, de Graaf RA, Shulman GI, Rothman DL (2007) Measurements of the anaplerotic rate in the human cerebral cortex using ^{13}C magnetic resonance spectroscopy and [1-^{13}C] and [2-^{13}C] glucose. J Neurochem 100:73–86

McCormick DA, Prince DA (1987) Post-natal development of electrophysiological properties of rat cerebral cortical pyramidal neurones. J Physiol 393:743–762

McKenna MC, Tildon JT, Stevenson JH, Huang X, Kingwell KG (1995) Regulation of mitochondrial and cytosolic malic enzymes from cultured rat brain astrocytes. Neurochem Res 20:1491–1501

McKenna MC, Sonnewald U, Huang X, Stevenson J, Zielke HR (1996) Exogenous glutamate concentration regulates the metabolic fate of glutamate in astrocytes. J Neurochem 66:386–393

McKenna MC, Stevenson JH, Huang X, Tildon JT, Zielke CL, Hopkins IB (2000) Mitochondrial malic enzyme activity is much higher in mitochondria from cortical synaptic terminals compared with mitochondria from primary cultures of cortical neurons or cerebellar granule cells. Neurochem Int 36:451–459

McKenna M, Gruetter R, Sonnewald U, Waagepetersen HS, Schousboe A (2012) Energy metabolism of the brain. In: Brady ST, Siegel GJ, Albers RW, Price DL (eds) Basic neurochemistry: principles of molecular, cellular, and medical neurobiology, 8th edn. Elsevier Academic, Oxford, pp 200–231

Melø TM, Nehlig A, Sonnewald U (2006) Neuronal-glial interactions in rats fed a ketogenic diet. Neurochem Int 48:498–507

Merle M, Bouzier-Sore AK, Canioni P (2002) Time-dependence of the contribution of pyruvate carboxylase versus pyruvate dehydrogenase to rat brain glutamine labelling from [1-(13)C] glucose metabolism. J Neurochem 82:47–57

Mohler H, Patel AJ, Balazs R (1974) Metabolic compartmentation in the brain: metabolism of a tricarboxylic acid cycle intermediate, (1,4-^{14}C)succinate, after intracerebral administration. J Neurochem 23:1281–1289

Morken TS, Brekke E, Haberg A, Wideroe M, Brubakk AM, Sonnewald U (2013) Neuron-astrocyte interactions, pyruvate carboxylation and the pentose phosphate pathway in the neonatal rat brain. Neurochem Res 39(3):556–569

Morken TS, Brekke E, Haberg A, Wideroe M, Brubakk AM, Sonnewald U (2014) Altered astrocyte-neuronal interactions after hypoxia-ischemia in the neonatal brain in female and male rats. Stroke 45:2777–2785

Nilsen LH, Witter MP, Sonnewald U (2014) Neuronal and astrocytic metabolism in a transgenic rat model of Alzheimer's disease. J Cereb Blood Flow Metab 34:906–914

Norenberg MD, Martinez-Hernandez A (1979) Fine structural localization of glutamine synthetase in astrocytes of rat brain. Brain Res 161:303–310

Oz G, Berkich DA, Henry PG, Xu Y, LaNoue K, Hutson SM, Gruetter R (2004) Neuroglial metabolism in the awake rat brain: CO_2 fixation increases with brain activity. J Neurosci 24:11273–11279

Patel MS (1974a) The effect of ketone bodies on pyruvate carboxylation by rat brain mitochondria. J Neurochem 23:865–867

Patel MS (1974b) The relative significance of CO_2-fixing enzymes in the metabolism of rat brain. J Neurochem 22:717–724

Puka-Sundvall M, Sandberg M, Hagberg H (1997) Brain injury after hypoxia-ischemia in newborn rats: relationship to extracellular levels of excitatory amino acids and cysteine. Brain Res 750(1–2):325–328

Qu H, Haberg A, Haraldseth O, Unsgard G, Sonnewald U (2000) (13)C MR spectroscopy study of lactate as substrate for rat brain. Dev Neurosci 22:429–436

Qu H, Eloqayli H, Unsgard G, Sonnewald U (2001) Glutamate decreases pyruvate carboxylase activity and spares glucose as energy substrate in cultured cerebellar astrocytes. J Neurosci Res 66:1127–1132

Rae C, Hansen JT, Bubb WA, Bröer S, Bröer A (2005) Alanine transport, metabolism and cycling in the brain. Proc Intl Soc Magn Reson Med 2481

Rae C, Nasrallah F, Bröer S (2009) Metabolic effects of blocking lactate transport in brain cortical tissue slices using an inhibitor specific to MCT1 and MCT2. Neurochem Res 34:1783–1791

Rao VR, Finkbeiner S (2007) NMDA and AMPA receptors: old channels, new tricks. Trends Neurosci 30:284–291

Rathman SC, Gregory JF III, McMahon RJ (2003) Pharmacological biotin supplementation maintains biotin status and function in rats administered dietary carbamazepine. J Nutr 133:2857–2862

Scafidi S, O'Brien J, Hopkins I, Robertson C, Fiskum G, McKenna M (2009) Delayed cerebral oxidative glucose metabolism after traumatic brain injury in young rats. J Neurochem 109(Suppl 1):189–197

Schiff M, Levrat V, Acquaviva C, Vianey-Saban C, Rolland MO, Guffon N (2006) A case of pyruvate carboxylase deficiency with atypical clinical and neuroradiological presentation. Mol Genet Metab 87:175–177

Schousboe A (1972) Development of potassium effects on ion concentrations and indicator spaces in rat brain-cortex slices during postnatal ontogenesis. Exp Brain Res 15:521–531

Schousboe A, Scafidi S, Bak LK, Waagepetersen HS, McKenna MC (2014) Glutamate metabolism in the brain focusing on astrocytes. Adv Neurobiol 11:13–30

Schousboe A, Walls A, Bak L, Waagepetersen H (2015) Astroglia and brain metabolism: focus on energy and neurotransmitter amino acid homeostasis. In: Verkhratsky A, Parpura V (eds) Colloquium series on neuroglia in biology and medicine from physiology to disease. Morgan & Claypool Life Sciences, San Rafael, pp 1–63

Shank RP, Bennett GS, Freytag SO, Campbell GL (1985) Pyruvate carboxylase: an astrocyte-specific enzyme implicated in the replenishment of amino acid neurotransmitter pools. Brain Res 329:364–367

Shank RP, Leo GC, Zielke HR (1993) Cerebral metabolic compartmentation as revealed by nuclear magnetic resonance analysis of D-[1-^{13}C]glucose metabolism. J Neurochem 61:315–323

Smeland OB, Hadera MG, McDonald TS, Sonnewald U, Borges K (2013) Brain mitochondrial metabolic dysfunction and glutamate level reduction in the pilocarpine model of temporal lobe epilepsy in mice. J Cereb Blood Flow Metab 33(7):1090–1097

Snead OC III, Stephens HI (1983) Ontogeny of cortical and subcortical electroencephalographic events in unrestrained neonatal and infant rats. Exp Neurol 82:249–269

Sonnewald U (2014) Glutamate synthesis has to be matched by its degradation—where do all the carbons go? J Neurochem 131(4):399–406

Varoqui H, Zhu H, Yao D, Ming H, Erickson JD (2000) Cloning and functional identification of a neuronal glutamine transporter. J Biol Chem 275:4049–4054

Vogel R, Hamprecht B, Wiesinger H (1998a) Malic enzyme isoforms in astrocytes: comparative study on activities in rat brain tissue and astroglia-rich primary cultures. Neurosci Lett 247:123–126

Vogel R, Jennemann G, Seitz J, Wiesinger H, Hamprecht B (1998b) Mitochondrial malic enzyme: purification from bovine brain, generation of an antiserum, and immunocytochemical localization in neurons of rat brain. J Neurochem 71:844–852

Voss CM, Pajecka K, Stridh MH, Nissen JD, Schousboe A, Waagepetersen HS (2015) AMPK activation affects glutamate metabolism in astrocytes. Neurochem Res 40(12):2431–2442. doi:10.1007/s11064-015-1558-5

Waagepetersen HS, Qu H, Schousboe A, Sonnewald U (2001a) Elucidation of the quantitative significance of pyruvate carboxylation in cultured cerebellar neurons and astrocytes. J Neurosci Res 66:763–770

Waagepetersen HS, Sonnewald U, Larsson OM, Schousboe A (2001b) Multiple compartments with different metabolic characteristics are involved in biosynthesis of intracellular and released glutamine and citrate in astrocytes. Glia 35:246–252

Wallace JC, Jitrapakdee S, Chapman-Smith A (1998) Pyruvate carboxylase. Int J Biochem Cell Biol 30:1–5

Weiss MD, Derazi S, Rossignol C, Varoqui H, Erickson JD, Kilberg MS, Anderson KJ (2003) Ontogeny of the neutral amino acid transporter SAT1/ATA1 in rat brain. Brain Res Dev Brain Res 143:151–159

Wilbur DO, Patel MS (1974) Development of mitochondrial pyruvate metabolism in rat brain. J Neurochem 22:709–715

Willis S, Stoll J, Sweetman L, Borges K (2010) Anticonvulsant effects of a triheptanoin diet in two mouse chronic seizure models. Neurobiol Dis 40:565–572

Xu Y, Oz G, LaNoue KF, Keiger CJ, Berkich DA, Gruetter R, Hutson SH (2004) Whole-brain glutamate metabolism evaluated by steady-state kinetics using a double-isotope procedure: effects of gabapentin. J Neurochem 90:1104–1116

Yager JY, Brucklacher RM, Vannucci RC (1992) Cerebral energy metabolism during hypoxia-ischemia and early recovery in immature rats. Am J Physiol 262:H672–H677

Yu AC, Drejer J, Hertz L, Schousboe A (1983) Pyruvate carboxylase activity in primary cultures of astrocytes and neurons. J Neurochem 41:1484–1487

Yudkoff M, Daikhin Y, Nissim I, Lazarow A (2004) Ketogenic diet, brain glutamate metabolism and seizure control. Prostaglandins Leukot Essent Fatty Acids 70:277–285

Zwingmann C (2007) The anaplerotic flux and ammonia detoxification in hepatic encephalopathy. Metab Brain Dis 22:235–249

Chapter 4
Enzyme Complexes Important for the Glutamate–Glutamine Cycle

Mary C. McKenna and Gustavo C. Ferreira

Abstract Transient multienzyme and/or multiprotein complexes (metabolons) direct substrates toward specific pathways and can significantly influence the metabolism of glutamate and glutamine in the brain. Glutamate is the primary excitatory neurotransmitter in brain. This neurotransmitter has essential roles in normal brain function including learning and memory. Metabolism of glutamate involves the coordinated activity of astrocytes and neurons and high affinity transporter proteins that are selectively distributed on these cells. This chapter describes known and possible metabolons that affect the metabolism of glutamate and related compounds in the brain, as well as some factors that can modulate the association and dissociation of such complexes, including protein modifications by acylation reactions (e.g., acetylation, palmitoylation, succinylation, SUMOylation, etc.) of specific residues. Development of strategies to modulate transient multienzyme and/or enzyme–protein interactions may represent a novel and promising therapeutic approach for treatment of diseases involving dysregulation of glutamate metabolism.

Keywords Transient multienzyme complexes • Metabolon • Glutamate • Brain • Transporters • Acylation

M.C. McKenna, Ph.D. (✉)
Department of Pediatrics, University of Maryland School of Medicine, 655 W. Baltimore Street, BRB 13-019, Baltimore, MD 21201, USA

Program in Neuroscience, University of Maryland School of Medicine, Baltimore, MD 21201, USA
e-mail: mmckenna@umaryland.edu

G.C. Ferreira
Department of Pediatrics, University of Maryland School of Medicine, 655 W. Baltimore Street, BRB 13-019, Baltimore, MD 21201, USA

Institute of Medical Biochemistry Leopoldo de Meis, Federal University of Rio de Janeiro, Rio de Janeiro, RJ 21941-902, Brazil

A. Schousboe, U. Sonnewald (eds.), *The Glutamate/GABA-Glutamine Cycle*, Advances in Neurobiology 13, DOI 10.1007/978-3-319-45096-4_4

Abbreviations

α-KG	α-Ketoglutarate
AAT	Aspartate aminotransferase
AMPA	α-Amino-3-hydroxy-5-methyl-4-isoxazolepropionic acid
BCAA	Branched-chain amino acid
BCAT	Branched-chain aminotransferase
BCATc	Cytosolic branched-chain aminotransferase
BCATm	Mitochondrial branched-chain aminotransferase
BCKD	Branched-chain ketoacid dehydrogenase
BCKDC	Branched-chain ketoacid dehydrogenase complex
GABA	Gamma-aminobutyric acid
GAD	Glutamic acid decarboxylase
GAPDH	Glyceraldehyde 3-phosphate dehydrogenase
GDH	Glutamate dehydrogenase
GLAST	EAAT1 or excitatory amino acid transporter 1
GLT-1	EAAT2 or excitatory amino acid transporter 2
hBCAT	Human branched-chain aminotransferase
hBCATm	Human mitochondrial branched-chain aminotransferase
ICDH	Isocitrate dehydrogenase
KGDH	α-Ketoglutarate dehydrogenase
MDH	Malate dehydrogenase
ME	Malic enzyme
mGluR	Metabotropic glutamate receptor
mMDH	Mitochondrial malate dehydrogenase
mTOR	Mammalian target of rapamycin
NCX	Na^+/Ca^{2+} exchanger
nNOS	Neuronal nitric oxide synthase
OAA	Oxaloacetate
PAG	Phosphate-activated glutaminase
PC	Pyruvate carboxylase
PDH	Pyruvate dehydrogenase
SCHAD	Short-chain 3-hydroxyacyl-CoA dehydrogenase
SIRTs or sirtuins	Silent information regulator proteins
SUMO	Small ubiquitin-like modifiers
TCA	Tricarboxylic acid
VDAC	Voltage-dependent anion channel
vGLUTs	Vesicular glutamate transporters

4.1 Introduction

There is increasing evidence that the formation of transient multienzyme complexes (metabolons), which can rapidly associate and dissociate, can greatly influence metabolism. The association and dissociation of these transiently formed heteroenzyme complexes is governed by fluctuations in the concentration of key metabolites in mitochondria and to some extent by protein modifications (Fahien et al. 1977, 1985, 1988; Teller et al. 1990; Fahien and Teller 1992; Malik et al. 1993; McKenna et al. 2000a, 2006b; McKenna 2007, 2011; Gibson et al. 2015; McKenna and Rae 2015; Yang et al. 2015; Robinson and Jackson 2016). Formation of these transient enzyme and/or enzyme–protein complexes occurs dynamically in response to localized changes in the concentration of metabolites and thus has the ability to modulate metabolism in "real time." Biochemical modifications of enzyme molecules can modulate activity and/or the ability of these proteins to associate into enzyme complexes. Given the importance of glutamate and glutamine metabolism in brain for neurotransmission and bioenergetics, an understanding of the intricate regulation of participating enzymes and associated proteins is crucial.

This chapter includes information about the enzymes involved in glutamate metabolism and summarizes reports on the formation and modulation of transient multienzyme complexes that could potentially influence glutamate and glutamine metabolism and cycling in astrocytes and neurons. We discuss the regulation of glutamate dehydrogenase as this large homohexameric enzyme may have a role in organizing multienzyme enzyme complexes (Fahien et al. 1989). Common post-translational modifications that influence the activity and/or properties of proteins directly or indirectly involved in glutamate metabolism are also discussed.

4.2 Overview of Glutamate and Glutamine Metabolism and Compartmentation in Brain

Glutamate is the primary excitatory neurotransmitter in brain and has essential roles in normal brain function including learning and memory (Schweigert et al. 2005; McKenna 2007, 2013; Schousboe et al. 2014; Connor and Wang 2015; Kumar 2015; Kennedy 2016). Astrocytes are key partners in brain function, and glutamatergic neurotransmission requires a continuous, dynamic interaction between astrocytes and neurons (McKenna 2013). After depolarization of glutamatergic neurons the concentration of glutamate in the synaptic cleft increases from a low resting concentration of ~0.1–1 μM to ~100 μM –1 mM (Bergles et al. 1999; Matsui et al. 2005). The high affinity glutamate transporters in astrocytes rapidly and efficiently remove glutamate from the synaptic cleft, which allows glutamatergic neurotransmission to continue and prevents excitotoxicity (Danbolt 2001; McKenna 2013). Glutamate, taken up by astrocytes is primarily oxidized for energy (McKenna 2007, 2013; Schousboe et al. 2014) or converted to glutamine, which can be released by astrocytes and taken up by neurons to replenish TCA cycle intermediates and neuronal glutamate (Leke and Schouboe 2016). Glutamate can also participate in the purine nucleotide cycle (Yudkoff et al. 1986) or be used for synthesis of proteins or the tripeptide glutathione (Dringen 2000).

Glutamine, which is formed primarily, if not exclusively in astrocytes due to the selective localization of glutamine synthetase in these cells (Norenberg and Martinez-Hernandez 1979), is taken up from the extracellular milieu by neurons and converted back to neurotransmitter glutamate, oxidized for energy via the tricarboxylic acid (TCA) cycle (Yudkoff et al. 1988), or decarboxylated to form gamma-aminobutyric acid (GABA), the main inhibitory neurotransmitter in brain (Schousboe 1981; Sonnewald et al. 1993b; Westergaard et al. 1995a; McKenna et al. 2011). De novo synthesis of glutamine occurs in astrocytes due to the compartmentation of the enzyme pyruvate carboxylase (PC) in these cells (Yu et al. 1983; Shank et al. 1985). By adding a carbon to pyruvate to form the 4-carbon TCA cycle intermediate oxaloacetate, metabolism via PC leads to formation of glutamine that is released by astrocytes and ultimately serves to replenish the glutamate and TCA cycle intermediates in neurons that would otherwise be depleted via neurotransmission (McKenna 2007, 2013; Schousboe et al. 2014; Sonnewald 2014 and refs therein).

Numerous transporters and enzymes are needed for glutamate, glutamine, and GABA metabolism, and a number of these proteins are differentially distributed between astrocytes and neurons (McKenna 2007, 2013; Schousboe et al. 2014). Such differential distribution or compartmentation of transporters, enzymes, and metabolic pathways is needed so that glutamatergic and GABAergic neurons and the astrocytes surrounding synapses can perform their highly specialized functions in brain (McKenna 2007, 2013; Schousboe et al. 2014) (Fig. 4.1 and Table 4.1). The phenomenon of compartmentation requires the trafficking or exchange of metabolites between different cell types in brain (Westergaard et al. 1995a; Schousboe et al. 1997; McKenna et al. 2000a). Key factors contributing to compartmentation are the heterogeneity in the localization of transporters and metabolic enzymes, *within* populations of astrocytes and neurons (McKenna et al. 1990, 2000a; Oh et al. 1991; Cesar and Hamprecht 1995), and differences *between* glia and neurons that lead to the unique metabolic capabilities of each cell type (Kvamme et al. 1985; Lai et al. 1989; Schousboe et al. 1997; Sonnewald et al. 1997, 1998; Hanu et al. 2000; McKenna et al. 2000b; Yudkoff et al. 2000; McKenna 2007). A number of the enzymes important in metabolism are regulated differently in astrocytes and neurons (Malik et al. 1993; McKenna et al. 2000a, 2006b; McKenna 2007, 2011 and references therein). *Intracellular* compartmentation of metabolism also occurs in both astrocytes and neurons (McKenna et al. 1990, 1995, 1996a, b, 2000a; Schousboe et al. 1993; Sonnewald et al. 1993a, b; Bakken et al. 1997; Waagepetersen et al. 1998, 1999 and references therein). In addition to compartmentation among different cell types and organelles, indication of cytoplasmic compartmentation of metabolites in neural cells has been reported (McKenna et al. 1994, 1995, 1996a, b, 2006a; Bakken et al. 1997; Waagepetersen et al. 1998). Indeed, the gel-like composition of the cytoplasm contains proteins and extensive membranous surfaces and that can restrict the diffusion of small molecules (Jones 1986; Pollack 2001). Interestingly, the metabolic state of the cell may regulate mobility of glycolytic enzymes (Pagliaro and Taylor 1992).

Since compartmentation leads to differences in neuronal and glial metabolism, recognizing the importance of this phenomenon is essential for understanding regulation of metabolism in brain (McKenna et al. 2000a; McKenna 2007). A knowledge of the enzymes involved in the formation and disposal of glutamate, glutamine, and related metabolites is essential for understanding glutamate metabolism in brain. Enzymes with a role in glutamate and glutamine metabolism in astrocytes and neurons are shown in Fig. 4.1 and Table 4.1.

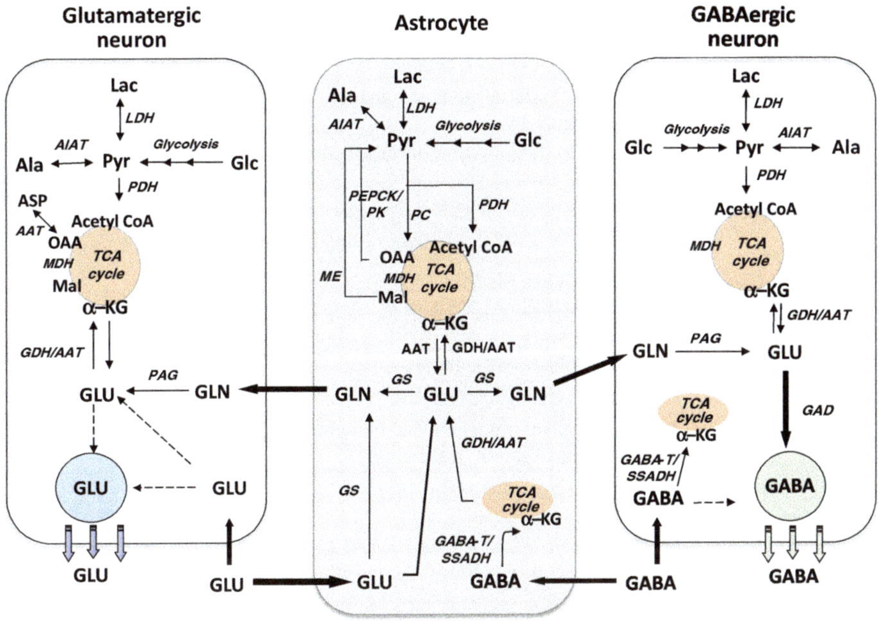

Fig. 4.1 Neurotransmitter glutamate (GLU) interacts with receptors and is rapidly taken up by the high affinity transporters on astrocytes surrounding the synaptic cleft. In the astrocyte, glutamate is either metabolized in the TCA cycle for energy or converted to glutamine (GLN) via the enzyme glutamine synthetase (GS), which forms part of an open glutamate–glutamine cycle. Glutamine is released by astrocytes and taken up by glutamatergic neurons to be used for energy or converted to glutamate via phosphate-activated glutaminase (PAG). Glutamate is converted to the TCA cycle intermediate α-ketoglutarate by the activity of GDH or via AAT or other aminotransferase enzymes. The carbon skeleton of glutamate can replenish the pool of TCA cycle intermediates in neurons and astrocytes. In GABAergic neurons, GABA is synthesized from GLU via the activity of glutamate decarboxylase (GAD). The carbon skeleton of glutamate can also be partially or fully oxidized via the pyruvate recycling pathway which involves malic enzyme (ME), or phosphoenol-pyruvate carboxykinase (PEPCK) and pyruvate kinase (PK). The de novo synthesis of glutamate and glutamine from glucose occurs via the combined action of PDH and pyruvate carboxylase (PC) in astrocytes leading to net synthesis of TCA cycle intermediates. Some of the enzymes discussed in this chapter are not shown in this figure. It should be noted that aspartate formation from OAA occurs in all of three cell types but is shown only in one cell type; and pyruvate recycling, shown only in astrocytes may also occur in neurons. *Ala* alanine, *ASP* aspartate, *Glc* glucose, *GLN* glutamine, *GLU* glutamate, *α-KG* α-ketoglutarate, *Lac* lactate, *Mal* malate, *OAA* oxaloacetate, *Pyr* pyruvate, *AlAT* alanine aminotransferase, *AAT* aspartate aminotransferase, *GDH* glutamate dehydrogenase, *GAD* glutamate decarboxylase, *GABA-T* GABA transaminase, *LDH* lactate dehydrogenase, *MDH* malate dehydrogenase, *ME* malic enzyme, *PC* pyruvate carboxylase, *PDH* pyruvate dehydrogenase, *SSADH* succinic semialdehyde dehydrogenase (Figure adapted from Schousboe et al. 2014). See references therein and this chapter for more details

Table 4.1 Compartmentation of a number of enzymes and proteins involved in glutamate metabolism in brain

Localization	Enzymes/transporters/proteins
Located in neurons (not in astrocytes)	cBCAT, cytosolic branched-chain aminotransferase
	BCKDC-E1α, branched-chain α-ketoacid dehydrogenase complex subunit E1α
Located only in GABAergic neurons	GAD, glutamic acid decarboxylase
Highly enriched in neurons	PDH, pyruvate dehydrogenase
	PAG, phosphate-activated glutaminase
	mME, mitochondrial malic enzyme
Neuronal glutamate transporters	EAAC1/EAAT3, EAAT4, EAAT5
Located in astrocytes (not in neurons)	GS, glutamine synthetase
	PC, pyruvate carboxylase
	hBCATm, mitochondrial branched-chain aminotransferase (in rodents, but not in humans)
Highly enriched in astrocytes	cME, cytosolic malic enzyme
Astrocytic glutamate transporters	GLAST/EAAT1, GLT1/EAAT2
Located in both astrocytes and neurons	AAT, aspartate aminotransferase (both mitochondrial and cytosolic isoforms)
	GDH, glutamate dehydrogenase
	Binding protein on the inner mitochondrial membrane that interacts with AAT, MDH, and mutienzyme complexes (unidentified but expected to be present); see section 4.3.3
Located in endothelial cells	BCKDC-E1α, branched-chain α-ketoacid dehydrogenase complex E1α subunit (in humans)
	hBCATm, mitochondrial branched-chain aminotransferase (in humans) (Hull et al. 2012)

Adapted from McKenna (2007)

4.3 Glutamate Dehydrogenase (GDH) and Aminotransferase Enzymes Have Essential Roles in Linking Glutamate Metabolism to Energy Metabolism

4.3.1 Metabolism of Glutamate to α-Ketoglutarate

As noted earlier glutamate taken up by astrocytes can be oxidatively metabolized via the TCA cycle. The first step in this process is conversion to the TCA cycle intermediate α-ketoglutarate either by deamination by the enzyme glutamate dehydrogenase (GDH) or transamination by one of the aminotransferase enzymes, aspartate aminotransferase, branched-chain amino acid transaminase, or alanine aminotransferase (McKenna 2011; Schousboe et al. 2014; McKenna et al. 2016 and refs therein).

The mitochondrial enzyme GDH functions in brain primarily to catalyze the deamination of glutamate to the TCA cycle intermediate α-ketoglutarate, releasing NH_3 and forming NAD(P)H (Cooper and Meister 1985; Yudkoff et al. 1990, 1991; Wysmyk-Cybula et al. 1991; McKenna et al. 1993, 1996a, 2000a; Westergaard et al. 1996; Zaganas et al. 2009; Cooper 2011). Although the thermodynamic equilibrium of GDH favors the energy requiring reductive amination of α-ketoglutarate to glutamate, the NAD^+/NADH ratio in the mitochondria and the high Km for ammonia favors the oxidative deamination in brain (Cooper and Meister 1985). Two forms of GDH are found in human brain, GDH1 which is widely expressed (Stanley 2004; Raizen et al. 2005; Spanaki et al. 2010), and GDH2 which is particularly high in astrocyte mitochondria and is also labeled in neuronal mitochondria (Spanaki et al. 2010). The structure and regulation of GDH1 and GDH2 have been reported (Plaitakis et al. 2000; Plaitakis and Zaganas 2001; Smith et al. 2001; Mastorodemos et al. 2005, 2009, 2015; Zaganas et al. 2009; Li et al. 2011; Stanley 2011; Botman et al. 2014; Karaca et al. 2015; Rasgado et al. 2015). Changes in GDH activity have been reported in several conditions leading to neurodegeneration (Plaitakis et al. 1982, 1984, 2003; Plaitakis and Zaganas 2001; Zaganas et al. 2009; Shashidharan and Plaitakis 2014).

Aminotransferase enzymes mediate the transfer of amino nitrogen in brain. These enzymes, which require the cofactor pyridoxal-5′-phosphate, catalyze the reversible transfer of amino groups between glutamate and α-ketoglutarate, in conjunction with another amino acids and their cognate keto acids (Cooper 1988; McKenna et al. 2000a). Aspartate aminotransferase (AAT) which has the highest transaminase activity in brain catalyzes the transamination of oxaloacetate to aspartate in conjunction with the interconversion of glutamate and α-ketoglutarate (Cooper and Meister 1985; Cooper 2011). GDH and AAT act together in metabolism in most tissues since GDH either adds an amino group to α-ketoglutarate or removes an amino group from glutamate (Yudkoff et al. 1990, 1991; Wysmyk-Cybula et al. 1991; McKenna et al. 1993, 1996a, 2000a; Westergaard et al. 1996), whereas aminotransferase enzymes, including AAT, transfer the amino groups to or from glutamate to convert keto acids into amino acids (Cooper and Meister 1985; Yudkoff et al. 1990, 1991; Wysmyk-Cybula et al. 1991; McKenna et al. 1993, 1996a, 2000a; Westergaard et al. 1996). The extent to which GDH and AAT or other aminotransferase enzymes work together in brain cells is not known. However, work from the Hutson lab recently reported that GDH and enzymes involved in branched-chain amino acid (BCAA) metabolism can form a functional complex in some tissues (Islam et al. 2010; Conway and Hutson 2016).

GDH and AAT have specific roles in astrocytes and neurons in brain (Moran and Rivera-Gaxiola 1992; Westergaard et al. 1996; Moran et al. 1999; McKenna et al. 2000a; Yudkoff et al. 2005). Formation of α-ketoglutarate from exogenous glutamate can be catalyzed by both GDH and AAT or other transaminases; however, many studies support a more predominant role for GDH in astrocytes (McKenna et al. 1996b, 2000a; Schousboe et al. 2014). The association of glutamate transporters, GDH and mitochondria into complexes in astrocytes would favor metabolism via GDH and subsequent oxidation for energy (Genda et al. 2011; Bauer et al. 2012;

Robinson and Jackson 2016). In contrast, the resynthesis of glutamate from α-ketoglutarate occurs primarily via AAT in astrocytes (Westergaard et al. 1996). It is well established that *exogenous* glutamate taken up from the extracellular milieu, and glutamate formed endogenously from glutamine via the action of glutaminase are metabolized differently in brain cells (McKenna et al. 1996b; Sonnewald and McKenna 2002; McKenna 2007; Schousboe et al. 2014). This is likely due to compartmentation and the close interaction of enzymes that can essentially channel metabolites through sequential reactions (McKenna et al. 2000a, 2006b; McKenna 2011 and references therein).

Interestingly, the entry of endogenous glutamate (formed from glutamine via glutaminase) into the TCA cycle is mediated only via GDH in synaptosomes, as the oxidation of glutamine by synaptosomes is not inhibited by aminooxyacetate which inhibits AAT and all aminotransferase enzymes (McKenna et al. 1993). The compartmentation and roles of AAT and GDH in brain have been reviewed by our group and others (McKenna et al. 1993, 1996b, 2000a, 2006b; Westergaard et al. 1996; McKenna 2007, 2011; Zaganas et al. 2012; Schousboe et al. 2014; Spanaki et al. 2014, 2015). Protein modifications such as acylation and palmitoylation can modify the kinetic parameters and binding properties of enzymes, including those involved in glutamate and glutamine metabolism (McKenna et al. 2006b; Zhao et al. 2010; McKenna 2011; Gibson et al. 2015; McKenna and Rae 2015).

The enzymes alanine aminotransferase (AlAT) and branched-chain amino transferase (BCAT) can also form α-ketoglutarate from glutamate via transamination with pyruvate or branched-chain keto acids, respectively (Yudkoff et al. 1989, 1991, 1993, 1996; Westergaard et al. 1995b; Yudkoff 1997; Waagepetersen et al. 2000; Hutson 2001; Hutson et al. 2001; Sweatt et al. 2004a; Bak et al. 2005, 2007; Islam et al. 2010). It has been proposed that both enzymes, particularly BCAT, function to transfer nitrogen between neurons and astrocytes to provide nitrogen for glutamate formation in brain (Cooper 2011; Hutson et al. 2011; Schousboe et al. 2014; Hull et al. 2015b; Conway and Hutson 2016). Studies by the Hutson lab have reported the compartmentation of BCAT in brain and interaction with other brain enzymes (Bixel et al. 2001; Garcia-Espinosa et al. 2007; Islam et al. 2010; Hutson et al. 2011; Cole et al. 2012; Hull et al. 2015b; Conway and Hutson 2016).

4.3.2 Regulation of GDH in Brain

Regulation of GDH is discussed as this enzyme may have a role in organizing heteroenzyme complexes (Fahien et al. 1993). Regulation of GDH has been extensively studied since this enzyme is so important in nitrogen metabolism in brain and other tissues (Cooper and Plum 1987; Plaitakis and Zaganas 2001; Zaganas et al. 2009, 2012; Cooper 2011; Plaitakis et al. 2011; Spanaki et al. 2012, 2014). GDH1 and GDH2 associate into homohexamers but are regulated differently (Stanley 2004; Raizen et al. 2005; Spanaki et al. 2010). GDH1 activity is controlled by the energy needs of cells and is allosterically inhibited by GTP (Stanley 2004; Raizen et al. 2005; Spanaki et al. 2010). GDH2 from human brain is not inhibited by GTP, and therefore can metabolize glutamate in the presence of an active TCA cycle and high energy levels (Smith and Stanley 2008; Spanaki et al. 2010; Li et al. 2011; Plaitakis et al. 2011; Stanley 2011).

The lipid modifications acetylation and palmitoylation can regulate GDH by altering activity and binding properties of the protein (Schlicker et al. 2008). Activity of GDH can also be modulated by mitochondrial silent information regulator proteins (sirtuins or SIRTs). Human SIRT3 binds and deacetylates GDH leading to activation (Schlicker et al. 2008); whereas SIRT4 inactivates GDH by ADP-ribosylation of the enzyme in liver and likely in brain (Haigis et al. 2006). Modulation of GDH activity by sirtuins is consistent with the roles of SIRT3 and SIRT4 in controlling many important mitochondrial processes including metabolism, intracellular signaling, energy production, and apoptosis (Verdin et al. 2010).

Overexpression of GDH1 in cultured neurons from mice led to altered synaptic plasticity, increased release of glutamate, and selective vulnerability of CA1 neurons (Bao et al. 2009). Interestingly, both decreased activity and gain-of-function mutations of GDH led to seizures and hyperexcitability of brain (Raizen et al. 2005; Malthankar-Phatak et al. 2006; McKenna 2011; Stanley 2011) and discussed by McKenna (2011 and references therein). Studies inducing decreased expression of GDH in astrocytes by siRNA and knockout mice with no GDH activity showed that reduction/loss of GDH activity impaired oxidation of glutamate in the TCA cycle (Skytt et al. 2012).

4.3.3 Aminotransferase Enzymes Can Be Subject to Regulation

Even though aminotransferase catalyzed reactions are considered freely reversible as the direction of the reaction is driven by the concentration of substrates and products, there is evidence that these enzymes may be subject to other levels of regulation (McKenna et al. 2000a, 2006b; Hull et al. 2015b; Yang et al. 2015). For example, Teller et al. (1990) reported that full activation of AAT requires the enzyme to associate with a binding protein that transfers it to lipids on the inner mitochondrial membrane. Palmitoylation enhances the interaction between GDH and AAT (Fahien and Kmiotek 1983). A report from McKenna et al. (2006b) demonstrating that addition of palmitate and other small molecules including malate, citrate, glutamate, bovine serum albumin, and Mg^{2+} modulated AAT activity in synaptic and nonsynaptic mitochondria from brain is consistent with regulation of this enzyme. Furthermore, it has recently been shown that acetylation of three lysines on mitochondrial AAT enhances the association between mAAT and mitochondrial malate dehydrogenase (mMDH; also known as MDH2) and promotes the transfer of the reducing equivalents from NADH into mitochondria via the malate-aspartate shuttle (Yang et al. 2015). Taken together, these studies demonstrate that AAT activity, and thus the metabolism of glutamate, may be regulated in part by protein modifications and metabolites that influence the formation of multienzyme complexes and/or binding of the enzyme to mitochondrial membranes and proteins. Interestingly, the association of AAT with its mitochondrial binding protein may facilitate the transfer of endogenously formed glutamate to AAT rather than to GDH in astrocytes, consistent with our demonstration that conversion of the glutamate formed endogenously from glutamine to α-ketoglutarate occurs primarily via AAT, as it can be blocked by the transaminase inhibitor aminooxyacetic acid (AOAA) (McKenna et al. 1993, 1996b, 2000a, 2006b). Our group recently reported that exogenous glutamate is oxidized in astrocytes at a rate several fold higher than endogenously formed glutamate (McKenna 2012). Constraints in mitochondrial binding of enzymes reported by Teller et al. (1990) can shed light on the compartmentation and differential metabolism of exogenous and endogenously formed glutamate. As discussed in depth by McKenna et al. (2000a), Teller showed that either AAT or MDH (alone or MDH associated with TCA cycle enzymes) can bind to the inner mitochondrial membrane (Teller et al. 1990). Although the binding of MDH to Complex I on the inner membrane has higher affinity than the association of AAT with its binding protein, the binding of AAT prevents the binding of MDH (Teller et al. 1990 and references therein). Given these constraints, it is logical that the endogenous glutamate formed from glutamine is not as rapidly oxidized as exogenous glutamate taken up from the extracellular milieu, since the AAT used to metabolize *endogenous* glutamate would have to dissociate from the binding protein and/or lipids, permitting MDH to bind, for oxidation via the TCA cycle to occur (McKenna et al. 1996b, 2000a).

In contrast to astrocytes, studies of metabolism in synaptosomes isolated from rat brain are consistent with the channeling of endogenous glutamate, formed from glutamine through a complex between GDH and phosphate-activated glutaminase (PAG) that facilitates neuronal oxidation of glutamine (McKenna et al. 1993). ^{13}C-NMR studies showing differential compartmentation of the metabolism of exogenous and endogenously formed glutamate also support the possibility of a PAG–GDH complex that provides substrate to a neuronal TCA cycle associated with GABA formation via glutamic acid decarboxylase (GAD) (Sonnewald and McKenna 2002). GDH has also been shown to associate with a complex between AAT and PC; MDH also associated with this ternary complex (Fahien et al. 1993). It is not known if this complex occurs in brain.

It should be emphasized that an essential, but sometimes overlooked, role of glutamate in both neurons and astrocytes is its role in the malate-aspartate shuttle that transfers reducing equivalents produced during glycolysis into the mitochondria so the energy can be harvested via the electron transport chain (McKenna et al. 2006c). This transfer of reducing equivalents is essential to maintain the NAD^+/NADH ratio required for continued glycolysis. Proteins that participate in this shuttle include both cytosolic and mitochondrial forms of AAT and MDH, the mitochondrial aspartate-glutamate carrier Aralar, and the malate-α-ketoglutarate carrier. This shuttle has a crucial role in brain, thus the formation of any complexes that facilitate either entry of glutamate into mitochondria or the activity of the malate-aspartate shuttle could have a major influence on metabolism.

The sections earlier provided an overview of glutamate and glutamine metabolism and compartmentation. In the section below we address some known and possible multienzyme/multiprotein complexes that can potentially affect metabolism of glutamate and related compounds, as well as factors that can modulate the association and dissociation of these complexes. The formation, dynamics, and metabolic consequences of these transient multiprotein complexes are likely to represent a further level of regulation of the metabolism of glutamate in brain.

4.4 Multienzyme Complexes in Metabolism

4.4.1 Evidence for and Consequences of Multienzyme Complex Formation

The metabolic outcome of the transient formation of these enzyme complexes, together with mitochondrial binding proteins that interact with enzymes and transporters (Teller et al. 1990; McKenna et al. 2000a), is variable. The activity of some enzymes and/or metabolism via certain pathways can be decreased by such interactions; whereas, metabolism through other pathways may be enhanced as the high affinity of participating enzymes for substrates can lead to 'channeling' of metabolites into specific pathways (Fahien et al. 1977, 1988, 1989; Beeckmans and Kanarek 1981; Beeckmans et al. 1990; Stanley 2011). Intrinsic mitochondrial properties, including the high density of enzymes, proteins, and lipids, restrict diffusion of both metabolites and proteins and increase the likelihood of multienzyme complex formation (Lopez-Beltran et al. 1996; McKenna et al. 2000a; Pollack 2001). Evidence supporting the formation of multienzyme complexes is from a wide variety of experimental tools, from simple studies assessing the in vitro binding capabilities of isolated/purified enzymes or purified mitochondrial membranes to other proteins and/or membranes (Fahien et al. 1977, 1985, 1988, 1989, 1990; Beeckmans and Kanarek 1981, 1987; Beeckmans et al. 1990, 1993; Teller et al. 1990; Fahien and Teller 1992; McKenna et al. 2000a) (and references therein), to more sophisticated approaches, using photobleaching (Haggie and Brindle 1999; Haggie and Verkman 2002) and proteomics using mass spectrometry (Rose et al. 2009; Li et al. 2010a; Genda et al. 2011). To date a number of transient multienzyme complexes that may directly or indirectly affect the metabolism of glutamate and glutamine have been described (Islam et al. 2007; Rose et al. 2009; McKenna 2011; Yang et al. 2015; Conway and Hutson 2016; Robinson and Jackson 2016; and references in these papers) (see Table 4.2). The dynamic assembly/disassembly of multienzyme complexes provides unique kinetic advantages that can coordinately impact multiple metabolic pathways, and the knowledge in this field will undoubtedly increase as novel technologies are developed and utilized to identify expected and unexpected interactions in living cells. Several studies reported evidence of heteroenzyme complex formation in vivo (Haigis et al. 2006; Li et al. 2011; Stanley 2011; Regenold et al. 2012; Shan et al. 2014).

Table 4.2 Known and likely enzyme complexes that may occur in brain

Enzymes interacting	Effect/predicted effect	References
CS+MDH	Increases TCA cycle activity	Fahien and Kmiotek (1983) and Morgunov and Srere (1998)
ICDH+KGDH+Complex I	Increases TCA cycle activity	Porpaczy et al. (1987)
CS+MDH+KGDH	Increases TCA cycle activity	Fahien et al. (1988)
KGDH+MDH+AAT	Facilitates metabolism via malate-aspartate shuttle	Fahien et al. (1988)
KGDH+MDH+AAT+GDH	Facilitates metabolism via malate-aspartate shuttle	Fahien et al. (1988)
AAT+GDH+ME	Transfers substrate and cofactor; increases pyruvate recycling pathway	McKenna et al. (1995)
AAT+inner mitochondrial binding protein+lipids	Activates mitochondrial AAT	Teller et al. (1990)
PAG+GDH	Facilitates entry of glutamine into TCA cycle for oxidation only via GDH in synaptosomes	McKenna et al. (1993)
PAG+GAD (glutamate decarboxylase)	Colocalized in thalamus; likely facilitates conversion of glutamine to GABA	Fisher (2007)
AAT+GDH	GDH adds or removes NH_3; AAT transfers amino group to oxaloacetate or from aspartate; increases TCA cycle activity	Fahien and Kmiotek (1983)
AAT+CS	Increases TCA cycle activity	Fahien and Kmiotek (1983)
MDH+GDH	Increases GDH activity (by removing NADH from the catalytic site)	Fahien and Kmiotek (1983)
GDH+AlAT	GDH adds or removes NH_3; AlAT transfers amino group to pyruvate or from alanine	Fahien et al. (1977)
GDH+BCAT	GDH adds or removes NH_3; BCAT transfers amino group to branched-chain ketoacids or from branched-chain amino acids (BCAA)	Islam et al. (2010)

GDH + BCAT + BCKD	Facilitates NH_3 transfer and oxidation of BCAA	Islam et al. (2010) and Hutson et al. (2011)
GDH + GLT1 (glutamate transporter)	Facilitates delivery of exogenous glutamate to mitochondria and oxidation in astrocytes	Genda et al. (2011)
GLT1 + hexokinase + mitochondria	Facilitates glutamate stimulation of glucose metabolism in astrocytes	Genda et al. (2011)
GLT1 + mitochondria	Facilitates delivery of exogenous glutamate to mitochondria and oxidation in astrocytes	Genda et al. (2011)
GLT1 + Na^+/K^+-ATPase	Drives Na^+-dependent glutamate uptake against an unfavorable gradient	Genda et al. (2011)
GLAST + mitochondria	Facilitates delivery of exogenous glutamate to mitochondria and oxidation in astrocytes	Bauer et al. (2012)
GLAST + Na^+/K^+-ATPase + HK + GAPDH + mitochondrial enzymes	Provides energy and drives Na^+-dependent glutamate uptake against an unfavorable gradient—supports local glutamate uptake	Bauer et al. (2012)
GLT1 + GDH	GDH inhibition leads to decreased Na^+-dependent L-glutamate uptake	Whitelaw and Robinson (2013)
KGDHC + PDH	KGDHC succinylates PDH which increases PDH activity; KGDHC can also acylate PDH which decreases activity	Gibson et al. (2015)
KGDHC + fumarase	KGDHC succinylates fumarase which increases activity	Gibson et al. (2015)
SCHAD + GDH	Inhibitory regulation of GDH by SCHAD—lost in SCHAD deficiency, leading to hyperinsulinism	Li et al. (2010a)

4.4.2 Many Factors Influence the Association and Dissociation of Transiently Formed Enzyme Complexes in Mitochondria

A number of different transient multienzyme complexes may be formed between mitochondrial enzymes, transporters, and proteins in brain. There are many effectors that influence the kinetics and binding of binary, ternary, and quaternary enzyme complexes between proteins (Fahien et al. 1977, 1985, 1988, 1989, 1990; Halper and Srere 1977; Beeckmans and Kanarek 1981; Moore et al. 1984; Sumegi and Srere 1984; Srere 1987; Beeckmans et al. 1990; Teller et al. 1990; Fahien and Teller 1992; McKenna et al. 2000a) (and references therein). As noted by McKenna (2011) "specific complexes are formed depending on (a) the binding affinities of enzymes for other enzymes, (b) the relative concentration of each enzyme in the mitochondrial matrix, (c) the affinity of binding to the mitochondrial membranes, (d) or to specific binding proteins in the mitochondrial membrane, (e) the order in which binding occurs, (f) the concentration of metabolites (e.g. citrate, oxaloacetate, α-ketoglutarate, malate, aspartate, fumarate, etc.) and ions (e.g. Mg^{2+}) that influence binding of specific enzymes (Fahien et al. 1977, 1985, 1988; Teller et al. 1990; Fahien and Teller 1992; Malik et al. 1993; McKenna et al. 2000a, 2006b; McKenna 2007), (g) the form in which enzymes are found e.g., AAT and MDH are dimers, GDH forms homohexamers when the concentration is sufficient, (h) the stoichiometry of enzyme binding, [e.g. α-ketoglutarate dehydrogenase (KGDH) complex can associate with AAT and MDH with a ratio of 10 total AAT or MDH bound per KGDH (Teller et al. 1990)], and (i) palmitoylation which alters protein conformation and allows proteins to associate and dissociate from specific subcellular microdomains (Fukata and Fukata 2010), e.g. palmitoyl-CoA has been shown to enhance the interactions between GDH and AAT (Fahien and Kmiotek 1983)." In addition, attachment of other types of acyl-moieties to proteins including acetylation (Zhao et al. 2010) and succinylation (Zhang et al. 2011; Gibson et al. 2015; McKenna and Rae 2015) can significantly alter the physicochemical properties and binding affinity as discussed later.

4.4.3 Fluctuations in the Concentration of Small Molecules Can Modulate Enzyme Activity and Complex Formation

Our group and others have reported that many small molecules can modulate the activity of key enzymes including MDH, mitochondrial AAT, and malic enzyme (ME) in nonsynaptic and synaptic mitochondria (McKenna et al. 1990, 1994, 1996a, b, 2000a, b, 2006b; Malik et al. 1993; McKenna and Sonnewald 2005; McKenna 2007; Gibson et al. 2015). A recent report from Gibson's group demonstrated KGDH-mediated succinylation modulates the activity of several brain enzymes including KGDH, fumarase, and pyruvate dehydrogenase (PDH) (Gibson et al. 2015;

McKenna and Rae 2015). The concentration of small molecule effectors fluctuates during normal brain metabolism and can change dramatically during pathological conditions (McKenna et al. 2006b; McKenna 2007). Compounds known to influence complex formation and/or enzyme activity include citrate, oxaloacetate, α-ketoglutarate, malate, aspartate, fumarate, glutamate, Ca^{2+}, and Mg^{2+} (McKenna et al. 2006b and references therein); Ca^{2+} also stimulates the activity of the mitochondrial aspartate-glutamate carrier (Contreras et al. 2007; Bak et al. 2012; Llorente-Folch et al. 2013). Small metabolite effectors are important since they can modulate the association and dissociation of transient multienzyme complexes that can rapidly influence metabolism (McKenna et al. 2006b; McKenna 2007).

Due to the viscous mitochondrial matrix, diffusion of enzymes away from other enzymes or proteins or away from membranes would be slow even when multienzyme complexes are dissociated (Srere 1987; Srere et al. 1987). With such close physical proximity, complexes could associate when the concentration of key effectors favors association, and dissociate when the concentration of effectors fluctuates under physiological or pathological conditions.

4.5 Transient Multienzyme Complexes That Affect Metabolism of Glutamate and Related Compounds

4.5.1 Possible Role of GDH and Other Large Enzymes in Organizing Complexes

Relatively little is known about the localization, regulation, and/or roles of transiently formed enzyme complexes that can influence metabolism in brain. Interestingly, many of the complexes formed contain GDH, leading Fahien and coworkers to propose that GDH "could play an organizational role" "in the quaternary complex between GDH, the aminotransferases, MDH, and pyruvate carboxylase" similar to the role performed by the KGDH complex in some tissues (Fahien et al. 1993) [see related discussion in (McKenna 2011)]. Such an organizational role for GDH would be extremely important as it would provide a mechanism to position enzymes that do not bind directly to each other (e.g., AAT and MDH) so that they are physically close enough to transfer substrates and cofactors (Fahien et al. 1993). The high concentration of GDH (up to 10 % of mitochondrial matrix protein in astrocyte processes) would allow for such an organizational role (Fahien and Teller 1992; Fahien et al. 1993; Rothe et al. 1994, 1995). In addition, a key role for GDH in organizing other enzymes into multienzyme complexes would help to explain the severe neurological consequences of mutations that cause conformational changes in GDH (Bahi-Buisson et al. 2008a, b).

Fahien and coworkers noted that the rate-limiting step in the GDH reaction is dissociation of NADH, and that binding of MDH to GDH in the quaternary complex (GDH, MDH, AAT, and fumarase) facilitated the removal of NADH from

GDH (Fahien and Kmiotek 1983; Fahien and Teller 1992; Fahien et al. 1993). KGDH, which like GDH is a large stable enzyme complex, can also associate with the AAT–GDH–MDH complex: however, as GDH is found in much higher concentration it would likely perform the organizational role of forming the complex with AAT and MDH in brain (Salganicoff and De Robertis 1965; Faff-Michalak and Albrecht 1991; Fahien et al. 1993; McKenna et al. 2000a, 2006b and references therein). Despite the fact that high GDH activity has been reported in neurons and synaptosomes, immunolabeling of GDH is lower in neurons than in astrocytes (Erecinska and Silver 1990; McKenna et al. 1993, 2000a; Rothe et al. 1994, 1995). Many groups have reported the activity of brain enzymes (Salganicoff and De Robertis 1965; Lai and Clark 1989; Faff-Michalak and Albrecht 1991; McKenna et al. 2000a, 2006b) but there is little information on the relative concentration of the different brain enzymes in different cell types and cellular compartments.

4.5.2 GDH Participates in Heteroenzyme Complexes with AAT and TCA Cycle Enzymes

As reviewed by McKenna et al. (2000a) and McKenna (2011) both AAT and MDH can form complexes with GDH, and the presence of AAT increases the affinity of MDH binding to GDH (Fahien and Kmiotek 1983; Fahien and Teller 1992); however, due to stoichiometry of binding, multiple complexes can be formed. The Fahein group observed that under the conditions used, the molar ratio of enzymes bound in vitro was always >1 AAT per GDH homohexamer (Fahien and Teller 1992). The only complexes formed were ternary complexes of GDH–AAT–MDH and GDH–MDH–(AAT)2 and the binary complex of GDH–AAT (Fahien and Teller 1992). A quaternary complex was formed by binding of fumarase to the ternary complex (Fahien and Teller 1992). It is important to note that even when such a complex is dissociated the physical proximity of the participating enzymes is so close that even very small crosslinker molecules (with only 12 Å span) can link all three of these enzymes or GDH and MDH (Fahien et al. 1985; Fahien and Teller 1992).

Binding of allosteric effectors induces conformational changes that impact enzyme activity (Fahien et al. 1989). Binding of substrate to AAT facilitates the reaction of this enzyme (Hayashi et al. 2003). Even though a complex between MDH and AAT is not actually formed (Fahien and Teller 1992), the ternary complex of AAT–GDH–MDH brings these two enzymes in close enough physical proximity to transfer the oxaloacetate from MDH to AAT (Fahien and Teller 1992). In addition, a binary complex between AAT and GDH can associate with malic enzyme (ME) and facilitates transfer of both substrate and cofactors (Teller et al. 1990; Fahien and Teller 1992) to ME, which has a higher Km for malate than mitochondrial MDH (McKenna et al. 1995, 2000b). This association "would facilitate conversion of TCA cycle metabolites into pyruvate" and thus increase metabolism via the pyruvate recycling pathway in brain. In liver, palmitoyl-CoA enhances the association of AAT and GDH into a complex (Fahien and Kmiotek 1983); whereas citrate enhances dissociation of AAT from GDH (Fahien et al. 1985), as discussed by McKenna (2011). The concentrations of mitochondrial enzymes in liver are an order of magnitude higher than the dissociation constants of the heteroenzyme complexes (Fahien and Teller 1992); however, such information is not readily available for brain.

4.5.3 Glutamate Transporters Can Form Complexes with Metabolic Enzymes and Mitochondria

The striking difference in glutamate concentration among distinct biological compartments within (e.g., vesicles, mitochondria, cytosol) or outside (e.g., synaptic cleft) neural cells is driven by glutamate transporters. As glutamate cannot be degraded extracellularly, GLAST (Excitatory amino acid transporter 1—EAAT1) and GLT-1 (Excitatory amino acid transporter 2—EAAT2), high affinity transporters on astrocytes have the key role of removing glutamate from the synaptic cleft to prevent excessive stimulation of glutamate receptors on neuronal plasma membranes (Danbolt 2001). Vesicular glutamate transporters (vGLUTs) are responsible for vesicular packing of this excitatory amino acid (Danbolt 2001).

Studies over the past decade, using a variety of preparations, have shown that glutamate transporters interact with a number of crucial proteins from different metabolic pathways of the CNS, including PSD-95 (Gonzalez-Gonzalez et al. 2008, 2009), Na^+/H^+ exchanger regulatory proteins (Lee et al. 2007; Sato et al. 2013), and glutamine transporters (SNAT3) (Martinez-Lozada et al. 2013). There is also evidence that glutamate transporters interact with proteins involved in brain bioenergetics, including Na^+/K^+-ATPase (Rose et al. 2009) and Na^+/Ca^{2+}exchanger (NCX) (Magi et al. 2012, 2013), as well as glycogen metabolizing enzymes, glycolytic enzymes, and mitochondrial proteins (Genda et al. 2011; Bauer et al. 2012). Some of these protein complexes facilitate metabolism, e.g., the macromolecular complex between GLAST and GDH facilitates the oxidative metabolism of transported glutamate in astrocyte mitochondria for energy production (Bauer et al. 2012), which offsets the energy requirements associated with glutamate uptake (McKenna 2013). The colocalization of GLT-1 with glycolytic enzymes may facilitate any glutamate and/or potassium stimulated increase in glucose metabolism that may occur after neurotransmitter release (Pellerin and Magistretti 1994; Peng et al. 1994, 2001), and would provide "a mechanism to spatially match energy and buffering capacity to the demands imposed by transport" as noted by Genda et al. (2011). Although the binding of GLT-1 to mitochondrial proteins may occur through GDH (Bauer et al. 2012), hexokinase was also hypothesized as a putative link between these proteins. However, the presence of membrane-permeant peptides, which are known to disrupt the interaction between hexokinase and the voltage-dependent anion channel (VDAC), did not affect coimmunoprecipitation and interactions between GLT-1 and the mitochondrial proteins adenine nucleotide translocator (ANT) or the ubiquinol–cytochrome-c reductase complex in a brain subfraction, suggesting that the hexokinase is not the protein that anchors GLT-1 to mitochondria (Jackson et al. 2015).

However, displacing hexokinase from VDAC diminished GLT-1-dependent glutamate uptake, implying that "this interaction is either required to fuel uptake or supports co-compartmentalization" as noted by Jackson et al. (2015).

Interactions of glutamate transporters and other proteins appear to have functional relevance. For instance, acute activation of adenosine A2A receptors, which were shown to interact with GLT-1 and the Na^+/K^+-ATPase α2 isoform in astrocytes, leads to decreased glutamate uptake and Na^+/K^+-ATPase activity, while glutamate uptake is enhanced following deletion of astrocytic A2A receptors (Matos et al. 2012a, b, 2013). In addition, it has recently been shown that the glutamate transporter EAAC1 (EAAT3) is found on both neuronal and astrocyte mitochondria, and that a physical and functional interaction of the EAAC1 and NCX1 has a modulatory effect on glutamate-enhanced ATP production in brain mitochondria (Magi et al. 2012, 2013). The functional coupling between these systems may support glutamate uptake, as well as be regulated by glutamate transport (Robinson and Jackson 2016).

Recent studies suggest that synaptic activity promotes the colocalization of astrocyte membrane transporters with mitochondria following synaptic activity, since colocalization between mitochondria and GLT1 was observed in mixed cultures of astrocytes and primary cortical neurons that contained a higher amount of mitochondria in astrocytic processes; whereas, no colocalization was found in pure astrocyte cultures (Ugbode et al. 2014). Interestingly, even though GLAST and GLT-1 are often colocalized on astrocyte membranes, they may take part in distinct multienzyme complexes, as there is no evidence for coimmunoprecipitable interactions between them (Haugeto et al. 1996). Understanding the nature, dynamics, and regulation of these protein complexes will enable us to better unravel their role(s) in the physiological functioning of different brain cells, as well as any role(s) that association/dissociation of macromolecular complexes may have in the pathophysiology of CNS diseases. Glutamate receptors can also form complexes with other proteins, for example glyceraldehyde 3-phosphate dehydrogenase (GAPDH), a crucial enzyme for glycolysis, was shown to interact with the GluR2 subunit of AMPA receptors. The inability to form GluR2/GAPDH interactions protected cells against AMPA receptor-mediated excitotoxicity (Wang et al. 2012; Zhai et al. 2013), and thus may represent an alternative therapeutic target for glutamate toxicity.

4.5.4 Branched-Chain Aminotransferase (BCAT) and Branched-Chain Ketoacid Dehydrogenase (BCKD) Participate in a Metabolon with GDH in Some Tissues

Leucine and other branched-chain amino acids (BCAA) may serve as a source of nitrogen in the brain and other tissues (Hutson et al. 2011). Plasma concentrations of BCAA are reflected in peripheral tissues as levels increase proportionately with higher intake of BCAAs (Hutson and Harper 1981; Neishabouri et al. 2015). BCAA levels in brain are moderately elevated following a high protein diet, since competition with other amino acids limits uptake across the blood–brain barrier (Hutson and Harper 1981; Neishabouri et al. 2015). The two first enzymes in their catabolic pathway, namely, the mitochondrial branched-chain aminotransferase (BCATm) and branched-chain ketoacid dehydrogenase (BCKD), can form a heteroenzyme metabolon that facilitates degradation of BCAA (Islam et al. 2007, 2010). Work from Hutson's group (Islam et al. 2010; Hutson et al. 2011) reported that this metabolon interacts with GDH in rat liver as well as with bovine GDH1. GDH binds to the pyridoxamine-5′-phosphate form of BCATm in vitro (Islam et al. 2010; Hutson et al. 2011). BCAA oxidation was stimulated by the formation of BCAT/BCKD complex and the provision of α-ketoglutarate for the transfer of nitrogen from the BCAT to glutamate was facilitated by the binding of GDH1 to the complex, ultimately enhancing BCKD enzyme activity. It is possible that GDH may have a role in assembling/facilitating this complex.

It was demonstrated that the cytosolic BCAT (BCATc) is the predominant isozyme of BCAT in the CNS (Hall et al. 1993; Sweatt et al. 2004a, b). While BCATc is expressed in "select populations of glutamatergic and GABAergic neurons" (Sweatt et al. 2004a; Garcia-Espinosa et al. 2007), BCATm is enriched in astrocytes of rodents (most often located in the neuropil and myelinated tracts, instead of in gray matter) and the "labeling for BCATm in these astrocytes extends into the cell processes, indicating a wide intracellular distribution of the enzyme in this cell type" (Cole et al. 2012). Labeling for the BCKDC-E1α subunit is present in several neuronal cell types, including glutamatergic, and cholinergic neurons, while astrocytes lack BCKDC labeling (Cole et al. 2012). "The segregation of BCATm to astrocytes and BCKDC to neurons provides further support for the existence of a BCAA-dependent glial-neuronal nitrogen shuttle since the data show that branched-chain keto acids produced by glial BCATm must be exported to neurons" (Cole et al. 2012). Since the complex between GDH and BCAT/BCKD identified by the Hutson group (Islam et al. 2010; Hutson et al. 2011) was restricted to BCATm, one can assume that neurons are devoid of such complexes. However, the expression of BCATm and the GDH-related glutamate disposal in astrocytes, leading to formation of α-ketoglutarate and subsequent oxidation suggest that a complex could exist in astrocytes. It should be noted that the conversion of α-ketoglutarate to glutamate in astrocytes is mediated by AAT, not by GDH, as noted earlier (McKenna et al. 1996a; Sonnewald et al. 1996; Westergaard et al. 1996).

Besides the ability to form a metabolon with BCKD and GDH, a novel functional role for human BCAT (hBCAT) in redox protein folding was recently proposed, particularly when hBCAT was S-glutathionylated (El Hindy et al. 2014). The formation of complexes between hBCAT with unfolded proteins, catalyzing their refolding, suggests that mitochondrial hBCAT (hBCATm) may also function as chaperone, by a thiol disulfide exchange mechanism or as a redox-regulated oxidoreductase (El Hindy et al. 2014).

In distinct contrast to rodent brain, hBCATm is not found in astrocytes in humans, where GDH is predominantly expressed (Hull et al. 2015a). A recent study from Hull and colleagues (2015a) reported an increase of hBCATm in the frontal and temporal cortex in AD subjects, relative to matched controls, suggesting that the expression of this enzyme may be directly related to the severity of the disease.

As initially hypothesized by Helms and colleagues (2012) and later discussed by Hull and colleagues (2012), the blood–brain barrier itself may play a role in the regulation of brain glutamate levels. Due to the high expression of BCAT throughout the endothelial layer in different areas of the human brain (Hull et al. 2012), one can assume that transamination is taking place in these cells, particularly when glutamate exceeds the Km of the Na^+-dependent transporters involved in its uptake (O'Kane et al. 1999), during periods of excitation or excitotoxicity (Dzubay and Jahr 1999; Moussawi et al. 2011). The potential source of branched-chain ketoacids remains uncertain but mRNA levels of the monocarboxylic acid transporters (MCT1 and MCT2), that are able to transport these monocarboxilic acids, were shown in rat brain endothelial cells (RBE4 cell line) (Mac and Nalecz 2003). The direction of BCAT reaction should be driven by the concentration of substrates and redox state of the cell. Therefore, if α-ketoglutarate is sufficiently provided by the TCA cycle and considering that these cells contain L-system transporter, which allow the transport of BCAAs (Smith et al. 1987; Shulkin et al. 1995; O'Kane and Hawkins 2003; O'Kane et al. 2004), it is also feasible that this reaction would occur in the opposite direction. GDH, if present, could provide an alternative source of α-ketoglutarate for this reaction and the association between BCATm and GDH could replenish this TCA intermediate when BCAT acts in the direction of glutamate formation (Hutson et al. 2011). Therefore, a novel role for hBCATm in glutamate homeostasis in the endothelial cells of the human brain vasculature as well as the possible occurrence of a BCAT/BCKD metabolon in these cells has been proposed (Hull et al. 2012, 2015b). The mammalian target of rapamycin (mTOR) complex 1 integrates many intra- and extracellular signals to control major processes in cell metabolism, including protein and lipid synthesis and autophagy (Laplante and Sabatini 2012). Dysregulation of mTOR complex 1 signaling was reported to be involved in the aging process (Gilley et al. 2013) and also in human diseases, including cancer (Menon and Manning 2008, 2013), Fragile X Syndrome (Richter et al. 2015), and depression (Li et al. 2010b). It was also recently reported that BCATc indirectly regulates mTOR complex 1 signaling, by modulating leucine concentration and glycolytic metabolism in immune cells ($CD4^+$ T cells) (Ananieva et al. 2014) and that the chronic mTOR complex 1 activation by BCAAs may result in hypertrophy in peripheral organs, including heart, kidney, and spleen (Neishabouri et al. 2015). However, it is uncertain if these effects also occur in the brain.

4.5.5 Interactions of Short-Chain 3-Hydroxyacyl-CoA Dehydrogenase (SCHAD) and Glutamate/Glutamine Metabolism Enzymes

It has recently been reported that an inherited deficiency of short chain 3-hydroxyacyl-CoA dehydrogenase (SCHAD) can lead to hyperinsulinemia in patients (Li et al. 2010a). Studies in SCHAD knockout mice provide evidence that in SCHAD deficiency, loss of the inhibitory interaction between SCHAD and GDH leads to activation of GDH in islet cells of the pancreas resulting in increased insulin release (Stanley 2011). This occurs because SCHAD, like many other multifunctional proteins, can act as a "moonlighting protein" which performs a role different than the primary role of the protein (Jeffery 2005). In this role SCHAD serves to inhibit GDH activity (Stanley 2011). The interaction between SCHAD and GDH was also later reported in brain, kidney, and liver of mice (Narayan et al. 2012). Other enzymes involved in glutamate metabolism known to have protein–protein interactions with SCHAD include AAT and glutamine synthetase (Narayan et al. 2012).

4.6 Posttranslational Modifications Can Influence Enzyme Activity and the Ability to Form Transient Multienzyme Complexes

Posttranslational covalent modifications to proteins can affect their stability and function, modulating their biological activity and interfering with major cellular pathways (Fischer 2014; Perluigi et al. 2014). The addition of each specific type of functional group confers unique properties to modified proteins and results in alterations in protein/protein interactions, membrane binding, targeting of proteins, and in intracellular signaling (Resh 1996; Baeza et al. 2016). The next section describes a number of covalent modifications affecting enzymes related to glutamate and glutamine metabolism, including covalent attachment of an acetyl-, palmitoyl-, succinyl-, or small ubiquitin-like modifiers (SUMO) moiety, and indicates the role(s) of sirtuins in these modifications of other proteins (Verdin et al. 2010; Osborne et al. 2014; Baeza et al. 2016) (Table 4.3).

4.6.1 Acetylation Is a Common Posttranslational Modification That Alters Enzyme Activity

Acetylation of lysine residues is a widespread posttranslational modification affecting metabolic enzymes and the underlying mechanism is believed to involve non-enzymatic attachment of an acetyl moiety from acetyl-CoA to reactive lysine residues. Reversible acetylation is one of the most extensively studied posttranslational modifications in mitochondria and is frequently associated to impaired functioning of key metabolic pathways involved in the integrity and redox homeostasis in this organelle (for an extensive review, see Baeza et al. 2016). Acetylation of mitochondrial enzymes is reversed by the NAD^+-dependent deacetylase SIRT3 and mitochondrial sirtuins were recently proposed to play an important role in the modulation of mitochondrial metabolism (Osborne et al. 2014). SIRT3 regulates protein deacetylation and is important for mitochondrial metabolism, cell survival, and longevity (Cimen et al. 2010). Acetylation affects many metabolic enzymes involved

Table 4.3 Common protein covalent modifications affecting glutamate and glutamine metabolism

Protein modification		References
Acetylation	Found on essentially all enzymes in glycolysis, gluconeogenesis, the tricarboxylic acid (TCA) cycle, the urea cycle, fatty acid metabolism, and glycogen metabolism in liver; less studies in brain	Zhao et al. (2010)
	Deacetylation of GDH by SIRT 3 activates its enzyme activity in mitochondrial matrix	Schlicker et al. (2008)
	SIRT 3 deacetylates and increases pyruvate dehydrogenase activity in cancer cells	Ozden et al. (2014)
	Acetylation activates MDH activity in human liver cells	Zhao et al. (2010)
	Acetylation of mitochondrial AAT enhances the association between mAAT and MDH2 and promotes the transfer of reducing equivalents from NADH into mitochondria via the malate-aspartate shuttle. Stimulates NADPH production in mitochondria	Yang et al. (2015)
	Deacetylation of succinate dehydrogenase flavoprotein subunit increases Complex II activity in liver of mice	Cimen et al. (2010)
ADP-Ribosylation	ADP-ribosylation of GDH by SIRT 4 leads to inactivation of GDH activity	Haigis et al. (2006)
Palmitoylation	Palmitoylation enhances interactions between GDH and AAT	Fahien and Kmiotek (1983)
	Palmitoylation of GAD65 is crucial for post-Golgi trafficking of the protein to presynaptic clusters in rat hippocampal neurons	Kanaani et al. (2008)
	Glutamate dehydrogenase isozymes are inhibited by palmitoylation	Fahien and Kmiotek (1981)
	Palmitoylation of GAD65 leads to its anchoring to membranes of small vesicles in rat pancreatic β-cells	Christgau et al. (1992)
Succinylation	Succinylation of isocitrate dehydrogenase decreases its activity in cultured neurons from embryonic C57BL/6 mice	Gibson et al. (2015)
	Succinylation of KGDHC, fumarase, and PDHC increases their activities in cultured neurons from embryonic C57BL/6 mice	Gibson et al. (2015)
	Succinate dehydrogenase is activated by succinylation in human embryonic kidney (HEK) 293T cultured cells	Park et al. (2013)
SUMOylation	SUMOylation of presynaptic proteins increases glutamate release in synaptosomes	Feligioni et al. (2009)
	SUMOylation of metabotropic glutamate receptors (mGluR) results in novel protein interactions and regulates their density, localization, and functional properties	Scheschonka et al. (2007)

in glutamate metabolism in the cell, notably impairing the activities of GDH, PDH (Lombard et al. 2007; Schlicker et al. 2008), and the TCA enzymes isocitrate dehydrogenase (Someya et al. 2010; Yu et al. 2012), the NDUFA9 subunit of Complex I (Ahn et al. 2008), and succinate dehydrogenase flavoprotein (SdhA), one of the Complex II subunits (Cimen et al. 2010; Finley et al. 2011). SIRT3 binds to GDH and leads to its deacylation and subsequent activation (Schlicker et al. 2008). A different sirtuin, SIRT4, uses NAD to inactivate GDH in the brain by ADP-ribosylating it (Haigis et al. 2006). SIRT4 has mitochondrial localization in rodent brain (Komlos et al. 2013), particularly in astrocytes in the postnatal brain and in radial glia during embryogenesis, and GDH inactivation may occur after a modification of a single subunit of an active homohexamer (Herrero-Yraola et al. 2001). In line with this, a repression of transcriptional production of SIRT4 by mTOR complex 1 leads to increased glutamine anaplerosis to the TCA cycle, indicating GDH activation (Csibi et al. 2013).

However, it should be underscored that acetylation does not always impair enzyme activity; for example, acetylation stimulates the activity of aconitase (Fernandes et al. 2015). Interestingly, MDH activity can be stimulated or inhibited depending on the specific lysine residue that is acetylated (Zhao et al. 2010; Hebert et al. 2013). Besides interfering with enzyme activities, acetylation can also modulate binding properties of proteins, as in the case of the triple acetylation of mitochondrial AAT that enhances the formation of a complex between mAAT and MDH2, that increases malate-aspartate shuttle activity, and the transfer of the reducing equivalents from cytosolic NADH into mitochondria (Yang et al. 2015). Production of NADPH to attenuate ROS formation and protect cells from oxidative damage is also enhanced by the formation of this multienzyme complex (Yang et al. 2015). Despite the function of thiamin as a cofactor for enzyme reactions, a novel role for this vitamin as an allosteric regulator of acetyl-CoA metabolism was recently described, ultimately affecting acetylation of proteins (Mkrtchyan et al. 2015). One should not underestimate the importance of this modification since it was recently stated that "reversible acetylation of mitochondrial proteins represents a new layer of protein regulation mediating acute and adaptive changes in mammalian metabolism" (Wagner and Payne 2011).

4.6.2 *Palmitoylation of Proteins Can Dynamically Influence Compartmentation, Localization, and Interactions of Proteins and Signaling*

Palmitoylation is the attachment of a 16-carbon fatty acid moiety to a protein via a thioester bond (Fukata and Fukata 2010). Unlike some modifications, palmitoylation occurs rapidly and allows proteins to shuttle between different intracellular compartments (e.g., Golgi and plasma membrane), and usually leads to a more rapid turnover of the protein (Fukata and Fukata 2010 and references therein). The relevance of palmitoylation and other lipid modifications to glutamate and glutamine metabolism was recently discussed and includes modulation of enzyme activities and binding properties (for details, see (McKenna 2011) and references therein). In this scenario, activation of glutamate receptors was shown to modulate palmitoylation of proteins and play a role in brain development and synaptic plasticity (Noritake et al. 2009; Fukata and Fukata 2010). McKenna and colleagues demonstrated that palmitate interfered with AAT activity in synaptic and nonsynaptic mitochondria (McKenna et al. 2006b), which is in line with previous reports indicating that AAT binds to the inner mitochondrial membrane, and must subsequently be transferred to lipids to become fully activated (Teller et al. 1990; Fahien and Teller 1992). Palmitoylation can occur spontaneously (Fukata and Fukata 2010) or by enzymatic reaction of palmitoyl acyltransferase enzymes, while palmitoyl acyl thioesterases mediate depalmitoylation (Fukata and Fukata 2010), and the regulation of these processes under physiological or pathophysiological context is still under debate (Noritake et al. 2009).

4.6.3 Succinylation of Lysine Residues in Enzymes Can Alter Activity

Succinylation of lysine residues in proteins represents a new level of metabolic regulation (Park et al. 2013; Choudhary et al. 2014), affecting key enzymes in glycolysis, the TCA cycle, fatty acid metabolism, and others (Papanicolaou et al. 2014). A recent report demonstrated that succinylation of multiple proteins is reduced when KGDH is inhibited in neural cells, even though KGDH is not the sole mechanism of attachment of this acyl moiety to proteins in the cell (Gibson et al. 2015). Identified targets for protein succinylation by KGDH include KGDH itself and other enzymes involved in brain bioenergetics, such as fumarase, isocitrate dehydrogenase, and pyruvate dehydrogenase (PDH). The catalytic outcome of this modification was varied: in contrast to the decreased isocitrate dehydrogenase activity observed, the activities of KGDH, fumarase, and PDH were found to be increased by succinylation (Gibson et al. 2015), which is in line with a previously reported activation of succinate dehydrogenase by succinylation (Park et al. 2013). Taken together, a putative role for KGDH inducing succinylation includes activation of TCA cycle steps from KGDH to oxaloacetate, and deceleration of the initial steps (from isocitrate dehydrogenase to KGDH), which would facilitate the oxidation of glutamate via a partial TCA cycle (McKenna and Rae 2015). It is worth noting that KGDH activity can also lead to acetylation of proteins in cytosol and mitochondria, including PDH (Gibson et al. 2015). In contrast to the effect of succinylation, glutathionylation (Cooper et al. 2011) and internally generated reactive oxygen species produced by the E3 subunit can inhibit KGDH activity (Bunik and Sievers 2002). In addition to effects on enzyme activity, the attachment of one or more large negatively charged succinyl groups can modify the physicochemical properties of proteins, and thus alter their ability to associate into transient multienzyme complexes (McKenna and Rae 2015), which can substantially influence metabolism

(McKenna et al. 2000a, 2006b; Genda et al. 2011; McKenna 2011; Bauer et al. 2012; Whitelaw and Robinson 2013; Jackson et al. 2014; McKenna and Rae 2015).

There is evidence that SIRT5 participates in the regulation of desuccinylation of proteins, with a weak deacetylation activity (Houtkooper et al. 2012; Newman et al. 2012; Park et al. 2013). Increased levels of protein succinylation are described in SIRT5 knock-out mice, that present with hyperammonemia probably due to the succinylation and inhibition of carbamoyl phosphate synthetase 1 (Nakagawa et al. 2009), but do not display any other evident metabolic abnormalities (Yu et al. 2013). Since succinylation occurs rapidly and dynamically within the cell (Gibson et al. 2015), it is feasible that the modulation elicited by such modification on TCA cycle oxidative capacity might be regulated in response to cellular demands.

4.6.4 SUMOylation of Proteins Can Affect the Glutamate–Glutamine Cycle

Sumoylation of proteins, the covalent addition of small ubiquitin-like modifiers (SUMO), provides an important posttranslational modification for many biological processes in the brain, particularly during development (Kerscher 2007). It may lead to alterations in the localization, activity, and stability of modified proteins (Geiss-Friedlander and Melchior 2007). Feligioni and colleagues demonstrated that SUMOylation of presynaptic proteins increases glutamate release (Feligioni et al. 2009). EAAT2 cleavage by the activation of caspase-3 in a mouse model of ALS results in the formation of a peptide fragment that can be SUMOylated and builds up along with disease course in the nucleus of astrocytes (Gibb et al. 2007), reaching up to 20 % of the GFAP$^+$ cells of the ventral horn of the spinal cord at later stages of disease (Foran et al. 2011). SUMOylation of proteins could also affect protein–protein interactions, by inhibiting, modifying, or enhancing such phenomena (Ulrich 2008). For instance, it was suggested that metabotropic glutamate receptors (mGluR) might be SUMOylated, resulting in specific protein interactions and ultimately regulating their density, localization, and functional properties (Scheschonka et al. 2007).

4.6.5 Protein Modifications Following Oxidative and/or Nitrosative Stress Can Modulate Glutamate Metabolism

Proteins can undergo many oxidative modifications, particularly under pathological conditions, that can alter their intrinsic activity, including carbonylation, nitration, and S-nitrosylation, which are frequently considered deleterious to cell survival (Pedersen et al. 1998; Beal 2002; Johnson and Johnson 2015). However, an interesting case is nitration of a tyrosine residue of glutamine synthetase that leads to the inactivation of the enzyme (Schliess et al. 2002; Haussinger et al. 2005). This apparent deleterious loss of enzyme function was in fact proposed to be protective during hyperammonemia, since the decrease in glutamine synthetase activity mitigated the metabolic changes contributing to ammonia toxicity including glutamine accumulation, cell swelling, Ca^{2+} influx, and the production of NO and reactive oxygen intermediates in rat astrocytes (Schliess et al. 2002).

A recent report also investigated the biological functions of protein S-nitrosylation in the tripartite glutamatergic synapse (Raju et al. 2015). Proteomic analyses indicated that mice lacking neuronal nitric oxide synthase ($nNOS^{-/-}$) activity had decreased S-nitrosylation of proteins that participate in the glutamate-glutamine cycle, and that brain extracts from $nNOS^{-/-}$ mice converted less glutamate to glutamine and oxidized more glutamate than those from mice of the other genotypes. This most likely is due to increased metabolism of glutamate to α-ketoglutarate via GDH and/or mAAT and subsequent oxidation, as well as a partial loss of glutamine synthetase activity. S-Nitrosylation of GLT-1 at Cys373 and Cys561 residues was absent in $nNOS^{-/-}$ mice, and synaptosomes from these animals showed decreased glutamate uptake sensitive to dihydrokainate (Raju et al. 2015), which inhibits GLT-1 (Robinson 1998), indicating a decrease in GLT-1 glutamate transport capacity. The finding that a functional mutant GLT-1 protein (containing serine instead of cysteine in the equivalent residues) did not have any nitrosative modifications of these residues in the presence of an S-nitrosylating agent further supports the nNOS-dependent S-nitrosylation of this transporter (Raju et al. 2015). These studies indicate that, under certain conditions, glutamatergic neurotransmission may be influenced by nNOS-dependent S-nitrosylation of crucial enzymes involved in glutamate homeostasis in the brain (Raju et al. 2015).

4.7 Clinical Relevance of Transient Multienzyme Complexes Formation

At this point, the clinical relevance of the possible association and dissociation of these multienzyme complexes in vivo is not fully understood. Recent reports provide evidence that the efficiency of brain metabolism may rely on physical interactions involving enzymes or transporters forming complexes with other proteins and/or mitochondria. The disruption of such interactions may be related to brain abnormalities, as demonstrated in patients with psychiatric diseases in which a lack of interaction between hexokinase and the outer mitochondrial membrane has been reported (Regenold et al. 2012), as well as in schizophrenia in which association of hexokinase 1 with GLT1 found in control brain was not found in brain tissue from patients (Shan et al. 2014). Increased expression of hBCATm was identified in specific regions in postmortem brain from patients affected by vascular dementia, dementia with Lewy body (Ashby et al. 2016), and Alzheimer's disease (Hull et al. 2015a). Interestingly, alterations leading to either increased or decreased GDH activity were related to seizures and hyperexcitability of brain (Raizen et al. 2005; Malthankar-Phatak et al. 2006; McKenna 2011; Stanley 2011) as discussed by McKenna (2011 and references therein). GDH insensitivity to

GTP inhibition caused by mutations in the *GLUD1* gene sequence leads to the neurological impairment in patients with hyperinsulinism–hyperammonemia syndrome (Bahi-Buisson et al. 2008b). Impaired GDH activity in the temporal cortex of patients with medial temporal lobe epilepsy was observed (Malthankar-Phatak et al. 2006). Both the expression and activity of GDH are increased in human glioma tissue, and higher GDH level is associated with poor outcome (Zhang et al. 2016).

Since a localized alteration in the amount of an enzyme in a particular cellular and/or mitochondrial domain can alter its binding capacity, which is driven by affinity, it is tempting to speculate that the transient formation and/or disassembly of some multienzyme complexes may play a role in the pathophysiological mechanisms leading to these conditions.

4.8 Concluding Remarks

The occurrence and dynamic control of transient multienzyme and/or multiprotein complexes (metabolons) can facilitate metabolism within cells. As noted in this chapter, there is compelling evidence that the formation of such transient metabolons can markedly influence cell bioenergetics and neurotransmission by channeling substrates into specific pathways including the TCA cycle, malate-asparte shuttle, and oxidative metabolism of glutamate and glutamine for energy in neurons and astrocytes. Although it was not extensively covered in this chapter, the influence of physiological (as well as pathological) fluctuations in the concentration of small molecules on the association/dissociation of complexes that impact glutamate and glutamine metabolism should not be underestimated. There is still much to learn about the extent and role of common protein modifications on the activity of enzymes and transporters involved in these pathways. As noted by Robinson and Jackson (2016), more efforts should be undertaken to determine how transporter/enzyme/mitochondria/proteins complexes are regulated. As novel technologies are developed and employed to identify expected and unexpected interactions in living cells and in vivo, a better understanding of these types of interactions will be reached. Modulation of such transient protein–protein interactions may represent a novel and promising therapeutic approach for treatment of diseases involving dysregulation of glutamate metabolism.

References

Ahn BH, Kim HS, Song S, Lee IH, Liu J, Vassilopoulos A, Deng CX, Finkel T (2008) A role for the mitochondrial deacetylase Sirt3 in regulating energy homeostasis. Proc Natl Acad Sci U S A 105:14447–14452

Ananieva EA, Patel CH, Drake CH, Powell JD, Hutson SM (2014) Cytosolic branched chain aminotransferase (BCATc) regulates mTORC1 signaling and glycolytic metabolism in CD4+ T cells. J Biol Chem 289:18793–18804

Ashby EL, Kierzkowska M, Hull J, Kehoe PG, Hutson SM, Conway ME (2016) Altered expression of human mitochondrial branched chain aminotransferase in dementia with Lewy bodies and vascular dementia. Neurochem Res doi:10.1007/s11064-016-1855-7

Baeza J, Smallegan MJ, Denu JM (2016) Mechanisms and dynamics of protein acetylation in mitochondria. Trends Biochem Sci 41:231–244

Bahi-Buisson N, El Sabbagh S, Soufflet C, Escande F, Boddaert N, Valayannopoulos V, Bellane-Chantelot C, Lascelles K, Dulac O, Plouin P, de Lonlay P (2008a) Myoclonic absence epilepsy with photosensitivity and a gain of function mutation in glutamate dehydrogenase. Seizure 17:658–664

Bahi-Buisson N, Roze E, Dionisi C, Escande F, Valayannopoulos V, Feillet F, Heinrichs C, Chadefaux-Vekemans B, Dan B, de Lonlay P (2008b) Neurological aspects of hyperinsulinism-hyperammonaemia syndrome. Dev Med Child Neurol 50:945–949

Bak LK, Sickmann HM, Schousboe A, Waagepetersen HS (2005) Activity of the lactate-alanine shuttle is independent of glutamate-glutamine cycle activity in cerebellar neuronal-astrocytic cultures. J Neurosci Res 79:88–96

Bak LK, Johansen ML, Schousboe A, Waagepetersen HS (2007) Among the branched-chain amino acids, only valine metabolism is up-regulated in astrocytes during glutamate exposure. J Neurosci Res 85:3465–3470

Bak LK, Obel LF, Walls AB, Schousboe A, Faek SA, Jajo FS, Waagepetersen HS (2012) Novel model of neuronal bioenergetics: postsynaptic utilization of glucose but not lactate correlates positively with Ca2+ signalling in cultured mouse glutamatergic neurons. ASN Neuro 4(3). pii: e00083

Bakken IJ, White LR, Aasly J, Unsgard G, Sonnewald U (1997) Lactate formation from [U-13C] aspartate in cultured astrocytes: compartmentation of pyruvate metabolism. Neurosci Lett 237:117–120

Bao X, Pal R, Hascup KN, Wang Y, Wang WT, Xu W, Hui D, Agbas A, Wang X, Michaelis ML, Choi IY, Belousov AB, Gerhardt GA, Michaelis EK (2009) Transgenic expression of Glud1 (glutamate dehydrogenase 1) in neurons: in vivo model of enhanced glutamate release, altered synaptic plasticity, and selective neuronal vulnerability. J Neurosci 29:13929–13944

Bauer DE, Jackson JG, Genda EN, Montoya MM, Yudkoff M, Robinson MB (2012) The glutamate transporter, GLAST, participates in a macromolecular complex that supports glutamate metabolism. Neurochem Int 61:566–574

Beal MF (2002) Oxidatively modified proteins in aging and disease. Free Radic Biol Med 32:797–803

Beeckmans S, Kanarek L (1981) Demonstration of physical interactions between consecutive enzymes of the citric acid cycle and of the aspartate-malate shuttle. A study involving fumarase, malate dehydrogenase, citrate synthesis and aspartate aminotransferase. Eur J Biochem 117:527–535

Beeckmans S, Kanarek L (1987) Enzyme-enzyme interactions as modulators of the metabolic flux through the citric acid cycle. Biochem Soc Symp 54:163–172

Beeckmans S, Van Driessche E, Kanarek L (1990) Clustering of sequential enzymes in the glycolytic pathway and the citric acid cycle. J Cell Biochem 43:297–306

Beeckmans S, Van Driessche E, Kanarek L (1993) Immobilized enzymes as tools for the demonstration of metabolon formation. A short overview. J Mol Recognit 6:195–204

Bergles DE, Diamond JS, Jahr CE (1999) Clearance of glutamate inside the synapse and beyond. Curr Opin Neurobiol 9:293–298

Bixel M, Shimomura Y, Hutson S, Hamprecht B (2001) Distribution of key enzymes of branched-chain amino acid metabolism in glial and neuronal cells in culture. J Histochem Cytochem 49:407–418

Botman D, Tigchelaar W, Van Noorden CJ (2014) Determination of glutamate dehydrogenase activity and its kinetics in mouse tissues using metabolic mapping (quantitative enzyme histochemistry). J Histochem Cytochem 62:802–812

Bunik VI, Sievers C (2002) Inactivation of the 2-oxo acid dehydrogenase complexes upon generation of intrinsic radical species. Eur J Biochem 269:5004–5015

Cesar M, Hamprecht B (1995) Immunocytochemical examination of neural rat and mouse primary cultures using monoclonal antibodies raised against pyruvate carboxylase. J Neurochem 64:2312–2318
Choudhary C, Weinert BT, Nishida Y, Verdin E, Mann M (2014) The growing landscape of lysine acetylation links metabolism and cell signalling. Nat Rev Mol Cell Biol 15:536–550
Christgau S, Aanstoot HJ, Schierbeck H, Begley K, Tullin S, Hejnaes K, Baekkeskov S (1992) Membrane anchoring of the autoantigen GAD65 to microvesicles in pancreatic beta-cells by palmitoylation in the NH2-terminal domain. J Cell Biol 118:309–320
Cimen H, Han MJ, Yang Y, Tong Q, Koc H, Koc EC (2010) Regulation of succinate dehydrogenase activity by SIRT3 in mammalian mitochondria. Biochemistry 49:304–311
Cole JT, Sweatt AJ, Hutson SM (2012) Expression of mitochondrial branched-chain aminotransferase and alpha-keto-acid dehydrogenase in rat brain: implications for neurotransmitter metabolism. Front Neuroanat 6:18
Connor SA, Wang YT (2015) A place at the table: LTD as a mediator of memory genesis. Neuroscientist 22(4):359–371
Contreras L, Gomez-Puertas P, Iijima M, Kobayashi K, Saheki T, Satrustegui J (2007) Ca2+ Activation kinetics of the two aspartate-glutamate mitochondrial carriers, aralar and citrin: role in the heart malate-aspartate NADH shuttle. J Biol Chem 282:7098–7106
Conway ME, Hutson SM (2016) BCAA metabolism and NH_3 homeostasis. In: Schousboe A, Sonnewald U (eds) The glutamate/GABA-glutamine cycle, vol 13, Advances in neurobiology. Springer, New York, pp 99–112
Cooper AJ (1988) L-glutamate(2-oxoglutarate) aminotransferases. In: Kvamme E (ed) Glutamine and glutamate in mammals. CRC, Boca Raton, pp 123–152
Cooper AJ (2011) (13)N as a tracer for studying glutamate metabolism. Neurochem Int 59(4):456–464
Cooper AJ, Meister A (1985) Metabolic significance of transamination. In: Christen P, Meltzer DE (eds) Transaminases. Wiley, New York, pp 500–580
Cooper AJ, Plum F (1987) Biochemistry and physiology of brain ammonia. Physiol Rev 67:440–519
Cooper AJL, Pinto JT, Callery PS (2011) Reversible and irreversible protein glutathionylation: biological and clinical aspects. Expert Opin Drug Metab Toxicol 7:891–910
Csibi A, Fendt SM, Li C, Poulogiannis G, Choo AY, Chapski DJ, Jeong SM, Dempsey JM, Parkhitko A, Morrison T, Henske EP, Haigis MC, Cantley LC, Stephanopoulos G, Yu J, Blenis J (2013) The mTORC1 pathway stimulates glutamine metabolism and cell proliferation by repressing SIRT4. Cell 153:840–854
Danbolt NC (2001) Glutamate uptake. Prog Neurobiol 65:1–105
Dringen R (2000) Metabolism and functions of glutathione in brain. Prog Neurobiol 62:649–671
Dzubay JA, Jahr CE (1999) The concentration of synaptically released glutamate outside of the climbing fiber-Purkinje cell synaptic cleft. J Neurosci 19:5265–5274
El Hindy M, Hezwani M, Corry D, Hull J, El Amraoui F, Harris M, Lee C, Forshaw T, Wilson A, Mansbridge A, Hassler M, Patel VB, Kehoe PG, Love S, Conway ME (2014) The branched-chain aminotransferase proteins: novel redox chaperones for protein disulfide isomerase—implications in Alzheimer's disease. Antioxid Redox Signal 20:2497–2513
Erecinska M, Silver IA (1990) Metabolism and role of glutamate in mammalian brain. Prog Neurobiol 35:245–296
Faff-Michalak L, Albrecht J (1991) Aspartate aminotransferase, malate dehydrogenase, and pyruvate carboxylase activities in rat cerebral synaptic and nonsynaptic mitochondria: effects of in vitro treatment with ammonia, hyperammonemia and hepatic encephalopathy. Metab Brain Dis 6:187–197
Fahien LA, Kmiotek E (1981) Regulation of glutamate dehydrogenase by palmitoyl-coenzyme A. Arch Biochem Biophys 212:247–253
Fahien LA, Kmiotek E (1983) Complexes between mitochondrial enzymes and either citrate synthase or glutamate dehydrogenase. Arch Biochem Biophys 220:386–397

Fahien LA, Teller JK (1992) Glutamate-malate metabolism in liver mitochondria. A model constructed on the basis of mitochondrial levels of enzymes, specificity, dissociation constants, and stoichiometry of hetero-enzyme complexes. J Biol Chem 267:10411–10422

Fahien LA, Hsu SL, Kmiotek E (1977) Effect of aspartate on complexes between glutamate dehydrogenase and various aminotransferases. J Biol Chem 252:1250–1256

Fahien LA, Kmiotek EH, Woldegiorgis G, Evenson M, Shrago E, Marshall M (1985) Regulation of aminotransferase-glutamate dehydrogenase interactions by carbamyl phosphate synthase-I, Mg2+ plus leucine versus citrate and malate. J Biol Chem 260:6069–6079

Fahien LA, Kmiotek EH, MacDonald MJ, Fibich B, Mandic M (1988) Regulation of malate dehydrogenase activity by glutamate, citrate, alpha-ketoglutarate, and multienzyme interaction. J Biol Chem 263:10687–10697

Fahien LA, MacDonald MJ, Teller JK, Fibich B, Fahien CM (1989) Kinetic advantages of hetero-enzyme complexes with glutamate dehydrogenase and the alpha-ketoglutarate dehydrogenase complex. J Biol Chem 264:12303–12312

Fahien LA, Teller JK, Macdonald MJ, Fahien CM (1990) Regulation of glutamate dehydrogenase by Mg2+ and magnification of leucine activation by Mg2+. Mol Pharmacol 37:943–949

Fahien LA, Davis JW, Laboy J (1993) Interactions between pyruvate carboxylase and other mitochondrial enzymes. J Biol Chem 268:17935–17942

Feligioni M, Nishimune A, Henley JM (2009) Protein SUMOylation modulates calcium influx and glutamate release from presynaptic terminals. Eur J Neurosci 29:1348–1356

Fernandes J, Weddle A, Kinter CS, Humphries KM, Mather T, Szweda LI, Kinter M (2015) Lysine acetylation activates mitochondrial aconitase in the heart. Biochemistry 54:4008–4018

Finley LW, Haas W, Desquiret-Dumas V, Wallace DC, Procaccio V, Gygi SP, Haigis MC (2011) Succinate dehydrogenase is a direct target of sirtuin 3 deacetylase activity. PLoS One 6:e23295

Fischer A (2014) Targeting histone-modifications in Alzheimer's disease. What is the evidence that this is a promising therapeutic avenue? Neuropharmacology 80:95–102

Fisher RS (2007) Co-localization of glutamic acid decarboxylase and phosphate-activated glutaminase in neurons of lateral reticular nucleus in feline thalamus. Neurochem Res 32:177–186

Foran E, Bogush A, Goffredo M, Roncaglia P, Gustincich S, Pasinelli P, Trotti D (2011) Motor neuron impairment mediated by a sumoylated fragment of the glial glutamate transporter EAAT2. Glia 59:1719–1731

Fukata Y, Fukata M (2010) Protein palmitoylation in neuronal development and synaptic plasticity. Nat Rev Neurosci 11:161–175

Garcia-Espinosa MA, Wallin R, Hutson SM, Sweatt AJ (2007) Widespread neuronal expression of branched-chain aminotransferase in the CNS: implications for leucine/glutamate metabolism and for signaling by amino acids. J Neurochem 100:1458–1468

Geiss-Friedlander R, Melchior F (2007) Concepts in sumoylation: a decade on. Nat Rev Mol Cell Biol 8:947–956

Genda EN, Jackson JG, Sheldon AL, Locke SF, Greco TM, O'Donnell JC, Spruce LA, Xiao R, Guo W, Putt M, Seeholzer S, Ischiropoulos H, Robinson MB (2011) Co-compartmentalization of the astroglial glutamate transporter, GLT-1, with glycolytic enzymes and mitochondria. J Neurosci 31:18275–18288

Gibb SL, Boston-Howes W, Lavina ZS, Gustincich S, Brown RH Jr, Pasinelli P, Trotti D (2007) A caspase-3-cleaved fragment of the glial glutamate transporter EAAT2 is sumoylated and targeted to promyelocytic leukemia nuclear bodies in mutant SOD1-linked amyotrophic lateral sclerosis. J Biol Chem 282:32480–32490

Gibson GE, Xu H, Chen HL, Chen W, Denton T, Zhang S (2015) Alpha-ketoglutarate dehydrogenase complex-dependent succinylation of proteins in neurons and neuronal cell lines. J Neurochem 134:86–96

Gilley R, Balmanno K, Cope CL, Cook SJ (2013) Adaptation to chronic mTOR inhibition in cancer and in aging. Biochem Soc Trans 41:956–961

Gonzalez-Gonzalez IM, Garcia-Tardon N, Cubelos B, Gimenez C, Zafra F (2008) The glutamate transporter GLT1b interacts with the scaffold protein PSD-95. J Neurochem 105:1834–1848

Gonzalez-Gonzalez IM, Garcia-Tardon N, Gimenez C, Zafra F (2009) Splice variants of the glutamate transporter GLT1 form hetero-oligomers that interact with PSD-95 and NMDA receptors. J Neurochem 110:264–274

Haggie PM, Brindle KM (1999) Mitochondrial citrate synthase is immobilized in vivo. J Biol Chem 274:3941–3945

Haggie PM, Verkman AS (2002) Diffusion of tricarboxylic acid cycle enzymes in the mitochondrial matrix in vivo. Evidence for restricted mobility of a multienzyme complex. J Biol Chem 277:40782–40788

Haigis MC, Mostoslavsky R, Haigis KM, Fahie K, Christodoulou DC, Murphy AJ, Valenzuela DM, Yancopoulos GD, Karow M, Blander G, Wolberger C, Prolla TA, Weindruch R, Alt FW, Guarente L (2006) SIRT4 inhibits glutamate dehydrogenase and opposes the effects of calorie restriction in pancreatic beta cells. Cell 126:941–954

Hall TR, Wallin R, Reinhart GD, Hutson SM (1993) Branched chain aminotransferase isoenzymes. Purification and characterization of the rat brain isoenzyme. J Biol Chem 268:3092–3098

Halper LA, Srere PA (1977) Interaction between citrate synthase and mitochondrial malate dehydrogenase in the presence of polyethylene glycol. Arch Biochem Biophys 184:529–534

Hanu R, McKenna M, O'Neill A, Resneck WG, Bloch RJ (2000) Monocarboxylic acid transporters, MCT1 and MCT2, in cortical astrocytes in vitro and in vivo. Am J Physiol Cell Physiol 278:C921–C930

Haugeto O, Ullensvang K, Levy LM, Chaudhry FA, Honore T, Nielsen M, Lehre KP, Danbolt NC (1996) Brain glutamate transporter proteins form homomultimers. J Biol Chem 271:27715–27722

Haussinger D, Gorg B, Reinehr R, Schliess F (2005) Protein tyrosine nitration in hyperammonemia and hepatic encephalopathy. Metab Brain Dis 20:285–294

Hayashi H, Mizuguchi H, Miyahara I, Nakajima Y, Hirotsu K, Kagamiyama H (2003) Conformational change in aspartate aminotransferase on substrate binding induces strain in the catalytic group and enhances catalysis. J Biol Chem 278:9481–9488

Hebert AS, Dittenhafer-Reed KE, Yu W, Bailey DJ, Selen ES, Boersma MD, Carson JJ, Tonelli M, Balloon AJ, Higbee AJ, Westphall MS, Pagliarini DJ, Prolla TA, Assadi-Porter F, Roy S, Denu JM, Coon JJ (2013) Calorie restriction and SIRT3 trigger global reprogramming of the mitochondrial protein acetylome. Mol Cell 49:186–199

Helms HC, Madelung R, Waagepetersen HS, Nielsen CU, Brodin B (2012) In vitro evidence for the brain glutamate efflux hypothesis: brain endothelial cells cocultured with astrocytes display a polarized brain-to-blood transport of glutamate. Glia 60:882–893

Herrero-Yraola A, Bakhit SM, Franke P, Weise C, Schweiger M, Jorcke D, Ziegler M (2001) Regulation of glutamate dehydrogenase by reversible ADP-ribosylation in mitochondria. EMBO J 20:2404–2412

Houtkooper RH, Pirinen E, Auwerx J (2012) Sirtuins as regulators of metabolism and healthspan. Nat Rev Mol Cell Biol 13:225–238

Hull J, Hindy ME, Kehoe PG, Chalmers K, Love S, Conway ME (2012) Distribution of the branched chain aminotransferase proteins in the human brain and their role in glutamate regulation. J Neurochem 123:997–1009

Hull J, Patel V, El Hindy M, Lee C, Odeleye E, Hezwani M, Love S, Kehoe P, Chalmers K, Conway M (2015a) Regional increase in the expression of the BCAT proteins in Alzheimer's disease brain: implications in glutamate toxicity. J Alzheimers Dis 45:891–905

Hull J, Patel VB, Hutson SM, Conway ME (2015b) New insights into the role of the branched-chain aminotransferase proteins in the human brain. J Neurosci Res 93:987–998

Hutson S (2001) Structure and function of branched chain aminotransferases. Prog Nucleic Acid Res Mol Biol 70:175–206

Hutson SM, Harper AE (1981) Blood and tissue branched-chain amino and alpha-keto acid concentrations: effect of diet, starvation, and disease. Am J Clin Nutr 34:173–183

Hutson SM, Lieth E, LaNoue KF (2001) Function of leucine in excitatory neurotransmitter metabolism in the central nervous system. J Nutr 131:846S–850S

Hutson SM, Islam MM, Zaganas I (2011) Interaction between glutamate dehydrogenase (GDH) and L-leucine catabolic enzymes: intersecting metabolic pathways. Neurochem Int 59:518–524

Islam MM, Wallin R, Wynn RM, Conway M, Fujii H, Mobley JA, Chuang DT, Hutson SM (2007) A novel branched-chain amino acid metabolon. Protein-protein interactions in a supramolecular complex. J Biol Chem 282:11893–11903

Islam MM, Nautiyal M, Wynn RM, Mobley JA, Chuang DT, Hutson SM (2010) Branched-chain amino acid metabolon: interaction of glutamate dehydrogenase with the mitochondrial branched-chain aminotransferase (BCATm). J Biol Chem 285:265–276

Jackson JG, O'Donnell JC, Takano H, Coulter DA, Robinson MB (2014) Neuronal activity and glutamate uptake decrease mitochondrial mobility in astrocytes and position mitochondria near glutamate transporters. J Neurosci 34:1613–1624

Jackson JG, O'Donnell JC, Krizman E, Robinson MB (2015) Displacing hexokinase from mitochondrial voltage-dependent anion channel impairs GLT-1-mediated glutamate uptake but does not disrupt interactions between GLT-1 and mitochondrial proteins. J Neurosci Res 93:999–1008

Jeffery CJ (2005) Mass spectrometry and the search for moonlighting proteins. Mass Spectrom Rev 24:772–782

Johnson DA, Johnson JA (2015) Nrf2—a therapeutic target for the treatment of neurodegenerative diseases. Free Radic Biol Med 88:253–267

Jones DP (1986) Intracellular diffusion gradients of O2 and ATP. Am J Physiol 250:C663–C675

Kanaani J, Patterson G, Schaufele F, Lippincott-Schwartz J, Baekkeskov S (2008) A palmitoylation cycle dynamically regulates partitioning of the GABA-synthesizing enzyme GAD65 between ER-Golgi and post-Golgi membranes. J Cell Sci 121:437–449

Karaca M, Frigerio F, Migrenne S, Martin-Levilain J, Skytt DM, Pajecka K, Martin-del-Rio R, Gruetter R, Tamarit-Rodriguez J, Waagepetersen HS, Magnan C, Maechler P (2015) GDH-dependent glutamate oxidation in the brain dictates peripheral energy substrate distribution. Cell Rep 13:365–375

Kennedy MB (2016) Synaptic signaling in learning and memory. Cold Spring Harb Perspect Biol 8 (a)016824 doi:10.1101/cshperspect.a016824

Kerscher O (2007) SUMO junction-what's your function? New insights through SUMO-interacting motifs. EMBO Rep 8:550–555

Komlos D, Mann KD, Zhuo Y, Ricupero CL, Hart RP, Liu AY, Firestein BL (2013) Glutamate dehydrogenase 1 and SIRT4 regulate glial development. Glia 61:394–408

Kumar A (2015) NMDA receptor function during senescence: implication on cognitive performance. Front Neurosci 9:473

Kvamme E, Svenneby G, Torgner IA, Drejer J, Schousboe A (1985) Postnatal development of glutamate developing enzymes in hippocampus from mice. Int J Dev Neurosci 3:359–364

Lai JC, Clark JB (1989) Isolation and characterization of synaptic and nonsynaptic mitochondria from the mammalian brain. In: Boulton AA, Baker GB (eds) Neuromethods, carbohydrates and energy metabolism. Humana Press, Clifton, pp 43–98

Lai JC, Murthy CR, Cooper AJ, Hertz E, Hertz L (1989) Differential effects of ammonia and beta-methylene-DL-aspartate on metabolism of glutamate and related amino acids by astrocytes and neurons in primary culture. Neurochem Res 14:377–389

Laplante M, Sabatini DM (2012) mTOR signaling in growth control and disease. Cell 149:274–293

Lee A, Rayfield A, Hryciw DH, Ma TA, Wang D, Pow D, Broer S, Yun C, Poronnik P (2007) Na+-H+ exchanger regulatory factor 1 is a PDZ scaffold for the astroglial glutamate transporter GLAST. Glia 55:119–129

Leke R, Schousboe A (2016) The glutamine transporters and their role in the glutamate/GABA–glutamine cycle. In: Schousboe A, Sonnewald U (eds) The glutamate/GABA-glutamine cycle, vol 13, Advances in neurobiology. Springer, New York, pp 223–258

Li C, Chen P, Palladino A, Narayan S, Russell LK, Sayed S, Xiong G, Chen J, Stokes D, Butt YM, Jones PM, Collins HW, Cohen NA, Cohen AS, Nissim I, Smith TJ, Strauss AW, Matschinsky FM, Bennett MJ, Stanley CA (2010a) Mechanism of hyperinsulinism in short-chain 3-hydroxyacyl-CoA dehydrogenase deficiency involves activation of glutamate dehydrogenase. J Biol Chem 285:31806–31818

Li N, Lee B, Liu RJ, Banasr M, Dwyer JM, Iwata M, Li XY, Aghajanian G, Duman RS (2010b) mTOR-dependent synapse formation underlies the rapid antidepressant effects of NMDA antagonists. Science 329:959–964

Li M, Li C, Allen A, Stanley CA, Smith TJ (2011) The structure and allosteric regulation of glutamate dehydrogenase. Neurochem Int 59(4):445–455

Llorente-Folch I, Rueda CB, Amigo I, del Arco A, Saheki T, Pardo B, Satrustegui J (2013) Calcium-regulation of mitochondrial respiration maintains ATP homeostasis and requires ARALAR/AGC1-malate aspartate shuttle in intact cortical neurons. J Neurosci 33:13957–13971, 13971a

Lombard DB, Alt FW, Cheng HL, Bunkenborg J, Streeper RS, Mostoslavsky R, Kim J, Yancopoulos G, Valenzuela D, Murphy A, Yang Y, Chen Y, Hirschey MD, Bronson RT, Haigis M, Guarente LP, Farese RV Jr, Weissman S, Verdin E, Schwer B (2007) Mammalian Sir2 homolog SIRT3 regulates global mitochondrial lysine acetylation. Mol Cell Biol 27:8807–8814

Lopez-Beltran EA, Mate MJ, Cerdan S (1996) Dynamics and environment of mitochondrial water as detected by 1H NMR. J Biol Chem 271:10648–10653

Mac M, Nalecz KA (2003) Expression of monocarboxylic acid transporters (MCT) in brain cells. Implication for branched chain alpha-ketoacids transport in neurons. Neurochem Int 43:305–309

Magi S, Lariccia V, Castaldo P, Arcangeli S, Nasti AA, Giordano A, Amoroso S (2012) Physical and functional interaction of NCX1 and EAAC1 transporters leading to glutamate-enhanced ATP production in brain mitochondria. PLoS One 7:e34015

Magi S, Arcangeli S, Castaldo P, Nasti AA, Berrino L, Piegari E, Bernardini R, Amoroso S, Lariccia V (2013) Glutamate-induced ATP synthesis: relationship between plasma membrane Na+/Ca2+ exchanger and excitatory amino acid transporters in brain and heart cell models. Mol Pharmacol 84:603–614

Malik P, McKenna MC, Tildon JT (1993) Regulation of malate dehydrogenases from neonatal, adolescent, and mature rat brain. Neurochem Res 18:247–257

Malthankar-Phatak GH, de Lanerolle N, Eid T, Spencer DD, Behar KL, Spencer SS, Kim JH, Lai JC (2006) Differential glutamate dehydrogenase (GDH) activity profile in patients with temporal lobe epilepsy. Epilepsia 47:1292–1299

Martinez-Lozada Z, Guillem AM, Flores-Mendez M, Hernandez-Kelly LC, Vela C, Meza E, Zepeda RC, Caba M, Rodriguez A, Ortega A (2013) GLAST/EAAT1-induced glutamine release via SNAT3 in Bergmann glial cells: evidence of a functional and physical coupling. J Neurochem 125:545–554

Mastorodemos V, Zaganas I, Spanaki C, Bessa M, Plaitakis A (2005) Molecular basis of human glutamate dehydrogenase regulation under changing energy demands. J Neurosci Res 79:65–73

Mastorodemos V, Kotzamani D, Zaganas I, Arianoglou G, Latsoudis H, Plaitakis A (2009) Human GLUD1 and GLUD2 glutamate dehydrogenase localize to mitochondria and endoplasmic reticulum. Biochem Cell Biol 87:505–516

Mastorodemos V, Kanavouras K, Sundaram S, Providaki M, Petraki Z, Kokkinidis M, Zaganas I, Logothetis DE, Plaitakis A (2015) Side-chain interactions in the regulatory domain of human glutamate dehydrogenase determine basal activity and regulation. J Neurochem 133:73–82

Matos M, Augusto E, Machado NJ, dos Santos-Rodrigues A, Cunha RA, Agostinho P (2012a) Astrocytic adenosine A2A receptors control the amyloid-beta peptide-induced decrease of glutamate uptake. J Alzheimers Dis 31:555–567

Matos M, Augusto E, Santos-Rodrigues AD, Schwarzschild MA, Chen JF, Cunha RA, Agostinho P (2012b) Adenosine A2A receptors modulate glutamate uptake in cultured astrocytes and gliosomes. Glia 60:702–716

Matos M, Augusto E, Agostinho P, Cunha RA, Chen JF (2013) Antagonistic interaction between adenosine A2A receptors and Na+/K+-ATPase-alpha2 controlling glutamate uptake in astrocytes. J Neurosci 33:18492–18502

Matsui K, Jahr CE, Rubio ME (2005) High-concentration rapid transients of glutamate mediate neural-glial communication via ectopic release. J Neurosci 25:7538–7547

McKenna MC (2007) The glutamate-glutamine cycle is not stoichiometric: fates of glutamate in brain. J Neurosci Res 85:3347–3358

McKenna MC (2011) Glutamate dehydrogenase in brain mitochondria: do lipid modifications and transient metabolon formation influence enzyme activity? Neurochem Int 59:525–533
McKenna MC (2012) Substrate competition studies demonstrate oxidative metabolism of glucose, glutamate, glutamine, lactate and 3-hydroxybutyrate in cortical astrocytes from rat brain. Neurochem Res 37:2613–2626
McKenna MC (2013) Glutamate pays its own way in astrocytes. Front Endocrinol (Lausanne) 4:191
McKenna MC, Rae CD (2015) A new role for alpha-ketoglutarate dehydrogenase complex: regulating metabolism through post-translational modification of other enzymes. J Neurochem 134:3–6
McKenna MC, Sonnewald U (2005) GABA alters the metabolic fate of [U-13C]glutamate in cultured cortical astrocytes. J Neurosci Res 79:81–87
McKenna MC, Stridh MH, McNair LF, Sonnewald U, Waagepetersen HS, Schousboe A (2016) Glutamate oxidation in astrocytes: Roles of glutamate dehydrogenase and aminotransferases. J Neurosci Res. 2016 Sep 15. doi: 10.1002/jnr.23908. [Epub ahead of print]
McKenna MC, Tildon JT, Couto R, Stevenson JH, Caprio FJ (1990) The metabolism of malate by cultured rat brain astrocytes. Neurochem Res 15:1211–1220
McKenna MC, Tildon JT, Stevenson JH, Boatright R, Huang S (1993) Regulation of energy metabolism in synaptic terminals and cultured rat brain astrocytes: differences revealed using aminooxyacetate. Dev Neurosci 15:320–329
McKenna MC, Tildon JT, Stevenson JH, Hopkins IB (1994) Energy metabolism in cortical synaptic terminals from weanling and mature rat brain: evidence for multiple compartments of tricarboxylic acid cycle activity. Dev Neurosci 16:291–300
McKenna MC, Tildon JT, Stevenson JH, Huang X, Kingwell KG (1995) Regulation of mitochondrial and cytosolic malic enzymes from cultured rat brain astrocytes. Neurochem Res 20:1491–1501
McKenna MC, Sonnewald U, Huang X, Stevenson J, Zielke HR (1996a) Exogenous glutamate concentration regulates the metabolic fate of glutamate in astrocytes. J Neurochem 66:386–393
McKenna MC, Tildon JT, Stevenson JH, Huang X (1996b) New insights into the compartmentation of glutamate and glutamine in cultured rat brain astrocytes. Dev Neurosci 18:380–390
McKenna MC, Stevenson JH, Huang X, Hopkins IB (2000a) Differential distribution of the enzymes glutamate dehydrogenase and aspartate aminotransferase in cortical synaptic mitochondria contributes to metabolic compartmentation in cortical synaptic terminals. Neurochem Int 37:229–241
McKenna MC, Stevenson JH, Huang X, Tildon JT, Zielke CL, Hopkins IB (2000b) Mitochondrial malic enzyme activity is much higher in mitochondria from cortical synaptic terminals compared with mitochondria from primary cultures of cortical neurons or cerebellar granule cells. Neurochem Int 36:451–459
McKenna MC, Gruetter R, Sonnewald U, Waagepetersen HS, Schousboe A, Siegel GJ, Albers RW, Brady ST, Price DL (2006a) Energy metabolism of the brain. In: Siegel GJ, Albers RW, Brady ST, Price DL (eds) Basic neurochemistry. Elsevier Academic, Burlington, MA, pp 531–557
McKenna MC, Hopkins IB, Lindauer SL, Bamford P (2006b) Aspartate aminotransferase in synaptic and nonsynaptic mitochondria: differential effect of compounds that influence transient hetero-enzyme complex (metabolon) formation. Neurochem Int 48:629–636
McKenna MC, Waagepetersen HS, Schousboe A, Sonnewald U (2006c) Neuronal and astrocytic shuttle mechanisms for cytosolic-mitochondrial transfer of reducing equivalents: current evidence and pharmacological tools. Biochem Pharmacol 71:399–407
McKenna MC, Dienel GA, Waagepetersen HS, Schousboe A, Sonnewald U (2011) Energy metabolism of the brain. In Basic Neurochemistry, (Brady S., Price, D., eds.), Elsevier, London. pp. 200–231
Menon S, Manning BD (2008) Common corruption of the mTOR signaling network in human tumors. Oncogene 27(Suppl 2):S43–S51
Menon S, Manning BD (2013) Cell signalling: nutrient sensing lost in cancer. Nature 498:444–445
Mkrtchyan G, Aleshin V, Parkhomenko Y, Kaehne T, Luigi Di Salvo M, Parroni A, Contestabile R, Vovk A, Bettendorff L, Bunik V (2015) Molecular mechanisms of the non-coenzyme action of thiamin in brain: biochemical, structural and pathway analysis. Sci Rep 5:12583

Moore GE, Gadol SM, Robinson JB Jr, Srere PA (1984) Binding of citrate synthase and malate dehydrogenase to mitochondrial inner membranes: tissue distribution and metabolite effects. Biochem Biophys Res Commun 121:612–618

Moran J, Rivera-Gaxiola M (1992) Effect of potassium and N-methyl-D-aspartate on the aspartate aminotransferase activity in cultured cerebellar granule cells. J Neurosci Res 33:239–247

Moran J, Alavez S, Rivera-Gaxiola M, Valencia A, Hurtado S (1999) Effect of NMDA antagonists on the activity of glutaminase and aspartate aminotransferase in the developing rat cerebellum. Int J Dev Neurosci 17:57–65

Morgunov I, Srere PA (1998) Interaction between citrate synthase and malate dehydrogenase. Substrate channeling of oxaloacetate. J Biol Chem 273:29540–29544

Moussawi K, Riegel A, Nair S, Kalivas PW (2011) Extracellular glutamate: functional compartments operate in different concentration ranges. Front Syst Neurosci 5:94

Nakagawa T, Lomb DJ, Haigis MC, Guarente L (2009) SIRT5 Deacetylates carbamoyl phosphate synthetase 1 and regulates the urea cycle. Cell 137:560–570

Narayan SB, Master SR, Sireci AN, Bierl C, Stanley PE, Li C, Stanley CA, Bennett MJ (2012) Short-chain 3-hydroxyacyl-coenzyme A dehydrogenase associates with a protein super-complex integrating multiple metabolic pathways. PLoS One 7:e35048

Neishabouri SH, Hutson SM, Davoodi J (2015) Chronic activation of mTOR complex 1 by branched chain amino acids and organ hypertrophy. Amino Acids 47:1167–1182

Newman JC, He W, Verdin E (2012) Mitochondrial protein acylation and intermediary metabolism: regulation by sirtuins and implications for metabolic disease. J Biol Chem 287:42436–42443

Norenberg MD, Martinez-Hernandez A (1979) Fine structural localization of glutamine synthetase in astrocytes of rat brain. Brain Res 161:303–310

Noritake J, Fukata Y, Iwanaga T, Hosomi N, Tsutsumi R, Matsuda N, Tani H, Iwanari H, Mochizuki Y, Kodama T, Matsuura Y, Bredt DS, Hamakubo T, Fukata M (2009) Mobile DHHC palmitoylating enzyme mediates activity-sensitive synaptic targeting of PSD-95. J Cell Biol 186:147–160

O'Kane RL, Hawkins RA (2003) Na+-dependent transport of large neutral amino acids occurs at the abluminal membrane of the blood-brain barrier. Am J Physiol Endocrinol Metab 285:E1167–E1173

O'Kane RL, Martinez-Lopez I, DeJoseph MR, Vina JR, Hawkins RA (1999) Na(+)-dependent glutamate transporters (EAAT1, EAAT2, and EAAT3) of the blood-brain barrier. A mechanism for glutamate removal. J Biol Chem 274:31891–31895

O'Kane RL, Vina JR, Simpson I, Hawkins RA (2004) Na+-dependent neutral amino acid transporters A, ASC, and N of the blood-brain barrier: mechanisms for neutral amino acid removal. Am J Physiol Endocrinol Metab 287:E622–E629

Oh YJ, Markelonis GJ, Oh TH (1991) Immunocytochemical localization of mitochondrial malate dehydrogenase in primary cultures of rat astrocytes and oligodendrocytes. J Histochem Cytochem 39:681–688

Osborne B, Cooney GJ, Turner N (2014) Are sirtuin deacylase enzymes important modulators of mitochondrial energy metabolism? Biochim Biophys Acta 1840:1295–1302

Ozden O, Park SH, Wagner BA, Yong Song H, Zhu Y, Vassilopoulos A, Jung B, Buettner GR, Gius D (2014) SIRT3 deacetylates and increases pyruvate dehydrogenase activity in cancer cells. Free Radic Biol Med 76:163–172

Pagliaro L, Taylor DL (1992) 2-Deoxyglucose and cytochalasin D modulate aldolase mobility in living 3T3 cells. J Cell Biol 118:859–863

Papanicolaou KN, O'Rourke B, Foster DB (2014) Metabolism leaves its mark on the powerhouse: recent progress in post-translational modifications of lysine in mitochondria. Front Physiol 5:301

Park J, Chen Y, Tishkoff DX, Peng C, Tan M, Dai L, Xie Z, Zhang Y, Zwaans BM, Skinner ME, Lombard DB, Zhao Y (2013) SIRT5-mediated lysine desuccinylation impacts diverse metabolic pathways. Mol Cell 50:919–930

Pedersen WA, Fu W, Keller JN, Markesbery WR, Appel S, Smith RG, Kasarskis E, Mattson MP (1998) Protein modification by the lipid peroxidation product 4-hydroxynonenal in the spinal cords of amyotrophic lateral sclerosis patients. Ann Neurol 44:819–824

Pellerin L, Magistretti PJ (1994) Glutamate uptake into astrocytes stimulates aerobic glycolysis: a mechanism coupling neuronal activity to glucose utilization. Proc Natl Acad Sci U S A 91:10625–10629

Peng L, Zhang X, Hertz L (1994) High extracellular potassium concentrations stimulate oxidative metabolism in a glutamatergic neuronal culture and glycolysis in cultured astrocytes but have no stimulatory effect in a GABAergic neuronal culture. Brain Res 663:168–172

Peng L, Swanson RA, Hertz L (2001) Effects of L-glutamate, D-aspartate, and monensin on glycolytic and oxidative glucose metabolism in mouse astrocyte cultures: further evidence that glutamate uptake is metabolically driven by oxidative metabolism. Neurochem Int 38:437–443

Perluigi M, Swomley AM, Butterfield DA (2014) Redox proteomics and the dynamic molecular landscape of the aging brain. Ageing Res Rev 13:75–89

Plaitakis A, Zaganas I (2001) Regulation of human glutamate dehydrogenases: implications for glutamate, ammonia and energy metabolism in brain. J Neurosci Res 66:899–908

Plaitakis A, Berl S, Yahr MD (1982) Abnormal glutamate metabolism in an adult-onset degenerative neurological disorder. Science 216:193–196

Plaitakis A, Berl S, Yahr MD (1984) Neurological disorders associated with deficiency of glutamate dehydrogenase. Ann Neurol 15:144–153

Plaitakis A, Metaxari M, Shashidharan P (2000) Nerve tissue-specific (GLUD2) and housekeeping (GLUD1) human glutamate dehydrogenases are regulated by distinct allosteric mechanisms: implications for biologic function. J Neurochem 75:1862–1869

Plaitakis A, Spanaki C, Mastorodemos V, Zaganas I (2003) Study of structure-function relationships in human glutamate dehydrogenases reveals novel molecular mechanisms for the regulation of the nerve tissue-specific (GLUD2) isoenzyme. Neurochem Int 43:401–410

Plaitakis A, Latsoudis H, Spanaki C (2011) The human GLUD2 glutamate dehydrogenase and its regulation in health and disease. Neurochem Int 59(4):495–509

Pollack GH (2001) Cells, gels and the engines of life; a new, unifying approach to cell function. Ebner and Sons, Seattle, pp 1–320

Porpaczy Z, Sumegi B, Alkonyi I (1987) Interaction between NAD-dependent isocitrate dehydrogenase, alpha-ketoglutarate dehydrogenase complex, and NADH: ubiquinone oxidoreductase. J Biol Chem 262:9509–9514

Raizen DM, Brooks-Kayal A, Steinkrauss L, Tennekoon GI, Stanley CA, Kelly A (2005) Central nervous system hyperexcitability associated with glutamate dehydrogenase gain of function mutations. J Pediatr 146:388–394

Raju K, Doulias PT, Evans P, Krizman EN, Jackson JG, Horyn O, Daikhin Y, Nissim I, Yudkoff M, Sharp KA, Robinson MB, Ischiropoulos H (2015) Regulation of brain glutamate metabolism by nitric oxide and S-nitrosylation. Sci Signal 8:ra68 doi:10.1126/scisignal.aaa4312

Rasgado LA, Reyes GC, Diaz FV (2015) Modulation of brain glutamate dehydrogenase as a tool for controlling seizures. Acta Pharm 65:443–452

Regenold WT, Pratt M, Nekkalapu S, Shapiro PS, Kristian T, Fiskum G (2012) Mitochondrial detachment of hexokinase 1 in mood and psychotic disorders: implications for brain energy metabolism and neurotrophic signaling. J Psychiatr Res 46:95–104

Resh MD (1996) Regulation of cellular signalling by fatty acid acylation and prenylation of signal transduction proteins. Cell Signal 8:403–412

Richter JD, Bassell GJ, Klann E (2015) Dysregulation and restoration of translational homeostasis in fragile X syndrome. Nat Rev Neurosci 16:595–605

Robinson MB (1998) Examination of glutamate transporter heterogeneity using synaptosomal preparations. Methods Enzymol 296:189–202

Robinson MB, Jackson JG (2016) Astroglial glutamate transporters coordinate excitatory signaling and brain energetics. Neurochem Int 98:56–71

Rose EM, Koo JC, Antflick JE, Ahmed SM, Angers S, Hampson DR (2009) Glutamate transporter coupling to Na, K-ATPase. J Neurosci 29:8143–8155

Rothe F, Brosz M, Storm-Mathisen J (1994) Quantitative ultrastructural localization of glutamate dehydrogenase in the rat cerebellar cortex. Neuroscience 62:1133–1146

Rothe F, Brosz M, Storm-Mathisen J (1995) Quantitative ultrastructural localization of glutamate dehydrogenase in the rat cerebellar cortex. Neuroscience 64:iii–xvi

Salganicoff L, De Robertis E (1965) Subcellular distribution of the enzymes of the glutamic acid, glutamine and gamma-aminobutyric acid cycles in rat brain. J Neurochem 12:287–309

Sato K, Otsu W, Otsuka Y, Inaba M (2013) Modulatory roles of NHERF1 and NHERF2 in cell surface expression of the glutamate transporter GLAST. Biochem Biophys Res Commun 430:839–845

Scheschonka A, Tang Z, Betz H (2007) Sumoylation in neurons: nuclear and synaptic roles? Trends Neurosci 30:85–91

Schlicker C, Gertz M, Papatheodorou P, Kachholz B, Becker CF, Steegborn C (2008) Substrates and regulation mechanisms for the human mitochondrial sirtuins Sirt3 and Sirt5. J Mol Biol 382:790–801

Schliess F, Gorg B, Fischer R, Desjardins P, Bidmon HJ, Herrmann A, Butterworth RF, Zilles K, Haussinger D (2002) Ammonia induces MK-801-sensitive nitration and phosphorylation of protein tyrosine residues in rat astrocytes. FASEB J 16:739–741

Schousboe A (1981) Transport and metabolism of glutamate and GABA in neurons are glial cells. Int Rev Neurobiol 22:1–45

Schousboe A, Westergaard N, Sonnewald U, Petersen SB, Huang R, Peng L, Hertz L (1993) Glutamate and glutamine metabolism and compartmentation in astrocytes. Dev Neurosci 15:359–366

Schousboe A, Westergaard N, Waagepetersen HS, Larsson OM, Bakken IJ, Sonnewald U (1997) Trafficking between glia and neurons of TCA cycle intermediates and related metabolites. Glia 21:99–105

Schousboe A, Scafidi S, Bak LK, Waagepetersen HS, McKenna MC (2014) Glutamate metabolism in the brain focusing on astrocytes. Adv Neurobiol 11:13–30

Schweigert ID, de Oliveira DL, Scheibel F, da Costa F, Wofchuk ST, Souza DO, Perry ML (2005) Gestational and postnatal malnutrition affects sensitivity of young rats to picrotoxin and quinolinic acid and uptake of GABA by cortical and hippocampal slices. Brain Res Dev Brain Res 154:177–185

Shan D, Mount D, Moore S, Haroutunian V, Meador-Woodruff JH, McCullumsmith RE (2014) Abnormal partitioning of hexokinase 1 suggests disruption of a glutamate transport protein complex in schizophrenia. Schizophr Res 154:1–13

Shank RP, Bennett GS, Freytag SO, Campbell GL (1985) Pyruvate carboxylase: an astrocyte-specific enzyme implicated in the replenishment of amino acid neurotransmitter pools. Brain Res 329:364–367

Shashidharan P, Plaitakis A (2014) The discovery of human of GLUD2 glutamate dehydrogenase and its implications for cell function in health and disease. Neurochem Res 39:460–470

Shulkin BL, Betz AL, Koeppe RA, Agranoff BW (1995) Inhibition of neutral amino acid transport across the human blood-brain barrier by phenylalanine. J Neurochem 64:1252–1257

Skytt DM, Klawonn AM, Stridh MH, Pajecka K, Patruss Y, Quintana-Cabrera R, Bolanos JP, Schousboe A, Waagepetersen HS (2012) siRNA knock down of glutamate dehydrogenase in astrocytes affects glutamate metabolism leading to extensive accumulation of the neuroactive amino acids glutamate and aspartate. Neurochem Int 61:490–497

Smith TJ, Stanley CA (2008) Untangling the glutamate dehydrogenase allosteric nightmare. Trends Biochem Sci 33:557–564

Smith QR, Momma S, Aoyagi M, Rapoport SI (1987) Kinetics of neutral amino acid transport across the blood-brain barrier. J Neurochem 49:1651–1658

Smith TJ, Peterson PE, Schmidt T, Fang J, Stanley CA (2001) Structures of bovine glutamate dehydrogenase complexes elucidate the mechanism of purine regulation. J Mol Biol 307:707–720

Someya S, Yu W, Hallows WC, Xu J, Vann JM, Leeuwenburgh C, Tanokura M, Denu JM, Prolla TA (2010) Sirt3 mediates reduction of oxidative damage and prevention of age-related hearing loss under caloric restriction. Cell 143:802–812

Sonnewald U (2014) Glutamate synthesis has to be matched by its degradation—where do all the carbons go? J Neurochem 131:399–406
Sonnewald U, McKenna M (2002) Metabolic compartmentation in cortical synaptosomes: influence of glucose and preferential incorporation of endogenous glutamate into GABA. Neurochem Res 27:43–50
Sonnewald U, Westergaard N, Hassel B, Muller TB, Unsgard G, Fonnum F, Hertz L, Schousboe A, Petersen SB (1993a) NMR spectroscopic studies of 13C acetate and 13C glucose metabolism in neocortical astrocytes: evidence for mitochondrial heterogeneity. Dev Neurosci 15:351–358
Sonnewald U, Westergaard N, Schousboe A, Svendsen JS, Unsgard G, Petersen SB (1993b) Direct demonstration by [13C]NMR spectroscopy that glutamine from astrocytes is a precursor for GABA synthesis in neurons. Neurochem Int 22:19–29
Sonnewald U, White LR, Odegard E, Westergaard N, Bakken IJ, Aasly J, Unsgard G, Schousboe A (1996) MRS study of glutamate metabolism in cultured neurons/glia. Neurochem Res 21:987–993
Sonnewald U, Westergaard N, Schousboe A (1997) Glutamate transport and metabolism in astrocytes. Glia 21:56–63
Sonnewald U, Hertz L, Schousboe A (1998) Mitochondrial heterogeneity in the brain at the cellular level. J Cereb Blood Flow Metab 18:231–237
Spanaki C, Zaganas I, Kleopa KA, Plaitakis A (2010) Human GLUD2 glutamate dehydrogenase is expressed in neural and testicular supporting cells. J Biol Chem 285:16748–16756
Spanaki C, Zaganas I, Kounoupa Z, Plaitakis A (2012) The complex regulation of human glud1 and glud2 glutamate dehydrogenases and its implications in nerve tissue biology. Neurochem Int 61:470–481
Spanaki C, Kotzamani D, Petraki Z, Drakos E, Plaitakis A (2014) Heterogeneous cellular distribution of glutamate dehydrogenase in brain and in non-neural tissues. Neurochem Res 39:500–515
Spanaki C, Kotzamani D, Petraki Z, Drakos E, Plaitakis A (2015) Expression of human GLUD1 and GLUD2 glutamate dehydrogenases in steroid producing tissues. Mol Cell Endocrinol 415:1–11
Srere PA (1987) Complexes of sequential metabolic enzymes. Annu Rev Biochem 56:89–124
Srere PA, Sumegi B, Sherry AD (1987) Organizational aspects of the citric acid cycle. Biochem Soc Symp 54:173–178
Stanley CA (2004) Hyperinsulinism/hyperammonemia syndrome: insights into the regulatory role of glutamate dehydrogenase in ammonia metabolism. Mol Genet Metab 81(Suppl 1):S45–S51
Stanley CA (2011) Two genetic forms of hyperinsulinemic hypoglycemia caused by dysregulation of glutamate dehydrogenase. Neurochem Int 59(4):465–472
Sumegi B, Srere PA (1984) Complex I binds several mitochondrial NAD-coupled dehydrogenases. J Biol Chem 259:15040–15045
Sweatt AJ, Garcia-Espinosa MA, Wallin R, Hutson SM (2004a) Branched-chain amino acids and neurotransmitter metabolism: expression of cytosolic branched-chain aminotransferase (BCATc) in the cerebellum and hippocampus. J Comp Neurol 477:360–370
Sweatt AJ, Wood M, Suryawan A, Wallin R, Willingham MC, Hutson SM (2004b) Branched-chain amino acid catabolism: unique segregation of pathway enzymes in organ systems and peripheral nerves. Am J Physiol Endocrinol Metab 286:E64–E76
Teller JK, Fahien LA, Valdivia E (1990) Interactions among mitochondrial aspartate aminotransferase, malate dehydrogenase, and the inner mitochondrial membrane from heart, hepatoma, and liver. J Biol Chem 265:19486–19494
Ugbode CI, Hirst WD, Rattray M (2014) Neuronal influences are necessary to produce mitochondrial co-localization with glutamate transporters in astrocytes. J Neurochem 130:668–677
Ulrich HD (2008) The fast-growing business of SUMO chains. Mol Cell 32:301–305
Verdin E, Hirschey MD, Finley LW, Haigis MC (2010) Sirtuin regulation of mitochondria: energy production, apoptosis, and signaling. Trends Biochem Sci 35:669–675

Waagepetersen HS, Bakken IJ, Larsson OM, Sonnewald U, Schousboe A (1998) Comparison of lactate and glucose metabolism in cultured neocortical neurons and astrocytes using 13C-NMR spectroscopy. Dev Neurosci 20:310–320
Waagepetersen HS, Sonnewald U, Larsson OM, Schousboe A (1999) Synthesis of vesicular GABA from glutamine involves TCA cycle metabolism in neocortical neurons. J Neurosci Res 57:342–349
Waagepetersen HS, Sonnewald U, Larsson OM, Schousboe A (2000) A possible role of alanine for ammonia transfer between astrocytes and glutamatergic neurons. J Neurochem 75:471–479
Wagner GR, Payne RM (2011) Mitochondrial acetylation and diseases of aging. J Aging Res 2011:234875. doi:10.4061/2011/234875
Wang M, Li S, Zhang H, Pei L, Zou S, Lee FJ, Wang YT, Liu F (2012) Direct interaction between GluR2 and GAPDH regulates AMPAR-mediated excitotoxicity. Mol Brain 5:13
Westergaard N, Sonnewald U, Petersen SB, Schousboe A (1995a) Glutamate and glutamine metabolism in cultured GABAergic neurons studied by 13C NMR spectroscopy may indicate compartmentation and mitochondrial heterogeneity. Neurosci Lett 185:24–28
Westergaard N, Sonnewald U, Schousboe A (1995b) Metabolic trafficking between neurons and astrocytes: the glutamate/glutamine cycle revisited. Dev Neurosci 17:203–211
Westergaard N, Drejer J, Schousboe A, Sonnewald U (1996) Evaluation of the importance of transamination versus deamination in astrocytic metabolism of [U-13C]glutamate. Glia 17:160–168
Whitelaw BS, Robinson MB (2013) Inhibitors of glutamate dehydrogenase block sodium-dependent glutamate uptake in rat brain membranes. Front Endocrinol (Lausanne) 4:123
Wysmyk-Cybula U, Faff-Michalak L, Albrecht J (1991) Effects of acute hepatic encephalopathy and in vitro treatment with ammonia on glutamate oxidation in bulk-isolated astrocytes and mitochondria of the rat brain. Acta Neurobiol Exp (Wars) 51:165–169
Yang H, Zhou L, Shi Q, Zhao Y, Lin H, Zhang M, Zhao S, Yang Y, Ling ZQ, Guan KL, Xiong Y, Ye D (2015) SIRT3-dependent GOT2 acetylation status affects the malate-aspartate NADH shuttle activity and pancreatic tumor growth. EMBO J 34(8):1110–1125
Yu AC, Drejer J, Hertz L, Schousboe A (1983) Pyruvate carboxylase activity in primary cultures of astrocytes and neurons. J Neurochem 41:1484–1487
Yu W, Dittenhafer-Reed KE, Denu JM (2012) SIRT3 protein deacetylates isocitrate dehydrogenase 2 (IDH2) and regulates mitochondrial redox status. J Biol Chem 287:14078–14086
Yu J, Sadhukhan S, Noriega LG, Moullan N, He B, Weiss RS, Lin H, Schoonjans K, Auwerx J (2013) Metabolic characterization of a Sirt5 deficient mouse model. Sci Rep 3:2806
Yudkoff M (1997) Brain metabolism of branched-chain amino acids. Glia 21:92–98
Yudkoff M, Nissim I, Hummeler K, Medow M, Pleasure D (1986) Utilization of [15N]glutamate by cultured astrocytes. Biochem J 234:185–192
Yudkoff M, Nissim I, Pleasure D (1988) Astrocyte metabolism of [15N]glutamine: implications for the glutamine-glutamate cycle. J Neurochem 51:843–850
Yudkoff M, Zaleska MM, Nissim I, Nelson D, Erecinska M (1989) Neuronal glutamine utilization: pathways of nitrogen transfer studied with [15N]glutamine. J Neurochem 53:632–640
Yudkoff M, Nissim I, Hertz L (1990) Precursors of glutamic acid nitrogen in primary neuronal cultures: studies with 15N. Neurochem Res 15:1191–1196
Yudkoff M, Nissim I, Nelson D, Lin ZP, Erecinska M (1991) Glutamate dehydrogenase reaction as a source of glutamic acid in synaptosomes. J Neurochem 57:153–160
Yudkoff M, Nissim I, Daikhin Y, Lin ZP, Nelson D, Pleasure D, Erecinska M (1993) Brain glutamate metabolism: neuronal-astroglial relationships. Dev Neurosci 15:343–350
Yudkoff M, Daikhin Y, Grunstein L, Nissim I, Stern J, Pleasure D, Nissim I (1996) Astrocyte leucine metabolism: significance of branched-chain amino acid transamination. J Neurochem 66:378–385
Yudkoff M, Daikhin Y, Nissim I (2000) Acidosis and astrocyte amino acid metabolism. Neurochem Int 36:329–339

Yudkoff M, Daikhin Y, Nissim I, Horyn O, Lazarow A, Luhovyy B, Wehrli S (2005) Response of brain amino acid metabolism to ketosis. Neurochem Int 47:119–128

Zaganas I, Kanavouras K, Mastorodemos V, Latsoudis H, Spanaki C, Plaitakis A (2009) The human GLUD2 glutamate dehydrogenase: localization and functional aspects. Neurochem Int 55:52–63

Zaganas I, Spanaki C, Plaitakis A (2012) Expression of human GLUD2 glutamate dehydrogenase in human tissues: functional implications. Neurochem Int 61:455–462

Zhai D, Li S, Wang M, Chin K, Liu F (2013) Disruption of the GluR2/GAPDH complex protects against ischemia-induced neuronal damage. Neurobiol Dis 54:392–403

Zhang Z, Tan M, Xie Z, Dai L, Chen Y, Zhao Y (2011) Identification of lysine succinylation as a new post-translational modification. Nat Chem Biol 7:58–63

Zhang J, Wang G, Mao Q, Li S, Xiong W, Lin Y, Ge J (2016) Glutamate dehydrogenase (GDH) regulates bioenergetics and redox homeostasis in human glioma. Oncotarget. doi: 10.18632/oncotarget.7657

Zhao S, Xu W, Jiang W, Yu W, Lin Y, Zhang T, Yao J, Zhou L, Zeng Y, Li H, Li Y, Shi J, An W, Hancock SM, He F, Qin L, Chin J, Yang P, Chen X, Lei Q, Xiong Y, Guan KL (2010) Regulation of cellular metabolism by protein lysine acetylation. Science 327:1000–1004

Chapter 5
BCAA Metabolism and NH_3 Homeostasis

M.E. Conway and S.M. Hutson

Abstract The branched chain amino acids (BCAA) are essential amino acids required not only for growth and development, but also as nutrient signals and as nitrogen donors to neurotransmitter synthesis and glutamate/glutamine cycling. Transamination and oxidative decarboxylation of the BCAAs are catalysed by the branched-chain aminotransferase proteins (BCATm, mitochondrial and BCATc, cytosolic) and the branched-chain α-keto acid dehydrogenase enzyme complex (BCKDC), respectively. These proteins show tissue, cell compartmentation, and protein–protein interactions, which call for substrate shuttling or channelling and nitrogen transfer for oxidation to occur. Efficient regulation of these pathways is mediated through the redox environment and phosphorylation in response to dietary and hormonal stimuli. The wide distribution of these proteins allows for effective BCAA utilisation. We discuss how BCAT, BCKDC, and glutamate dehydrogenase operate in supramolecular complexes, allowing for efficient channelling of substrates. The role of BCAAs in brain metabolism is highlighted in rodent and human brain, where differential expression of BCATm indicates differences in nitrogen metabolism between species. Finally, we introduce a new role for BCAT, where a change in function is triggered by oxidation of its redox-active switch. Our understanding of how BCAA metabolism and nitrogen transfer is regulated is important as many studies now point to BCAA metabolic dysregulation in metabolic and neurodegenerative conditions.

Keywords BCAT • BCAAs • Glutamate • Nitrogen shuttling • Metabolon • Redox-switch • MSUD

M.E. Conway (✉)
Department of Applied Sciences, University of the West of England,
Coldharbour Lane, Bristol BS16 1QY, UK
e-mail: myra.conway@uwe.ac.uk

S.M. Hutson
Department of Human Nutrition, Foods and Exercise, Integrated Life Sciences Building (0913), 1981 Kraft Drive, Blacksburg, VA 24060, USA
e-mail: susanh5@vt.edu

A. Schousboe, U. Sonnewald (eds.), *The Glutamate/GABA-Glutamine Cycle*,
Advances in Neurobiology 13, DOI 10.1007/978-3-319-45096-4_5

Abbreviations

AGC	Malate/aspartate carrier
ALT	Alanine aminotransferase
BCAA	Branched chain amino acids
BCAT	Branched chain aminotransferase
BCKA	Branched chain α-keto acid
BCKDC	Branched-chain α-keto acid dehydrogenase enzyme complex
E1	Branched chain α-keto acid dehydrogenase
E2	Dihydrolpoyl transacylase subunits
E3	Dihydrolipoyl dehydrogenase
GABA	γ-Amino butyric acid
GDH	Glutamate dehydrogenase
GS	Glutamine synthetase
GSNO	S-nitrosoglutathione
KG	α-Ketoglutarate
KIC	α-Ketoisocaproate
KIV	α-Ketoisovalerate
KMV	α-Keto-β-methylvalerate
MDH	Malate dehydrogenase
ME	Malic enzyme
MSUD	Maple syrup urine disease
mTOR	The mechanistic target of rapamycin
PLP	Pyridoxal phosphate
PMP	Pyridoxamine
TPP	Thiamine pyrophosphate

5.1 Introduction

The branched chain amino acids (BCAA) (leucine, isoleucine, and valine) are three of the nine nutritionally essential amino acids that cannot be synthesized endogenously and must therefore be obtained through a dietary source. Dietary BCAA are required for normal growth and development (Hutson et al. 2005). These amino acids play a key role as nitrogen donors, for the major nitrogen carriers, alanine and glutamine, where nitrogen is shuttled between muscle and liver. In the brain, leucine is an important nitrogen donor for the synthesis of the neurotransmitter glutamate and the inhibitory neurotransmitter γ-amino butyric acid (GABA). Leucine also plays a role as an anabolic nutrient signal regulating insulin secretion and protein synthesis through the mechanistic target of rapamycin (mTOR) pathway in skeletal muscle (Hutson et al. 1980). However, toxic levels of BCAAs and the α-keto acids observed in conditions such as Maple Syrup Urine Disease (MSUD) highlight the importance of regulating their metabolism (Chuang et al. 2006). Control of BCAA transport and oxidation is governed by the cell and tissue expression of the branched-chain aminotransferase (BCAT) proteins, the branched-chain α-keto acid dehydrogenase enzyme complex (BCKDC), glutamate dehydrogenase (GDH), and the redox state of the cell coupled with phosphorylation/dephosphorylation of branched chain α–keto dehydrogenase enzyme (E1) of BCKDC.

The BCAT proteins catalyze the transamination of the BCAA, leucine, isoleucine, and valine with α-ketoglutarate (α-KG) releasing their respective branched chain α-keto acids (BCKA: α-ketoisocaproate (KIC), α-keto-β-methylvalerate (KMV), and α-ketoisovalerate (KIV)) and glutamate, regenerating the pyridoxal phosphate (PLP) form of the enzyme (Ichihara 1975). The second step in BCAA catabolism is catalyzed by the BCKDC in which irreversible oxidative decarboxylation of the BCKA is the committed step in oxidation of the BCAAs (Harris et al. 2004). Glutamate dehydrogenase, a mitochondrial matrix enzyme, is responsible for the oxidative deamination of glutamate to α-KG and free NH_4^+, using either NAD^+ or $NADP^+$ as a co-factor. Whole body tissue expression and compartmentation of these proteins drives intra- and inter-organ exchange of nitrogen and carbon. In addition to being the major excitatory neurotransmitter, glutamate is also the substrate for Glutamine Synthetase which forms glutamine. These two amino acids are part of the Glutamine/Glutamate Cycle. BCAA are not transaminated in liver hepatocytes and transamination exceeds oxidation in skeletal muscle, which accounts for 40 % of total body weight and is one of the main sites of BCAA uptake. Because skeletal muscle transaminase activity exceeds the oxidative capacity, BCAAs act as nitrogen donors in peripheral metabolism. In the central nervous system, transamination and oxidation also exhibit cell-specific localization and participate in nitrogen and carbon shuttles. In this chapter, we discuss the rodent BCAA-dependent astroglial-neuronal nitrogen shuttle and how the absence of BCATm from astrocytes in human brain indicates distinct differences between species.

BCATm and BCKDC and BCATm and GDH also form metabolons, which facilitate effective substrate channelling (Islam et al. 2010; Hutson et al. 2011). These interactions are regulated through changes in the redox environment, where a reduced environment favours binding and oxidation prevents metabolon formation (Islam et al. 2007). The unique redox-active CXXC motif of the mammalian BCAT becomes oxidised and transamination is terminated. We discuss the potential for this redox switch to behave as a signal for change in function from a metabolic role to operate as a redox chaperone. How this would impact BCAA catabolism and nitrogen transfer remains to be elucidated, but is important to answer, given the role of these amino acids in nutrient cell signalling and protein synthesis.

5.2 Transamination and Oxidative Decarboxylation of the BCAA

The BCAT isozymes are encoded by BCAT1 (located on the short arm of chromosome 12) and BCAT2 (located in chromosome 19) (Hall et al. 1993; Bledsoe et al. 1997). The resulting (predominant) isoforms include a cytosolic isozyme, BCATc, and a mitochondrial isozyme, BCATm, which show both cell and tissue compartmentalization. These proteins operate as homodimers with a molecular mass of 44,650 Da and 41,730, respectively, and are structurally very similar, where they share 58 % sequence homology (Yennawar et al. 2002, 2006; Goto et al. 2005). However, they possess distinct differences in catalytic efficiency and regulation (Davoodi et al. 1998; Conway and Hutson 2000). The mitochondrial isozyme is ubiquitous and found in most tissues, whereas the cytosolic isozyme is limited to the brain, peripheral nervous system, ovaries, and testes (Suryawan et al. 1998; Hall et al. 1993; Hutson 1988; Hutson et al. 1998; Sweatt et al. 2004a, b; García-Espinosa et al. 2007). Although these two isozymes predominate, two other splice variants have been identified, namely, a novel alternatively spliced PP18b variant found in placental tissue and a novel co-repressor for thyroid hormone nuclear receptors (P3) (Lin et al. 2001; Than et al. 2001). Both spliced variants are homologous to BCATm. Although the function of the PP18b variant is unknown, P3 was reported to act as a co-repressor for thyroid hormone nuclear receptors. The biological significance of these variants remains to be determined, but may have significance under pathological conditions such as Alzheimer's disease (Hull et al. 2015). Therefore, based on existing evidence, with the exception of the brain and peripheral nervous system, BCAA metabolism occurs in the mitochondria.

The mammalian BCAT proteins are PLP-dependent enzymes that belong to the fold-type IV class of aminotransferase proteins where the proton is abstracted from the C4 atom of the coenzyme-imine or external aldimine on the Re face instead of the Si face of the PLP cofactor (Yoshimura et al. 1996). This is unlike most aminotransferases, which fall into the fold type I family type I (Jansonius 1998; Schneider et al. 2000). Other key aminotransferases include the alanine aminotransferase (ALT) [glutamate pyruvate transaminase or α-keto-glutarate E.C. 2.6.1.2] and the aspartate aminotransferase proteins (AST) [glutamic oxaloacetic transaminase or L-aspartate: α-keto-glutarate aminotransferase, E.C. 2.6.1.1] (Mehta and Christen 2000; Salzmann et al. 2000). These enzymes follow the same basic transamination reaction and exhibit ping-pong kinetic mechanisms that have two half-reactions consisting of three main stages (Fig. 5.1). This reversible transamination step is initiated in the first half-reaction, where the PLP form of BCAT reacts with the α-amino group of the BCAAs. Here, the PLP cofactor is bound to BCAT by an internal aldimine, with a Schiff base between the active site lysine and the PLP cofactor. As the BCAA enters the active site, a transaldimation reaction occurs, where the substrate amino group displaces the amino groups of the active site lysine, forming an external aldimine between the BCAA and the PLP cofactor. This external aldimine is subsequently converted into its ketamine intermediate, where lysine acts as the catalytic base. The first-half reaction is completed with the hydrolysis of the ketamine to give the pyridoxamine phosphate (PMP) form of the enzyme and the respective BCKA. In the second half-reaction, the PMP cofactor binds to α-KG following a reversal of the three main stages. Here, the net result is that the PLP-form of BCAT is regenerated together with the release of glutamate. Transamination in the reverse direction results in the BCKAs receiving the amino group of glutamate to generate BCAAs and α-KG. Although changes in the redox environment do not directly play a role in catalysis, oxidation of the human BCAT proteins inacti-

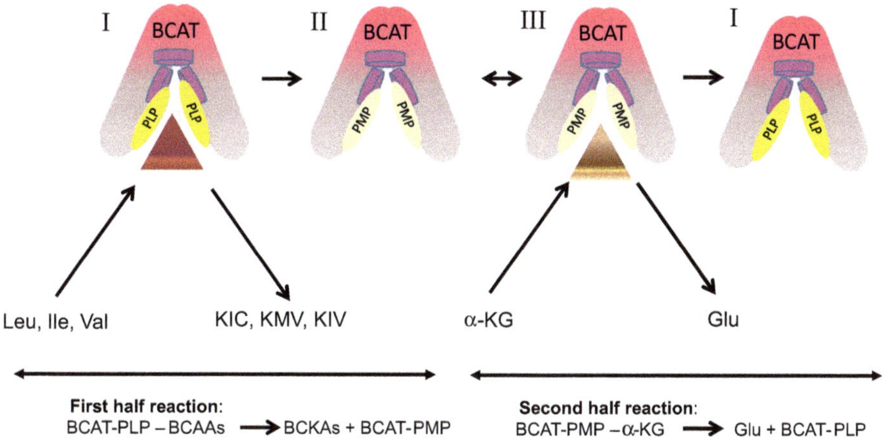

Fig. 5.1 Ping-pong kinetics of the BCAT protein. Each half reaction is divided into three stages. (*I*) Interaction of BCAT-PLP with their respective substrates involves the nucleophillic attack of the α-amino group of amino acid 1 (e.g. isoleucine) with the BCAT-PLP Schiff base carbon atom to form an amino acid-PLP Schiff base (external aldimine) with release of the BCAT. (*II*) After transamination the second step is keto-enol tautomerism which involves the interconversion between the keto-enol form resulting in the formation of an α-keto acid-PMP Schiff base *III*. This is subsequently hydrolysed to PMP and an α-keto acid, the final step in the first half-reaction

vates the enzymes disrupting transamination (discussed later). This redox switch may offer a control point during conditions where oxidative stress prevails.

The BCAT proteins show substrate specificity for the BCAA, with substrate preferences for isoleucine ≥ leucine > valine ⋙ glutamate (Wallin et al. 1990; Hall et al. 1993; Davoodi et al. 1998). Moreover, the kinetic data indicates that the BCKAs and glutamate are the favoured products of the BCAT reaction. In vivo studies using ^{15}N-labeled leucine showed that the majority the ^{15}N was found in glutamate and other amino acids formed through transamination with glutamate, such as alanine and aspartate (Forslund et al. 1998; Kanamori et al. 1998). Cycling of BCAA nitrogen and labelling of the amino acid pool can occur through transamination and reamination. Transfer of nitrogen from the BCAA between tissues happens when BCKAs are released from one tissue, or cells in a tissue and reaminated in other tissues or different cells within a tissue. However, only oxidation and removal of the carbon skeleton will result in a net transfer of BCAA nitrogen to body dispensable amino acids. The widespread expression of BCATs in cells and tissues promotes nitrogen cycling, transfer, and net transfer.

As described, the net transfer of BCAA nitrogen into the nonessential amino acid pool requires oxidation of the BCAA carbon skeleton. Complete oxidation is catalysed by the BCKDC complex (4.5 MDa) located in the mitochondrial matrix and associated with inner mitochondrial membrane (reviewed in Wynn et al. 2004). The activity of the BCKDC determines the rate of oxidation of the BCAA (irreversible loss), which is equivalent to the rate of net nitrogen transfer from BCAAs to glutamate. The BCKDC shares significant functional homology to the pyruvate dehydrogenase enzyme complex and the α-ketoglutarate dehydrogenase enzyme complex (Reed et al. 1985). It is composed of multiple copies of three enzyme proteins including the branched chain α-keto acid dehydrogenase [E1, heterotetramer of 2E1α and 2E1β subunits] (12 copies) and dihydrolipoyl dehydrogenase [E3] (six copies), which non-covalently surround a core of 24 dihydrolpoyl transacylase enzymes [E2 homodimer]. Also, at the N-terminal end of E2 is a flexible lipoyl moiety, which manoeuvres between the active sites of the enzymes and an inner core domain at the COOH-terminal, important for oligomeric core formation and catalysis of the acyltransferase reaction. For catalysis, several cofactors are required including thiamine pyrophosphate (E1), Coenzyme A (E2), lipoamide, FAD, and NAD (E3) and activity is tightly regulated by phosphorylation and dephosphorylation at Ser 293 of the E1α subunit of the E1 enzyme (Harris et al. 1990; Popov et al. 1992).

Oxidative decarboxylation of the α-keto acid (e.g. KIC) is initiated when the E1β subunit binds thiamine pyrophosphate and catalyses decarboxylation forming 2-methylpropanol-TPP (Wynn et al. 2004). TPP is regenerated when 2-methylpropanol-TPP is oxidised to form an acyl group, while it is simultaneously transferred to the lipoyl cofactor covalently attached to E2. The flexible lipoyl arm carrying the S-acyldihydrolipoamide now swings to E2 where acyl transfer to Coenzyme A forming acyl-CoA occurs. Finally, oxidation of the dihydrolipoyl moiety on E2 occurs via the FAD bound to the E3 enzyme, where NAD^+ acts as the ultimate electron acceptor. The end products of BCKDC-mediated catalysis include isovaleryl-CoA, α-methylbutyryl-CoA, and isobutryl-CoA that undergo further enzymatic reactions and either enter the Krebs cycle or act as precursors for lipogenesis (Ile and Leu). Therefore, the overall end products of BCAA catabolism are acetoacetate and succinyl-CoA (Summarized in Fig. 5.2). Mutations in one or more proteins in the BCKDC result in the inborn error of metabolism, MSUD, characterised by elevated plasma BCAAs and BCKAs (Chuang and Chuang 2000; Chuang et al. 2006).

5.3 Peripheral BCAA Metabolism: Inter-Organ Nitrogen Shuttling

Whole body distribution of the BCAT proteins was first described in rats, followed by human and nonhuman primates (Shinnick and Harper 1976; Hall et al. 1993; Suryawan et al. 1998). In rat, the mitochondrial isoform is expressed in most tissues, but its expression tends to be cell-specific (e.g. in astrocytes in brain, acinar cells of the pancreas, etc.) (Sweatt et al. 2004a, b; Cole et al. 2012). Expression of BCATc is, however, limited to brain (neuronal) and the peripheral nervous system, ovary, and placenta (Hall et al. 1993; García-Espinosa et al. 2007; Sweatt et al. 2004a, b; Hull et al. 2012). Neither BCAT was reported in rat liver, indicating that transamination occurs largely outside the liver and it is not in liver hepatocytes in primates. On the other hand, rat BCKDC shows high activity in the liver, low in skeletal muscle and moderate levels in the kidney and heart, indicating that transamination and oxidative decarboxylation differ between tissues (Goodwin and Harris et al. 1988; Harris et al. 2004). In rat skeletal muscle, however, the ratio of

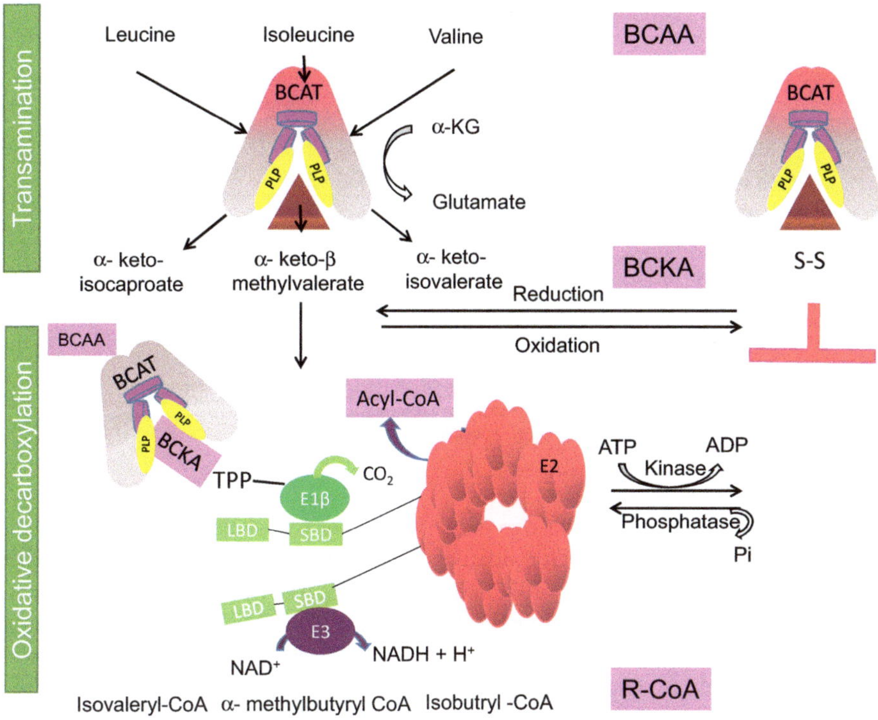

Fig. 5.2 BCAA catabolism. BCAT catalyze the transamination of the BCAA, leucine, isoleucine, and valine to α-ketoglutarate (α-KG) releasing their respective branched chain α-keto acids (BCKAs: α-ketoisocaproate (KIC), α-keto-β-methylvalerate (KMV) and α-ketoisovalerate (KIV)) and glutamate, regenerating the enzyme. The committed step is catalyzed by the second enzyme in the catabolic pathway, the branched-chain α-keto acid dehydrogenase complex (BCKDC) where the respective BCKAs undergo oxidative decarboxylation to generate Isovaleryl-CoA, α-methylbutryl CoA, and Isobutryl-CoA, respectively. Transamination is regulated through a peroxide-sensitive redox switch and oxidation is regulation through phosphorylation/dephosphorylation of the BCKDC

BCAT activity to actual BCKDC activity is ≥90 % and ≥25 % when you take into account the total potential BCKD activity in its unphosphorylated form, i.e. the oxidative capacity. Expression of BCAT and BCKDC in humans showed that although the ratio of BCAT to BCKDC expression in each tissue was comparable with rat, the activity (Table 5.1) and expression pattern showed differences, indicating that transaminase and oxidative capacity is regulated by specific pathways in each species. In humans, overall measured transaminase activity and oxidative capacity, factoring in tissue size, was greatest in muscle; however, high activity reported for brain and kidney indicated that these two organs may also contribute to whole body BCAA metabolism and nitrogen shuttling (Suryawan et al. 1998). Still liver can handle BCAA oxidation in humans (see MSUD section). Although not measured in these studies, based on primate analysis, it is anticipated that the human pancreas will also show high levels of BCAT and BCKDC.

Table 5.1 Distribution of BCAT activity in rat, human, and monkey tissues[a]

Tissue	Rat	Human	Monkey
mU/g wet weight			
Heart	4894±216	387±216	–
Muscle	1599±60	124±216	245±216
Brain	1944±94	510±216	434±216
liver	78±5	248±216	250±216
Kidney	3486±142	880±216	1215±216
Pancreas	11,088±1187	–	1790±216
Stomach	5842±415	447±216	559±216
Small intestine	489±22	241±216	383±216
Colon	894±37	254±216	–
Adipose	166±16	84±216	98±216

[a]Data are from Suryawan et al. (1998)

The high ratio of transaminase to oxidative capacity in skeletal muscle of the rat and absence of BCAT and high oxidative capacity in liver led to the prediction of extensive inter-organ shuttling of BCAA metabolites, first postulated by Harper and collaborators (Shinnick and Harper 1976; Hutson et al. 1978; Harper and Benjamin 1984). In rat hepatocytes, the absence of the BCATm isozyme and high activity of BCKDC favour liver oxidation of circulating BCKAs (Suryawan et al. 1998). This would account for the low concentrations of BCKAs found in liver. The liver is the first organ that receives dietary amino acids, where essential amino acids, with exception to the BCAA, undergo irreversible degradation primarily in this organ. As described, in humans, skeletal muscle is the main site for BCAA transamination, generating BCKAs and glutamate. Although skeletal muscle due to its mass also reflected the highest relative value of potential oxidative capacity compared to other sites measured, transaminase activity exceeds the oxidative capacity, indicating that the fate of glutamate and the BCKAs are not coordinated in this tissue. It has been established that BCAA provide nitrogen via glutamate for glutamine and NH_4 production, as well as alanine synthesis (Harper and Benjamin 1984; Elia and Livesey 1983). Glutamine and alanine are major sources of fuel during gluconeogenesis and they are released from muscle and are taken up by the liver, kidney, and small intestine. Alanine and glutamine also are major mechanisms for transferring nitrogen from peripheral amino acid oxidation to liver for urea synthesis. Liver, intestine, and skeletal muscle are the major organs affecting blood glutamine concentrations. Supplementation studies with BCAA have demonstrated that the concentrations of NH_3 and glutamine increase during exercise; however, performance was suggested to be affected, possibly due to a compromised TCA cycle (van Hall et al. 1995a, b). The role of BCAA metabolism in shuttling nitrogen is ultimately linked to the fate of glutamate. The fate of glutamate can be to exit the cell, act as a precursor for alanine or glutamine synthesis, nitrogen donor to other transaminases, or undergo oxidation by GDH. Mitochondrially generated Glu must first exit the mitochondria via the glutamate/hydroxyl carrier, which depletes TCA cycle α-KG. This would need to be replenished through anaplerosis or amino acid oxidation from protein breakdown. Anaplerosis through pyruvate carboxylase would be a candidate, but it is not abundant in muscle, but is high in brain and liver. Oxidation of glutamate by GDH would generate α-KG and NH_4 and not cause loss of TCA cycle intermediates. GDH1 is found at varying levels in tissues. Coupling of BCAA metabolism to GDH activity would generate the needed NH_4 for glutamine synthesis, without compromising the TCA cycle. Thus, the high ratios of BCAT/BCKDC activities favour (1) release of BCKAs rather than their oxidation and (2) efficient transfer of BCAA nitrogen. This led to the hypothesis that BCAAs play an important role in body nitrogen metabolism, particularly as nitrogen donors for the major nitrogen carriers alanine and glutamine (Odessey and Goldberg 1979; Yudkoff et al. 1994).

Distribution studies in rat were extended to tissues of the digestive tract, which showed BCATc was expressed in peripheral nerves, without the expression of other BCAA catabolic enzymes (Sweatt et al. 2004b). Moreover, the expression of BCATm was pronounced throughout the secretory epithelial cells that secrete acid, enzymes, or enzyme precursors rather than in the absorptive epithelial cells of the intestinal mucosa. This would indicate that during absorption leucine oxidation would be limited, which correlates with previous studies indicating that although transaminase activity was measured in these areas, oxidative capacity was low (Suryawan et al. 1998). The high immunopositive staining of BCATm and BCKDC found in the pancreatic acini rather than the islets was suggested to serve as a regulatory mechanism for leucine signalling in the islets, allowing leucine to activate GDH (Sweatt et al. 2004a, b). Low expression in the islets has been proposed to promote activation of GDH1, related to insulin secretion (Cole et al. 2012). A study by Zhou et al. in 2010 demonstrated that KIC, but not leucine-induced insulin secretion, was abolished in mice deficient in BCATm. The increase in ATP and $NADH/NAD^+$ ratios normally reported with KIC stimulation was reduced in these deficient mice, and malate concentrations were also attenuated. These studies indicate that transamination in the direction of leucine production is important for insulin release when KIC is the substrate provided to the β-cells, where a role for leucine in stimulating GDH activity, rather than transamination, is important. Recent studies detailing BCATm-induced enhancement of GDH1 metabolism indicate that this reaction involves several metabolic proteins, described in detail later in this chapter (Islam et al. 2010).

The compartmentalization of BCAT isozymes and BCKDC within organs and tissues demonstrates that, for complete oxidation to occur, expression of BCAT with BCKDC is required, but not necessarily in the same tissue or cell. Skeletal muscle clearly has the highest transaminase capacity, which in both rat and human exceeds the oxidative capacity, supporting the role for BCAA as nitrogen donors in peripheral metabolism. Although the expression of BCKDC is high in rat liver, highlighting a role for liver as the main site for BCKA oxidation, the relatively lower levels reported in human does question if the fate of the BCKAs is the same across species. Nevertheless, liver transplantation is being used to treat MSUD, so the liver can serve as the major site of BCAA oxidation in individuals with a defect in BCKDC, hence BCAA oxidation. However, what is certain is that both alanine and glutamine are metabolized in the liver and kidney, in particular during gluconeogenesis.

This concept of nitrogen and BCAA carbon shuttling originally described in peripheral tissues was extended to the Central Nervous System where it has been studied extensively by a number of laboratories including those of Yudkoff and co-workers and Hutson and co-workers. It is generally accepted that BCAAs donate nitrogen to brain glutamate neurotransmitter and are involved in regulating brain glutamate levels.

5.4 Brain Metabolism of the BCAA and the BCAA/BCKA Nitrogen Shuttle (Rodent Model)

In the brain, glutamate is the major excitatory neurotransmitter, a precursor for glutamine as well as the precursor for the synthesis of the major inhibitory neurotransmitter, GABA. Glutamate cannot easily traverse the blood brain barrier, which is important because excess accumulation of glutamate can induce excitotoxicity, leading to neuronal damage (Danbolt 2001). The glutamate/glutamine cycle is the predominant pathway which regulates neuronal glutamate stores (Berl and Clarke 1983; Yudkoff et al. 1993). Replenishment of neuronal glutamate is important because, during excitatory neurotransmission, not all glutamate released from the pre-synaptic neuron is recovered. Under normal excitatory conditions, excess glutamate not involved in transmission is taken up by astrocytes, which express high concentrations of the glutamate-specific transporters (GLAST/EAAT and GLT1/EAAT2) (Chaudhry et al. 1995; Lehre et al. 1995; Rothstein et al. 1994). Astrocytes specifically express the microsomal enzyme glutamine synthetase (GS), which catalyses the conversion of glutamate to glutamine through ATP-dependent amidation. As described for the peripheral generation of glutamine in skeletal muscle, this reaction requires NH_3, supplied from blood or brain metabolism. Glutamine is a non-neuroactive amino acid and can be released to the extracellular fluid for subsequent uptake by the pre-synaptic neuronal cells. Deamidation of glutamine by mitochondrial phosphate-dependent glutaminase regenerates neuronal glutamate, closing the cycle (Yudkoff et al. 2005). Cycling of glutamate between neuronal cells and astrocytes supports the provision of high levels of glutamate in neuronal cells and prevents toxicity in the synaptic space. However, metabolic studies using cultured rat neonatal astrocytes showed that up to 30 % of glutamate removed from the synaptic space by astrocytes was not converted to glutamine (Sonnewald et al. 1993; Hutson et al. 1998). Instead, these studies illustrated that glutamate was transaminated to α-KG (or oxidatively deaminated by GDH) oxidised for energy, especially when glutamate concentrations are high (McKenna et al. 1996a, b). Further metabolism to malate and oxaloacetate and decarboxylation to pyruvate will depend on substrate availability and the distribution and regulation of key metabolic enzymes including Malate dehydrogenase (MDH), Malic enzyme (ME), and pyruvate carboxylase (PC) or whether the source of glutamate is exogenous or endogenous (Shank et al. 1985; Teller et al. 1990; McKenna et al. 1993). For example, when levels of external glutamate are low, the glial glutamate synthetase pathway is favoured, whereas in glutamate excess considerable oxidation occurs (McKenna et al. 1996a). Glutamate is also a source of carbon for the production of purines and reducing equivalents such as glutathione (Yudkoff 1997). The fate of glutamate other than regeneration through the glutamate/glutamine cycle highlighted a gap in our metabolic understanding of how this 'lost' glutamate carbon could be replaced.

Pyruvate carboxylase, expressed solely in astrocytes, can replenish this lost glutamate, but is dependent on the supply of nitrogen (Gamberino et al. 1997; Hutson et al. 1998). As NH_3 was not considered to contribute to this reaction, other pathways catalysed by the transaminases were suggested, where the BCAA and aspartate serve as potential nitrogen donors (Shank et al. 1985; Brookes 1993; Hutson et al. 1998, 2001; Lieth et al. 2001; Bixel and Hamprecht 1995; Yudkoff et al. 1996a, b, Yudkoff 1997; Kanamori et al. 1998). Several lines of evidence support the role of BCAA as a nitrogen donor in rat brain metabolism. First, they can easily traverse the blood brain barrier, particularly leucine, where a substantial uptake by brain of BCAAs has been reported (Oldendorf 1973; Smith et al. 1987). In rat brain slices, rather than simply providing energy, their metabolism was faster than their incorporation into proteins (Chaplin et al. 1976). As described for skeletal muscle, the overall BCAT activity exceeds that of BCKDC catalysed oxidation in rat brain (≥50). In rat brain, reflected through metabolic studies, transamination of leucine is preferred over complete oxidation. By measuring the incorporation of [^{15}N]leucine into glutamate, these studies illustrated that approximately 30 % of the nitrogen of glutamate/glutamine was derived from leucine alone (Brand 1981; Brand and Hauschildt 1984). In another study, 9 h of continuous intra-gastric feeding in rats estimated that approximately 50 % of brain glutamate was derived from leucine with significant reamination of α-KIC (Sakai et al. 2004). In rat retina models, gabapentin, an inhibitor of BCATc, impacted glutamate synthesis by 30 % in retina cells, also supporting a role for BCAT metabolism in brain glutamate regulation (Lieth et al. 2001; LaNoue et al. 2001). Thus, these studies among others support the flow of metabolites between neuronal and astrocytic cells, complementing the glutamate/glutamine cycle (Yudkoff et al. 1996b; Yudkoff 1997; Daikhin and Yudkoff 2000). As the BCAT and BCKDC activities in human brain are second to muscle, it indicates that this tissue could be an important site of BCAA metabolism.

The role of BCAA metabolism in the CNS was supported further through enzyme distribution studies of rat brain and led to the development of the BCAA/BCKA nitrogen shuttle hypothesis (Fig. 5.3). Early studies showed that BCATc mRNA was expressed in whole rat brain and regionally in the cerebellum and hippocampus (Hutson et al. 1995). In astroglia-rich primary cultures derived from newborn rat brains, BCATc was found expressed in oligodendroglial and O2A progenitor cells. Using immunohistochemistry, BCATc was found only in neuronal cells, i.e. primarily glutamatergic and GABAergic cells of the hippocampus and cerebellum (Sweatt et al. 2004a). No astrocytic staining was observed. Analysis of these images showed that intense staining was observed in the processes rather than the cell bodies of glutamatergic neurons (e.g. granule cells of the cerebellum and dentate gyrus), indicating a role for BCAA metabolism in glutamate production during excitation. Conversely, in GABAergic neurons, the cell body was heavily stained (e.g. cerebellar Purkinje cells and pyramidal basket cells in the dentate gyrus), suggesting a role

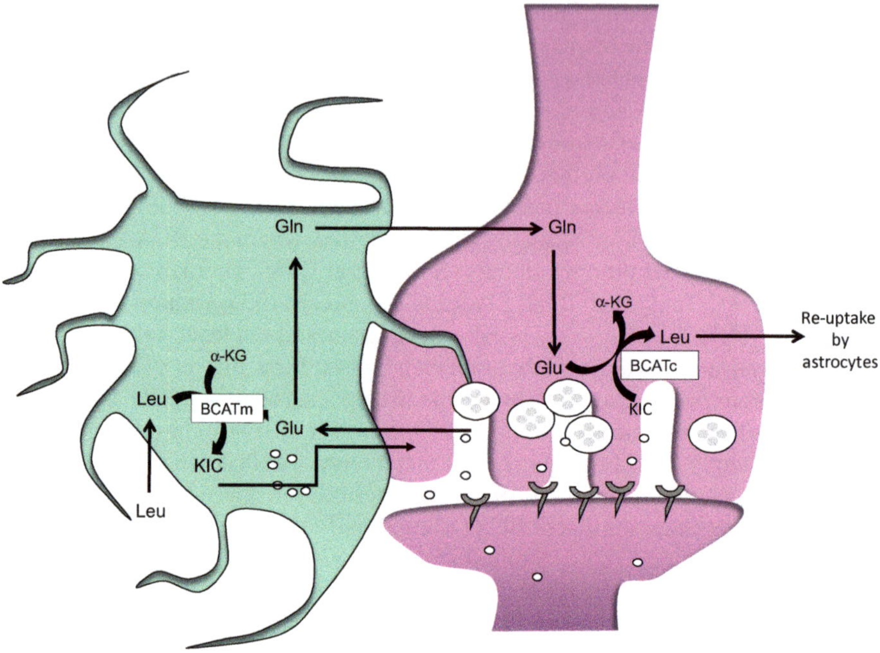

Fig. 5.3 Nitrogen transfer in rodent brain. Leucine is actively taken up by astrocyes where it undergoes transamination via BCATm forming glutamate and α-ketoisocaproate (KIC). The glutamate formed can enter the glutamate-glutamine cycle, whereas KIC, which is poorly metabolised by astrocytes, is metabolized further in neurons cells. Subsequently, the α-keto acid is transaminated with glutamate to regenerate leucine, which can be returned to the astrocyte to complete the cycle. These cycles also operate with the TCA cycle and GDH metabolism (Yudkoff et al. 1990; Hutson et al. 2001)

in maintaining the glutamate storage pool rather than contributing to excitation. In the cerebellar cortex, there were regional differences in the pattern of BCATc immunoreactivity, where intense staining was reported in the glutamatergic inputs to most of the cortex but limited in the vestibule cerebellum, which showed preferential expression in GABAergic cells. The varied expression and intensity of BCATc in these regions suggested differential regulatory controls mechanisms between subtypes of neuronal cells. The pattern of staining in the hippocampus was more consistent, where again staining in glutamatergic and GABAergic cells was reported (Sweatt et al. 2004a). However, more extensive distribution studies of rat brain showed that intracellular staining was not particular to neuronal transmitter phenotype (for example, sertonergic neurons of brainstem contained BCATc), where cell body staining throughout glutamatergic neurons was reported. BCATc was found to be expressed across all regions of rat brain in neurons of different phenotypes, where proposed roles range from generating glutamate and GABA to regulating feeding behaviour through the mTOR pathway (García-Espinosa et al. 2007). In a separate study, BCATc, considered to be expressed solely in neuronal cells, were found in the peripheral nerves of the salivary gland, the pancreas, and in the Auerbach's and Meissner's plexuses of the gut (Sweatt et al. 2004b). The principal neurotransmitters in this region are acetylcholine, the catecholamines, and neuroactive peptides. Although the role of the BCAT proteins in this region is not known, the authors speculate that leucine may act as an anabolic signal. As BCKDC was not found in this region, oxidation cannot occur, and therefore, the BCKAs would be released.

BCATm expression was initially reported to be highly expressed in astro-glial primary cultures derived from the brains of newborn Wistar rats and also in microglia (Bixel et al. 1997, 2001). However, it was not until recently that BCATm was mapped to specific regions of rat brain (Cole et al. 2012). Consistent with these cellular studies, expression of BCATm was found throughout astrocytes in rat brain. Immunopositive staining of astrocytes was reported in the white matter tracts and in the distal processes of the Bergmann glia of the cerebellar cortex. In the hippocampal region, positive staining was reported in the molecular, granule cell, and polymorphic layers of the dentate gyrus. Labelling of the astrocytic processes surrounding blood vessels along the hippocampal fissure and in the subpial region also showed concentrated BCATm expression. Thus, in rat brain the clear labelling of BCATm to astrocytic processes and cell bodies in the spinal cord and the neuronal specific expression of BCATc supported the proposed BCAT-dependent leucine/glutamine cycle (Yudkoff et al. 1996a, b; Yudkoff 1997) and the BCAA/BCKA nitrogen shuttle (Hutson et al. 2001; Lieth et al. 2001). As described above, for glutamate to be replenished in neuronal cells, additional support networks are needed, in particular a supply of nitrogen. The BCAAs were proposed as a major source of nitrogen required for this pathway, since transfer of BCAA nitrogen to glutamate and glutamine was shown to occur in primary cultures of rat astroglia and in preparations of synaptosomes (Yudkoff et al. 1996a, b; Hutson et al. 1998). Further support using ^{13}C-NMR demonstrated that leucine contributes 25–50 %, of glutamate nitrogen, respectively in rat brain (Kanamori et al. 1998; Sakai et al. 2004). In the BCAA/BCKA nitrogen shuttle model, it was proposed that leucine is taken up by astrocytes, which undergo transamination with α-KG forming glutamate and the respective BCKAs. Glutamate enters the glutamate/glutamine cycle and the BCKAs, which are not oxidized by astrocytes due to lack of BCKDC, are shuttled to neuronal cells. In neuronal cells, BCATc catalyses the transamination of these BCKAs with glutamate, producing the respective BCAAs and α-KG. Leucine is released by neuronal cells for re-uptake by astrocytes, completing the BCAA/BCKA nitrogen cycle (Fig. 5.3). α-KG can enter the TCA cycle or undergo reductive amination to glutamate by GDH. However, as the flux in the direction of reductive amination is low, this is a less likely outcome (Hutson et al. 2001; Yudkoff et al. 1990). Therefore, the role of the BCAT proteins in brain metabolism does not only facilitate a role in energy metabolism, but also in providing essential nitrogen for the anaplerotic regeneration of glutamate.

In rodent brain, BCATm and BCKDC do not co-localize. Immunolocalization of BCATm and BCKDC in rats revealed that BCATm is present in astrocytes in white matter and in neuropil, while BCKDC is essentially found in neuronal populations (Cole et al. 2012). In the rat, BCATm appears uniformly distributed in astrocyte cell bodies throughout the brain. The segregation of BCATm to astrocytes and BCKDC to neurons provides further support for the existence of a BCAA-dependent astroglial-neuronal nitrogen shuttle and support the hypothesis that BCATm association with GDH1 in astrocytes could provide ammonia for glutamine synthesis (Hutson et al. 2011). In rat brain, GDH1 has been found expressed in many nerve cells, including neurons, astrocytes, the Bergman glia, oligodendrocytes, and satellite cells, but immunoreactivity is predominantly associated with astrocytes (Zaganas et al. 2009). Because BCKDC is also localized in neurons, though BCKDC and BCATc are not necessarily found in the same neurons, net nitrogen transfer occurs when the BCKA product is oxidized. The localization of these key enzymes provides the potential for significant shuttling of BCAA metabolites between different cell types in the rodent brain.

5.5 Branched Chain Amino Acid Metabolon: Substrate Channeling

To further explore the integrated metabolism of the BCAAs and nitrogen transfer, the physical interaction of BCAT with other key metabolic enzymes involved in BCAA oxidation was pursued. Here, overexpressed human BCATm (hBCATm) was reported to bind to the E1α subunit of the E1 enzyme of BCKDC forming a BCAA metabolon (Islam et al. 2007). The association of hBCATm with E1 was governed by several factors including, (1) Only hBCATm in its PLP form could bind (2) Open structure of both proteins during catalysis was required (3) The reduced form of the reactive cysteine residues of hBCATm was necessary. Binding of the PLP-form of hBCAT to E1 increased the kinetic rate of decarboxylation of the BCKAs, whereas no binding occurred when hBCATm was in the PMP form. The K_m values for BCAAs for both enzymes are at (Leu and Ile) or above (Val) the physiological range of plasma and tissue BCAAs with BCATc having lower K_m values and higher k_{cat} than BCATm (Table 5.2). For both enzymes, transamination is responsive to changes in plasma and tissue BCAAs. The K_m for Glu is above tissue concentrations, with the exception of neurons where Glu is concentrated in the nerve terminals in the mM range. In addition, the rate of transamination exceeds that of oxidative decarboxylation by BCKDC. For BCATm, metabolon formation with BCKDC not only increases the k_{cat} for oxidative decarboxylation (DCPIP assay), but also the overall reaction of the complex (Overall assay), but lowers the K_m values for BCAA compared to BCATm alone. Thus, K_m values for BCAA are now within the range observed after a meal. This is especially true for Val. Metabolon formation would also limit BCKA release and facilitate oxidation. Lack of colocalization of hBCATm and BCKDC in the CNS actually promotes metabolite shuttling. Metabolon formation may occur in the capillary endothelium facilitating oxidation.

During transamination, the BCAA substrate binds to the PLP cofactor at the bottom of the active site and donates its nitrogen to PLP-hBCATm. This is stabilised by the hydrophobic core, where unlike other transaminases, domain closure does not occur (Yennawar et al. 2002). This open structure has been proposed to facilitate the

Table 5.2 Kinetic constants for branched chain amino acid substrates of hBCATm and hBCATc

	Isoleucine		Leucine		Valine		Glutamate	
Enzyme	k_{cat}/K_m ($\times 10^3$ M^{-1} s^{-1})	K_m (mM)	k_{cat}/K_m ($\times 10^3$ M^{-1} s^{-1})	K_m (mM)	k_{cat}/K_m ($\times 10^3$ M^{-1} s^{-1})	K_m (mM)	k_{cat}/K_m ($\times 10^3$ M^{-1} s^{-1})	K_m (mM)
hBCATm	372±2	0.62±0.03	242±1	0.80±0.02	39±0	4.80±0.25	11±0	12±0.42
hBCATc	3359±2	0.32±0.04	2494±1	0.36±0.02	618±0	2.0±0.45	108±0	9.9±0.42

substrate channelling of the α-keto acid to the E1 subunit of the BCKDC complex, where the active site also stays exposed during catalysis. Also important to this interaction is the redox-active –CXXC– motif, located about 10 Å from the active site (Conway et al. 2002, 2004; Yennawar et al. 2006). The redox sensitivity of the hBCAT proteins is unique to these aminotransferase proteins (Conway et al. 2002). This motif acts as a peroxide-mediated redox-switch, which regulates hBCAT activity, where the oxidised and reduced forms represent the inactive and active enzyme, respectively (Conway et al. 2002). When the hBCATm oxidised or mutant form of the –CXXC– motif was used, complex formation with BCKDC did not occur (Islam et al.). This is because disulphide bond formation increases the net dipole moment of the β-turn and β-sheet residues, which destabilises the substrate and PLP conformation (Yennawar et al. 2006). These changes subsequently interfere with the next step in BCKDC association. Therefore, the thiols in their reduced form and the net dipole charge are essential for association of hBCATm with BCKDC. Once transamination has proceeded, the BCKAs are channelled to the E1 subunit and conversion of PLP-hBCATm to PMP-hBCATm signals release from the E1 subunit. E1 catalysed decarboxylation and reductive acylation. This reaction was enhanced with the addition of α-KG, considered to promote the conversion of PMP-hBCATm to PLP-hBCATm, which can then initiate another cycle (Islam et al. 2007). The interaction was regulated not only by redox, but signalling pathways governed by phosphorylation. When E1 was phosphorylated, hBCATm could not bind. This is interesting because phosphorylation of the E1 subunit renders it inactive and therefore cannot bind BCKAs for decarboxylation. As demonstrated from the distribution studies, skeletal muscle has high expression of the BCKDC kinase, enhancing exit of the BCKAs (Suryawan et al. 1998). Moreover, high ratios of NADH/NAD also prevent BCATm/BCKDC binding and BCKA release or recycling is favoured. Therefore, metabolon formation and its regulation are key events that regulate nitrogen transfer and BCAA oxidation. This switch from a reduced to an oxidised BCATm may also signal a change in function, i.e. from the role of hBCAT in metabolism to its redox-chaperone role, discussed later.

Not only was the BCKDC found to associate with hBCATm, but so too were other key metabolic proteins including GDH, pyruvate carboxylase, 4-nitrophenylphosphatase domain, and non-neuronal SNAP25-like protein homolog 1 (Islam et al. 2010). Pyruvate carboxylase catalyses the formation of oxaloacetate from pyruvate and bicarbonate. Although this protein is important for de novo synthesis of glutamate, its role in this metabolon is unclear. On the other hand, binding of hBCATm with GDH was found to occur in a BCKDC-dependent or BCKDC-independent manner. GDH catalyses the oxidative deamination of glutamate to α-KG, using NAD^+ or NADH as a coenzyme. Following the first half-reaction of transamination and after dissociation with the E1 subunit, the PMP-hBCATm form requires conversion to its PLP form. Binding occurs between PMP-hBCATm and GDH, whereupon the α-KG produced by oxidative deamination of glutamate is channelled to the BCATm active site to generate glutamate and re-establish the PLP form of hBCATm and release GDH. Oxidation of the BCKAs by BCKDC generates the NADH which is a cofactor in the GDH reaction.

Thus, wherever BCATm and GDH are expressed, both are mitochondrial enzymes; α-KG formed from the GDH reaction (preferred direction) is reaminated by metabolon formation, resulting in no net change in glutamate concentration. If BCKDC is also found in the same cell as GDH and BCATm, formation of the metabolon would facilitate BCKA oxidation. However, the ultimate products of the metabolons are the branched chain acyl-CoAs and NH_3. If BCKDC is not present, as is found in rodent astrocytes, then BCKAs would be released and could be reaminated or oxidised in another cell type in brain such as neurons or released into the blood. The ammonia produced could be used for glutamine synthesis in astrocytes, otherwise the ammonia would be released and used in other cells. Another factor other than expression and relative activities of GDH1 and hBCATm (PLP-form v/s PMP-form) is the redox state of the cells (NAD+/NADH). If only GDH and BCKDC are present, the kinetics of both reactions are not affected, because α-KG is not a substrate for BCKDC and GDH does not bind to BCKDC (Hutson et al. 2011). However, although GDH1 alone did not have a significant effect on BCKDC activity, the inclusion of hBCATm and glutamate enhanced the k_{cat} by 150 % when compared to BCKDC and BCKAs alone and by about 40 % compared to the addition of hBCATm alone. Equally important, binding of hBCATm to GDH enhances leucine-induced activation of GDH, where the interaction with hBCATm and leucine promotes nitrogen shuttling and ammonia formation. However, addition of ADP to the above reaction did not enhance oxidative deamination by GDH (Islam et al. 2010). Interestingly, although GTP is an allosteric inhibitor of GDH, addition of PMP-hBCATm protected GDH from inhibition. Conversely, addition of GDH to PMP-hBCATm enhanced the k_{cat} of PMP-hBCATm reamination of α-KG by 15-fold and reduced the k_m by 4.5 fold (Hutson et al. 2011). Therefore, actual net transfer of BCAA nitrogen occurs when BCKDC is active (BCAAs are oxidised) and this is initiated by PLP-hBCATm/E1 complex formation. Together, these metabolons offer a complex yet efficient mechanism for the transfer of nitrogen between enzyme systems providing a possible role for regulating NH_3. However, metabolon formation and the functional significance of the metabolon are ultimately dependent on the expression of these enzymes in tissues and cell types within tissues.

Thus, knowledge of the distribution and relative ratios of these enzymes in cells offers insight into the direction of metabolism, but also the fate of the BCKAs. In rodents, where BCKDC is not co-expressed with BCATm, in tissues such as the liver and brain, this metabolon does not function. In this forum, compartmentation of BCATm to astrocytes (or endothelial cells in humans) and BCKDC to neuronal cells facilitates the shuttling of metabolites between cells, where transamination is favoured in astrocytes or endothelial cells and oxidation in neuronal cells. Also, in skeletal muscle, although BCKDC is expressed, the high transaminase activity and high BCKDC kinase activity will favour BCKA production, preserving the carbon skeleton, which can be reaminated or oxidised in other tissues. The role of these metabolons is also likely to expand to other perhaps to be described functions, where insight into their role in hormone secretion has already been described (Zhou et al. 2010).

5.6 Patterns of BCAA Metabolic Enzyme Expression in Human Brain

BCAT expression studies in the human brain were expected to reflect that of the rodent model, where the BCAT cycle model would support the glutamate/glutamine cycle. Indeed, for hBCATc, consistency between models was reported where immunopositive labelling was predominant in neuronal cells (Fig. 5.4) (Hull et al. 2012). Similar to reports from the rodent models, the majority of hBCATc-labelled neurons in the human brain were either GABAergic or glutamatergic, with GABAergic neurons shown to be more strongly immunopositive than glutamatergic neurons (Table 5.3). However, neurons that were neither glutamatergic nor GABAergic also labelled for hBCATc. Each of the rat distribution studies reported staining that was more intense in some areas over others, and whereas cell body staining was clear in all neuronal cells labelled, staining of processes was more cell-type-specific. For example, in the hippocampus neuronal staining was more intense than the CA1 region. This supports earlier findings, where differential mRNA expression of hBCATc was reported in postnatal and adult brain of mice (Castellano et al. 2007). Compared with axonal staining in the areas of the supraoptic tract, it was more intense in the cell bodies of neuronal cells within the temporal and hippocampus relative to the dendrite regions that had more pronounced staining. Thus, it appears that depending on the neuronal sub-type, hBCATc may have a role in either contributing to the glutamate metabolic pool or the glutamate pool used during excitation (Table 5.2). Therefore, in these neuronal sub-types it was proposed that the role of

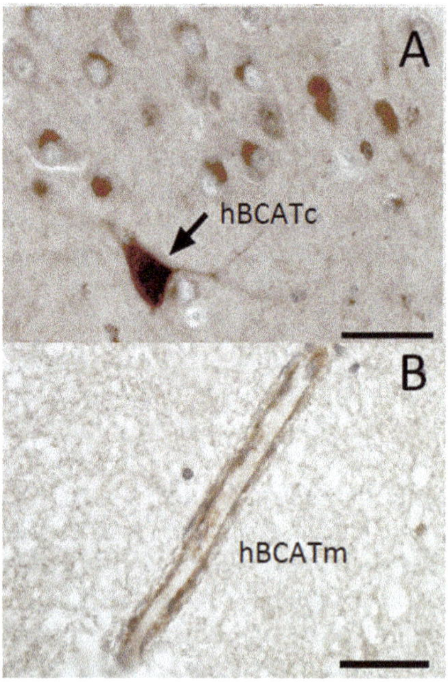

Fig. 5.4 Immunostaining of hBCATc and hBCATm in the temporal lobe and cerebellum. (**a**) The temporal neocortex (inferior temporal gyrus) showing hBCATc immunopositive neurons. (**b**) hBCATm staining of the vasculature surrounding the hippocampus (Hull et al. 2012)

Table 5.3 An overview of hBCATc immunoreactivity throughout the human brain[a]

Area	Immunopositive neuronal cells	Intensity of staining
Temporal lobe		
• *Hippocampus*		
– GABAergic interneurons	+++	+++
– Pyramidal neurons	+	++++
– Dentate gyrus (neurons)	+	+
• *Cortex*		
– Neurons	+++	++
– Subiculum sub-population of neurons	+	++
– Lamina II neurons	++	++
Cortex and white matter		
• Cortical pyramidal cells	+	+++
• Cortical neurons	++	++
• Axonal staining	–/+	++
Putamen and Basal ganglia		
• Large neurons	+	+++++
• Small neurons	+++	+
• Insular cortex (small neurons)	+	++
• Thalamus (neurons)	+	++
• Caudate nucleus (neurons)	+	+
• Lateral geniculate nucleus (neurons)	+	+
Midbrain		
• Widespread neuronal staining	++++	+++
• Periaqueductal grey matter (neurons)	++	++
• Neuropil staining	++	+
• Inferior colliculus (nerve cells)	+	+++
• Supraoptic nucleus of the hypothalamus	+	++++
• Paraventricular neurons of the hypothalamus	+	++++
Cerebellum		
• Oligodendrocyte staining in the white matter.	+	++
• Purkinje cells	+	+
• Swollen axon terminals	++	++
• Interneurons	++	+
• Neurons in the dentate nucleus	++	+
Pons		
• Pontine nuclei (neurons)	+	++
• Tegmental neurons	+	++
• Neuronal processes (nigro striatal processes)	++	+
• Raphe nuclei (neurons)	+	++
• Nucleus basalis of Meynert (cholinergic neurons)	+	++++
Medulla		
• Hypoglossal nucleus (neurons)	+	++
• Dorsal motor nucleus (neurons)	+	++
• Nucleus ambiguus (neurons)	++	+++
• Gracile nucleus	++	++
• Inferior olivary nuclei (neurons)	++	++
• Inferior olivary nuclei neuropil	++	++

[a]Date from Hull et al. (2012)

BCAAs was to produce glutamate, which is corroborated by historical studies that show BCAAs can label a significant proportion of the glutamate pool in neuronal cell culture, which has also been shown in vivo in rats and ex vivo rat retina models (Yudkoff et al. 1983, 1994; Kanamori et al. 1998; LaNoue et al. 2001).

Conversely, unlike the rodent model, which showed expression of BCATm in astrocytes, there is extensive immunopositive staining for hBCATm in the vasculature of the brain with no evidence of staining within astrocytes, suggesting a different role for hBCATm in glutamate regulation in the human brain (Fig. 5.5). Cerebral endothelial cells have a high density of mitochondria (Oldendorf et al. 1977), giving greater capacity for energy production than other capillary endothelial cells and explaining the punctuate staining of mitochondrial hBCAT in these cells. This result was surprising; however, discrepancies between the rat model and humans have previously been described, in particular with respect to transaminase activity and oxidative capacity (Suryawan et al. 1998). If hBCATm is not expressed in human astrocytes, transamination cannot occur indicating that the BCAA/BCKA shuttle is particular to rodents and that in human brain its clear staining in the endothelial layer of the vasculature supports a role in metabolite channelling across the blood

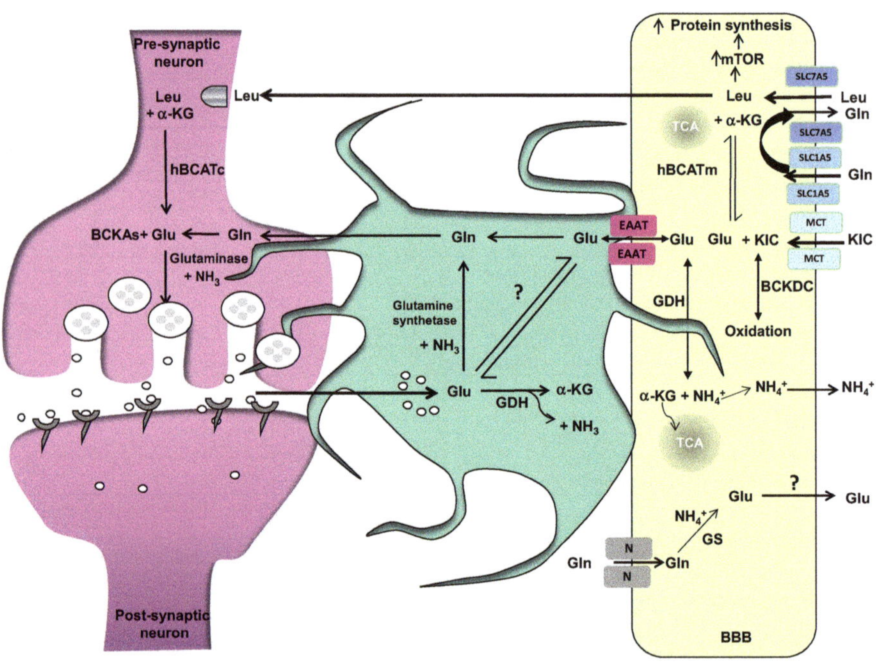

Fig. 5.5 Nitrogen transfer in human brain. Leucine, readily taken up by the blood brain barrier by the L1 system in exchange for glutamine, where it can be taken up by neuronal cells and transaminated by hBCATc or be used as a nutrient signal in endothelial cells. Excess glutamate generated during excitation is taken up by astrocytes, where glutamate is regenerated through the glutamate/glutamine cycle. Glutamate may also be removed by endothelial cells of the vasculature. Here, the fate of glutamate could be (1) exit into blood; (2) oxidative deamination by GDH; and (3) reamination by hBCATm generating leucine for reuptake by neuronal cells (Hull et al. 2012)

brain barrier. The reason behind these differences is not clear. However, hBCATm expression is induced under conditions that promote cell proliferation, which may explain why in cell models hBCATm expression is increased (Pérez-Villaseñor et al. 2005) or that the pattern of expression may be altered by placing cells in culture outside the normal milieu. It may also be due to the age of the rat brain compared with the brains used in the human study, reflecting a younger model relative to older brains used in the human study. Nevertheless, it is most likely the difference in BCAT expression between species.

So, what is the role of hBCATm in the endothelial layer of the vasculature? Based on structural and metabolic studies, we would anticipate that its role will be dependent on the expression of other key metabolic proteins, such as the GDH and BCKDC and the inter-relationships of metabolism to transport key amino acids. Moreover, we anticipate that the redox environment will also feature as it is known to regulate and potentially redirect the function of hBCATs. In humans, hGDH is expressed in two isoforms, hGDH1, that is widely expressed but high in the liver, whereas hGDH2 is expressed in the brain, testis, and kidney (Lai et al. 1986; Schmitt and Kugler 1999; Zaganas et al. 2009, 2010; Spanaki et al. 2010). In human brain, the GDH-specific activity is fourfold less than that described for liver (Spanaki et al. 2010). Similar to reports from rat, in human brain clear astrocyte staining is evident for GDH1 and GDH2, though GDH2 is highly expressed in astrocytes (Spanaki et al. 2014); however, punctuate staining throughout the vasculature was also observed. Here, the authors proposed that this was due to the astrocytic end-feet surrounding the vasculature. However, unpublished work from our lab shows vasculature staining, but indicates that it is the endothelial layer of the vasculature rather than just the astrocytic end-feet. As GDH is found together with hBCATm, then it is possible that the hBCATm/GDH1 metabolon can form and hBCATm does not bind to GDH2 (Hutson et al. 2011). Excess glutamate taken up by EAAT transporters (EAAT1, 2 and 3), the rate of which is thought to be influenced by blood glutamate, could be regulated by the hBCATm/GDH metabolon (Chaudhry et al. 1995; O'Kane et al. 2004; Hosoya et al. 1999; Gottlieb et al. 2003). Glutamate in the ECF must be maintained at low concentrations to avoid excitotoxicity. Glutamine is also transported, primarily by the N transporter, and can be metabolised by glutaminase generating glutamate and ammonium adding to the glutamate pool in endothelial cells (Lee et al. 1998). It is expected that when levels of intra-cellular glutamate rise, net transport across the luminal membrane occurs. Removal of glutamine is also important, given the absence of the urea cycle. Here, an opportunity for the BBB to regulate ammonia levels is important. Accumulation of glutamine in the brain would result in an increased osmolarity and swelling (Hawkins et al. 2006). Oxidation of glutamate by GDH would release α-KG, forming ammonia and NAD^+. If the ammonia produced in the endothelial cell could enter the astrocyte end-feet, it could provide ammonia for glutamine synthesis or would be free to diffuse across the luminal membrane into blood. In the absence of metabolon formation, transamination of glutamate with BCKA would form α-KG that could enter the TCA cycle, generating leucine for neuronal uptake (dependent on the availability of α-keto-acids). Brain transport of α-keto-acids is facilitated by the monocarboxylate transporter expressed in the BBB and neuronal cells (Mac et al. 2000). Provision of blood α-keto-acids

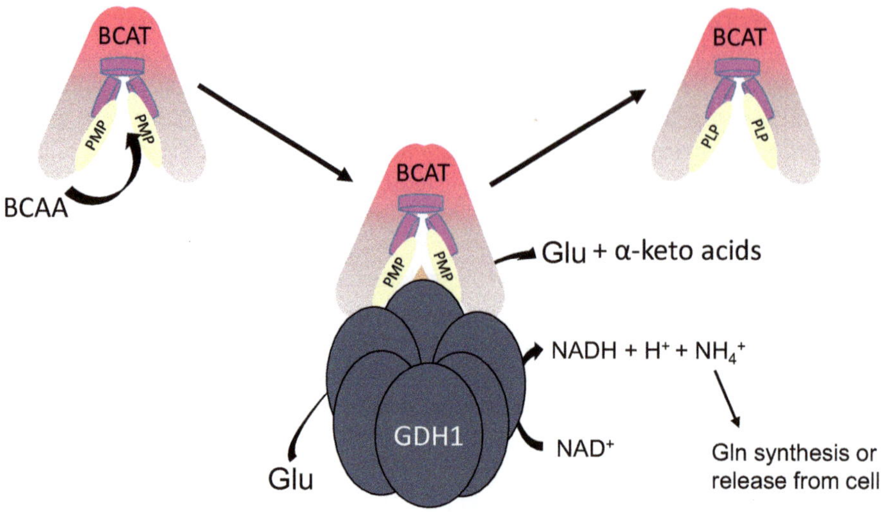

Fig. 5.6 BCATm and GDH1 form a metabolon. The PMP form of BCATm binds to GDH1. The product of oxidative deamination, α-KG, is channeled from GDH1 to PMP BCATm followed by formation of glutamate (Glu) and release of the PLP form of BCATm, which is then ready to transaminate with a BCAA. There is no net change in cellular glutamate. The products of the reaction are NADH and ammonia (NH_4^+). Thus, the metabolon facilitates formation of ammonia and NADH

could drive metabolism in the direction of leucine synthesis. Metabolism of brain glutamate by hBCATm or hBCATm/GDH could therefore serve as an auxillary part to the glutamate/glutamine cycle, the predominant system that regulates brain glutamate between neurons and astrocytes (Fig. 5.6).

Because leucine is easily taken up by specific transporters on the BBB, and as hBCATm is expressed in these endothelial cells, transamination can occur. These amino acid transporters play a key role in the maintenance of free amino acid concentrations in cells and also regulating brain nitrogen. It is the coupling of these amino acid transporters with metabolism that enables the cell to respond to nutrient signals through activation of increased protein translation and growth. Conversely, cell survival through autophagy is stimulated when amino acids are limiting (Chen et al. 2014). Entry of large neutral amino acids such as leucine and tryptophan is facilitated by the membrane transport L1 system, found both on the luminal and abluminal side of the BBB (Oldendorf 1973; Smith et al. 1987; O'Kane et al. 2004; O'Kane and Hawkins 2003). The plasma concentration of tryptophan and competition with other large neutral amino acids (LNAA) has been shown to reflect serotonin levels in the brain and will thus dictate metabolic signalling in the cell (Fernstrom and Wurtman 1972). Recent studies have indicated that in order for leucine to be imported by the solute carrier family 7 member 5 (SLC7A5), glutamine exchange must occur (Nicklin et al. 2009). So, when glutamine is taken up by the solute carrier family 1 member 5 (SLC1A5), it is concomitantly exchanged for leucine (within 1–2 min). This bidirectional transport has been proven to be important for leucine-mediated mTOR activation. Studies indicate that p70S6K was not activated until glutamine was exchanged for leucine and reactivation of starved cells was dependent on glutamine uptake (Chen et al. 2014). Moreover, knockdown of the SLC1A5 transporter impaired mTor activation, further supporting the synergistic role of glutamine with leucine in regulating the anabolic response (Nicklin et al. 2009).

In contrast to the facilitative transporters on the luminal face, Na^+-dependent transport systems are required on the abluminal side, which allows amino acid removal from the brain via the Na^+ gradient that exists between the ECF and the endothelial cells of brain capillaries (reviewed in Hawkins et al. 2006). In theory, amino acids that both pass the endothelial cell membrane and enter the basement membrane space could be actively and selectively pumped back across the abluminal membrane. It is this gradient system and abluminal expression of the Na^+-dependent carriers that favours removal of nonessential amino acids and toxic amino acids such as glutamate. As L1 transporters also exist on the abluminal side, this indicates that transport can occur in either direction, the impact and kinetics of which remains to be determined (O'Kane and Hawkins 2003). BCKA produced by BCATm could be oxidized by endothelial cell BCKDC (Conway unpublished observations) and enter the blood or the brain. However, further knowledge of the relative expression/activities of these enzyme systems is required, which would inform the BCAA metabolic capacity of these cells. Moreover, we anticipate that as neuronal stores of glutamate require constant replacement, leucine uptake by endothelial cells will be preferentially for this role and mTOR regulation rather than oxidation. Thus, the specific expression of transporters on the respective membranes, together with the metabolic complexes expressed in the BBB, orchestrates the fine tuning of amino acid homeostasis and nitrogen balance, which aims to protect the brain from excitotoxicity. If as we have proposed that in human brain the direction of transamination in neuronal cells is towards glutamate production catalysed by the cytosolic BCATc isozyme, transamination functions in both contributing to the neuronal pool and also excitation, dependent on the neuronal sub-type. On the other hand, if glutamate concentrations are high and BCKAs are available, BCATc could attenuate excitation. As BCKDC is widely expressed in neuronal cells (Cole et al. 2012), oxidation of the BCKAs can occur, but again this will depend on the relative activities of these two enzymes and transport of BCKA either into the mitochondria or release from the cell. As shown in CD4(+) T cells, activation of the T cell receptor induces expression of BCATc and the primary fate of BCKA is released from the T cells (Ananieva et al. 2014). Conversely, in the endothelial layer of the vasculature although the role of hBCATm may work with GDH to metabolise excess glutamate released at the synaptic region, it is also possible that BCATm is detrimental, particularly in conditions such as Alzheimer's disease, where glutamate excitotoxicity prevails. hBCATm was found to be increased by 160 % in the brain of patients with AD relative to matched control (Hull et al. 2015). This was correlated with increased Braak stage, indicating a link with disease severity. On the other hand, multiple forms of BCATm between 41 and 52 kDa were also observed. Redox regulation on BCATm, which inhibits BCAT activity, may alter the function of these proteins and their role in glutamate metabolism.

5.7 The Redox-Active hBCAT Proteins: The Nano-switch

When discussing the role of the hBCAT proteins in metabolon structures, we mentioned the importance of the redox-active –CXXC– motif. Each isoform is differentially regulated through changes in the redox environment, where the mitochondrial isoform can undergo complete reversible oxidation, losing all activity (Conway et al. 2002). The cytosolic isoform is also susceptible to oxidation, where the formation of a disulphide bond between its reactive cysteine residues results in the loss of 50 % hBCATc activity. Interestingly, unlike hBCATm, further oxidation did not affect activity (Conway et al. 2008). The N-terminal cysteine residue of each –CXXC– motif was the most reactive, forming a cysteine sulphenic acid intermediate that quickly undergoes further oxidation to produce a disulphide bond (Conway et al. 2002, 2004). Loss in activity occurs due to the re-orientation of the substrate-binding channel compromising stability, which impacts the second half reaction rather than the first half reaction (Conway et al. 2003; Yennawar et al. 2006). The reactive cysteines of both isoforms are also targets for the S-nitrosylating agent, S-nitroso glutathione (GSNO), where a transition between S-nitrosation and S-glutathionylation was reported dependent on the level of RNS exposure (Coles et al. 2009). The glutaredoxin/glutathione system reversed this inactive form, supporting a role for hBCAT in cellular redox control. We therefore had characterised a reversible redox-sensitive switch, which controlled transaminase activity.

We have already seen that oxidation of hBCATm influences its role in forming a metabolon with the E1 subunit of BCKD, where oxidation prevents metabolite channeling. We propose that this nano-switch has wider implications, such that during periods of oxidative stress the role of hBCAT switches from BCAA metabolism to a redox chaperone (Conway et al. 2008; Conway and Lee 2015). Evidence in support of this stems from recent work where a novel functional role for hBCAT in redox protein folding was reported (El Hindy et al. 2014). Here, these proteins were found to catalyse the refolding of reduced and denatured RNase, the activity of which was regulated by the oxidising environment. When treated with GSNO, high molecular weight (HMW) structures were observed and the rate of refolding increased. Generation of HMW structures during chaperone activity has been reported for other chaperones such as human peroxyredoxin and also Hsp33 (when overoxidised), which are important to their role in protein folding (reviewed in Conway and Lee 2015). We therefore speculate that hBCAT can form oligomers that prevent aggregation and promote correct protein folding. Moreover, as hBCATm also forms a redox complex with protein disulphide isomerase (PDI), influencing the rate of refolding, it is likely that hBCATm is a novel redox chaperone for PDI, operating through a thiol-disulphide exchange mechanism. Or these transaminases may operate independently as an oxidoreductase regulated through the redox environment. In summary, we propose that, under reduced conditions, the role of hBCAT favours transamination and metabolon formation with the BCKDC and/or GDH, metabolic role in brain neurotransmitter metabolism. In an oxidizing environment such as increased ROS or RNS, oxidation changes the structural conformation opening the active site to novel substrates to promote protein folding and chaperone

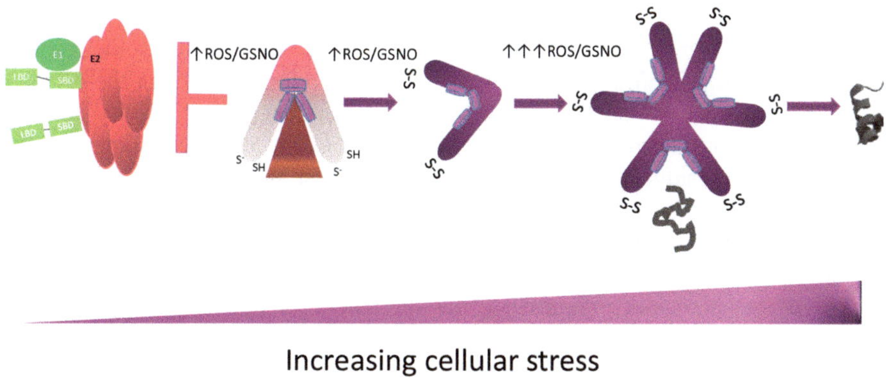

Fig. 5.7 Thiol redox sensor–dual role for BCAT. Oxidation results in the inactivation of BCAT and the prevention of BCAT/BCKDC metabolon formation. With increased oxidation, the BCAT protein can form multimers and function as oxidoreductases, where a role in protein folding has been implicated (El Hindy et al. 2014; Conway and Lee 2015)

activity, supported by the three hydrophobic pockets in the active site. This change in function will be further accelerated as the levels of ATP decrease and the NADH/NAD+ ratios are compromised, impacting the functional state of the metabolons. The physiological role of this interaction has not yet been elucidated, but new data has isolated a protein-folding metabolon, offering exciting new roles for these metabolic proteins as redox chaperones (unpublished observations) (Fig. 5.7).

5.8 Inborn Errors of BCAA Metabolism

A recent case involving mutations in BCATm (BCAT2), which resulted in reduced activity of this enzyme, has been reported (Wang et al. 2015). In this patient, the observed hypervalinemia and hyperleucinemia were associated with brain white matter lesions. This case provides evidence that blocking BCAA catabolism has consequences on brain structure and function that are independent of other BCAA metabolites. Nevertheless, by far the most common and deleterious inborn error of BCAA metabolism is Maple Syrup Urine disease (MSUD; OMIM 248600) caused by deficiency of the mitochondrial BCKDC, the second step in the catabolic pathway. This disease results in the accumulation of BCAA and their corresponding BCKA in tissues and plasma. The disorder typically manifests with potentially lethal episodes of intoxication presenting with acute neurological deterioration, feeding problems, weight loss, and a maple syrup odor to the urine (Morton et al. 2002). These episodes usually occur during states where the body is catabolic such as fasting or intercurrent illnesses. Toxicity is associated with the increase in plasma concentrations of leucine and its α-keto acid KIC. There is little apparent toxicity associated with increases in isoleucine or valine (Korein et al. 1994). Based on its severity, MSUD has been classified into five clinical subtypes: a “classic” neonatal severe form, an “intermediate” form, an “intermittent” form, a “thiamine-responsive” form, and an “E3-deficient with lactic acidosis” form. Although the correlation between clinical severity and degree of residual enzymatic activity is often inconsistent, the intermediate and intermittent forms are usually associated with some degree of residual activity with later clinical onset, while the classic form usually exhibits extremely low activity (Dancis et al. 1967; Schulman et al. 1970). Current treatment is based on dietary manipulations with protein restriction and a synthetic formula with reduced BCAA content (Morton et al. 2002). More recently, total liver transplant has become a popular alternative for MSUD patients, and they appear to be able to tolerate a normal diet after transplantation. From a recent review that used our published data and another study, it was concluded that the effectiveness of liver transplantation in MSUD indicates that providing 9–13 % of the bodies’ BCKDC activity is sufficient to restore BCAA homeostasis (Suryawan et al. 1998; Burrage et al. 2014). However, even with early detection and dietary restriction or liver transplantation, some mental and social impairments are still present in the majority of these individuals (Simon et al. 2007; Muelly et al. 2013).

Accumulation of leucine and, particularly its α-keto acid KIC, is likely responsible for the brain pathology (Funchal et al. 2005; Ribeiro et al. 2008). Indeed, the BCATm knockout mouse, which exhibits elevated BCAAs without subsequent elevation of BCKAs, does not exhibit an MSUD phenotype (She et al. 2007), although the human clinical case indicated white matter abnormality (Wang et al. 2015a, b). The BCATm KO mouse with concentrations of BCAA seen only in patients undergoing metabolic decompensation exhibited increased energy expenditure due in part to increased protein turnover supporting the hypothesis that the toxic metabolites are the BCKA. It has been reported that BCKA treatment triggers elevation of reactive oxygen species in cells (Funchal et al. 2005), rising BCKA and loss of BCKDC activity increases oxidative stress in MSUD patients (Guerreiro et al. 2015; Sitta et al. 2014). Because oxidative stress can trigger apoptosis in neurons, this could be one mechanism for neuronal cell damage. Neurons are characterized by lower glycolytic rate compared to astrocytes and so are thought to be less resistant to cellular stresses (Bolaños and Almeida 2010). A study by Klee et al. (2013) reported changes in white matter microstructural organization in MSUD patients and interpreted the data as evidence of structural abnormality and reduction in amounts of fibre tracts in these patients. It is known that both mTORC1 and mTORC2 are involved in cytoskeletal polymerization and reorganization (Jacinto et al. 2004; Gulhati et al. 2011; Li et al. 2014). Our unpublished data showed significant reduction of β3-tubulin as well as F-actin in the brain of newborn E1α (E1 enzyme protein) knockout mouse brain and in cultured neurons, when compared to the WT animals. These data suggest either neuronal death or a major cytoskeletal alteration. There was increased depolymerization of actin and β3-tubulin and increased β3 tubulin cleavage in BCKA plus BCAA-treated neurons.

Finally, there are consequences to the brain when BCAA are limiting. Regulation of BCKDC enzymatic activity depends on the phosphorylation status of the E1α subunit of the E1 enzyme that is specified by a kinase (BDK), which inactivates the E1α subunit, and by a mitochondrial matrix resident type 2C phosphatase gene, *PP2Cm*, that activates it. When phosphorylated, the E1 enzyme is inactive preventing decarboxylation of the BCKA. Mutations in the BCKDC kinase have now been reported (García-Cazorla et al. 2014; Novarino et al. 2012). Impaired kinase activity results in uncontrolled oxidation of BCAA. Therefore, BCAA can become limiting for protein synthesis, which may also alter mitochondrial metabolism (energy metabolism). These mutations have been associated with mild autism, epilepsy, and intellectual disability (Novarino et al. 2012) and developmental delay, microcephaly, and neurobehavioral abnormalities (García-Cazorla et al. 2014). Fortunately, protein-rich diets or oral BCAA supplementation can be used to treat BDK disorders.

5.9 Conclusions

Nitrogen transfer within and between cells and tissues illustrates a fascinating interplay of metabolic pathways that play key roles in protein metabolism, nutrient signalling, and neurotransmitter regulation. Changes in cell signalling and the redox homeostasis clearly define the direction of metabolism and may signal new roles for these metabolic proteins under oxidative stress. Our current understanding of these mechanisms supports a clear role for the BCAA in the supply of nitrogen for glutamate synthesis. This is of particular importance given that the BCAA, particularly leucine, can cross the blood brain barrier and provide a nitrogen source for the synthesis of the neurotransmitter, glutamate. As these aminotransferase proteins play a key role in facilitating the anaplerotic generation of glutamate, it is highly likely that their metabolism will be altered in neurodegenerative disease conditions. Therefore, understanding how these suggested shuttles are altered in disease will offer possible targets for novel therapeutic treatment to either delay onset or prevent further neuronal destruction.

References

Ananieva EA, Patel CH, Drake CH, Powell JD, Hutson SM (2014) Cytosolic branched chain aminotransferase (BCATc) regulates mTORC1 signaling and glycolytic metabolism in CD4+ T cells. J Biol Chem 289(27):18793–18804

Berl S, Clarke DD (1983) The metabolic compartmentation concept. In: Hertz L, Kvamme E, McGeer EG, Schousboe A (eds) Glutamine, glutamate and GABA in the central nervous system, Neurology and neurobiology. Liss, New York, pp 215–217

Bixel MG, Hamprecht B (1995) Generation of ketone bodies from leucine by cultured astroglial cells. J Neurochem 65(6):2450–2461

Bixel MG, Hutson SM, Hamprecht B (1997) Cellular distribution of branched-chain amino acid aminotransferase isoenzymes among rat brain glial cells in culture. J Histochem Cytochem 45(5):685–694

Bixel M, Shimomura Y, Hutson S, Hamprecht B (2001) Distribution of key enzymes of branched-chain amino acid metabolism in glial and neuronal cells in culture. J Histochem Cytochem 49(3):407–418

Bledsoe RK, Dawson PA, Hutson SM (1997) Cloning of the rat and human mitochondrial branched chain aminotransferases (BCATm). Biochim Biophys Acta 1339(1):9–13
Bolaños JP, Almeida A (2010) The pentose-phosphate pathway in neuronal survival against nitrosative stress. IUBMB Life 62(1):14–18
Brand K (1981) Metabolism of 2-oxoacid analogues of leucine, valine and phenylalanine by heart muscle, brain and kidney of the rat. Biochim Biophys Acta 677(1):126–132
Brand K, Hauschildt S (1984) Metabolism of 2-oxo-acid analogues of leucine and valine in isolated rat hepatocytes. Hoppe Seylers Z Physiol Chem 365(4):463–468
Brookes N (1993) Interaction between the glutamine cycle and the uptake of large neutral amino acids in astrocytes. J Neurochem 60(5):1923–1928
Burrage LC, Nagamani SC, Campeau PM, Lee BH (2014) Branched-chain amino acid metabolism: from rare Mendelian diseases to more common disorders. Hum Mol Genet 23(R1):R1–R8
Castellano S, Casarosa S, Sweatt AJ, Hutson SM, Bozzi Y (2007) Expression of cytosolic branched chain aminotransferase (BCATc) mRNA in the developing mouse brain. Gene Expr Patterns 7(4):485–490
Chaplin ER, Goldberg AL, Diamond I (1976) Leucine oxidation in brain slices and nerve endings. J Neurochem 26(4):701–707
Chaudhry FA, Lehre KP, van Lookeren Campagne M, Ottersen OP, Danbolt NC, Storm-Mathisen J (1995) Glutamate transporters in glial plasma membranes: highly differentiated localizations revealed by quantitative ultrastructural immunocytochemistry. Neuron 15(3):711–720
Chen R, Zou Y, Mao D, Sun D, Gao G, Shi J, Liu X, Zhu C, Yang M, Ye W, Hao Q, Li R, Yu L (2014) The general amino acid control pathway regulates mTOR and autophagy during serum/glutamine starvation. J Cell Biol 206(2):173–182
Chuang JL, Chuang DT (2000) Diagnosis and mutational analysis of maple syrup urine disease using cell cultures. Methods Enzymol 324:413–423
Chuang DT, Chuang JL, Wynn RM (2006) Lessons from genetic disorders of branched-chain amino acid metabolism. J Nutr 136(1 Suppl):243S–249S, Review
Cole JT, Sweatt AJ, Hutson SM (2012) Expression of mitochondrial branched-chain aminotransferase and α-keto-acid dehydrogenase in rat brain: implications for neurotransmitter metabolism. Front Neuroanat 6:18
Coles SJ, Easton P, Sharrod H et al (2009) S-Nitrosoglutathione inactivation of the mitochondrial and cytosolic BCAT proteins: S-nitrosation and S-thiolation. Biochemistry 48(3):645–656
Conway ME, Hutson SM (2000) Mammalian branched-chain aminotransferases. Methods Enzymol 324:355–365
Conway ME, Lee C (2015) The redox switch that regulates molecular chaperones. Biomol Concepts 6(4):269–284
Conway ME, Yennawar N, Wallin R et al (2002) Identification of a peroxide-sensitive redox switch at the CXXC motif in the human mitochondrial branched chain aminotransferase. Biochemistry 41(29):9070–9078
Conway ME, Yennawar N, Wallin R, Poole LB, Hutson SM (2003) Human mitochondrial branched chain aminotransferase: structural basis for substrate specificity and role of redox active cysteines. Biochim Biophys Acta 1647(1-2):61–65, Review
Conway ME, Poole LB, Hutson SM (2004) Roles for cysteine residues in the regulatory CXXC motif of human mitochondrial branched chain aminotransferase enzyme. Biochemistry 43(23):7356–7364
Conway ME, Coles SJ, Islam MM et al (2008) Regulatory control of human cytosolic branched-chain aminotransferase by oxidation and S-glutathionylation and its interactions with redox sensitive neuronal proteins. Biochemistry 47(19):5465–5479
Daikhin Y, Yudkoff M (2000) Compartmentation of brain glutamate metabolism in neurons and glia. J Nutr 130(4S Suppl):1026S–1031S, Review
Danbolt NC (2001) Glutamate uptake. Prog Neurobiol 65(1):1–105, Review

Dancis J, Hutzler J, Rokkones T (1967) Intermittent branched-chain ketonuria. Variant of maple-syrup-urine disease. N Engl J Med 276:84–89

Davoodi J, Drown PM, Bledsoe RK et al (1998) Overexpression and characterization of the human mitochondrial and cytosolic branched-chain aminotransferases. J Biol Chem 273(9):4982–4989

El Hindy M, Hezwani M, Corry D, Hull J, El Amraoui F, Harris M, Lee C, Forshaw T, Wilson A, Mansbridge A, Hassler M, Patel VB, Kehoe PG, Love S, Conway ME (2014) The branched-chain aminotransferase proteins: novel redox chaperones for protein disulfide isomerase-implications in Alzheimer's disease. Anti oxid Redox Signal 20(16):2497–2513

Elia M, Livesey G (1983) Effects of ingested steak and infused leucine on forelimb metabolism in man and the fate of the carbon skeletons and amino groups of branched-chain amino acids. Clin Sci (Lond) 64(5):517–526

Fernstrom JD, Wurtman RJ (1972) Brain serotonin content: physiological regulation by plasma neutral amino acids. Science 178(4059):414–416

Forslund AH, Hambraeus L, Olsson RM, El-Khoury AE, Yu YM, Young VR (1998) The 24-h whole body leucine and urea kinetics at normal and high protein intakes with exercise in healthy adults. Am J Physiol 275(2 Pt 1):E310–E320

Funchal C, Gottfried C, de Almeida LM, dos Santos AQ, Wajner M, Pessoa-Pureur R (2005) Morphological alterations and cell death provoked by the branched-chain alpha-amino acids accumulating in maple syrup urine disease in astrocytes from rat cerebral cortex. Cell Mol Neurobiol 25(5):851–867

Gamberino WC, Berkich DA, Lynch CJ, Xu B, LaNoue KF (1997) Role of pyruvate carboxylase in facilitation of synthesis of glutamate and glutamine in cultured astrocytes. J Neurochem 69(6):2312–2325

García-Cazorla A, Oyarzabal A, Fort J, Robles C, Castejón E, Ruiz-Sala P, Bodoy S, Merinero B, Lopez-Sala A, Dopazo J, Nunes V, Ugarte M, Artuch R, Palacín M, Rodríguez-Pombo P, Alcaide P, Navarrete R, Sanz P, Font-Llitjós M, Vilaseca MA, Ormaizabal A, Pristoupilova A, Agulló SB (2014) Two novel mutations in the BCKDK (branched-chain keto-acid dehydrogenase kinase) gene are responsible for a neurobehavioral deficit in two pediatric unrelated patients. Hum Mutat 35(4):470–477

García-Espinosa MA, Wallin R, Hutson SM, Sweatt AJ (2007) Widespread neuronal expression of branched-chain aminotransferase in the CNS: implications for leucine/glutamate metabolism and for signaling by amino acids. J Neurochem 100(6):1458–1468

Goodwin GW, Zhang B, Paxton R, Harris RA (1988) Determination of activity and activity state of branched-chain alpha-keto acid dehydrogenase in rat tissues. Methods Enzymol 166: 189–201

Goto M, Miyahara I, Hirotsu K et al (2005) Structural determinants for branched-chain aminotransferase isozyme-specific inhibition by the anticonvulsant drug gabapentin. J Biol Chem 280(44):37246–37256

Gottlieb M, Wang Y, Teichberg VI (2003) Blood-mediated scavenging of cerebrospinal fluid glutamate. J Neurochem 87(1):119–126

Guerreiro G, Mescka CP, Sitta A, Donida B, Marchetti D, Hammerschmidt T, Faverzani J, Coelho Dde M, Wajner M, Dutra-Filho CS, Vargas CR (2015) Urinary biomarkers of oxidative damage in Maple syrup urine disease: the L-carnitine role. Int J Dev Neurosci 42:10–14

Gulhati P, Bowen KA, Liu J, Stevens PD, Rychahou PG, Chen M, Lee EY, Weiss HL, O'Connor KL, Gao T, Evers BM (2011) mTORC1 and mTORC2 regulate EMT, motility, and metastasis of colorectal cancer via RhoA and Rac1 signaling pathways. Cancer Res 71(9):3246–3256

Hall TR, Wallin R, Reinhart GD et al (1993) Branched chain aminotransferase isoenzymes. Purification and characterization of the rat brain isoenzyme. J Biol Chem 268(5):3092–3098

Harper AE, Benjamin E (1984) Relationship between intake and rate of oxidation of leucine and alpha-ketoisocaproate in vivo in the rat. J Nutr 114(2):431–440

Harris RA, Joshi M, Jeoung NH (2004) Mechanisms responsible for regulation of branched-chain amino acid catabolism. Biochem Biophys Res Commun 313(2):391–396, Review

Harris RA, Zhang B, Goodwin GW, Kuntz MJ, Shimomura Y, Rougraff P, Dexter P, Zhao Y, Gibson R, Crabb DW (1990) Regulation of the branched-chain alpha-ketoacid dehydrogenase

and elucidation of a molecular basis for maple syrup urine disease. Adv Enzyme Regul 30:245–263
Hawkins RA, O'Kane RL, Simpson IA, Viña JR (2006) Structure of the blood-brain barrier and its role in the transport of amino acids. J Nutr 136(1 Suppl):218S–226S, Review
Hosoya K, Sugawara M, Asaba H, Terasaki T (1999) Blood-brain barrier produces significant efflux of L-aspartic acid but not D-aspartic acid: in vivo evidence using the brain efflux index method. J Neurochem 73(3):1206–1211
Hull J, Hindy ME, Kehoe PG et al (2012) Distribution of the branched chain aminotransferase proteins in the human brain and their role in glutamate regulation. J Neurochem 123(6):997–1009
Hull J, Patel V, El Hindy M et al (2015) Regional increase in the expression of the BCAT proteins in Alzheimer's disease brain: Implications in glutamate toxicity. J Alzheimers Disease 45(3): 891–905.
Hutson SM (1988) Subcellular distribution of branched-chain aminotransferase activity in rat tissues. J Nutr 118(12):1475–1481
Hutson SM, Cree TC, Harper AE (1978) Regulation of leucine and alpha-ketoisocaproate metabolism in skeletal muscle. J Biol Chem 253(22):8126–8133
Hutson SM, Zapalowski C, Cree TC et al (1980) Regulation of leucine and alpha-ketoisocaproic acid metabolism in skeletal muscle. Effects of starvation and insulin. J Biol Chem 255(6):2418–2426
Hutson SM, Bledsoe RK, Hall TR, Dawson PA (1995) Cloning and expression of the mammalian cytosolic branched chain aminotransferase isoenzyme. J Biol Chem 270(51):30344–30352
Hutson SM, Berkich D, Drown P et al (1998) Role of branched-chain aminotransferase isoenzymes and gabapentin in neurotransmitter metabolism. J Neurochem 71(2):863–874
Hutson SM, Lieth E, LaNoue KF (2001) Function of leucine in excitatory neurotransmitter metabolism in the central nervous system. J Nutr 131(3):846S–850S
Hutson SM, Sweatt AJ, Lanoue KF (2005) Branched-chain [corrected] amino acid metabolism: implications for establishing safe intakes. J Nutr 135(6 Suppl):1557S–1564S, Review
Hutson SM, Islam MM, Zaganas I (2011) Interaction between glutamate dehydrogenase (GDH) and L-leucine catabolic enzymes: intersecting metabolic pathways. Neurochem Int 59(4):518–524
Ichihara A (1975) Isozyme patterns of branched-chain amino acid transaminase during cellular differentiation and carcinogenesis. Ann N Y Acad Sci 259:347–354
Islam MM, Wallin R, Wynn RM, Conway M, Fujii H, Mobley JA, Chuang DT, Hutson SM (2007) A novel branched-chain amino acid metabolon. Protein-protein interactions in a supramolecular complex. J Biol Chem 282(16):11893–11903
Islam MM, Nautiyal M, Wynn RM et al (2010) Branched-chain amino acid metabolon: interaction of glutamate dehydrogenase with the mitochondrial branched-chain aminotransferase (BCATm). J Biol Chem 285(1):265–276
Jacinto E, Loewith R, Schmidt A, Lin S, Rüegg MA, Hall A, Hall MN (2004) Mammalian TOR complex 2 controls the actin cytoskeleton and is rapamycin insensitive. Nat Cell Biol 6(11):1122–1128
Jansonius JN (1998) Structure, evolution and action of vitamin B6-dependent enzymes. Curr Opin Struct Biol 8(6):759–769
Kanamori K, Ross BD, Kondrat RW (1998) Rate of glutamate synthesis from leucine in rat brain measured in vivo by 15N NMR. J Neurochem 70(3):1304–1315
Klee D, Thimm E, Wittsack HJ, Schubert D, Primke R, Pentang G, Schaper J, Mödder U, Antoch A, Wendel U, Cohnen M (2013) Structural white matter changes in adolescents and young adults with maple syrup urine disease. J Inherit Metab Dis 36(6):945–953
Korein J, Sansaricq C, Kalmijn M, Honig J, Lange B (1994) Maple syrup urine disease: clinical, EEG, and plasma amino acid correlations with a theoretical mechanism of acute neurotoxicity. Int J Neurosci 79:21–45

Lai JC, Sheu KF, Kim YT, Clarke DD, Blass JP (1986) The subcellular localization of glutamate dehydrogenase (GDH): is GDH a marker for mitochondria in brain? Neurochem Res 11(5):733–744

LaNoue KF, Berkich DA, Conway M, Barber AJ, Hu LY, Taylor C, Hutson S (2001) Role of specific aminotransferases in de novo glutamate synthesis and redox shuttling in the retina. J Neurosci Res 66(5):914–922

Lee WJ, Hawkins RA, Viña JR, Peterson DR (1998) Glutamine transport by the blood-brain barrier: a possible mechanism for nitrogen removal. Am J Physiol 274(4 Pt 1):C1101–C1107

Lehre KP, Levy LM, Ottersen OP, Storm-Mathisen J, Danbolt NC (1995) Differential expression of two glial glutamate transporters in the rat brain: quantitative and immunocytochemical observations. J Neurosci 15(3 Pt 1):1835–1853

Li H, Zhang B, Liu Y, Yin C (2014) EBP50 inhibits the migration and invasion of human breast cancer cells via LIMK/cofilin and the PI3K/Akt/mTOR/MMP signalling pathway. Med Oncol 31(9):162

Lieth E, LaNoue KF, Berkich DA et al (2001) Nitrogen shuttling between neurons and glial cells during glutamate synthesis. J Neurochem 76(6):1712–1723

Lin HM, Kaneshige M, Zhao L, Zhang X, Hanover JA, Cheng SY (2001) An isoform of branched-chain aminotransferase is a novel co-repressor for thyroid hormone nuclear receptors. J Biol Chem 276(51):48196–48205

Mac M, Nehlig A, Nałecz MJ, Nałecz KA (2000) Transport of alpha-ketoisocaproate in rat cerebral cortical neurons. Arch Biochem Biophys 376(2):347–353

McKenna MC, Tildon JT, Stevenson JH, Boatright R, Huang S (1993) Regulation of energy metabolism in synaptic terminals and cultured rat brain astrocytes: differences revealed using aminooxyacetate. Dev Neurosci 15(3–5):320–329

McKenna MC, Sonnewald U, Huang X, Stevenson J, Zielke HR (1996a) Exogenous glutamate concentration regulates the metabolic fate of glutamate in astrocytes. J Neurochem 66(1):386–393

McKenna MC, Tildon JT, Stevenson JH, Huang X (1996b) New insights into the compartmentation of glutamate and glutamine in cultured rat brain astrocytes. Dev Neurosci 18(5–6):380–390

Mehta PK, Christen P (2000) The molecular evolution of pyridoxal-5′-phosphate-dependent enzymes. Adv Enzymol Relat Areas Mol Biol 74:129–184, Review

Morton DH, Strauss KA, Robinson DL, Puffenberger EG, Kelley RI (2002) Diagnosis and treatment of maple syrup disease: a study of 36 patients. Pediatrics 109:999–1008

Muelly ER, Moore GJ, Bunce SC, Mack J, Bigler DC, Morton DH, Strauss KA (2013) Biochemical correlates of neuropsychiatric illness in maple syrup urine disease. J Clin Invest 123(4):1809–1820

Nicklin P, Bergman P, Zhang B, Triantafellow E, Wang H, Nyfeler B, Yang H, Hild M, Kung C, Wilson C, Myer VE, MacKeigan JP, Porter JA, Wang YK, Cantley LC, Finan PM, Murphy LO (2009) Bidirectional transport of amino acids regulates mTOR and autophagy. Cell 136(3):521–534

Novarino G, El-Fishawy P, Kayserili H, Meguid NA, Scott EM, Schroth J, Silhavy JL, Kara M, Khalil RO, Ben-Omran T, Ercan-Sencicek AG, Hashish AF, Sanders SJ, Gupta AR, Hashem HS, Matern D, Gabriel S, Sweetman L, Rahimi Y, Harris RA, State MW, Gleeson JG (2012) Mutations in BCKD-kinase lead to a potentially treatable form of autism with epilepsy. Science 338(6105):394–397

O'Kane RL, Hawkins RA (2003) Na+-dependent transport of large neutral amino acids occurs at the abluminal membrane of the blood-brain barrier. Am J Physiol Endocrinol Metab 285(6):E1167–E1173

O'Kane RL, Viña JR, Simpson I, Hawkins RA (2004) Na+-dependent neutral amino acid transporters A, ASC, and N of the blood-brain barrier: mechanisms for neutral amino acid removal. Am J Physiol Endocrinol Metab 287(4):E622–E629

Odessey R, Goldberg AL (1979) Leucine degradation in cell-free extracts of skeletal muscle. Biochem J 178(2):475–489

Oldendorf WH (1973) Stereospecificity of blood-brain barrier permeability to amino acids. Am J Physiol 224(4):967–969

Oldendorf WH, Cornford ME, Brown WJ (1977) The large apparent work capability of the blood-brain barrier: a study of the mitochondrial content of capillary endothelial cells in brain and other tissues of the rat. Ann Neurol 1(5):409–417

Pérez-Villaseñor G, Tovar AR, Moranchel AH, Hernández-Pando R, Hutson SM, Torres N (2005) Mitochondrial branched chain aminotransferase gene expression in AS-30D hepatoma rat cells and during liver regeneration after partial hepatectomy in rat. Life Sci 78(4):334–339

Popov KM, Zhao Y, Shimomura Y, Kuntz MJ, Harris RA (1992) Branched-chain alpha-ketoacid dehydrogenase kinase. Molecular cloning, expression, and sequence similarity with histidine protein kinases. J Biol Chem 267(19):13127–13130

Reed LJ, Damuni Z, Merryfield ML (1985) Regulation of mammalian pyruvate and branched-chain alpha-keto acid dehydrogenase complexes by phosphorylation-dephosphorylation. Curr Top Cell Regul 27:41–49

Ribeiro CA, Sgaravatti AM, Rosa RB, Schuck PF, Grando V, Schmidt AL, Ferreira GC, Perry ML, Dutra-Filho CS, Wajner M (2008) Inhibition of brain energy metabolism by the branched-chain amino acids accumulating in maple syrup urine disease. Neurochem Res 33(1):114–124

Rothstein JD, Martin L, Levey AI, Dykes-Hoberg M, Jin L, Wu D, Nash N, Kuncl RW (1994) Localization of neuronal and glial glutamate transporters. Neuron 13(3):713–725

Sakai R, Cohen DM, Henry JF, Burrin DG, Reeds PJ (2004) Leucine-nitrogen metabolism in the brain of conscious rats: its role as a nitrogen carrier in glutamate synthesis in glial and neuronal metabolic compartments. J Neurochem 88(3):612–622

Salzmann D, Christen P, Mehta PK, Sandmeier E (2000) Rates of evolution of pyridoxal-5′-phosphate-dependent enzymes. Biochem Biophys Res Commun 270(2):576–580

Schmitt A, Kugler P (1999) Cellular and regional expression of glutamate dehydrogenase in the rat nervous system: non-radioactive in situ hybridization and comparative immunocytochemistry. Neuroscience 92(1):293–308

Schneider G, Käck H, Lindqvist Y (2000) The manifold of vitamin B6 dependent enzymes. Structure 8(1):R1–R6, Review

Schulman JD, Lustberg TJ, Kennedy JL, Museles M, Seegmiller JE (1970) A new variant of maple syrup urine disease (branched chain ketoaciduria). Clinical and biochemical evaluation. Am J Med 49:118–124

Shank RP, Bennett GS, Freytag SO, Campbell GL (1985) Pyruvate carboxylase: an astrocyte-specific enzyme implicated in the replenishment of amino acid neurotransmitter pools. Brain Res 329(1–2):364–367

She P, Reid TM, Bronson SK, Vary TC, Hajnal A, Lynch CJ, Hutson SM (2007) Disruption of BCATm in mice leads to increased energy expenditure associated with the activation of a futile protein turnover cycle. Cell Metab 6(3):181–194

Shinnick FL, Harper AE (1976) Branched-chain amino acid oxidation by isolated rat tissue preparations. Biochim Biophys Acta 437(2):477–486

Simon E, Schwarz M, Wendel U (2007) Social outcome in adults with maple syrup urine disease (MSUD). J Inherit Metab Dis 30:264

Sitta A, Ribas GS, Mescka CP, Barschak AG, Wajner M, Vargas CR (2014) Neurological damage in MSUD: the role of oxidative stress. Cell Mol Neurobiol 34(2):157–165

Smith QR, Momma S, Aoyagi M, Rapoport SI (1987) Kinetics of neutral amino acid transport across the blood-brain barrier. J Neurochem 49(5):1651–1658

Sonnewald U, Westergaard N, Petersen SB, Unsgård G, Schousboe A (1993) Metabolism of [U-13C]glutamate in astrocytes studied by 13C NMR spectroscopy: incorporation of more label into lactate than into glutamine demonstrates the importance of the tricarboxylic acid cycle. J Neurochem 61(3):1179–1182

Spanaki C, Zaganas I, Kleopa KA, Plaitakis A (2010) Human GLUD2 glutamate dehydrogenase is expressed in neural and testicular supporting cells. J Biol Chem 285(22):16748–16756

Spanaki C, Kotzamani D, Petraki Z, Drakos E, Plaitakis A (2014) Heterogeneous cellular distribution of glutamate dehydrogenase in brain and in non-neural tissues. Neurochem Res 39(3): 500–515

Suryawan A, Hawes JW, Harris RA et al (1998) A molecular model of human branched-chain amino acid metabolism. Am J Clin Nutr 68(1):72–81

Sweatt AJ, Garcia-Espinosa MA, Wallin R et al (2004a) Branched-chain amino acids and neurotransmitter metabolism: expression of cytosolic branched-chain aminotransferase (BCATc) in the cerebellum and hippocampus. J Comp Neuro 477(4):360–370

Sweatt AJ, Wood M, Suryawan A et al (2004b) Branched-chain amino acid catabolism: unique segregation of pathway enzymes in organ systems and peripheral nerves. Am J Physiol Endocrinol Metab 286(1):E64–E76

Teller JK, Fahien LA, Valdivia E (1990) Interactions among mitochondrial aspartate aminotransferase, malate dehydrogenase, and the inner mitochondrial membrane from heart, hepatoma, and liver. J Biol Chem 265(32):19486–19494

Than NG, Sümegi B, Than GN, Bellyei S, Bohn H (2001) Molecular cloning and characterization of placental tissue protein 18 (PP18a)/human mitochondrial branched-chain aminotransferase (BCATm) and its novel alternatively spliced PP18b variant. Placenta 22(2–3):235–243

van Hall G, Raaymakers JS, Saris WH, Wagenmakers AJ (1995a) Ingestion of branched-chain amino acids and tryptophan during sustained exercise in man: failure to affect performance. J Physiol 486(Pt 3):789–794

van Hall G, van der Vusse GJ, Söderlund K, Wagenmakers AJ (1995b) Deamination of amino acids as a source for ammonia production in human skeletal muscle during prolonged exercise. J Physiol 489(Pt 1):251–261

Wallin R, Hall TR, Hutson SM (1990) Purification of branched chain aminotransferase from rat heart mitochondria. J Biol Chem 265(11):6019–6024

Wang XL, Li CJ, Xing Y, Yang YH, Jia JP (2015a) Hypervalinemia and hyperleucine-isoleucinemia caused by mutations in the branched-chain-amino-acid aminotransferase gene. J Inherit Metab Dis 38(5):855–861

Wang S, Tsun ZY, Wolfson RL, Shen K, Wyant GA, Plovanich ME, Yuan ED, Jones TD, Chantranupong L, Comb W, Wang T, Bar-Peled L, Zoncu R, Straub C, Kim C, Park J, Sabatini BL, Sabatini DM (2015b) Metabolism. Lysosomal amino acid transporter SLC38A9 signals arginine sufficiency to mTORC1. Science 347(6218):188–194

Wynn RM, Kato M, Machius M, Chuang JL, Li J, Tomchick DR, Chuang DT (2004) Molecular mechanism for regulation of the human mitochondrial branched-chain alpha-ketoacid dehydrogenase complex by phosphorylation. Structure 12(12):2185–2196

Yennawar NH, Conway ME, Yennawar HP et al (2002) Crystal structures of human mitochondrial branched chain aminotransferase reaction intermediates: ketimine and pyridoxamine phosphate forms. Biochemistry 41(39):11592–11601

Yennawar NH, Islam MM, Conway M et al (2006) Human mitochondrial branched chain aminotransferase isozyme: structural role of the CXXC center in catalysis. J Biol Chem 281(51):39660–39671

Yoshimura T, Jhee KH, Soda K (1996) Stereospecificity for the hydrogen transfer and molecular evolution of pyridoxal enzymes. Biosci Biotechnol Biochem 60(2):181–187, Review

Yudkoff M (1997) Brain metabolism of branched-chain amino acids. Glia 21(1):92–98, Review

Yudkoff M, Nissim I, Hertz L (1990) Precursors of glutamic acid nitrogen in primary neuronal cultures: studies with 15N. Neurochem Res 15(12):1191–1196

Yudkoff M, Nissim I, Kim S, Pleasure D, Hummeler K, Segal S (1983) [15N] leucine as a source of [15N] glutamate in organotypic cerebellar explants. Biochem Biophys Res Commun 115(1):174–179

Yudkoff M, Nissim I, Daikhin Y, Lin ZP, Nelson D, Pleasure D, Erecinska M (1993) Brain glutamate metabolism: neuronal-astroglial relationships. Dev Neurosci 15(3–5):343–350, Review

Yudkoff M, Daikhin Y, Nissim I, Pleasure D, Stern J, Nissim I (1994) Inhibition of astrocyte glutamine production by alpha-ketoisocaproic acid. J Neurochem 63(4):1508–1515

Yudkoff M, Daikhin Y, Grunstein L, Nissim I, Stern J, Pleasure D, Nissim I (1996a) Astrocyte leucine metabolism: significance of branched-chain amino acid transamination. J Neurochem 66(1):378–385

Yudkoff M, Daikhin Y, Nelson D, Nissim I, Erecińska M (1996b) Neuronal metabolism of branched-chain amino acids: flux through the aminotransferase pathway in synaptosomes. J Neurochem 66(5):2136–2145

Yudkoff M, Daikhin Y, Nissim I, Horyn O, Luhovyy B, Lazarow A, Nissim I (2005) Brain amino acid requirements and toxicity: the example of leucine. J Nutr 135(6 Suppl):1531S–1538S, Review

Zaganas I, Kanavouras K, Mastorodemos V, Latsoudis H, Spanaki C, Plaitakis A (2009) The human GLUD2 glutamate dehydrogenase: localization and functional aspects. Neurochem Int 55(1–3):52–63

Zhou Y, Jetton TL, Goshorn S, Lynch CJ, She P (2010) Transamination is required for {alpha}-ketoisocaproate but not leucine to stimulate insulin secretion. J Biol Chem 285(44):33718–33726. doi:10.1074/jbc.M110.136846, Epub 2010 Aug 24

Chapter 6
Glutaminases

Javier Márquez, José M. Matés, and José A. Campos-Sandoval

Abstract Mammalian glutaminases catalyze the stoichiometric conversion of L-glutamine to L-glutamate and ammonium ions. In brain, glutaminase is considered the prevailing pathway for synthesis of the neurotransmitter pool of glutamate. Besides neurotransmission, the products of glutaminase reaction also fulfill crucial roles in energy and metabolic homeostasis in mammalian brain. In the last years, new functional roles for brain glutaminases are being uncovered by using functional genomic and proteomic approaches. Glutaminases may act as multifunctional proteins able to perform different tasks: the discovery of multiple transcript variants in neurons and glial cells, novel extramitochondrial localizations, and isoform-specific proteininteracting partners strongly support possible moonlighting functions for these proteins. In this chapter, we present a critical account of essential works on brain glutaminase 80 years after its discovery. We will highlight the impact of recent findings and thoughts in the context of the glutamate/glutamine brain homeostasis.

Keywords Glutamate • Glutamine • Glutaminase-interacting proteins • Astrocytes • Neurons

Abbreviations

AGC	Aspartate–glutamate carrier
BBB	Blood-brain barrier
EM	Electron microscopy
GA	Phosphate-activated glutaminase
GAB	*Gls2*-encoded long GA isoform
GAC	*Gls*-encoded short GA isoform

J. Márquez (✉) • J.M. Matés • J.A. Campos-Sandoval
Departamento de Biología Molecular y Bioquímica, Canceromics Lab,
Facultad de Ciencias, Universidad de Málaga, Málaga 29071, Spain

Instituto de Investigación Biomédica de Málaga (IBIMA), Málaga, Spain
e-mail: marquez@uma.es

A. Schousboe, U. Sonnewald (eds.), *The Glutamate/GABA-Glutamine Cycle*,
Advances in Neurobiology 13, DOI 10.1007/978-3-319-45096-4_6

GAD	Glutamic acid decarboxylase
GDH	Glutamate dehydrogenase
GFAP	Glial fibrillary acidic protein
GIP	Glutaminase-interacting protein
GS	Glutamine synthetase
IMM	Inner mitochondrial membrane
KGA	*Gls*-encoded long GA isoform
LGA	*Gls2*-encoded short GA isoform
NBT	Nitroblue tetrazolium
NMR	Nuclear magnetic resonance
NPCs	Neural progenitor cells
PDZ	PSD95/Dlg/ZO1 domains
TCA	Tricarboxylic acid cycle

6.1 Introduction

Glutamate (Glu) is the most important neurotransmitter in normal brain function: about 80–90 % of all brain synapses release this nonessential amino acid (Braitenberg and Schüz 1998; Attwell and Laughlin 2001). Glu does not cross the blood-brain barrier (BBB) (Hawkins 2009) and therefore must be synthesized in neurons from local precursors. The enzyme phosphate-activated glutaminase (GA) catalyzes the hydrolytic deamidation of glutamine (Gln) to Glu and ammonium ions and appears as the major Gln metabolizing enzyme in brain (Hertz 1979; Kvamme 1984). This enzyme has been considered as the principal pathway for neurotransmitter Glu production (Nicklas et al. 1987); accordingly, the bulk of Glu released at the synapses derives from precursor Gln (Bradford et al. 1978; Hamberger et al. 1979; Thanki et al. 1983; Laake et al. 1995; Hertz and Zielke 2004), while inhibition of GA depletes the stores of Glu (Conti and Minelli 1994). Nevertheless, presynaptic Glu is not always necessarily derived from Gln: synthesis from tricarboxylic acid cycle (TCA) intermediates also contributes to generation of excitatory Glu (Waagepetersen et al. 2005; Takeda et al. 2012), although the exact contribution of each source is a matter of debate and remains to be established (Kam and Nicoll 2007).

Enzymes that break the amide bond of Gln and transfer the ammonia to other specific substrates in biosynthetic pathways are known as glutamine amidotransferases (EC 2.4.2). However, these enzymes should not be confused with GA, which is a glutamine amidohydrolase (EC 3.5.1.2), a genuine hydrolytic enzyme releasing the γ-glutamyl group and ammonium into the media without transfer to any other acceptor molecule. The catalytic reaction requires no energy and is essentially irreversible: the K'_{eq} is 322 at pH 7.0 and 25 °C (Benzinger et al. 1959), which indicates that synthesis of Glu is strongly favored. Mammalian glutaminases only show reactivity toward free Gln and cannot deamidate internal Gln residues in proteins, although such activity has been reported for microbial protein glutaminases (Yamaguchi et al. 2001). On the other hand, prokaryotic asparaginases may also deamidate free Gln and this has been used as a chemotherapeutic treatment to deplete plasma Gln levels in children suffering leukemia (Holcenberg et al. 1979). However, mammalian asparaginases (L-asparaginase-amidohydrolase, EC 3.5.1.1) do not utilize Gln as substrates (Schalk et al. 2014).

Besides the transmitter pool of Glu in excitatory synapses, there is a metabolic pool with great relevance because this amino acid is a primary metabolic fuel for brain cells (Erecinska and Silver 1990) and a precursor of many other important metabolites (Fonnum 1984). In fact, the first step in the mitochondrial breakdown of glutamine to pyruvate/lactate, a pathway known as glutaminolysis, is initiated by GA and has been largely recognized as a fundamental process for energy supply in the bioenergetics of many normal and transformed cell types (Kovacevic and McGivan 1983; Márquez et al. 2015). Furthermore, Gln can be used through glutaminolysis as an anaplerotic substrate to replenish TCA intermediates (DeBerardinis et al. 2007).

6.2 Historical Account: Pioneer Works

Krebs, in his pioneer studies on amino acids metabolism, discovered GA while studying the enzymatic hydrolysis of Gln in animal tissues (Krebs 1935). Krebs first noticed the existence of two different types of GA, classifying them as "brain type" (also present in kidney) and "hepatic type." Lately, Errera and Greenstein (1949) coined the term "phosphate-activated" GA to describe the enzymatic activity which was strongly stimulated in tissue extracts after addition of inorganic phosphate (Pi). The enzyme is widely distributed in mammalian tissues where it fulfills essential tasks related to tissue-specific function (Curthoys and Watford 1995).

The traditional terms, kidney- and liver-type isoenzymes, denoted marked differences in molecular structures, kinetic, regulatory, and immunological properties that were early detected and justified their study separately (Kovacevic and McGivan 1983). Historically, kidney-type enzymes were the first GAs isolated and characterized in mammalian tissues. Kvamme's group leads the way in this area by isolating to apparent homogeneity pig renal and pig brain GAs (Kvamme et al. 1970; Svenneby et al. 1973). Later rat kidney (Curthoys et al. 1976), cow brain (Chiu and Boeker 1979), pig brain (Nimmo and Tipton 1980), and rat brain (Haser et al. 1985; Kaneko et al. 1987) GAs were also purified. Kidney-type enzymes were isolated in highly purified states from cancer cells indeed (Quesada et al. 1988; Segura et al. 1995).

Most kidney-type enzymes undergo specific and extensive polymerization in the presence of phosphate and phosphate–borate buffers. The reversible association–dissociation equilibrium has been advantageously used to purify this GA isoform. In sharp contrast, the liver-type isoenzymes do not show aggregation and self-assembly in high molecular mass polymers. This fact, along with a remarkable susceptibility to proteolytic degradation, has hampered their purification; thus, liver-type isozymes have been considerably more refractory to purification than their kidney-type counterparts. Actually, the isolation of pure rat liver isoform was achieved almost two decades after the first kidney-type enzyme was isolated (Heini et al. 1987; Smith and Watford 1988), although partial purifications were also previously reported (Huang and Knox 1976; Patel and McGivan 1984). Finally, it is worthy to mention that the classical terms kidney(K)-type and liver(L)-type are currently obsolete and should be avoided for the sake of coherence: the new expression data for mammalian GA do not support any more such kind of terminology (see Matés et al. 2013 for an update on GA nomenclature).

6.3 Mammalian Glutaminase Genes

The mammalian GA family members are encoded by two paralogous genes, *Gls* and *Gls2*, presumably derived from a common ancestral gene by duplication and divergent evolution (Porter et al. 2002; Pérez-Gómez et al. 2003). In humans, the *GLS* gene is located on chromosome 2 and encodes GA isozymes classically referred to as kidney-type (K-type), while *GLS2* is located on chromosome 12 and codes for liver-type (L-type) isozymes (Aledo et al. 2000). Orthologous genes have been described in rat (Chung-Bok et al. 1997) and mouse (Mock et al. 1989). The human *GLS* gene (cited in many publications with the incorrect term *GLS1*: see HUGO Gene nomenclature, http://www.genenames.org/) spans 82 kb and splits into 19 exons (Porter et al. 2002). Two different transcripts arise from this gene (Fig. 6.1): the KGA transcript, originally found in kidney, composed by 18 exons and formed by joining exons 1–14 and 16–19, and the GAC transcript which appears by alternative splicing and uses only the first 15 exons (Elgadi et al. 1999; Porter et al. 2002) (Fig. 6.1).

The human *GLS2* gene has a length of approximately 18 kb and splits into 18 exons (Pérez-Gómez et al. 2003) (Fig. 6.1). Two transcripts have been identified from the mammalian *Gls2* gene: the canonical long transcript termed GAB, formed by joining all 18 exons of the gene (Pérez-Gómez et al. 2003), and the short transcript LGA that lacks exon 1 and was originally identified in rat liver (Smith and Watford 1990) (Fig. 6.1). Human GAB transcript was isolated as a cDNA clone from ZR-75 breast cancer cells encoding a protein of 602 amino acids, which is 67 amino acids longer than rat liver LGA protein (Gómez-Fabre et al. 2000) (Fig. 6.1).

Although located in different chromosomes, the human *GLS* and *GLS2* genes share a considerable degree of sequence similarity. Apart from the additional exon present in the human *GLS* gene, the main differences in the coding sequences of both genes are located at exons 1 and 18. Exon 1 shares 62.5 % similarity, but it codes for 129 amino acids in KGA and only for 61 amino acids in GAB, accounting for most of the 67 extra amino acids of KGA protein at the N-terminal. The sequences encoded by exon 1 contain the signals involved in mitochondrial targeting and translocation processes (Shapiro et al. 1991; Gómez-Fabre et al. 2000) (Fig. 6.2). Likewise, exon 18, which codes for the C-terminal region of both proteins, shows the lowest sequence similarity (29.4 %). This region of human

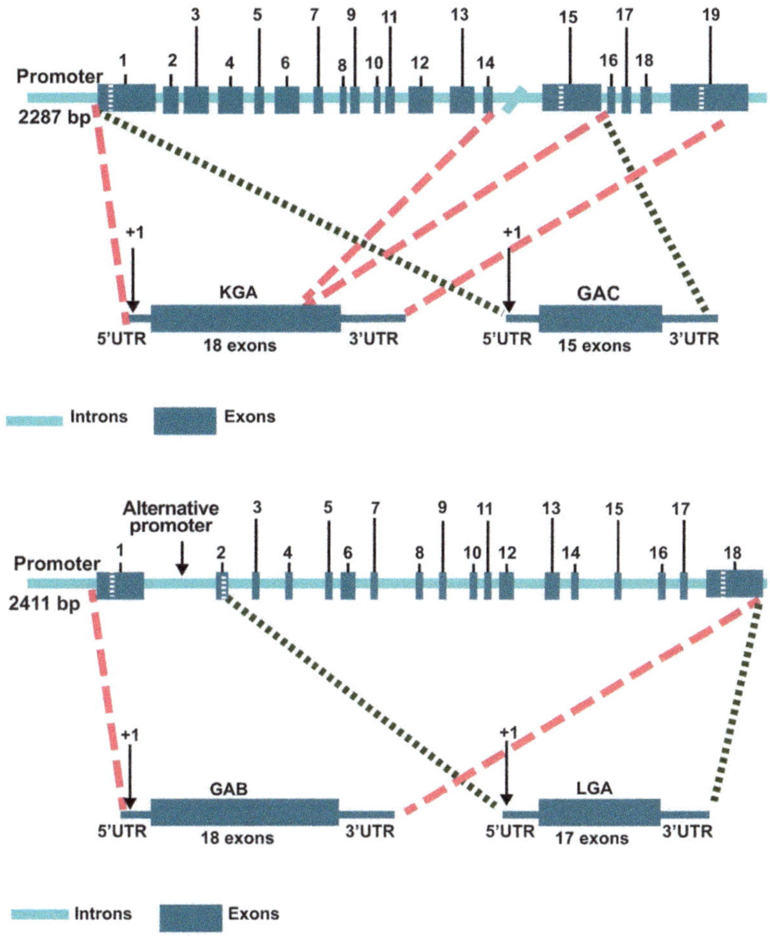

Fig. 6.1 Human glutaminase genes and mRNA transcripts. (*Top panel*) Human glutaminase *GLS* gene and alternative transcripts KGA and GAC. (*Bottom panel*) Human glutaminase *GLS2* gene and alternative transcripts GAB and LGA. Each gene is shown with introns depicted as solid light blue lines and exons as numbered dark blue boxes. The promoter regions are also indicated on the 5′-end of each gene, including the alternative promoter of the *GLS2* gene on intron 1. *Dashed red lines* indicate the exons forming KGA and GAB mRNA transcripts, while *dotted dark green lines* comprise exons involved in the generation of transcripts GAC and LGA. The transcription start site is marked by an arrow and numbered as +1

GLS2 proteins has been demonstrated to be involved in the recognition of PDZ (PSD95\Dlg\ZO1 domains) interaction modules (Olalla et al. 2001) (Fig. 6.2). Therefore, the most significant differences between human *GLS* and *GLS2* exons are located in regions involved with organelle targeting and protein–protein interactions, which may help to explain their differential function and regulation. Conversely, exons 3–17 of GLS and GLS2 mRNA transcripts have the same length and show a high sequence similarity.

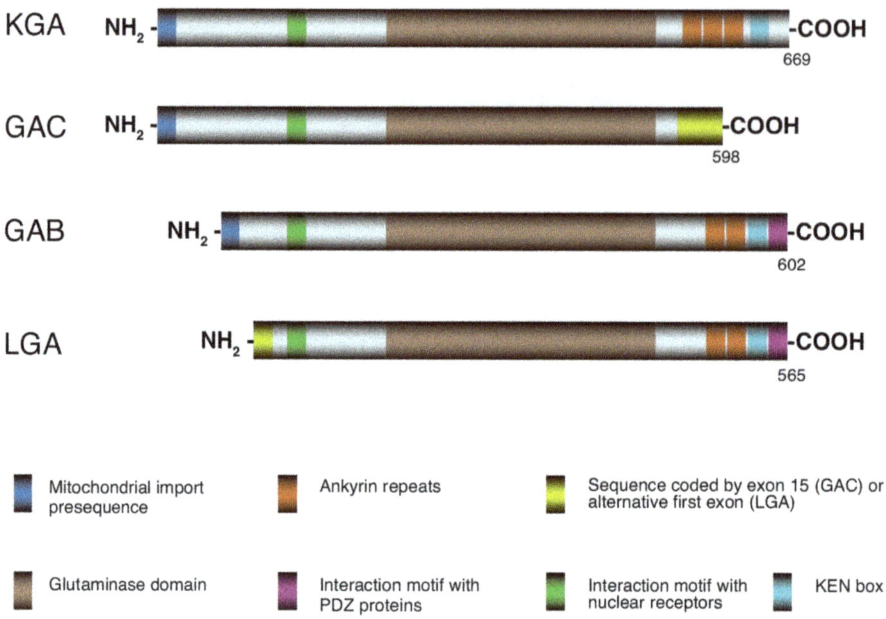

Fig. 6.2 Schematic illustration of the four human GA isoforms. The main signature sequences, motifs, and domains characterized or identified by sequence analysis are highlighted. Starting from the N-terminal end, the following domains are shown: mitochondrial import presequence, nuclear receptor interaction motif (LXXLL), glutaminase domain, ankyrin repeats, APC/C-Cdh1 (anaphase-promoting complex/cyclosome and activator protein Cdh1)-specific recognition motif (KEN) or KEN box, and PDZ protein recognition module (ESMV). The sequence coded by unique exon 15 (GAC isoform) and alternative first exon (LGA isoform) is also highlighted in *yellow* (For further information see text.)

The GLS isoform KGA is found in all tissues with GA activity with the exception of postnatal liver (Curthoys and Watford 1995; Aledo et al. 2000), although expression in liver endothelial cells has also been reported (Lohmann et al. 1999). The GAC splice variant, first isolated from a human HT-29 colon carcinoma cDNA library, is expressed predominantly in human cardiac muscle and pancreas, appreciably in placenta, kidney, and lung, but not in brain and liver (Elgadi et al. 1999). Other authors also found GAC expression in rat kidney and pig renal cells (Porter et al. 2002). On the other hand, GA transcripts derived from the *Gls2* gene were originally thought to be present in adult liver tissue and absent in extrahepatic tissues (Smith and Watford 1990; Curthoys and Watford 1995). This restricted pattern of expression was changed after new findings demonstrated the expression of GLS2 transcripts in extrahepatic tissues like brain, pancreas, and breast cancer cells (Gómez-Fabre et al. 2000). Moreover, simultaneous expression of both GLS and GLS2 isoenzymes is a more frequent event than previously thought, as has been demonstrated in human tissues (Aledo et al. 2000) and cancer cells (Turner and McGivan 2003; Pérez-Gómez et al. 2005). Therefore, the molecular portrait of GA isoform expression in mammalian tissues involves multiple transcripts even in a single cell type. The abundance of a particular GA mRNA species may significantly change depending upon the tissue type or the developmental or metabolic state of the tissue. The physiological meaning of the existence of different species of mRNA coding for GA is not fully understood. However, it is tempting to speculate that each transcript may represent a specific target for different stimuli, the overall GA expression being the balance between these stimuli.

6.4 Glutaminase Expression in Brain

6.4.1 GA Transcripts and Proteins Expressed in Brain Cells

As mentioned before, the *Gls*-encoded KGA isoform was the first GA enzyme isolated and characterized. Furthermore, very similar molecular and kinetic properties (see next section) were found for brain and kidney KGA (Haser et al. 1985; Kvamme 1998) while, in parallel, brain KGA cDNA sequences were also available after their cloning from rat and human brain (Banner et al. 1988; Nagase et al. 1998). These early studies established the classical pattern of GA expression in mammals, where KGA was originally thought as the only glutaminase present in brain (Haser et al. 1985; Curthoys and Watford 1995). Nowadays, the pattern of GA expression in mammalian tissues has been shown to be considerably more complex. Concretely, we first reported that human brain expresses GLS2 transcripts, in addition to the KGA isoform; these transcripts were ubiquitously expressed in brain regions with the strongest signal appearing in cerebral cortex (Aledo et al. 2000; Gómez-Fabre et al. 2000). Shortly after, Northern analysis and immunocytochemistry in brain of diverse mammalian species (human, monkey, rat, cow, mouse and rabbit) confirmed simultaneous expression of GLS (KGA) and GLS2 (GAB and/or LGA) isoenzymes; interestingly, both isoforms colocalize in numerous cells throughout the brain (Olalla et al. 2002).

The existence of alternative transcripts of the *Gls2* gene was recently demonstrated in brain and liver of three mammalian species: human, rat, and mouse (Martín-Rufián et al. 2012). Two GLS2 transcript variants showing alternative first exons were amplified: the long GAB transcript, previously cloned from ZR-75 breast cancer cells (Gómez-Fabre et al. 2000), and the short LGA transcript first characterized in rat liver (Smith and Watford 1990; Chung-Bok et al. 1997). Cloning and primer extension analysis of the human brain LGA transcript, to map its transcription start site (TSS), unambiguously demonstrated that this short GLS2 isoform arises by a combination of two mechanisms of transcriptional regulation: alternative transcription initiation and alternative promoter. The LGA variant contains both the TSS and the alternative promoter in the first intron of the *Gls2* gene, which is located 7 kb away from the canonical promoter of the GAB isoform (Martín-Rufián et al. 2012) (Fig. 6.1). Therefore, three main conclusions can be drawn from these studies: (1) two alternative GLS2 transcripts, GAB and LGA, are expressed in mammalian brain (and liver); (2) GAB is a new GA isozyme and not the human orthologous of the classical rat liver LGA isozyme; and (3) the short transcript variant LGA is not liver specific and is also expressed in brain.

Quantitative measurements of GLS2 transcript variants by real-time qRT-PCR demonstrated that the ratio of the two products varied very widely between tissues and species (Martín-Rufián et al. 2012). Thus, while LGA was slightly more abundant than GAB in mouse brain, the opposite was found in rat brain where GAB mRNA level was fourfold the amount of LGA. Despite this, GLS isoforms (KGA + GAC) showed the greatest abundance in mouse and rat brain, accounting for more than 90 % of total GA transcripts and becoming the predominant GA in mammalian brain (Martín-Rufián et al. 2012). In human brain tissue and cultured neurons and astrocytes, KGA and GAC transcripts showed similar expression levels (Szeliga et al. 2008). In liver, there was also a dramatic change in the relative abundance of GLS2 transcripts, which showed to be absolutely dependent upon the species: GAB was the predominant isoform in mouse, while the short LGA transcript was the most prominent in rat. These strikingly different expression patterns observed for GLS2 transcript variants imply selective transcriptional regulatory mechanisms being operative in distinct tissues and mammalian species.

Alternative mRNA isoforms not always translate into different protein isoforms. In fact, they can be involved in the regulation of translation initiation, RNA editing, and other processes. Therefore, additional experimental evidence at the protein level is required to assess the functionality of GLS and GLS2 mRNA variants. To our knowledge, evidence supporting expression, at the protein level, of the four different GA mRNAs detected in brain tissue (KGA, GAC, GAB, and LGA) have been obtained for all of them except for GAC isozyme. Isoform-specific antibodies against whole proteins, truncated proteins, and peptide sequences were widely used to reveal KGA and GAB isoforms in mammalian brain (Kaneko et al. 1987; Aoki et al. 1991; Laake et al. 1999; Olalla et al. 2002, 2008). The human brain LGA cDNA was in vitro transcribed and translated in a rabbit reticulocyte lysate system and GA activity was consistently detected for the whole LGA translated product. Moreover, immunoblot analysis of rat brain tissue, isolated rat brain mitochondria, and SHSY-5Y human neuroblastoma cells were consistent with co-expression of both GAB and LGA proteins (Martín-Rufián et al. 2012).

6.4.2 Molecular and Kinetic Properties

The subunit composition of rat brain (and kidney) KGA has been characterized: a heterotetrameric enzyme having two different subunits of 66 kDa and 68 kDa with a 3:1 stoichiometry (Haser et al. 1985; Shapiro et al. 1987; Perera et al. 1991). These two GA polypeptides are produced in vivo from a common precursor of 74 kDa (Shapiro et al. 1991; Holcomb et al. 2000) (Table 6.1). On the other hand, the KGA protomer of pig brain has been reported to contain identical subunits of 64 kDa (Svenneby et al. 1973) or 73 kDa (Nimmo and Tipton 1980). The full length human variant GAC cDNA codes for a protein of 598 amino acids with a molecular mass of about 65 kDa (Elgadi et al. 1999) (Table 6.1). Both KGA and GAC isoforms undergo specific and extensive polymerization in the presence of phosphate and

Table 6.1 Mammalian GA genes, transcripts, and protein isoforms

Transcript/protein name	KGA/GLS	GAC/GLS	GAB/GLS2	LGA/GLS2
Gene	*Gls*	*Gls*	*Gls2*	*Gls2*
Transcript lengths	18 exons/4348 nt	15 exons/3183 nt	18 exons/2408 nt	17 exons/2026 nt
Transcriptional mechanism	Alternative splicing	Alternative splicing	Canonical	Alternative transcription initiation and alternative promoter
Transcriptional regulation	c-Myc, c-Jun	–	p53, TAp63, TAp73	–
Expression	Ubiquitous in most nonhepatic mammalian tissues, human cancer cells	Cardiac muscle, pancreas, placenta, kidney, lung, human cancer cells	Brain, pancreas, immune cells, human cancer cells	Liver, brain, human cancer cells
Human proteins	669 aa	598 aa	602 aa	565 aa
Mature subunit molecular mass	68 and 66 kDa	58 kDa	63 kDa	58 kDa
Pi dependence	High	–	Low	Low
Ammonia activation	No	No	Very low	Strong
Glutamate inhibition	Strong	–	Moderate	No
Interacting partners	Caytaxin, Bmcc1s	–	GIP, SNT	–
Drug inhibitors	BPTES, 968, ebselen, chelerythrine, apomorphine, B839, zaprinast		Alkylbenzoquinones, apomorphine	

Data are for human GA except for kinetic and expression data that were also collected from pig and rat GA. (–) Not determined

phosphate–borate buffers. Thus, the purified native enzyme may appear in three different states: (1) a protomeric and inactive state in Tris–HCl buffer formed by two subunits (dimeric form); (2) an active form by combination of two protomers in the presence of Pi (four subunits, tetrameric form); and (3) a polymeric state with molecular masses higher than 2,000,000 in the presence of phosphate–borate buffer for KGA (Godfrey et al. 1977) or with Pi concentrations as low as 20 mM for GAC (Ferreira et al. 2013). These enzymes require a polyvalent anion for activity. In vitro, inorganic phosphate (Pi) is the most prominent stimulator of GA. Furthermore, there is a correlation between tetramer formation induced by Pi and enzyme activation (Svenneby 1971, 1972; Godfrey et al. 1977; Morehouse and Curthoys 1981). The tetramer–dimer equilibrium is concentration dependent; concretely, for the GAC isoform, it has been shown to occur also in the absence of Pi just by raising the enzyme concentration, which argues in favor of GAC having a greater tendency to oligomerize than KGA: the unique C-terminal sequence of GAC seems to be an important determinant for oligomerization (Cassago et al. 2012).

Purified brain KGA is an allosteric enzyme highly sensitive to changes in the level of Pi (Haser et al. 1985). For the purified enzyme, the activity was completely dependent on added Pi; the phosphate activation curve was sigmoidal with half-maximal activation at concentrations of 13.5 mM (Svenneby 1971) to 25 mM (Haser et al. 1985) and a Hill index of 1.4–1.5. In contrast, rat and pig brain KGA showed hyperbolic kinetics with regard to their dependence on Gln (Haser et al. 1985; Nimmo and Tipton 1981), although other authors have pointed out that this Michaelis–Menten behavior only occurs at pH 8, whereas at higher pH values the plots become concave upward indicating positive cooperativity (Kvamme 1984). Anyway, the K_M for Gln was dependent on the Pi concentration: increasing the Pi concentration from 10 mM to 150 mM decreased the K_M for Gln from 14 mM to 5 mM (Haser et al. 1985).

GLS isoforms are very sensitive to changes in the levels of Pi, but whether Pi is the true physiological stimulator of GA in brain remains to be determined. However, considering its brain concentration and the fact that it can be rapidly altered during neuronal activity, its candidature as an important physiological regulator of brain GA in vivo was postulated (Erecinska and Silver 1990). For example, in synaptosomes, the Pi activation curve of KGA was very steep and hyperbolic (Bradford and Ward 1976), instead of the sigmoidal curve seen in homogenates and purified enzyme preparations. This distinct kinetic behavior with Pi has been explained by the different intrasynaptosomal Pi concentration compared to the medium (Erecinska and Silver 1990). In addition, synaptosomes treated with KCl and veratridine, to simulate a state of increased neuronal activity in vivo, showed enhanced GA activity concomitantly with the rise in Pi, thereby authors postulated that increase in Pi was the main factor responsible for GA activation under depolarizing conditions (Erecinska et al. 1990).

The concentration of Pi in brain is 1–2 mM (Siesjö 1978; Erecinska and Silver 1989) but a sharp rise might be expected during periods of intense neuronal activation, due to the lowering of the cell energy charge after hydrolysis of high-energy compounds including nucleotide triphosphates. Interestingly, intramitochondrial levels of Pi can significantly increase in a rapid way in cellular models, like cancer cells, characterized by a strong induction of KGA/GAC activity: we found that tumor cell's mitochondria incubated with 0.5 mM Gln double their Pi concentration one minute after Gln exposure (Medina et al. 1988). Additional support for the candidature of Pi as a relevant in vivo effector of *Gls*-encoded GAs has recently come from structural data, obtained by X-ray analysis of truncated GAC proteins containing the GA domain (Cassago et al. 2012) (Fig. 6.2). They showed that Pi binds inside the catalytic pocket, resulting in allosteric stabilization of tetramers and facilitating substrate entry by outcompeting with the product, Glu, to guarantee enzyme cycling. This competence of Pi versus Glu for binding to the active site nicely explains the decreased inhibitory effect of Glu in the presence of increasing Pi concentrations (Shapiro et al. 1982). This competitive behavior between Pi and Glu was early postulated from kinetic analysis of KGA in human brain synaptosomes (Svenneby et al. 1986) and rat cortex homogenates (Bradford et al. 1984).

Main effectors of brain GA activity are, in fact, compounds that alter the Pi activation. For example, calcium activates the hydrolysis of Gln in brain synaptosomes (Kvamme et al. 1983), but it seems to affect GA indirectly by modulating the level of free intramitochondrial Pi (Erecinska et al. 1990). In fact, Ca^{2+} fails to stimulate purified brain KGA (Kvamme et al. 1983), reinforcing the notion that activation is rather an indirect effect. Another effector of possible physiological relevance is H^+. The enzyme's optimum pH is between 8 and 9; therefore, brain metabolism of Gln increases at alkaline pH (Erecinska et al. 1990), while strong reduction in GA activity was noted after lowering the pH of the incubation medium of synaptosomes (Kvamme and Olsen 1981). Of note, during increased neuronal activity the intracellular pH decreases and, hence, pH microdomains are generated in active regions, like dendrites, where these acid shifts are particularly larger (Willoughby and Schwiening 2002). Thus, it is tempting to speculate a decrease on neuronal GA function during periods of intense electrical activity. In contrast, an intracellular alkalinization is observed in mammalian astrocytes in response to repetitive neuronal activity (Chesler and Kraig 1989). Most important, in cultured astrocytes alkalinization induces Glu uptake and promotes a threefold increase in Gln content (Brookes 1997). Therefore, concrete populations of glial cells containing functional mitochondrial GA (see next section) might be activated through astrocytic alkalinization to enhance the production of endogenous Gln-derived Glu. Therefore, pH fluctuations during synaptic activity may also influence GA activity and cerebral function, being a physiologically important process during periods of intense neuronal electrical activity.

In brain, a wide variety of endogenous and exogenous effectors can modulate GA activity in the range of mM concentrations: sulphate, chloride, carboxylic acids, nucleotide triphosphates, riboflavin phosphate, acyl-CoA derivatives and the dye bromothymol blue (Erecinska and Silver 1990; Kvamme et al. 2000). The main inhibitor of brain KGA (and all GLS isoforms) was the first to be discovered: Glu, one product of the reaction (Krebs 1935). The inhibition is competitive (Chiu and Boeker 1979; Haser et al. 1985) and strongly dependent on relative Pi levels: the inhibitory effect of Glu sharply decreases at higher Pi concentration (Svenneby 1971; Shapiro et al. 1982; Campos et al. 1998), in agreement with the active site's mechanism explained before. The relative concentrations of Gln and Glu in glutamatergic terminals suggest that GA can be strongly inhibited in nerve cells as in their terminals (Kvamme et al. 2000). Furthermore, in vitro and in vivo measurements of GA activity in brain support this notion. Assuming an extracellular Gln concentration of 0.5 mM, the activity of pig brain GA in vivo with 5–8 mM Pi was calculated to be at most 5–10 % of that obtained at saturating concentrations of Gln and Pi (Kvamme 1998). Accordingly, in synaptosomal homogenates the GA rate was found to be 124 nmol/mg per min at 37 °C, whereas in intact synaptosomes it was only 7 nmol/mg per min (Ward and Bradford 1979). This dramatic decrease can be explained by Glu inhibition, since Glu concentration is at least 10 mM in intact nerve endings (Erecinska and Silver 1990) and 10 mM Glu inhibits GA by over 90 % in broken synaptosomes (Bradford et al. 1978). Moreover, a low rate of GA activity (4–5 nmol/mg per min at 28 °C) was calculated from the flux of ^{15}N from

0.5 mM [^{15}N]-Gln in nerve endings incubated at 2 mM Pi in the medium (Yudkoff et al. 1989). Finally, the in vivo GA activity in intact brain of hyperammonemic rats measured by ^{15}N nuclear magnetic resonance (NMR) is about 1 % of the reported in vitro activity shown by rat brain homogenates (Kanamori and Ross 1995). Taking together, these results suggest that GA activity in intact brain is maintained at a low level by a short-term regulation of its key effectors, the most likely ones being Pi, Glu and H^+. In conclusion, the situation is considerably more complex in vivo: GA activity will largely depend on the interplay of Gln, Glu, and Pi true concentrations in neurons and glial cells, as well as transient pH fluctuations induced by neural activity.

Conflicting data were reported for ammonium ions, the second end product of GA reaction, as effectors of brain GLS isoforms. In partially purified KGA protein isolated from pig brain (Nimmo and Tipton 1981) and cow brain (Chiu and Boeker 1979) ammonium showed no effect, as well as in cultured astrocytes and neurons (Kvamme et al. 1982; Hogstad et al. 1988). Nonetheless, inhibition of GA was reported in particulate preparations (Bradford and Ward 1976; Wallace and Dawson 1992) and homogenates (Benjamin 1981), while an activation effect was found for the purified pig brain enzyme (Svenneby 1971). In any case, the low concentration of ammonia in the CNS (<0.2 mM) makes unlikely a relevant effect of ammonia on GA activity, at least under normal physiological conditions. Other nonspecific inhibitors are compounds that react with thiol groups as well as Gln analogs like 6-diazo-5-oxonorleucine (DON) (Curthoys and Watford 1995; Campos et al. 1998).

A recently discovered GLS-specific inhibitor is BPTES [bis-2-(5-phenylacetamido-1,2,4-thiadiazol-2-yl)ethyl sulfide], a small molecule which prevents the formation of large phosphate-induced oligomers (Robinson et al. 2007). The inhibition mechanism was elucidated after structural data were recently obtained from crystallized mammalian GLS isoforms. Protein crystals were grown alone and in combination with Gln, Glu, Pi, and novel inhibitors. BPTES inhibition is achieved through an uncompetitive mechanism interfering with the Pi-induced allosteric activation: the GA domain (Fig. 6.2) of human KGA forms a highly symmetrical tetramer containing two molecules of BPTES that are positioned at the dimer/dimer interfaces; binding of BPTES triggers a major conformational change of the active site residues and also freezes the enzyme in a stable inactive tetramer (Thangavelu et al. 2012). Further, X-ray diffraction data of crystallized human and mouse GAC proteins demonstrated that BPTES inhibits this isoenzyme indeed, locking the GAC tetramer in a non-catalytic state and avoiding formation of larger active polymers (DeLaBarre et al. 2011; Ferreira et al. 2013). Other small molecules which are potent inhibitors of GLS isoenzymes are dibenzophenanthridines; in concrete, the compound 968, a bromo-benzophenanthridinone, is a potent specific inhibitor of the GAC isoform now receiving much attention as anticancer drug (Wang et al. 2010). It is also an allosteric inhibitor, but with a different mechanism to that of BPTES: it preferentially binds to an inactive monomeric state of GAC and prevents it from undergoing the required conformational changes to achieve activation (Stalnecker et al. 2015).

Much less is known with regard to the regulation of GLS2 isoforms in mammalian brain. The classical rat liver LGA isozyme has been characterized at the molecular and kinetic levels. The distinct kinetic behavior of mammalian GLS and GLS2 isozymes has been a hallmark frequently used to distinguish between GA isoforms. The main kinetic differences have been observed in the dependence of the activator Pi, low for GLS2, high for GLS; the relative affinity for the substrate Gln, higher in GLS than in GLS2 isoforms; and the inhibitory effect of Glu, a unique characteristic only reported for GLS isozymes (Kovacevic and McGivan 1983; Curthoys and Watford 1995). Also, rat liver LGA was completely dependent on the presence of ammonia, which is an obligatory activator at physiological concentrations (Verhoeven et al. 1983; Patel and McGivan 1984; Snodgrass and Lund 1984). This activation effect was largely dependent on the pH and ammonia (NH_3), rather than ammonium ions, as the activating species (McGivan and Bradford 1983). Rat liver LGA has a subunit molecular mass of 58 kDa (Heini et al. 1987; Smith and Watford 1988) (Table 6.1) and a flat pH curve with an optimum between 7.8 and 8.2.

The only GLS2 isoenzyme expressed in brain that has been characterized in a purified form is the GAB isoform. The recombinant human GAB protein was expressed in Sf9 insect cells (infected with recombinant baculovirus) and affinity purified (Campos-Sandoval et al. 2007). GAB is expressed as a precursor protein of 66 kDa and a mature, processed form, of about 63 kDa, after cleavage of the first 38-39 amino acids at the N-terminal sequence (Table 6.1). Its kinetic characterization demonstrated an allosteric behavior (Hill index of 2.7) with low affinity for Gln ($S_{0.5}$ values of 32 and 64 mM for high (150 mM) and low (5 mM) Pi, respectively), and low dependence for Pi as expected for a GLS2 isoenzyme. Surprisingly, GAB was inhibited by Glu, a characteristic only shown by GLS isoforms (Table 6.1). The inhibition has an IC_{50} value of 50 mM at low Pi concentrations (5 mM) and suboptimal Gln concentration (20 mM). These data are in the range of Ki values reported for Glu competitive inhibition of GLS isoenzymes at high Pi concentrations (Shapiro et al. 1982). The Glu inhibition of GAB disappears at high Pi concentrations (up to 100 mM Glu). This was the first report of inhibition by Glu of a purified *Gls2*-encoded GA isoform; previous studies using partially purified (Patel and McGivan 1984) or pure homogeneous (Smith and Watford 1988) LGA preparations from rat liver, as well as mitochondria isolated from human liver (Snodgrass and Lund 1984), always confirmed that liver LGA is not inhibited by Glu (up to 50 mM).

Another remarkable difference for human GAB was seen in relation with its main activators: it is scarcely activated by ammonia, unlike the liver LGA enzyme which is strictly dependent on ammonia. Partially purified LGA from rat and human liver, as well as LGA in intact and disrupted liver mitochondria, have an absolute requirement for ammonia as an obligatory activator (Häussinger and Sies 1979; Patel and McGivan 1984 and references therein). In fact, the ammonia activation has been considered the particular distinguishing feature of LGA isoforms. In liver, the physiological significance of this effect was considered in the context of feed-forward activation of hepatic GA by ammonia produced in the intestine (Häussinger and Sies 1979); thus, ammonia levels are amplified inside the mitochondria to increase flux through urea cycle via carbamoyl phosphate synthetase I (EC 6.3.4.16) activation (Häussinger 1983). However, as mentioned before, the ammonia levels in brain are kept low under normal circumstances, whereby activation of cerebral GA isozymes by this effector does not seem to play a relevant physiological role.

The existence of GLS2 isoforms with distinctive kinetic properties different from classical K- and L-type isozymes were also noted by other authors, although using homogenates or crude cell extracts instead of purified enzymes. For example, McGivan's group found mRNAs species identical to human GAB in several colorectal tumors, but without any evidence for the presence of an enzyme with LGA-type kinetics (Turner and McGivan 2003). In conclusion, more studies are needed to further characterize GLS2 protein isoforms in mammalian brain. Tissues like brain, which express both GLS and GLS2 isoforms, are endowed with a powerful system to guarantee biosynthesis of transmitter Glu and glutaminolysis under many different conditions, including a wide range of substrate, activator, and inhibitor concentrations.

Unlike GLS isoforms, relevant structural information from X-ray diffraction data and/or NMR studies at atomic resolution has not yet been published for GLS2 isoforms; hence, the search for specific compounds that block their activity has been hampered. Nonetheless, some selective inhibitors have been recently found for GLS2 isoforms: apomorphine showed greater affinity than BPTES for GA enzymes but with similar activities toward GLS and GLS2 isoforms (Thomas et al. 2013). More recently, a drug discovery screening with natural products discovered a series of alkyl benzoquinones that preferentially inhibit GLS2 isoenzymes and display antitumor activity (Lee et al. 2014).

6.4.3 Regional, Cellular, and Subcellular Locations

One of the main histochemical markers used for identification of excitatory glutamatergic neurons has been the enzyme GA (reviewed by Kaneko 2000), as expected from its role as a main source of transmitter Glu. Nonetheless, Glu also plays important roles in brain general metabolism and it is also a precursor of inhibitory neurotransmitter γ-aminobutyric acid (GABA); for these reasons, immunoreactivity against the amino acid Glu cannot be considered specific for identification of glutamatergic neurons. Traditionally, it was thought that the only cerebral GA isoform was KGA, whereby antibodies against this isoform were the unique tools employed in most immunocytochemical studies dealing with regional, cellular, and subcellular localization of brain GA.

A survey on relevant studies dealing with the immunocytochemical localization of brain KGA (most studies done in rat) indicates expression in cerebral cortex, glutamatergic pyramidal neurons, and GABAergic interneurons (Donoghue et al. 1985; Akiyama et al. 1990; Aoki et al. 1991; Kaneko et al. 1992, 1995; Kaneko and Mizuno 1994, 1996); hippocampus, glutamatergic pyramidal and granular neurons (Altschuler et al. 1985); basal telencephalon, striatum, nucleus accumbens, cholinergic and GABAergic neurons (Aoki et al. 1991; Kaneko and Mizuno 1992;

Manns et al. 2001); cerebellum, glutamatergic granular cells (Wenthold et al. 1986; Kaneko et al. 1989; Laake et al. 1999); thalamus, glutamatergic neurons (Kaneko and Mizuno 1996); mesencephalon and rhombencephalon, catecholaminergic and serotonergic neurons, cochlear nucleus (Kaneko et al. 1990; Godfrey et al. 1994); and brainstem, spinal cord, respiratory, and motor neurons (Magnusson et al. 1986; Senba et al. 1991; Yezierski et al. 1993; Turman and Chandler 1994; Pilowsky et al. 1997).

The first reports demonstrating expression of GLS2 isoforms in mammalian brain were published 15 years ago: Northern analysis of GA transcripts in human brain indicated simultaneous expression of GLS and GLS2 mRNAs (Aledo et al. 2000; Gómez-Fabre et al. 2000). Shortly after, co-expression was also demonstrated in brain of other mammalian species such as cow, mouse, rabbit, and rat, while chicken only displayed GLS transcript (Olalla et al. 2002). This pattern of expression was in conflict with that previously reported for the rat (Smith and Watford 1990). The co-expression of both types of GA isoforms was confirmed at the protein level by biochemical and immunological approaches in rat and monkey brain (Olalla et al. 2002) (Fig. 6.3). Both isozymes, KGA and GAB/LGA (GAB antibodies do not discriminate between GLS2 isoforms and would recognize both GAB and LGA proteins), are ubiquitously expressed in brain regions with the strongest signal appearing in cerebral cortex (Olalla et al. 2002). Since Gln serves as an oxidative substrate in most cells, the wide distribution of GA throughout the mammalian brain was not unexpected. Nevertheless, GA was highly concentrated in the cerebral cortex, which is consistent with Glu as a major excitatory neurotransmitter of projection neurons in this area. An excellent correlation between the KGA and GLS2 expression patterns was observed at the regional and cellular level.

Most of the total brain GA activity is found in the crude P2 fraction, which consists of both free mitochondria and synaptosomes (Bradford and Ward 1976; Svenneby et al. 1986) and 40–60 % is retained in purified synaptosomes (Ward and Bradford 1979). GAs have been traditionally considered as mitochondrial enzymes: it was first proposed by Errera and Greenstein (1949) in liver and later shown experimentally (Guha 1961). In brain, the isoenzyme KGA has been convincingly demonstrated to be expressed in mitochondria by immunocytochemistry (Aoki et al. 1991; Laake et al. 1999; Olalla et al. 2002) (Fig. 6.3).

However, the submitochondrial localization of GA has been a very controversial matter and has not been determined conclusively. This issue has profound physiological implications in cerebral metabolism and energy homeostasis. To illustrate the discrepancies, we can say that almost all possible submitochondrial localizations have been proposed for GA, including matrix soluble (Kalra and Brosnan 1974), matrix side of the inner mitochondrial membrane (IMM) (Shapiro et al. 1985), and simultaneous presence in both halves of the IMM existing two populations of GA: one oriented in the intermembrane space (c-side) and the other in the matrix space (m-side) (Kvamme and Olsen 1979). The last authors observed an incomplete blockade of GA activity in rat brain mitochondria by Glu, ammonia and the SH-group reagent *N*-ethylmaleimide (NEM) (Kvamme and Olsen 1981; Kvamme et al. 1983) and used this evidence as the basis to propose that GA exists

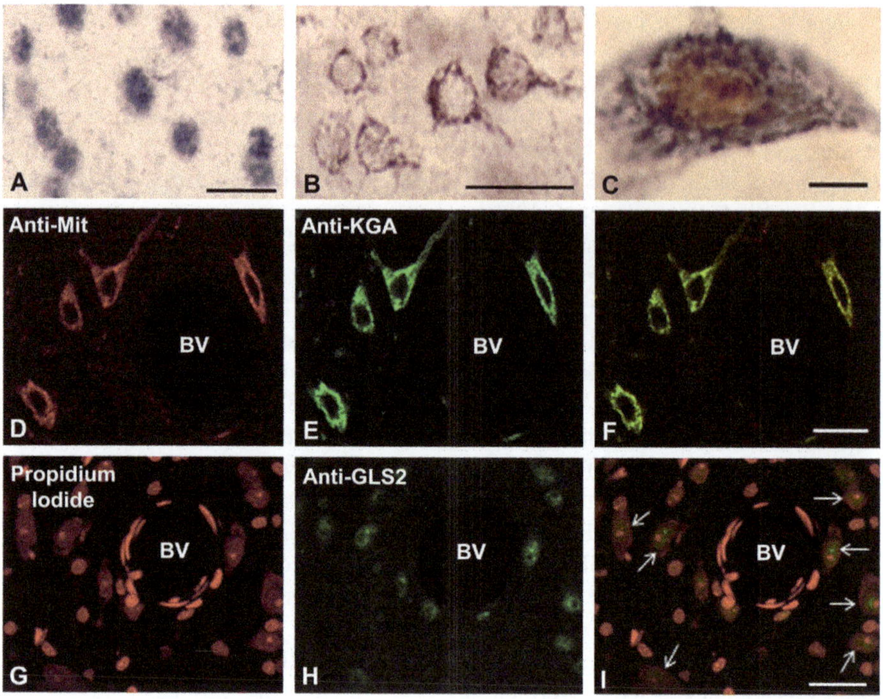

Fig. 6.3 Summary of confocal and immuno-EM results for KGA and GLS2 in neurons from mammalian brain. (**a**) Rat cortical neuronal nuclei (layer II) immunolabeled with anti-GAB isoform-specific antibody; (**b**) monkey cortical neurons displaying cytoplasmic particulate immunoreaction deposits with anti-KGA antibody; (**c**) double-labeled neuron showing granular cytoplasmic localization for KGA protein (*blue*) and nuclear localization for GLS2 protein (*brown*); (**d–f**) double-labeling immunofluorescence with anti-mitochondria (*red*, **d**), and anti-KGA (*green*, **e**) reveals a mitochondrial localization for this isoform (*yellow*, **f**); (**g–i**) double labeling with propidium iodide (*red*, **g**) and anti-GAB (*green*, **h**) shows that many cells, but not all, express nuclear GLS2 protein (**i**, *white arrows* show double-labeled cells). From (**d–i**), confocal laser scanning microscopy of rat brain sections. *BV* blood vessel. Scale bars = 20 μm (**a**), 25 μm (**b**), 10 μm (**c**), 40 μm (**d–i**). *Mit* mitochondria (Adapted from research originally published by Olalla et al. (2002) in Journal of Biological Chemistry, © the American Society for Biochemistry and Molecular Biology)

in two forms located in separate sites: the NEM-sensitive form was located at the outer surface of the IMM (c-side), whereas the insensitive form was situated at the inner surface (m-side) of the IMM. Pig brain mitochondria was also shown to contain two major forms of GA: a readily extractable soluble enzyme located in the matrix and a membrane-bound enzyme located in the IMM; only this latter enzyme was associated with mitochondria in nerve endings (Nimmo and Tipton 1979).

Nonetheless, other authors have questioned these results (Erecinska and Silver 1990), arguing that almost complete (98 %) inhibition of GA by Glu has been demonstrated in particulate preparations of brain (Krebs 1935; Bradford et al. 1978), as also happens in isolated mitochondria from kidney and tumor cells (Shapiro et al. 1982;

Campos et al. 1998). In addition, NEM crosses membranes easily and, hence, it is also hard to explain how this sulfhydryl reagent would not inhibit GA completely wherever located. For example, the inhibition pattern of Ehrlich tumor cell KGA with SH-reagents of different permeability clearly indicated that the IMM becomes a barrier to access to essential KGA residues (Aledo et al. 1997): in particulate fractions, mitochondria and mitoplasts, the membrane-permeant reagent NEM strongly inhibited KGA activity, but almost no effect was detected with membrane-impermeant reagents such as mersalyl and p-chloromercuriphenylsulphonic acid (PCMPS). However, at the same concentrations, mersalyl and PCMPS, as well as NEM, were very effective inhibitors of the soluble KGA (Aledo et al. 1997).

Later on, Kvamme and associates have suggested predominant c-side localization for the GA isoenzymes from kidney and brain, based on the inactivation by non-permeant SH-reagents and the lack of mixing between the Gln-derived Glu and the endogenous matrix Glu (Kvamme et al. 1991; Roberg et al. 1995). This location would not necessarily require a mitochondrial Gln transport while, at the same time, GA would be under the influence of the cytoplasmic Glu, Pi and Gln concentrations. Most notably, Gln-derived Glu should be imported through the IMM to be engaged in TCA reactions. In sharp contrast, all the KGA activity in kidney and tumor cell mitochondria was found to be localized on the inner surface of the IMM, and most of their functionally sensitive domain was located on the m-side of the IMM, as inferred from convergent enzymatic, immunological and chemical modification studies (Curthoys and Weiss 1974; Shapiro et al. 1985; Aledo et al. 1997).

Although submitochondrial localization of GA might differ from organism to organism, recent data support association with the inner surface of the IMM and, most notably, that the active site of GA must be facing the matrix side. Metabolic studies employing [U-^{13}C]Gln and NMR spectroscopy in cultured neurons have established that almost 50 % of the Gln metabolism involves TCA reactions (Waagepetersen et al. 2005). This result means that part of the transmitter Glu pool could be ascribed to transamination pathways but, at the same time, it is not compatible with most of the GA-derived Glu being generated outside the matrix. Similar findings were obtained using isolated rat brain mitochondria (Bak et al. 2008). In addition, an interesting observation also emerged from this last study: a 50 % decrease in the mitochondrial Glu level was detected in the presence of histidine, an inhibitor of Gln uptake in cerebral mitochondria (Albrecht et al. 2000). The authors interpreted these results as evidence of GA acting from the mitochondrial matrix; actually, Gln transport was a prerequisite for Glu formation and further metabolism in the TCA cycle (Bak et al. 2008). In fact, the location of GA's active site in the c-side of the IMM, being accessible to the cytosol, poses several questions difficult to ascertain, such as the necessity for very active mitochondrial Gln carriers, as those reported for rat kidney and tumor cells (see Matés et al. 2009 and references therein). In contrast, the generation of Glu into the matrix would facilitate their further metabolism for bioenergetics or anaplerotic purposes.

There was another noteworthy and unexpected finding concerning the subcellular location of GLS2 isoforms in brain: immunocytochemistry in monkey and rat brain revealed that GLS2-type GA immunoreactivity was mostly concentrated in neuronal nuclei (Olalla et al. 2002) (Fig. 6.3). For the first time, an extramitochondrial localization for a mammalian GA was reported, because GA enzymes were considered to be exclusively mitochondrial enzymes (Kovacevic and McGivan 1983; Curthoys and Watford 1995). Nevertheless, GLS2 was not the first neurotransmitter-synthesizing enzyme being located in the cell nucleus. In 1999, the group of Jane Rylett reported the nuclear localization of the 82 kDa form of human choline acetyltransferase [EC 2.3.1.6] (Resendes et al. 1999). We found clear isoform segregation in rat and monkey brain: KGA protein was detected in mitochondria, whereas GLS2-type GA protein was mostly localized in neuronal nuclei, albeit a minor cytoplasmic immunolabeling was also detected (Olalla et al. 2002) (Fig. 6.3). Of note, both isoforms colocalize in numerous cells throughout the brain and many neuronal cells, but not all, expressed nuclear GLS2-type GA protein (Fig. 6.3). Furthermore, the nuclear GLS2 was catalytically active, although showing kinetic characteristics atypical for classical L-type isozymes. As discussed above, the novel GAB isozyme in purified form shows mixed kinetic characteristics of GLS and GLS2-type isoforms (Campos-Sandoval et al. 2007) and seems the most plausible candidate for brain nuclear GA (de la Rosa et al. 2009). It is noteworthy that human functional GAB is targeted to both mitochondria and nucleus in Sf9 cells, and in both locations the protein was catalytically active (Campos-Sandoval et al. 2007), reinforcing the view that this isoform may be targeted to different subcellular locations, including the cell nucleus.

The expression in vivo of functional GA enzymes in astrocytes is also controversial and has been a matter of debate fueled by puzzled results. Primary cultures of astrocytes displayed efficient Gln uptake, strong GA activity (Schousboe et al. 1979; Kvamme et al. 1982), and expression of GA mRNA transcripts (Szeliga et al. 2008). However, these in vitro results were questioned arguing that GA may somehow be induced by the Gln present in the growth media and by the length of culturing (Erecinska and Silver 1990; Borg et al. 1985). On the other hand, contradictory results were obtained for the in situ expression of glial GA in brain tissue: immunohistochemical studies showed expression of KGA protein and GA activity in rat brain astrocytes (Aoki et al. 1991; Würdig and Kugler 1991), but others failed to detect immunoreactivity in non-neuronal elements such as glia and blood vessels (Kaneko et al. 1987; Akiyama et al. 1990; Laake et al. 1999). Therefore, the prevalent and currently accepted view in mammalian brain states that GA is almost exclusively confined in neuronal cells and shows a limited or no expression in glial cells, with only a few exceptions: in vitro cultured astrocytes and activated microglia cells that produce Glu in an autocrine manner through upregulation of GA expression (Takeuchi et al. 2006). Interestingly, previous studies suggested the possible existence of two GA enzyme forms in cultured astrocytes and neurons by the differences observed in kinetic properties of their main effectors (Hogstad et al. 1988).

The occurrence of GA in astrocytes in vivo has not been conclusively proved. Hence, a comprehensive study was devised to elucidate expression of GA in neuroglia and, more concretely, in astrocytes (Cardona et al. 2015) (Fig. 6.4). We demonstrated by in vivo and in vitro approaches that astrocytes express at least two GA isoforms: KGA and GLS2 (GAB/LGA), essentially considered as exclusive neuronal enzymes. Immunocytochemistry in rat and human brain tissues employing isoform-specific antibodies revealed expression of both GLS (KGA) and GLS2 (GAB/LGA) isozymes in glutamatergic and GABAergic neuronal populations as well as in astrocytes. Nevertheless, there was a different and clear subcellular distribution: KGA isoform was always present in mitochondria while GAB/LGA was preferentially detected in both nucleus and mitochondria (Fig. 6.4). Immunocytochemistry in cultured astrocytes confirmed the same pattern previously seen in brain tissue samples (Fig. 6.4). Astrocytic GA expression was also assessed

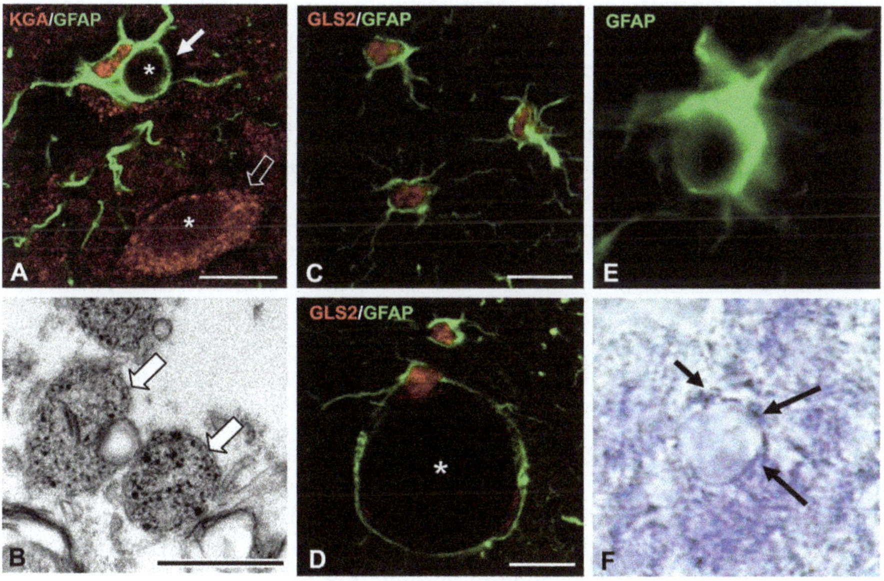

Fig. 6.4 Summary of confocal and immuno-EM microscopy results for KGA and GLS2 in astrocytes from mammalian brain. (**a**) Double immunofluorescence for KGA (*red*) and glial fibrillary acidic protein (GFAP) (*green*) in human brain temporal lobe sections. Punctate KGA immunostaining is clearly seen in the cytoplasm of neurons (*open arrows*) and astrocytes (*white arrows*). *Asterisks* (*) indicate the nuclei of neurons and astrocytes. (**b**) Immuno-EM detection of KGA in the mitochondria of rat brain astrocytes. KGA labeling was restricted to the mitochondria in astrocytes (immunoreactive mitochondria indicated with *white arrows*). (**c**, **d**) GLS2 in rat brain astrocytes: confocal microscopy and double immunofluorescence for GFAP (*green*) and GLS2 (*red*) in rat hippocampus. Merge images show the location of GAB in the nuclei (**c**, **d**) and specialized foot processes of perivascular astrocytes (**d**). *Asterisk* indicates a blood vessel. (**e**, **f**) GA activity in rat brain astrocytes. Double labeling in rat cerebellum sections for GFAP (*green*, **e**) and GA activity staining (blue formazan precipitate, **f**). In vivo GA activity appears as a granular or punctate staining (*black arrows*) in the cytoplasm of astrocytes (**f**). Scale bars = 10 μm (**a**), 0,5 μm (**b**), 20 μm (**c**–**f**) (Adapted from Cardona et al. (2015), GLIA 63:365–82)

at the mRNA level by real-time qRT-PCR: transcripts of four GA isozymes were detected, but with marked differences on their absolute copy number. The predominance of GLS isoforms over GLS2 transcripts was remarkable (ratio of 144:1). Finally, we proved that astrocytic GA proteins possess enzymatic activity by in situ activity staining: discrete populations of astrocytes were labeled in the cortex, cerebellum, and hippocampus of rat brain, demonstrating functional catalytic activity which can be ascribed to mitochondrial GA. In conclusion, subpopulations of astroglial cells possess GA activity in vivo; they are capable of generating Glu through GA pathway by utilizing Gln either internally generated or produced in neighboring cells or taken from the extracellular fluid (ECF, Fig. 6.5). Nonetheless, available experimental data predict an in vivo GA activity considerably lower in astrocytes as compared with that shown by neurons.

The presence of a non-mitochondrial brain GA in the neuronal cytoplasm has also been suggested by immunocytochemistry studies (Aoki et al. 1991). Nevertheless, the relevance of extramitochondrial GA to Gln/Glu metabolism deserves further investigations and is presently unknown, although interaction of GA with newly identified protein-interacting partners might give rise to transient sporadic appearance of GA in cytosol of neurons and astrocytes (see Fig. 6.5 and Sect. 6.6). On the other hand, it should be emphasized that the existence of a cytosolic GA in astrocytes seems unlikely taking into account the strong inhibition that synaptic Glu may exert. Most important, the simultaneous operation of GA and GS in the cytosol will give rise to a futile cycle with high energy expenditure in the form of ATP.

6.5 Role of GA in Glutamatergic Neurotransmission

The importance of GA in glutamatergic synaptic function has been largely recognized. Physiological, biochemical, immunological, and NMR spectroscopic data indicate that neurotransmitter Glu is mainly generated through GA reaction (Hertz 2004), although the relative contribution of each GA isoform to the transmitter pool is presently unknown. Quantitation of absolute mRNA levels of GA isoforms in total brain reveals a clear predominance of GLS isoforms versus GLS2 transcripts (Martín-Rufián et al. 2012). Although quantitative assessment of individual protein levels of neuronal GA isozymes is unknown, immunocytochemical results seem to support a greater abundance of KGA versus GLS2 isoforms (Olalla et al. 2002; Cardona et al. 2015). A knockout (KO) mice model for the *Gls* gene has been generated (Masson et al. 2006). Mice lacking the *Gls* gene die shortly after birth due to altered functioning of key glutamatergic neural networks, stressing the importance of GLS isozymes (mainly KGA) in glutamatergic transmission. Nevertheless, there was persistence of glutamatergic activity in null mutant mice and the quantal size was not reduced in cultured *Gls*$^{-/-}$ cortical neurons, suggesting that there were adequate levels of intrasynaptic Glu under conditions of basal activity; in fact, synaptic transmission was only impaired under intense stimulation (Masson et al. 2006).

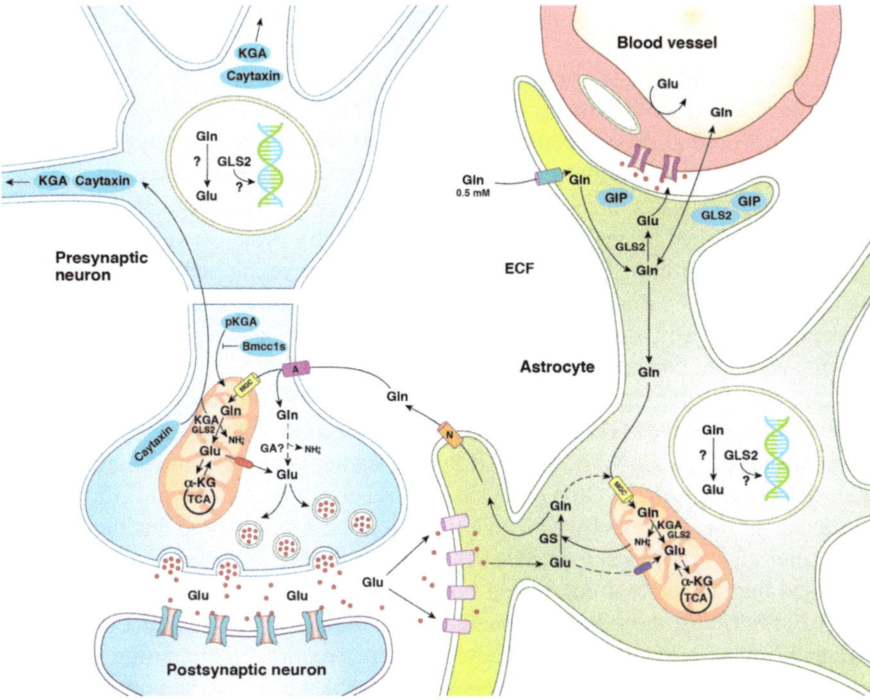

Fig. 6.5 Schematic illustration of the Glu/Gln cycle between neurones and astrocytes at the tripartite synapsis. This is a simplified model highlighting the main functions of GA isozymes, even if not enough evidence exists for a particular task. In the latter case, a question mark means "not enough evidence." For example, in neuronal nuclei the presence of GLS2 has been demonstrated; however, the nuclear function of this GA has not been fully ascertained. Two alternative nuclear functions appear: regulation of the Gln/Glu levels or transcriptional regulation (*arrow* pointing toward DNA). The relative contribution of GLS and GLS2 isoforms in the synthesis of the neurotransmitter Glu pool remains to be clarified. In neuronal body, Gln is converted to Glu by mitochondrial KGA and GLS2, whereas the existence of a cytosolic GA, which may contribute to the Glu transmitter pool, has not yet been confirmed. Synaptic Glu is primarily taken up by astrocytes, mostly converted to Gln by GS in the cytosol, and then cycled back to neurones (through N and A carriers) where GA regenerates the transmitter Glu. Exogenous Glu may be transported to glial mitochondria (*dashed arrow*) and converted to α-KG by GDH or transamination, followed by oxidation into the TCA cycle. Astroglial Gln may be exchanged with blood. Some Gln can be oxidatively degraded by astrocytic GA and TCA cycle after being transported into mitochondria through the uncharacterized MGC; the ammonium generated might be channeled to cytosolic Gln synthesis. Experimental evidence support expression of both KGA and GLS2 isoforms in astrocyte. GLS2 and GIP may interact in vivo and astrocytes are a likely anatomic substrate for their coupling in brain. The colocalization of GIP and GLS2 in perivascular end feet surrounding capillaries might be related to the regulation of the vascular tone through Glu receptors shown in endothelial cells. Caytaxin interacts with KGA and relocalizes it from mitochondria to neurites, while Bmcc1s blocks the targeting of precursor KGA (pKGA) to the mitochondria. The detailed anatomy of synapses, astrocytes, and blood vessels is not portrayed. MGC, Mitochondrial Gln Carrier; ECF, extracellular fluid; GLS2, GAB and/or LGA isoforms; KGA, *Gls*-encoded long GA isoform; α-KG, α-ketoglutarate

These results strongly suggest that alternative sources of vesicular Glu were activated in *Gls* null mutants: *Gls2*-encoded GA isoenzymes are the most suitable candidates to maintain releasable Glu stores after induction through a compensatory mechanism. On the other hand, the persistence of glutamatergic transmission in these mice would also be accounted for upregulation of other Glu-synthetic pathways, such as transamination reactions or by direct neuronal Glu reuptake.

The homeostasis of Glu and Gln in brain should be carefully regulated due to the toxic effects elicited by an excess of Glu. The Glu/Gln cycle between neurons and astrocytes is a central pathway for neurotransmitter recycling and to deal with Glu toxicity (Hertz 1979; Berl and Clarke 1983) (Fig. 6.5). The Glu released by neurons is taken up by nearby astrocytes at the synaptic cleft through efficient Glu transport systems and then converted to Gln by glutamine synthetase (GS; EC 6.3.1.2), an enzyme exclusively located in astrocytes (Norenberg and Martínez-Hernández 1979). The Glu-derived Gln is finally exported back to neurons where GA generates neurotransmitter Glu. In vivo ^{13}C-NMR kinetic studies, metabolic modeling and flux analysis to distinguish the Glu/Gln cycle from other sources of isotopic Gln labeling (particularly Gln synthesis by glial anaplerosis), have demonstrated that the Glu/Gln cycle between astrocytes and neurones is a major pathway for neuronal Glu repletion in rat and human cerebral cortex and hippocampus (Shen et al. 1999; Lebon et al. 2002). Furthermore, these studies concluded that the rate of the Glu/Gln cycle is very high and similar in magnitude to the rate of glucose oxidation, supporting the model proposed by Magistretti and coworkers which couples neuronal activity to glucose utilization and where glial glycolytic ATP represents the major source of energy for neurotransmission (Magistretti et al. 1999; Magistretti and Pellerin 1999).

The existence of GA isoforms in astrocytes is of great importance to fully understand the bioenergetics of glial cells and, particularly, to shed light into transmitter recycling and short-term regulation of the Glu/Gln cycle. The mitochondrial location of KGA and GLS2 isoforms in astrocytes may be a control mechanism allowing broad and fine tuning of Glu production depending on their energetic needs and/or synaptic activity. In this context, it is noteworthy the significant increase in mitochondrial GLS2 detected in cultured astrocytes, using a standard culture medium containing Gln, compared with the scarce mitochondrial label observed in rat brain tissue (Cardona et al. 2015). This fact easily illustrates how astrocytes can modulate expression of GA isoforms, depending on the environmental conditions to which they are exposed, to maintain their energy and metabolic homeostasis. Mammalian KGA and GLS2 proteins possess different kinetic and regulatory properties (Table 6.1); therefore, they reach full catalytic activity at different cellular concentrations of Gln, Glu, Pi, and other effectors.

At first sight, recent data demonstrating functional expression of two GA isoforms in astrocytes are in apparent conflict with the compartmentation of the key enzymes in the Glu/Gln cycle (Norenberg and Martínez-Hernández 1979; Donoghue et al. 1985; Erecinska and Silver 1990; Kaneko et al. 1995) and seem hard to reconcile with the constant supply of synaptic Glu being received by these glial cells, which may render superfluous the existence of astrocytic GA. Nonetheless, a GA activity in astrocytes could fit well with the Glu/Gln cycle, because it would endorse astrocytes with an endogenous mitochondrial source of Glu (and ammonium) whose synthesis from Gln does not require energy. This endogenous, non-synaptic, Gln-derived Glu could be converted through glutamate dehydrogenase (GDH) or transamination to α-ketoglutarate and then proceeds via TCA cycle to satisfy astrocytic energetic needs (Fig. 6.5). Catabolism of this endogenous Gln-derived Glu may cooperate with the strong energy and biosynthetic needs required by astrocytes, thus preserving most of the exogenous synaptic Glu to be transformed back to Gln and redirected toward neurons. In this way, astrocytes will not significantly deplete synaptic Glu stores saving most of it for Gln synthesis in the cytosol; such mechanism might be particularly relevant in periods of great synaptic activity. In addition, the ammonium generated in mitochondria by this glutaminolytic process could be channeled to Gln synthesis in cytosol, providing an additional source of nitrogen needed to recycle Glu in Gln through GS (Fig. 6.5).

The Glu/Gln cycle is a key mechanism for homeostatic control of Glu, Gln, and GABA concentrations; however, it is not always a stoichiometric pathway (McKenna 2007) because part of the exogenous Glu taken by astrocytes does not end as Gln but used to fulfill other energy and metabolic purposes instead (Sonnewald et al. 1993). Actually, it has been estimated that 10–30 % of synaptic-derived Glu is oxidized by astrocytes and must be replenished to keep the carbon balance of the Glu/Gln cycle (McKenna et al. 1996; Gamberino et al. 1997; Hutson et al. 1998). In vivo studies, using 13C-NMR spectroscopy, indicated that glial anaplerosis from glucose, through the astrocyte-specific enzyme pyruvate carboxylase (PC; EC 6.4.1.1), is the main contributor for replacing the oxidatively degraded Glu (Rothman et al. 1999; Lebon et al. 2002). In this sense, a GA activity will increase the metabolic versatility of astrocytes in such way that they do not have to exclusively rely on PC for *de novo* synthesis of Glu. The existence of a functional GA in astrocytes will diminish their anaplerotic pressure, which is important because Glu synthesis by anaplerosis would be detrimental if it is not matched by degradation through cataplerotic routes (Sonnewald 2014). Furthermore, the fast induction of GS protein in active synapse, which rapidly converts the captured synaptic Glu into Gln, means that the flux of Glu to the TCA cycle became negligible as compared with the flux through GS (Fonseca et al. 2005). Actually, these authors conclude that astrocytes adapted to a constant supply of Glu by increasing Glu uptake and GS activity in such a way that Glu was converted almost exclusively through GS. Under these conditions, the supply of endogenous Glu by GA would be a valuable resource for astrocyte bioenergetics.

An additional advantage of the existence of astrocyte GA can be related to the mitochondrial import of Glu. Generation of mitochondrial Glu in situ avoids another problem faced by synaptic cytosolic Glu: its import into astrocytic mitochondria. It has been shown that brain astrocytes do not express the main aspartate–glutamate carrier (Aralar/AGC1) (Ramos et al. 2003; Berckich et al. 2007); therefore, exogenous Glu must enter into astrocytes uniquely by the Glu/hydroxyl carrier. Studies with Aralar KO mice underscored the importance of this carrier for Glu transport in brain and skeletal muscle mitochondria (Jalil et al. 2005), while further evidence has shown that no other Glu carrier can substitute for Aralar in those tissues (Satrústegui et al. 2007 and references therein). Notwithstanding, it should be stated that evidence for anatomic and physical linkages between the astroglial Na^+-dependent Glu transporters (GLT-1/EAAT2, GLAST/EAAT1) and mitochondria has been recently reported (Whitelaw and Robinson 2013 and references therein). On the other hand, Gln can be actively concentrated into astrocyte mitochondria using a high-affinity mitochondrial Gln carrier (Roberg et al. 1999) and then be released into the mitochondrial matrix near the GA catalytic site. The generation of Glu into the matrix facilitates its further catabolism for bioenergetics or biosynthetic purposes. Thus, Gln-derived Glu is generated into the mitochondrial matrix without the need for an additional carrier, as required for exogenous Glu.

Finally, it is noteworthy to mention the role of neuronal GA in the synthesis of GABA. The Gln-derived Glu may also contribute to the synthesis of GABA at inhibitory synapses. We have recently found expression of both GLS and GLS2 isoforms in a scarcity of Purkinje cells and in many GABAergic neurons at the molecular layer of the cerebellum, which point toward GA as an important biosynthetic source of Glu at inhibitory synapses indeed (Cardona et al. 2015). Early studies did not detect KGA immunoreactivity in GABAergic neurons of neocortex from rodents and monkey (Altschuler et al. 1984, 1985; Donoghue et al. 1985; Kaneko et al. 1992; Kaneko and Mizuno 1994). However, more recent studies presented immunocytochemical evidence of the localization of KGA in GABAergic neurons in the cat visual cortex (Van der Gucht et al. 2003) and thalamus (Fisher 2007), thereby suggesting that Gln can be a metabolic precursor for GABA synthesis. Interestingly, co-expression of KGA and glutamic acid decarboxylase (GAD; EC 4.1.1.15) was also reported in rat hippocampus and basal forebrain neurons (Kaneko and Mizuno 1994; Manns et al. 2001). GABA is produced by GAD, and two GAD isoforms, encoded by different independently regulated genes, are also expressed in brain (Bu et al. 1992). Both forms can synthesize transmitter GABA, but have different roles in the coding of information by GABA-containing neurones and different subcellular localizations (Soghomonian and Martin 1998). Thus, each of the two main neurotransmitters in the CNS—Glu and GABA—can be synthesized by two isozymes coded by distinct genes, a unique situation different from all the other neurotransmitters. Glu and GABA production must be a process exquisitely regulated to ensure a proper Glu homeostasis. Although the need for several brain GA isoforms is not completely understood, it may represent the biochemical basis to achieve this fine modulation under different physiological circumstances.

6.6 New Functions Beyond Their Role in Glutamatergic Transmission

The mammalian *Gls* and *Gls2* genes are presumed to have been derived by gene duplication of a common ancestral gene, followed by gradual changes in the sequences of both copies (Chung-Bok et al. 1997; Pérez-Gómez et al. 2003). As paralogous proteins, GLS and GLS2 isoforms may acquire different functions during their evolution while maintaining a strong similarity in sequence and three-dimensional structure (Fig. 6.2). Furthermore, GA has been endowed with consensus protein motifs and domains that might support its role as multifunctional proteins (Fig. 6.2). We strongly support this view for GA in mammalian tissues and, with special relevance, for the nuclear localization of GLS2 isoforms in neurons and astrocytes, whose significance remains to be fully determined. The nuclear function has been related with transcriptional regulation associated to differentiation (Olalla et al. 2002; Márquez et al. 2006). First indications about this role for GLS2 isoenzymes came from cancer studies. We postulated a completely different role for GLS and GLS2 isoforms in cancer based on their relative expression patterns in human leukemia, breast cancer cells, and hepatocellular transformation: the data suggest that upregulation of GLS isoforms correlates with increased rates of proliferation, whereas prevalence of the GLS2 isoforms seems to be related with differentiated and quiescent cell states (Pérez-Gómez et al. 2005). Therefore, we proposed that the process of malignant transformation shifts the pattern of GA expression in such way that GLS becomes upregulated while GLS2 is frequently repressed. In accordance with this, human T98G glioma cells showed almost null expression of GLS2, but overexpression of human GAB induced a marked change in the cell's transcriptome correlated with a reversion of their transformed phenotype (Szeliga et al. 2009, see also chapter 9– in this book). Taking into account its presence in cell's nuclei, it has been speculated that GAB overexpression may contribute in altering the transcriptional program of glioma cells, yielding a less malignant and more differentiated phenotype, but the concrete molecular mechanisms underpinning this phenotypical change are unknown.

In this regard, important insights into GLS2 function in neuronal differentiation have been recently published (Velletri et al. 2013). The authors show that GLS2 is under the control of TAp73, a p53 family member, during neuronal differentiation of neuroblastoma cells induced by retinoic acid (Table 6.1). Indeed, GLS2 overexpression increased neurite outgrowth. Furthermore, inhibition of GA activity in cultures of mouse embryonary cortical neurons resulted in impaired neuritogenesis. In order to gain insights into the role of GLS2 in vivo, they also analyzed GLS2 expression in developing cerebellum. Interestingly, GLS2 transcript significantly increased between postnatal day 0 (P0) and P8, a development stage characterized by a massive expansion of granule cell progenitors (Velletri et al. 2013). Convergent pieces of evidence were obtained from primary human keratinocytes, where it was shown that GLS2 is a target gene regulated by the transcription factor TAp63, also belonging to the p53 family (Table 6.1). GLS2 and TAp63 expression increases during the in vitro differentiation of keratinocytes, while depletion of GLS2 inhibits skin differentiation both at molecular and cellular levels (Giacobbe et al. 2013). Altogether, these data reinforce our hypothesis relating GLS2 with cell differentiation. With regard to GLS, experimental evidence supporting a key role in the proliferation of neural cells has been reported. Thus, both GLS isoforms (KGA and GAC) were upregulated during neurogenesis of human neural progenitor cells (NPCs), and their expression pattern positively correlated with the neuronal marker MAP-2 (microtubule-associated protein 2) (Wang et al. 2014). Most important, studies of cultured human NPCs after siRNA silencing of *GLS* suggest a critical role of GLS isoforms for proliferation and survival of NPCs (Wang et al. 2014).

Another potential nuclear function for GA could be the regulation of Gln/Glu levels (Olalla et al. 2002), taking into account that Gln is a signal molecule involved in gene expression (Bungard and McGivan 2004; Häussinger and Schliess 2007). Therefore, the significance of its nuclear localization could be as simple as being an enzyme controlling in situ the Gln levels in the nucleoplasm and, thus, being indirectly involved in the expression of Gln-regulated genes (Fig. 6.5). Actually, GA activity was found in nuclear fractions isolated from rat brain, although with kinetic properties distinct from the classical *Gls2*-encoded LGA isoform (Olalla et al. 2002). Interestingly, purified human GAB isozyme showed mixed kinetic characteristics of GLS and GLS2 isoforms (Campos-Sandoval et al. 2007), in accord with the results obtained from nuclear rat brain extracts. Strikingly, no GA activity was revealed in astrocytic nuclei despite the clear immunoreactivity revealed for GLS2 isoforms; however, this result may arise because the NBT-based activity assay was not sensitive enough for low activity levels found in nucleus (Cardona et al. 2015).

In the search for discovering new physiological functions for brain GA, proteomics approaches, such as two-hybrid genetic screenings and immunoprecipitations (IP) coupled to mass spectrometry (MS) analysis, have proven to be very useful strategies for isolation of potential protein-interacting partners of GA. A yeast two-hybrid genetic assay was performed by screening a human brain cDNA library with the C-terminal region of human GAB. Two PDZ (PSD95/Dlg/ZO1) domain-containing proteins were isolated: α-1-syntrophin (SNT) and glutaminase-interacting protein (GIP), also known as tax-interacting binding protein-1 (TIP-1) (Olalla et al. 2001). One of the most common protein modules involved in scaffolding interactions are the PDZ domains, which are responsible for a wide array of protein–protein interactions in the CNS and other tissues (Kornau et al. 1995). These modules bind to short sequences at the extreme C-terminus of target proteins, thereby nucleating the formation of specific functional complexes. The last four amino acids of the C-terminal end of human GAB, ESMV, conforms a consensus sequence required for interaction with PDZ proteins (Fig. 6.2) (Olalla et al. 2001). The relevance of these residues for the human GAB–GIP binding was demonstrated by NMR and fluorescence spectroscopy studies, allowing determination of the dissociation constant (K_d) for their binding: 1.66 mM, which means a moderate affinity suitable for regulatory functions (Banerjee et al. 2008).

The physiological relevance of the GLS2–GIP interaction was first assessed by immunocytochemical studies in brain, in order to ascertain whether both proteins may be interacting partners in vivo (Olalla et al. 2008). GIP colocalizes with GLS2 in cerebral cortex astrocytes from rat brain. Double-label studies and confocal microscopy confirmed colocalization in astrocytes cell bodies and processes, including their perivascular end feet (Fig. 6.5). The presence of GIP in astrocytes was also confirmed by immuno-electron microscopy (EM), which revealed immunoreactive astrocytic end feet surrounding perivascular endothelial cells. These results strengthen the view that GA and GIP may interact in vivo and point to astrocytes as a likely anatomic substrate for their coupling in brain (Fig. 6.5).

Analysis of rat brain neurones by EM also found GIP immunoreactivity in the nuclear envelope (Olalla et al. 2008). This nuclear location agrees with previous studies on GIP which described its participation in the regulation of transcription: GIP physically interacts with beta-catenin inhibiting its transcriptional activity, being a critical component of the beta-catenin regulatory network (Kanamori et al. 2003). While human GLS2 has structural determinants needed for mitochondrial targeting (Gómez-Fabre et al. 2000), it does not possess a discernible nuclear localization signal (Fig. 6.2); therefore, the mechanism by which it reaches the nucleus is unclear. The PDZ-protein recognition motif would be implicated in GLS2 specific targeting to selective cellular locations (Olalla et al. 2008; Márquez et al. 2009). Of note, PDZ proteins have also been reported to play an important role in the targeting of protein-interacting partners to concrete subcellular localizations, including cell nucleus (Kausalya et al. 2004), and several PDZ proteins are known to localize to the nucleus (Hsueh et al. 2000). Therefore, we cannot exclude that GIP may be involved in the targeting of GLS2 to neuronal nuclei.

In addition to PDZ domains, human GAB has other sequence motifs and conserved modules that may be essential for its nuclear import (Márquez et al. 2006). For example, it has a consensus motif for interaction with nuclear receptors: the sequence LGDLL on exon 2 (residues 72–76) conform to the consensus nuclear receptor box LXXLL (Fig. 6.2). This sequence motif allows for specific interaction with nuclear receptors and is primarily found in coactivators of nuclear receptors (Heery et al. 1997). Nuclear translocation of a mitochondrial enzyme containing a mitochondrial targeting sequence, but lacking a specific nuclear targeting signal, is not without precedent: mitochondrial 3-hydroxy-3-methylglutarylCoA synthase has been detected in nuclei and its nuclear translocation seems to involve interactions with nuclear hormone receptors through its LXXLL motif (Meertens et al. 1998).

Other protein–protein interaction motifs present in the primary structure of GLS2 proteins (and KGA, but not in the GAC isoform) are two ankyrin repeats in the C-terminal region (Fig. 6.2). Ankyrin repeats are about 33 amino acids long consisting of a β-turn and two antiparallel α-helices and have been found in proteins of diverse function such as transcriptional initiators, cell cycle regulators, cytoskeletal, ion transporters and signal transducers (Sedgwick and Smerdon 1999). The ankyrin-repeat proteins mediate many important protein–protein interactions in virtually all species. Of particular relevance in this context is the existence of transcription factors (GA-binding protein) and transcriptional regulators (IkB protein family, ANCO proteins) whose ankyrin repeats are essential for proper transcriptional activity (Sedgwick and Smerdon 1999; Zhang et al. 2004). Furthermore, ankyrin-repeat motifs are critically required for nuclear localization of signaling enzymes (Hozumi et al. 2003) and transcriptional cofactors (Sedgwick and Smerdon 1999), being able to functionally substitute for a classical nuclear localization signal (Sachdev et al. 1998). Therefore, the involvement of this modular motif in the nuclear import of GLS2 will deserve further attention.

A second PDZ protein isolated as protein-interacting partner of human GAB was SNT. In brain, the PDZ domain of SNT is known to bind to protein kinases, to the neural nitric oxide synthase (nNOS), and to the water channel aquaporin-4 (AQP4) as a mechanism for selective targeting to unique subcellular sites (Adams et al. 2001). In astrocytes, it has been shown that SNT is a central organizer of the dystrophin complex, an important molecular scaffold for localization of AQP4 at the BBB (Bragg et al. 2006). SNT is the best characterized PDZ scaffold in astrocytes and is concentrated in the perivascular end feet of astrocytes anchoring the AQP4 water channel to these membranes (Neely et al. 2001). The coupling between AQP4 and SNT may be indirect and studies with KO mice for SNT have suggested that additional anchoring proteins (GIP?) may be needed for clustering AQP4 in the end feet of brain astrocytes (Amiry-Moghaddam et al. 2004). At present, the functional role of GIP in the perivascular end feet is unknown and whether GIP or SNT may act as scaffold proteins for GLS2 in astrocytes remains to be determined. Interestingly, GLS2 possesses the consensus C-terminal –SXV sequence, like the potassium channel Kir4.1, required for interaction with the dystrophin complex.

On the other hand, the localization of GIP and GLS2 in astrocytes, particularly in perivascular end feet surrounding blood vessels and capillaries, might be related to the regulation of the vascular tone (Fig. 6.5). Of note, a perivascular and pial localization of GA in rat brain was previously reported (Kaneko and Mizuno 1988). Indeed, we have found GLS2 in concrete perivascular end feet (Cardona et al. 2015), as well as colocalization of GIP and GLS2 in astrocytic processes (Olalla et al. 2008). Thus, it is tempting to speculate that GLS2 targeting to specialized foot processes of concrete perivascular astrocytes might be a plausible mechanism implicated in the regulation of the vascular function, considering the key role of astrocytes in cerebrovascular regulation (Koehler et al. 2006). The fact that Glu is a vasoactive compound (Fergus and Lee 1997) and the existence of Glu receptors in perivascular glia and vascular endothelial cells (Collard et al. 2002; Nedergaard et al. 2002) would implicate GLS2–GIP interaction in regulating the vascular tone (Fig. 6.5).

GLS isoforms can also traffic in brain and unexpected locations have been lately reported for them. The first interacting partner for a GLS protein was shown to be the brain-specific protein BNIP-H (for BNIP-2 homology) or caytaxin, a protein exclusively expressed in neural tissues and encoded by a gene associated with human cerebellar Cayman ataxia (Buschdorf et al. 2006). This protein contains a novel protein–protein interaction domain known as the BNIP-2 and Cdc42GAP homology (BCH) domain. The regional distribution of caytaxin in mouse brain broadly matched the pattern of expression previously known for KGA. In PC12 cells stimulated with nerve growth factor for 24 h to induce differentiation to neuronal phenotype, the co-transfection of KGA and caytaxin induced a specific redistribution of KGA to neurite terminals independently of, and away from, mitochondria (Fig. 6.5); therefore, caytaxin relocalized KGA from cell body to neurite terminals. Neuronal trafficking of KGA was specifically ascribed to its interaction with caytaxin, which also reduces the steady-state levels of Glu by inhibiting KGA activity (Buschdorf et al. 2006). Thus, the interaction KGA–caytaxin seems relevant for regulating the homeostasis of Glu synthesis important for proper neurotransmission and neuronal cell growth (Fig. 6.5). For example, absence of functional caytaxin would increase Glu levels in the cell bodies of neurons leading to neurotoxicity and/or abnormal neuronal growth. In agreement with these findings, a whole-brain proteomic analysis of *Gls*-deficient mice (*Gls*$^{-/-}$ and *Gls*$^{+/-}$) has revealed a link between GLS isoforms and ATPases, ion channels, nucleic acid and lipid metabolism, calcium, Akt and retinol signaling, cytoskeletal components and protein synthesis machinery, among others. This protein network involves key players of the cellular assembly and organization, cell signaling and cell cycle (Bae et al. 2013). Thus, reduction of GLS expression by genetic knockdown of *Gls* gene in mice suggests a crucial role of these proteins in neuronal maturation and development.

A second member of the BCH domain-containing family of proteins (related to caytaxin) has been identified as direct interacting partner of KGA isoform. BMCC1 (**B**cl2, the adenovirus E1B 19 kDa interacting protein 2 and the Cdc42 GAP homology BCH **m**otif-**c**ontaining molecule at the **c**arboxy-terminal region **1**) is a large polypeptide expressed in brain which encodes several isoforms; one of them, BMCC1s, is a novel short isoform of BMCC1 predominantly expressed in the mouse brain and involved in the regulation of cytoskeleton dynamics and cell survival (Arama et al. 2012). Proteomic analysis of BMCC1s-binding proteins identified KGA as an interacting partner. When primary cultures of cortical neurons were contransfected with both proteins, the subcellular distribution of KGA was drastically modified: almost half of transfected cells lost its unique KGA mitochondrial profile, being also detected in the cytoplasm (Boulay et al. 2013). The authors concluded that BMCC1s preferentially interacts with the 74 kDa cytosolic monomeric precursor of KGA (pKGA) and, thus, it cannot reach the mitochondria for being processed to the mature catalytically active form (Fig. 6.5).

The hypothesis of brain GA isoforms being regulated by protein-interacting partners is appealing because it may provide a mechanism for both control of Glu synthesis and targeting of cerebral GA to concrete cellular compartments. In fact, GIP has been shown to inhibit GLS2-type GA activity in crude extracts of rat liver (Aledo et al. 2001), whereas two binding partners recently discovered for KGA, caytaxin and BMCC1s, were also able to inhibit the catalytic activity of KGA, although by different mechanisms: direct interaction or blocking its path to the mitochondria.

6.7 Summary

Eighty years after its discovery, the knowledge accumulated on brain GA is considerable. Significant progress has been made in this field stressing the relevance of GA, but still a number of sound questions remain to be answered. Based on novel and growing pieces of evidence summarized in this chapter, the roles of GA in cerebral function are multiple and spreading out. Nowadays, the pattern of GA expression has been shown to be considerably more complex: the old concept of only one type of GA expressed in mammalian brain (KGA) was changed at the light of overwhelming evidence demonstrating that, at least, four GA isoenzymes are expressed. The concept of an exclusively mitochondrial location for GA was also demonstrated to be wrong as new subcellular locations (e.g., nucleus, cytosol) have been described for these proteins. Even more, the concept of GA as exclusive neuron-specific enzyme has been challenged by recent findings reporting in vivo expression of functional GA isoforms in astrocytes. The interactome of brain GA is starting to be uncovered. To reach their final destinations in brain cells, GA may interact with newly discovered scaffold proteins. Such arrangements could provide the molecular basis for selective and regulated targeting to concrete cellular locations and control of Glu synthesis by blocking GA activity. These unanticipated findings open a new avenue of research on how GA may affect the homeostasis of Glu/Gln in the tripartite synapsis. Besides their classical role in glutamatergic transmission, novel emerging roles for GA include transcriptional control, neuronal growth and differentiation, and cerebrovascular regulation. Thus, GA may be added to the growing list of moonlighting proteins, that is, proteins having a second (or third) function.

There are several key issues about GA function in brain that remain unresolved. Given our new knowledge about the presence of GA in astrocytes, it is expected that in the next years we can determine the function of GA in glial cells in physiological conditions and whether astrocytic GA contribute to the changes in brain function under pathological conditions. Further studies will undoubtedly reveal insights into the nuclear role of GA; particularly the role of GLS2 and/or Gln in the regulation of gene expression. The physiological relevance of the interaction between GA- and protein-binding partners, as well as their implications in neurological disorders, still remain to be elucidated. We expect that the function of KGA in neurogenesis and neuronal growth and development can be fully addressed in the next future as well. The expression of, at least, four GA in mammalian brain is so far unexplained. The source of neurotransmitter Glu and what is the relative contribution of each GA isoform to the transmitter pool of Glu are outstanding questions for future work, as well as elucidation of the factors which regulate GA activity in vivo, and hence Glu supply, during neuronal stimulation.

Acknowledgements We are grateful to everybody in the Canceromics group for creating most of the own work described in this chapter, particularly to our colleagues F.J. Alonso, J.A. Segura, M. Martín-Rufián, C. Cardona, C. Lobo, A. Peñalver, and former members L. Olalla, P. Gómez-Fabre, C. Pérez-Gómez, and J.C. Aledo. The excellent collaboration of Dr. Antonia Gutiérrez and her team is also gratefully acknowledged. We are in debt with Dr. Alfonso Gutiérrez-Adán, Dr. Fernando Rodríguez de Fonseca, and Dr. Emilio Ambrosio for helpful discussions and collaborations with animal models. This work was supported by Excellence Grant CVI-6656 (Regional Andalusian government) and Grant RD12/0028/0013 of the RTA RETICS network from the Spanish Health Institute Carlos III. Thanks are due to Mrs. Kristina Pačesaitė for the design of the graphic work and to Mrs. Laura Castilla for excellent technical work.

References

Adams ME, Mueller HA, Froehner SC (2001) *In vivo* requirement of the alphasyntrophin PDZ domain for the sarcolemmal localization of nNOS and aquaporin-4. J Cell Biol 155:113–122

Akiyama H, Kaneko T, Mizuno N, McGeer PL (1990) Distribution of phosphate-activated glutaminase in the human cerebral cortex. J Comp Neurol 297:239–252

Albrecht J, Dolinska M, Hilgier W, Lipkowski AW, Nowacki J (2000) Modulation of glutamine uptake and phosphate-activated glutaminase activity in rat brain mitochondria by amino acids and their synthetic analogues. Neurochem Int 36:341–347

Aledo JC, de Pedro E, Gómez-Fabre PM, Núñez de Castro I, Márquez J (1997) Submitochondrial localization and membrane topography of Ehrlich ascitic tumour cell glutaminase. Biochim Biophys Acta 1323:173–184

Aledo JC, Gómez-Fabre PM, Olalla L, Márquez J (2000) Identification of two human glutaminase loci and tissue-specific expression of the two related genes. Mamm Genome 11:1107–1110

Aledo JC, Rosado A, Olalla L, Campos JA, Márquez J (2001) Overexpression, purification, and characterization of glutaminase-interacting protein, a PDZ domain protein from human brain. Protein Expr Purif 23:411–418

Altschuler RA, Wenthold RJ, Schwartz AM, Haser WG, Curthoys NP, Parakkal MH et al (1984) Immunocytochemical localization of glutaminase-like immunoreactivity in the auditory nerve. Brain Res 291:173–178

Altschuler RA, Monaghan DT, Haser WG, Wenthold RJ, Curthoys NP, Cotman CW (1985) Immunocytochemical localization of glutaminase-like and aspartate aminotransferase-like immunoreactivities in the rat and guinea pig hippocampus. Brain Res 330:225–233

Amiry-Moghaddam M, Frydenlund DS, Ottersen OP (2004) Anchoring of aquaporin-4 in brain: molecular mechanisms and implications for the physiology and pathophysiology of water transport. Neuroscience 129:999–1010

Aoki C, Kaneko T, Starr A, Pickel VM (1991) Identification of mitochondrial and non-mitochondrial glutaminase within select neurons and glia of rat forebrain by electron microscopic immunocytochemistry. J Neurosci Res 28:531–548

Arama J, Boulay AC, Bosc C, Delphin C, Loew D, Rostaing P et al (2012) Bmcc1s, a novel brain-isoform of Bmcc1, affects cell morphology by regulating MAP6/STOP functions. PLoS One 7:e35488

Attwell D, Laughlin SB (2001) An energy budget for signaling in the grey matter of the brain. J Cereb Blood Flow Metab 21:1133–1145

Bae N, Wang Y, Li L, Rayport S, Lubec G (2013) Network of brain protein level changes in glutaminase deficient fetal mice. J Proteomics 80:236–249

Bak LK, Zieminska E, Waagepetersen HS, Schousboe A, Albrecht J (2008) Metabolism of [U-13C] Glutamine and [U-13C]Glutamate in isolated rat brain mitochondria suggests functional phosphate-activated glutaminase activity in matrix. Neurochem Res 33:273–278

Banerjee M, Huang C, Marquez J, Mohanty S (2008) Probing the structure and function of human glutaminase-interacting protein: a possible target for drug design. Biochemistry 47:9208–9219

Banner C, Hwang JJ, Shapiro RA, Wenthold RJ, Nakatani Y, Lampel KA, Thomas JW, Huie D, Curthoys NP (1988) Isolation of a cDNA for rat brain glutaminase. Brain Res 427:247–254

Benjamin AM (1981) Control of glutaminase activity in rat brain cortex *in vitro*: influence of glutamate, phosphate, ammonium, calcium and hydrogen ions. Brain Res 208:363–377

Benzinger T, Kitzinger C, Hems R, Burton K (1959) Free-energy changes of the glutaminase reaction and the hydrolysis of the terminal pyrophosphate bond of adenosine triphosphate. Biochem J 71:400–407

Berckich DA, Ola MS, Cole J, Sweatt AJ, Hutson SM (2007) Mitochondrial transport proteins of the brain. J Neurosci Res 85:3367–3377

Berl S, Clarke DD (1983) The metabolic compartmentation concept. In: Hertz L, Kvamme E, McGeer EG, Schousboe A (eds) Glutamine, glutamate and GABA in the central nervous system. Liss, New York, pp 205–217

Borg J, Spitz B, Hamel G, Mark J (1985) Selective culture of neurons from rat cerebral cortex: morphological characterization, glutamate uptake and related enzymes during maturation in various culture media. Brain Res 350:37–49

Boulay AC, Burbassi S, Lorenzo HK, Loew D, Ezan P, Giaume C et al (2013) Bmcc1s interacts with the phosphate-activated glutaminase in the brain. Biochimie 95:799–807

Bradford HF, Ward HK (1976) On glutaminase activity in mammalian synaptosomes. Brain Res 110:115–125

Bradford HF, Ward HK, Thomas AJ (1978) Glutamine: a major substrate for nerve endings. J Neurochem 30:1453–1459
Bradford HF, Ward HK, Sandberg M (1984) Kinetic properties of glutaminase from cerebral cortex. Neurochem Res 9:751–757
Bragg AD, Amiry-Moghaddam M, Ottersen OP, Adams ME, Froehner SC (2006) Assembly of a perivascular astrocyte protein scaffold at the mammalian blood–brain barrier is dependent on alpha-syntrophin. Glia 53:879–890
Braitenberg V, Schüz A (1998) Cortex: statistics and geometry of neuronal connectivity, 2nd edn. Springer, Berlin
Brookes N (1997) Intracellular pH as a regulatory signal in astrocyte metabolism. Glia 21:64–73
Bu DF, Erlander MG, Hitz BC, Tillakaratne NJ, Kaufman DL, Wagner-McPherson CB, Evans GA, Tobin AJ (1992) Two human glutamate decarboxylases, 65-kDa GAD and 67-kDa GAD, are each encoded by a single gene. Proc Natl Acad Sci U S A 89:2115–2119
Bungard CI, McGivan JD (2004) Glutamine availability up-regulates expression of the amino acid transporter protein ASCT2 in HepG2 cells and stimulates the ASCT2 promoter. Biochem J 382:27–32
Buschdorf JP, Chew LL, Zhang B, Cao Q, Liang FY, Liou YC et al (2006) Brain-specific BNIP-2-homology protein Caytaxin relocalises glutaminase to neurite terminals and reduces glutamate levels. J Cell Sci 119:3337–3350
Campos JA, Aledo JC, del Castillo-Olivares A, del Valle AE, Núñez de Castro I, Márquez J (1998) Involvement of essential cysteine and histidine residues in the activity of isolated glutaminase from tumour cells. Biochim Biophys Acta 1429:275–283
Campos-Sandoval JA, López de la Oliva AR, Lobo C, Segura JA, Matés JM, Alonso FJ, Márquez J (2007) Expression of functional human glutaminase in baculovirus system: affinity purification, kinetic and molecular characterization. Int J Biochem Cell Biol 39:765–773
Cardona C, Sánchez-Mejías E, Dávila JC, Martín-Rufián M, Campos-Sandoval JA, Vitorica J et al (2015) Expression of Gls and Gls2 glutaminase isoforms in astrocytes. Glia 63:365–382
Cassago A, Ferreira AP, Ferreira IM, Fornezari C, Gomes ER, Greene KS et al (2012) Mitochondrial localization and structure-based phosphate activation mechanism of Glutaminase C with implications for cancer metabolism. Proc Natl Acad Sci U S A 109:1092–1097
Chesler M, Kraig RP (1989) Intracellular pH transients of mammalian astrocytes. J Neurosci 9:2011–2019
Chiu JF, Boeker EA (1979) Cow brain glutaminase: partial purification and mechanism of action. Arch Biochem Biophys 196:493–500
Chung-Bok M-I, Vincent N, Jhala U, Watford M (1997) Rat hepatic glutaminase: identification of the full coding sequence and characterization of a functional promoter. Biochem J 324:193–200
Collard CD, Park KA, Montalto MC, Alapati S, Buras JA, Stahl GL et al (2002) Neutrophil-derived glutamate regulates vascular endothelial barrier function. J Biol Chem 277:14801–14811
Conti F, Minelli A (1994) Glutamate immunoreactivity in rat cerebral cortex is reversibly abolished by 6-diazo-5-oxo-L-norleucine (DON), an inhibitor of phosphate-activated glutaminase. J Histochem Cytochem 42:717–726
Curthoys NP, Watford M (1995) Regulation of glutaminase activity and glutamine metabolism. Annu Rev Nutr 15:133–159
Curthoys NP, Weiss RF (1974) Regulation of renal ammoniagenesis. Subcellular localization of rat kidney glutaminase isoenzymes. J Biol Chem 249:3261–3266
Curthoys NP, Kuhlenschmidt T, Godfrey SS (1976) Regulation of renal ammoniagenesis. Purification and characterization of phosphate-dependent glutaminase from rat kidney. Arch Biochem Biophys 174:82–89
de la Rosa V, Campos-Sandoval JA, Martín-Rufián M, Cardona C, Matés JM, Segura JA, Alonso FJ, Márquez J (2009) A novel glutaminase isoform in mammalian tissues. Neurochem Int 55:76–84
DeBerardinis RJ, Mancuso A, Daikhin E, Nissim I, Yudkoff M, Wehrli S et al (2007) Beyond aerobic glycolysis: transformed cells can engage in glutamine metabolism that exceeds the requirement for protein and nucleotide synthesis. Proc Natl Acad Sci U S A 104:19345–19350

DeLaBarre B, Gross S, Fang C, Gao Y, Jha A, Jiang F, Song JJ, Wei W, Hurov JB (2011) Full-length human glutaminase in complex with an allosteric inhibitor. Biochemistry 50:10764–10770
Donoghue JP, Wenthold RJ, Altschuler RA (1985) Localization of glutaminase-like and aspartate aminotransferase-like immunoreactivity in neurons of cerebral neocortex. J Neurosci 5:2597–2608
Elgadi KM, Meguid RA, Qian M, Souba WW, Abcouwer SF (1999) Cloning and analysis of unique human glutaminase isoforms generated by tissue-specific alternative splicing. Physiol Genomics 1:51–62
Erecinska M, Silver IA (1989) ATP and brain function. J Cereb Blood Flow Metab 9:12–19
Erecinska M, Silver IA (1990) Metabolism and role of glutamate in mammalian brain. Prog Neurobiol 35:245–296
Erecinska M, Zaleska MM, Nelson D, Nissim I, Yudkoff M (1990) Neuronal glutamine utilization: glutamine/glutamate homeostasis in synaptosomes. J Neurochem 54:2057–2069
Errera M, Greenstein JP (1949) Phosphate-activated glutaminase in kidney and other tissues. J Biol Chem 178:495–502
Fergus A, Lee KS (1997) Regulation of cerebral microvessels by glutamatergic mechanisms. Brain Res 754:35–45
Ferreira AP, Cassago A, Gonçalves Kde A, Dias MM, Adamoski D, Ascenção CF et al (2013) Active glutaminase C self-assembles into a supratetrameric oligomer that can be disrupted by an allosteric inhibitor. J Biol Chem 288:28009–28020
Fisher RS (2007) Co-localization of glutamic acid decarboxylase and phosphate-activated glutaminase in neurons of lateral reticular nucleus in feline thalamus. Neurochem Res 32:177–186
Fonnum F (1984) Glutamate: a neurotransmitter in mammalian brain. J Neurochem 42:1–11
Fonseca LL, Monteiro MAR, Alves PM, Carrondo MJT, Santos H (2005) Cultures of rat astrocytes challenged with a steady supply of glutamate: new model to study flux distribution in the glutamate-glutamine cycle. Glia 51:286–296
Gamberino WC, Berkich DA, Lynch CJ, Xu B, LaNoue KF (1997) Role of pyruvate carboxylase in facilitation of synthesis of glutamate and glutamine in cultured astrocytes. J Neurochem 69:2312–2325
Giacobbe A, Bongiorno-Borbone L, Bernassola F, Terrinoni A, Markert EK, Levine AJ et al (2013) p63 regulates glutaminase 2 expression. Cell Cycle 12:1395–1405
Godfrey S, Kuhlenschmidt T, Curthoys P (1977) Correlation between activation and dimer formation of rat renal phosphate-dependent glutaminase. J Biol Chem 252:1927–1931
Godfrey DA, Ross CD, Parli JA, Carlson L (1994) Aspartate aminotransferase and glutaminase activities in rat olfactory bulb and cochlear nucleus; comparisons with retina and with concentrations of substrate and product amino acids. Neurochem Res 19:693–703
Gómez-Fabre PM, Aledo JC, del Castillo-Olivares A, Alonso FJ, Núñez de Castro I, Campos JA, Márquez J (2000) Molecular cloning, sequencing and expression studies of the human breast cancer cell glutaminase. Biochem J 345:365–375
Guha SR (1961) Intercellular localization of glutaminase I in rat liver. Enzymologia 23:94–100
Hamberger AC, Chiang GH, Nylén ES, Scheff SW, Cotman CW (1979) Glutamate as a CNS transmitter. I. Evaluation of glucose and glutamine as precursors for the synthesis of preferentially released glutamate. Brain Res 168:513–530
Haser WG, Shapiro RA, Curthoys NP (1985) Comparison of the phosphate-dependent glutaminase obtained from rat brain and kidney. Biochem J 229:399–408
Häussinger D (1983) Hepatocyte heterogeneity in glutamine and ammonia metabolism and the role of an intercellular glutamine cycle during ureogenesis in perfused rat liver. Eur J Biochem 133:269–275
Häussinger D, Schliess F (2007) Glutamine metabolism and signaling in the liver. Front Biosci 12:371–391
Häussinger D, Sies H (1979) Hepatic glutamine metabolism under the influence of the portal ammonia concentration in the perfused rat liver. Eur J Biochem 101:179–184
Hawkins RA (2009) The blood-brain barrier and glutamate. Am J Clin Nutr 90(Suppl):867S–874S
Heery DM, Kalkhoven E, Hoare S, Parker MG (1997) A signature motif in transcriptional coactivators mediates binding to nuclear receptors. Nature 387:733–736

Heini HG, Gebhardt R, Brecht A, Mecke D (1987) Purification and characterization of rat liver glutaminase. Eur J Biochem 162:541–546
Hertz L (1979) Functional interactions between neurons and astrocytes I. Turnover and metabolism of putative amino acid transmitters. Prog Neurobiol 13:277–323
Hertz L (2004) Intercellular metabolic compartmentation in the brain: past, present and future. Neurochem Int 45:285–296
Hertz L, Zielke HR (2004) Astrocytic control of glutamatergic activity: astrocytes as stars of the show. Trends Neurosci 27:735–743
Hogstad S, Svenneby G, Torgner IA, Kvamme E, Hertz L, Schousboe A (1988) Glutaminase in neurons and astrocytes cultured from mouse brain: kinetic properties and effects of phosphate, glutamate, and ammonia. Neurochem Res 13:383–388
Holcenberg JS, Borella LD, Camitta BM, Ring BJ (1979) Human pharmacology and toxicology of succinylated Acinetobacter glutaminase-asparaginase. Cancer Res 39:3145–3151
Holcomb T, Taylor L, Trohkimoinen J, Curthoys NP (2000) Isolation, characterization and expression of a human brain mitochondrial glutaminase cDNA. Brain Res Mol Brain Res 76:56–63
Hozumi Y, Ito T, Nakano T, Nakagawa T, Aoyagi M, Kondo H et al (2003) Nuclear localization of diacylglycerol kinase zeta in neurons. Eur J Neurosci 18:1448–1457
Hsueh YP, Wang TF, Yang FC, Sheng M (2000) Nuclear translocation and transcription regulation by the membrane-associated guanylate kinase CASK/LIN-2. Nature 404:298–302
Huang YZ, Knox WE (1976) A comparative study of glytaminase isozymes in rat tissues. Enzyme 21:408–426
Hutson SM, Berkich D, Drown P, Xu B, Aschner M, LaNoue KF (1998) Role of branched-chain aminotransferase isoenzymes and gabapentin in neurotransmitter metabolism. J Neurochem 71:863–874
Jalil MA, Begum L, Contreras L, Pardo B, Iijima M, Li MX et al (2005) Reduced N-acetylaspartate levels in mice lacking aralar, a brain- and muscle-type mitochondrial aspartate-glutamate carrier. J Biol Chem 280:31333–31339
Kalra J, Brosnan JT (1974) The subcellular localization of glutaminase isoenzymes in rat kidney cortex. J Biol Chem 249:3255–3260
Kam K, Nicoll R (2007) Excitatory synaptic transmission persists independently of the glutamate-glutamine cycle. J Neurosci 27:9192–9200
Kanamori K, Ross BD (1995) *In vivo* activity of glutaminase in the brain of hyperammonaemic rats measured by 15N nuclear magnetic resonance. Biochem J 305:329–336
Kanamori M, Sandy P, Marzinotto S, Benetti R, Kai C, Hayashizaki Y et al (2003) The PDZ protein tax-interacting protein-1 inhibits beta-catenin transcriptional activity and growth of colorectal cancer cells. J Biol Chem 278:38758–38764
Kaneko T (2000) Enzymes responsible for glutamate synthesis and degradation. In: Ottersen OP, Storm-Mathisen J (eds) Handbook of chemical neuroanatomy. Elsevier, Amsterdam, pp 203–230
Kaneko T, Mizuno N (1988) Immunohistochemical study of glutaminase-containing neurons in the cerebral cortex and thalamus of the rat. J Comp Neurol 267:590–602
Kaneko T, Mizuno N (1992) Mosaic distribution of phosphate-activated glutaminase-like immunoreactivity in the rat striatum. Neuroscience 49:329–345
Kaneko T, Mizuno N (1994) Glutamate-synthesizing enzymes in GABAergic neurons of the neocortex: a double immunofluorescence study in the rat. Neuroscience 61:839–849
Kaneko T, Mizuno N (1996) Spiny stellate neurons in layer VI of the rat cerebral cortex. Neuroreport 7:2331–2335
Kaneko T, Urade Y, Watanabe Y, Mizuno N (1987) Production, characterization, and immunohistochemical application of monoclonal antibodies to glutaminase purified from rat brain. J Neurosci 7:302–309
Kaneko T, Itoh K, Shigemoto R, Mizuno N (1989) Glutaminase-like immunoreactivity in the lower brainstem and cerebellum of the adult rat. Neuroscience 32:79–98
Kaneko T, Akiyama H, Nagatsu I, Mizuno N (1990) Immunohistochemical demonstration of glutaminase in catecholaminergic and serotoninergic neurons of rat brain. Brain Res 507: 151–154

Kaneko T, Nakaya Y, Mizuno N (1992) Paucity of glutaminase-immunoreactive nonpyramidal neurons in the rat cerebral cortex. J Comp Neurol 322:181–190
Kaneko T, Kang Y, Mizuno N (1995) Glutaminase-positive and glutaminase-negative pyramidal cells in layer VI of the primary motor and somatosensory cortices: a combined analysis by intracellular staining and immunocytochemistry in the rat. J Neurosci 15:8362–8377
Kausalya PJ, Phua DC, Hunziker W (2004) Association of ARVCF with zonula occludens (ZO)-1 and ZO-2: binding to PDZ-domain proteins and cell-cell adhesion regulate plasma membrane and nuclear localization of ARVCF. Mol Biol Cell 15:5503–5515
Koehler RC, Gebremedhin D, Harder DR (2006) Role of astrocytes in cerebrovascular regulation. J Appl Physiol 100:307–317
Kornau HC, Schenker LT, Kennedy MB, Seeburg PH (1995) Domain interaction between NMDA receptor subunits and the postsynaptic density protein PSD-95. Science 269:1737–1740
Kovacevic Z, McGivan JD (1983) Mitochondrial metabolism of glutamine and glutamate and its physiological significance. Physiol Rev 63:547–605
Krebs HA (1935) The synthesis of glutamine from glutamate and ammonia, and the enzymatic hydrolysis of glutamine in animal tissues. Biochem J 29:1951–1969
Kvamme E (1984) Enzymes of cerebral glutamine metabolism. In: Häussinger D, Sies H (eds) Glutamine metabolism in mammalian tissues. Springer, Berlin, pp 32–48
Kvamme E (1998) Synthesis of glutamate and its regulation. Prog Brain Res 116:73–85
Kvamme E, Olsen BE (1979) Evidence for two species of mammalian phosphate-activated glutaminase having different regulatory properties. FEBS Lett 107:33–36
Kvamme E, Olsen BE (1981) Evidence for compartmentation of synaptosomal phosphate-activated glutaminase. J Neurochem 36:1916–1923
Kvamme E, Tveit B, Svenneby G (1970) Glutaminase from pig renal cortex. I. Purification and general properties. J Biol Chem 245:1871–1877
Kvamme E, Svenneby G, Hertz L, Schousboe A (1982) Properties of phosphate activated glutaminase in astrocytes cultured from mouse brain. Neurochem Res 7:761–770
Kvamme E, Svenneby G, Torgner IA (1983) Calcium stimulation of glutamine hydrolysis in synaptosomes from rat brain. Neurochem Res 8:25–38
Kvamme E, Torgner IA, Roberg B (1991) Evidence indicating that pig renal phosphate-activated glutaminase has a functionally predominant external localization in the inner mitochondrial membrane. J Biol Chem 266:13185–13192
Kvamme E, Roberg B, Torgner IA (2000) Phosphate-activated glutaminase and mitochondrial glutamine transport in the brain. Neurochem Res 25:1407–1419
Laake JH, Slyngstad TA, Haug FM, Ottersen OP (1995) Glutamine from glial cells is essential for the maintenance of the nerve terminal pool of glutamate: immunogold evidence from hippocampal slice cultures. J Neurochem 65:871–881
Laake JH, Takumi Y, Eidet J, Torgner IA, Roberg B, Kvamme E, Ottersen OP (1999) Postembedding immunogold labelling reveals subcellular localization and pathway-specific enrichment of phosphate activated glutaminase in rat cerebellum. Neuroscience 88:1137–1151
Lebon V, Petersen KF, Cline GW, Shen J, Mason GF, Dufour S et al (2002) Astroglial contribution to brain energy metabolism in humans revealed by 13C nuclear magnetic resonance spectroscopy: elucidation of the dominant pathway for neurotransmitter glutamate repletion and measurement of astrocytic oxidative metabolism. J Neurosci 22:1523–1531
Lee Y-Z, Yang C-W, Chang HY, Hsu H-Y, Chen I-S, Chang H-S et al (2014) Discovery of selective inhibitors of Glutaminase-2, which inhibit mTORC1, activate autophagy and inhibit proliferation in cancer cells. Oncotarget 5:6087–6101
Lohmann R, Souba WW, Bode BP (1999) Rat liver endothelial cell glutamine transporter and glutaminase expression contrast with parenchymal cells. Am J Physiol 276:G743–G750
Magistretti PJ, Pellerin L (1999) Astrocytes couple synaptic activity to glucose utilization in the brain. News Physiol Sci 14:177–182
Magistretti PJ, Pellerin L, Rothman DL, Shulman RG (1999) Energy on demand. Science 283:496–497

Magnusson KR, Larson AA, Madl JE, Altschuler RA, Beitz AJ (1986) Co-localization of fixative-modified glutamate and glutaminase in neurons of the spinal trigeminal nucleus of the rat: an immunohistochemical and immunoradiochemical analysis. J Comp Neurol 247:477–490

Manns ID, Mainville L, Jones BE (2001) Evidence for glutamate, in addition to acetylcholine and GABA, neurotransmitter synthesis in basal forebrain neurons projecting to the entorhinal cortex. Neuroscience 107:249–263

Márquez J, López de la Oliva AR, Matés JM, Segura JA, Alonso FJ (2006) Glutaminase: a multifaceted protein not only involved in generating glutamate. Neurochem Int 48:465–471

Márquez J, Tosina M, de la Rosa V, Segura JA, Alonso FJ, Matés JM et al (2009) New insights into brain glutaminases: beyond their role on glutamatergic transmission. Neurochem Int 55:64–70

Márquez J, Matés JM, Alonso FJ, Martín-Rufián M, Lobo C, Campos-Sandoval JA (2015) Canceromics studies unravel tumor's glutamine addiction after metabolic reprogramming. In: Mazurek S, Shoshan M (eds) Tumor cell metabolism. Pathways, regulation and biology. Springer, Wien, pp 257–287

Martín-Rufián M, Tosina M, Campos-Sandoval JA, Manzanares E, Lobo C, Segura JA et al (2012) Mammalian glutaminase Gls2 gene encodes two functional alternative transcripts by a surrogate promoter usage mechanism. PLoS One 7, e38380

Masson J, Darmon M, Conjard A, Chuhma N, Ropert N, Thoby-Brisson M et al (2006) Mice lacking brain/kidney phosphate-activated glutaminase have impaired glutamatergic synaptic transmission, altered breathing, disorganized goal-directed behavior and die shortly after birth. J Neurosci 26:4660–4671

Matés JM, Segura JA, Campos-Sandoval JA, Lobo C, Alonso L, Alonso FJ, Márquez J (2009) Glutamine homeostasis and mitochondrial dynamics. Int J Biochem Cell Biol 41:2051–2061

Matés JM, Segura JA, Martín-Rufián M, Campos-Sandoval JA, Alonso FJ, Márquez J (2013) Glutaminase isoenzymes as key regulators in metabolic and oxidative stress against cancer. Curr Mol Med 13:514–534

McGivan JD, Bradford NM (1983) Characteristics of the activation of glutaminase by ammonia in sonicated rat liver mitochondria. Biochim Biophys Acta 759:296–302

McKenna MC (2007) The glutamate-glutamine cycle is not stoichiometric: fates of glutamate in brain. J Neurosci Res 85:3347–3358

McKenna MC, Sonnewald U, Huang X, Hopkins IB (1996) Exogenous glutamate concentration regulates the metabolic fate of glutamate in astrocytes. J Neurochem 66:386–393

Medina MA, Quesada AR, Márquez FJ, Sánchez-Jiménez F, Núñez de Castro I (1988) Inorganic phosphate and energy charge compartmentation in Ehrlich ascites tumor cells in the presence of glucose and/or glutamine. Biochem Int 16:713–718

Meertens LM, Miyata KS, Cechetto JD, Rachubinski RA, Capone JP (1998) A mitochondrial ketogenic enzyme regulates its gene expression by association with the nuclear hormone receptor PPARα. EMBO J 17:6972–6978

Mock B, Kozak C, Seldin MF, Ruff N, D'Hoostelaere L, Szpirer C, Levan G, Seuanez H, O'Brien S, Banner C (1989) A glutaminase (gls) gene maps to mouse chromosome 1, rat chromosome 9, and human chromosome 2. Genomics 5:291–297

Morehouse RF, Curthoys NP (1981) Properties of rat renal phosphate-dependent glutaminase coupled to Sepharose. Evidence that dimerization is essential for activation. Biochem J 193:709–716

Nagase T, Ishikawa K, Suyama M, Kikuno R, Hirosawa M, Miyajima N et al (1998) Prediction of the coding sequences of unidentified human genes. XII. The complete sequences of 100 new cDNA clones from brain which code for large proteins *in vitro*. DNA Res 5:355–364

Nedergaard M, Takano T, Hansen AJ (2002) Beyond the role of glutamate as a neurotransmitter. Nat Rev Neurosci 3:748–755

Neely JD, Amiry-Moghaddam M, Ottersen OP, Froehner SC, Agre P, Adams ME (2001) Syntrophin-dependent expression and localization of aquaporin-4 water channel protein. Proc Natl Acad Sci U S A 98:14108–14113

Nicklas WJ, Zeevalk G, Hyndman A (1987) Interactions between neurons and glia in glutamate/glutamine compartmentation. Biochem Soc Trans 15:208–210

Nimmo GA, Tipton KF (1979) The distribution of soluble and membrane-bound forms of glutaminase in pig brain. J Neurochem 33:1083–1084

Nimmo GA, Tipton KF (1980) Purification of soluble glutaminase from pig brain. Biochem Pharmacol 29:359–367
Nimmo GA, Tipton KF (1981) Kinetic comparisons between soluble and membrane-bound glutaminase preparations from pig brain. Eur J Biochem 117:57–64
Norenberg MD, Martínez-Hernández A (1979) Fine structural localization of glutamine synthetase in astrocytes of rat brain. Brain Res 161:303–310
Olalla L, Aledo JC, Bannenberg G, Márquez J (2001) The C-terminus of human glutaminase L mediates association with PDZ domain-containing proteins. FEBS Lett 488:116–122
Olalla L, Gutiérrez A, Campos JA, Khan ZU, Alonso F, Segura JA, Márquez J, Aledo JC (2002) Nuclear localization of L-glutaminase in mammalian brain. J Biol Chem 277:38939–38944
Olalla L, Gutiérrez A, Jiménez AJ, López-Téllez JF, Khan ZU, Pérez J, Alonso FJ, de la Rosa V, Campos-Sandoval JA, Segura JA, Aledo JC, Márquez J (2008) Expression of scaffolding PDZ protein GIP (glutaminase-interacting-protein) in mammalian brain. J Neurosci Res 86:281–292
Patel M, McGivan JD (1984) Partial purification and properties of rat liver glutaminase. Biochem J 220:583–590
Perera SY, Voith DM, Curthoys NP (1991) Biosynthesis and processing of mitochondrial glutaminase in HTC hepatoma cells. Biochem J 273:265–270
Pérez-Gómez C, Matés JM, Gómez-Fabre PM, del Castillo-Olivares A, Alonso FJ, Márquez J (2003) Genomic organization and transcriptional analysis of the human L-glutaminase gene. Biochem J 370:771–784
Pérez-Gómez C, Campos-Sandoval JA, Alonso FJ, Segura JA, Manzanares E, Ruiz-Sánchez P, González ME, Márquez J, Matés JM (2005) Co-expression of glutaminase K and L isoenzymes in human tumour cells. Biochem J 386:535–542
Pilowsky P, Sun QJ, Llewellyn-Smith I, Arnolda L, Chalmers J, Minson J (1997) Phosphate-activated glutaminase immunoreactivity in brainstem respiratory neurons. J Auton Nerv Syst 63:85–90
Porter LD, Ibrahim H, Taylor L, Curthoys NP (2002) Complexity and species variation of the kidney-type glutaminase gene. Physiol Genomics 9:57–66
Quesada AR, Sánchez-Jiménez F, Pérez-Rodríguez J, Márquez J, Medina MA, Núñez de Castro I (1988) Purification of phosphate-dependent glutaminase from isolated mitochondria of Ehrlich ascites-tumour cells. Biochem J 255:1031–1035
Ramos M, del Arco A, Pardo B, Martínez-Serrano A, Martínez-Morales JR, Kobayashi K et al (2003) Developmental changes in the Ca^{2+}-regulated mitochondrial aspartate–glutamate carrier aralar1 in brain and prominent expression in the spinal cord. Dev Brain Res 143:33–46
Resendes MC, Dobransky T, Ferguson SS, Rylett RJ (1999) Nuclear localization of the 82-kDa form of human choline acetyltransferase. J Biol Chem 274:19417–19421
Roberg B, Torgner IA, Kvamme E (1995) The orientation of phosphate activated glutaminase in the inner mitochondrial membrane of synaptic and non-synaptic rat brain mitochondria. Neurochem Int 27:367–376
Roberg B, Torgner IA, Kvamme E (1999) Inhibition of glutamine transport in rat brain mitochondria by some amino acids and tricarboxylic acid cycle intermediates. Neurochem Res 24:811–816
Robinson MM, McBryant SJ, Tsukamoto T, Rojas C, Ferraris DV, Hamilton SK et al (2007) Novel mechanism of inhibition of rat kidney-type glutaminase by bis-2-(5-phenylacetamido-1,2,4-thiadiazol-2-yl)ethyl sulfide (BPTES). Biochem J 406:407–414
Rothman DL, Sibson NR, Hyder F, Shen J, Behar KL, Shulman RG (1999) *In vivo* nuclear magnetic resonance spectroscopy studies of the relationship between the glutamate-glutamine neurotransmitter cycle and functional neuroenergetics. Phil Trans R Soc Lond B 354:1165–1177
Sachdev S, Hoffmann A, Hannink M (1998) Nuclear localization of IkappaB alpha is mediated by the second ankyrin repeat: the IkappaB alpha ankyrin repeats define a novel class of cis-acting nuclear import sequences. Mol Cell Biol 18:2524–2534
Satrústegui J, Contreras L, Ramos M, Marmol P, del Arco A, Saheki T et al (2007) Role of aralar, the mitochondrial transporter of aspartate–glutamate, in brain N-acetylaspartate formation and Ca^{2+} signaling in neuronal mitochondria. J Neurosci Res 85:3359–3366
Schalk AM, Nguyen HA, Rigouin C, Lavie A (2014) Identification and structural analysis of an L-asparaginase enzyme from guinea pig with putative tumor cell killing properties. J Biol Chem 289:33175–33186

Schousboe A, Hertz L, Svenneby G, Kvamme E (1979) Phosphate activated glutaminase activity and glutamine uptake in primary cultures of astrocytes. J Neurochem 32:943–950

Sedgwick SG, Smerdon SJ (1999) The ankyrin repeat: a diversity of interactions on a common structural framework. Trends Biochem Sci 24:311–316

Segura JA, Aledo JC, Gómez-Biedma S, Núñez de Castro I, Márquez J (1995) Tumor glutaminase purification. Protein Expr Purif 6:343–351

Senba E, Kaneko T, Mizuno N, Tohyama M (1991) Somato-, branchio- and viscero-motor neurons contain glutaminase-like immunoreactivity. Brain Res Bull 26:85–97

Shapiro RA, Morehouse RF, Curthoys NP (1982) Inhibition by glutamate of phosphate-dependent glutaminase of rat kidney. Biochem J 207:561–566

Shapiro RA, Haser WG, Curthoys NP (1985) The orientation of phosphate-dependent glutaminase on the inner membrane of rat renal mitochondria. Arch Biochem Biophys 243:1–7

Shapiro RA, Haser WG, Curthoys NP (1987) Immunoblot analysis of glutaminase peptides in intact and solubilized mitochondria isolated from various rat tissues. Biochem J 242:743–747

Shapiro RA, Farrell L, Srinivasan M, Curthoys NP (1991) Isolation, characterization, and *in vitro* expression of a cDNA that encodes the kidney isoenzyme of the mitochondrial glutaminase. J Biol Chem 266:18792–18796

Shen J, Petersen KF, Behar KL, Brown P, Nixon TW, Mason GF, Petroff OA, Shulman GI, Shulman RG, Rothman DL (1999) Determination of the rate of the glutamate/glutamine cycle in the human brain by in vivo 13C NMR. Proc Natl Acad Sci U S A 96:8235–8240

Siesjö BK (1978) Brain energy metabolism. Wiley, New York

Smith EM, Watford M (1988) Rat hepatic glutaminase: purification and immunochemical characterization. Arch Biochem Biophys 260:740–751

Smith EM, Watford M (1990) Molecular cloning of a cDNA for rat hepatic glutaminase. Sequence similarity to kidney-type glutaminase. J Biol Chem 265:10631–10636

Snodgrass PJ, Lund P (1984) Allosteric properties of phosphate-activated glutaminase of human liver mitochondria. Biochim Biophys Acta 798:21–27

Soghomonian JJ, Martin DL (1998) Two isoforms of glutamate decarboxylase: why? Trends Pharmacol Sci 19:500–505

Sonnewald U (2014) Glutamate synthesis has to be matched by its degradation—where do all the carbons go? J Neurochem 131:399–406

Sonnewald U, Westergaard N, Petersen SB, Unsgard G, Schousboe A (1993) Metabolism of [U-13C]glutamate in astrocytes studied by 13C NMR spectroscopy: incorporation of more label into lactate than into glutamine demonstrates the importance of the tricarboxylic acid cycle. J Neurochem 61:1179–1182

Stalnecker CA, Ulrich SM, Li Y, Ramachandran S, McBrayer MK, DeBerardinis RJ, Cerione RA et al (2015) Mechanism by which a recently discovered allosteric inhibitor blocks glutamine metabolism in transformed cells. Proc Natl Acad Sci U S A 112:394–399

Svenneby G (1971) Activation of pig brain glutaminase. J Neurochem 18:2201–2208

Svenneby G (1972) Time- and temperature-dependent activation of pig brain glutaminase. J Neurochem 19:165–174

Svenneby G, Torgner I, Kvamme E (1973) Purification of phosphate-dependent pig brain glutaminase. J Neurochem 20:1217–1224

Svenneby G, Roberg B, Hogstad S, Torgner IAA, Kvamme E (1986) Phosphate-activated glutaminase in the crude mitochondrial fraction (P2 fraction) from human brain cortex. J Neurochem 47:1351–1355

Szeliga M, Matyja E, Obara M, Grajkowska W, Czernicki T, Albrecht J (2008) Relative expression of mRNAs coding for glutaminase isoforms in CNS tissues and CNS tumors. Neurochem Res 33:808–813

Szeliga M, Obara-Michlewska M, Matyja E, Lazarczyk M, Lobo C, Hilgier W et al (2009) Transfection with liver-type glutaminase (LGA) cDNA alteres gene expression and reduces viability, migration and proliferation of T98G glioma cells. Glia 57:1014–1023

Takeda K, Ishida A, Takahashi K, Ueda T (2012) Synaptic vesicles are capable of synthesizing the VGLUT substrate glutamate from α-ketoglutarate for vesicular loading. J Neurochem 121:184–196

Takeuchi H, Jin S, Wang J, Zhang G, Kawanokuchi J, Kuno R et al (2006) Tumor necrosis factor-a induces neurotoxicity via glutamate release from hemichannels of activated microglia in an autocrine manner. J Biol Chem 281:21362–21368

Thangavelu K, Pan CQ, Karlberg T, Balaji G, Uttamchandani M, Suresh V et al (2012) Structural basis for the allosteric inhibitory mechanism of human kidney-type glutaminase (KGA) and its regulation by Raf-Mek-Erk signaling in cancer cell metabolism. Proc Natl Acad Sci U S A 109:7705–7710

Thanki CM, Sugden D, Thomas AJ, Bradford HF (1983) *In vivo* release from cerebral cortex of [14C]glutamate synthesized from [U-14C]glutamine. J Neurochem 41:611–617

Thomas AG, Rojas C, Tanega C, Shen M, Simeonov A, Boxer MB et al (2013) Kinetic characterization of ebselen, chelerythrine and apomorphine as glutaminase inhibitors. Biochem Biophys Res Commun 438:243–248

Turman JE Jr, Chandler SH (1994) Immunohistochemical localization of glutamate and glutaminase in guinea pig trigeminal premotoneurons. Brain Res 634:49–61

Turner A, McGivan JD (2003) Glutaminase isoform expression in cell lines derived from human colorectal adenomas and carcinomas. Biochem J 370:403–408

Van der Gucht E, Jacobs S, Kaneko T, Vandesande F, Arckens L (2003) Distribution and morphological characterization of phosphate-activated glutaminase-immunoreactive neurons in cat visual cortex. Brain Res 988:29–42

Velletri T, Romeo F, Tucci P, Peschiaroli A, Annicchiarico-Petruzzelli M, Niklison-Chirou MV et al (2013) GLS2 is transcriptionally regulated by p73 and contributes to neuronal differentiation. Cell Cycle 12:1–10

Verhoeven AJ, van Iwaarden JF, Joseph SK, Meijer AJ (1983) Control of rat-liver glutaminase by ammonia and pH. Eur J Biochem 133:241–244

Waagepetersen HS, Qu H, Sonnewald U, Shimamoto K, Schousboe A (2005) Role of glutamine and neuronal glutamate uptake in glutamate homeostasis and synthesis during vesicular release in cultured glutamatergic neurons. Neurochem Int 47:92–102

Wallace DR, Dawson R Jr (1992) Ammonia regulation of phosphate-activated glutaminase displays regional variation and impairment in the brain of aged rats. Neurochem Res 17:1113–1122

Wang JB, Erickson JW, Fuji R, Ramachandran S, Gao P, Dinavahi R et al (2010) Targeting mitochondrial glutaminase activity inhibits oncogenic transformation. Cancer Cell 18:207–219

Wang Y, Huang Y, Zhao L, Li Y, Zheng J (2014) Glutaminase 1 is essential for the differentiation, proliferation, and survival of human neural progenitor cells. Stem Cells Dev 23:2782–2790

Ward HK, Bradford HF (1979) Relative activities of glutamine synthetase and glutaminase in mammalian synaptosomes. J Neurochem 33:339–342

Wenthold RJ, Skaggs KK, Altschuler RA (1986) Immunocytochemical localization of aspartate aminotransferase and glutaminase immunoreactivities in the cerebellum. Brain Res 363:371–375

Whitelaw BS, Robinson MB (2013) Inhibitors of glutamate dehydrogenase block sodium-dependent glutamate uptake in rat brain membranes. Front Endocrinol 4:123

Willoughby D, Schwiening CJ (2002) Electrically evoked dendritic pH transients in rat cerebellar Purkinje cells. J Physiol 544:487–499

Würdig S, Kugler P (1991) Histochemistry of glutamate metabolizing enzymes in the rat cerebellar cortex. Neurosci Lett 130:165–168

Yamaguchi S, Jeenes DJ, Archer DB (2001) Protein-glutaminase from Chryseobacterium proteolyticum, an enzyme that deamidates glutaminyl residues in proteins. Purification, characterization and gene cloning. Eur J Biochem 268:1410–1421

Yezierski RP, Kaneko T, Miller KE (1993) Glutaminase like immunoreactivity in rat spinomesencephalic tract cells. Brain Res 624:304–308

Yudkoff M, Zaleska MM, Nissim I, Nelson D, Erecinska M (1989) Neuronal glutamine utilization: pathways of nitrogen transfer studied with [^{15}N]glutamine. J Neurochem 53:632–640

Zhang A, Yeung PL, Li CW, Tsai SC, Dinh GK, Wu X et al (2004) Identification of a novel family of ankyrin repeats containing cofactors for p160 nuclear receptor coactivators. J Biol Chem 279:33799–33805

Chapter 7
Vesicular Glutamate Uptake

Tetsufumi Ueda

Abstract Glutamate is an excitatory neurotransmitter widely used in the vertebrate central nervous systems. The synaptic transmission process is characterized by three steps: (1) presynaptic vesicular transmitter uptake, (2) presynaptic release, and (3) postsynaptic receptor activation. Presynaptic vesicular glutamate uptake plays an initial pivotal role in glutamate transmission by concentrating glutamate in the vesicular lumen prior to its release. This active glutamate transport harnesses energy derived from ATP hydrolysis, and intra- or extravesicular chloride, and is highly specific to glutamate. The uptake system consists of a vesicular glutamate transporter (VGLUT) and v-type proton-pump ATPase, which generates an electrochemical proton gradient, the driving force of the transport. The major source of ATP is likely to be supplied by glycolytic vesicle-bound enzymes, glyceraldehyde 3-phosphate dehydrogenase, and 3-phosphoglycerate kinase, rather than by mitochondrial ATP synthase. The VGLUT substrate glutamate is proposed to be synthesized by vesicle-bound aspartate amino transferase from α-ketoglutarate, not directly from glutamine. VGLUT has three isoforms, and gaged by their distributions they perform different physiological functions. The mechanism and regulation of vesicular glutamate uptake are discussed. The pharmacology of vesicular glutamate uptake is a developing field of inquiry.

Keywords Glutamate uptake • Synaptic vesicles • VGLUT • Glycolytic ATP • Glyceraldehyde-3-phophate dehydrogenase • Aspartate aminotransferase

---

T. Ueda (✉)
Molecular and Behavioral Neuroscience Institute, The University of Michigan, 109 Zina Pitcher Place, Ann Arbor, MI 48109, USA

Department of Pharmacology, University of Michigan Medical School, 2301 MSRB III, 1150 W. Medical Center Dr., Ann Arbor, MI 48109-5632, USA

Department of Psychiatry, Medical School, The University of Michigan, 109 Zina Pitcher Place, Ann Arbor, MI 48109, USA
e-mail: tueda@umich.edu

A. Schousboe, U. Sonnewald (eds.), *The Glutamate/GABA-Glutamine Cycle*, Advances in Neurobiology 13, DOI 10.1007/978-3-319-45096-4_7

Abbreviations

AAT	Aspartate aminotransferase
t-ACPD	*trans*-1-amino-1,3-cyclopentanedicarboxylate
CCCP	Carbonyl cyanide *m*-chlorophenylhydrazone
CNS	Central nervous system
FCCP	Carbonyl cyanide *p*-trifluoromethoxyphenylhydrazone
GABA	γ-Amino butyric acid
GAPDH	Glyceraldehyde 3-phosphate dehydrogenase
α-KGA	α-Ketoglutarate
IPF	Inhibitory protein factor
3-PGK	3-Phosphoglycerate kinase
VGLUT	Vesicular glutamate transporter

Brain function derives from properly integrated neural communication systems. Neuronal communication in the central nervous system (CNS) relies heavily on glutamate synaptic transmission, which is excitatory, triggering neuronal firing via activation of glutamate receptors. Intense glutamate transmission results in activation of various types of glutamate receptors, leading to synaptic modification, which is involved in memory and learning, among other functions. Preparation for glutamate transmission begins with the accumulation of this acidic amino acid into synaptic vesicles, as well as with its synthesis, in the nerve terminal prior to its release into the synaptic cleft (Fig. 7.1). Glutamate uptake into synaptic vesicles is a pivotal step in a broad definition of glutamate transmission. Important to unraveling this narrative was the discovery of the glutamate uptake system in presynaptic vesicles and of its unique properties, which contributed to the recognition of glutamate as a neurotransmitter in the central nervous system. This discovery led to the identification of vesicular glutamate transporter (VGLUT) proteins, which, in combination with the availability of their antibodies, triggered an explosion in this field of neurobiology, largely because of their implication in a variety of physiological functions. A review of VGLUTs will be followed by the discussion of the mechanism and regulation of vesicular glutamate uptake, and a proposed efficient synthesis of the VGLUT substrate and of ATP required for VGLUT function. The development of VGLUT inhibitors will be summarized. The vesicular aspartate transporter will be evaluated, as well as evidence for and against VGLUT in astrocytes. Implications of aberrant VGLUT and vesicular glutamate accumulation in various types of neurological disorders will be described. Relevant reviews (Ueda 1986; Maycox et al. 1990; Ozkan and Ueda 1998; Fremeau et al. 2004a; Thompson et al. 2005; Edwards 2007; Erickson et al. 2006; Ueda and Ikemoto 2007; Chaudhry et al. 2008; El Mestikawy et al. 2011; Omote et al. 2011; Hackett and Ueda 2015) deal with some of the advances in this field.

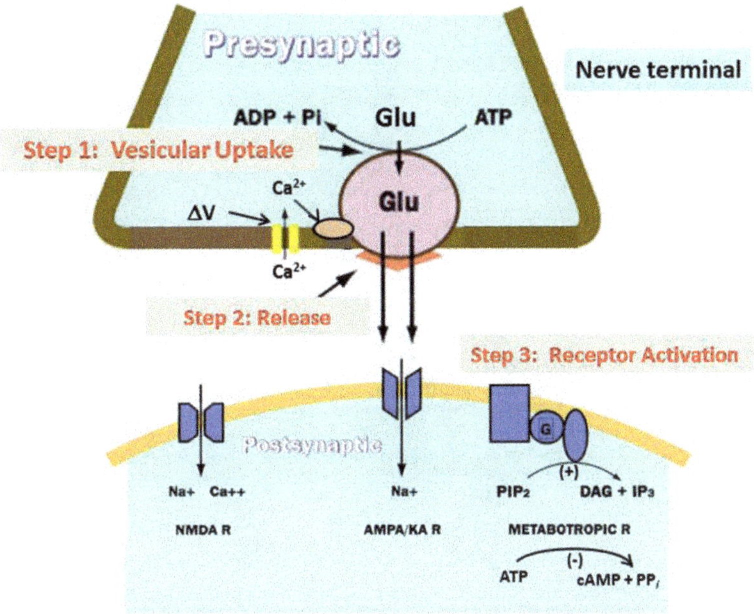

Fig. 7.1 Model of glutamate synaptic transmission. The first step in glutamate transmission is to accumulate glutamate into synaptic vesicles in an energy-dependent manner prior to its release. Upon arrival of the electrical impulse at the nerve terminal, depolarization (ΔV)-induced Ca^{2+} influx occurs. Increased rapid intracellular Ca^{2+} triggers rapid glutamate release into the synaptic cleft (step 2). Increased glutamate in the synaptic cleft activates various types of glutamate receptors, opening the Na^{+}/Ca^{2+} channels and leading to activation of various enzymes, and thereby serves as a neurotransmitter (step 3). *AMPA*, α-amino-3-hydroxy-5-methyl-4-isoxazolepropionic acid; *DAG*, diacyl glyceride; *PIP*$_2$, phosphatidyl inositol bisphosphate; *Glu*, glutamate; *NMDA*, *N*-methyl-aspartate; *IP*$_3$, inositol triphosphate; *R*, receptor

7.1 Discovery and Significance of Vesicular Glutamate Uptake

The discovery of glutamate uptake into synaptic vesicles can be traced back to the discovery of synapsin I (formerly referred to as Protein I). Synapsin Ia and Ib were discovered, purified, and characterized as brain-specific, major synaptic endogenous substrates for cyclic AMP-dependent protein kinase by Ueda and Greengard (1997), in an investigation to understand the molecular mechanism underlying the action of monoamine neurotransmitters via cyclic AMP production at the synapse (Greengard 1976). Their primary structures were deduced by Sudhof et al. (1989, 1990). Initial immunocytochemical and subcellular distribution studies suggested that synaptic vesicles were enriched with Protein I (Bloom et al. 1979; Ueda et al. 1979). Subsequent studies by De Camilli et al. (1983) indicated that Protein I was specifically localized to synaptic vesicles, hence the name "synapsin I." Based upon this knowledge, Naito and Ueda (1982, 1983) purified brain synaptic vesicles by use

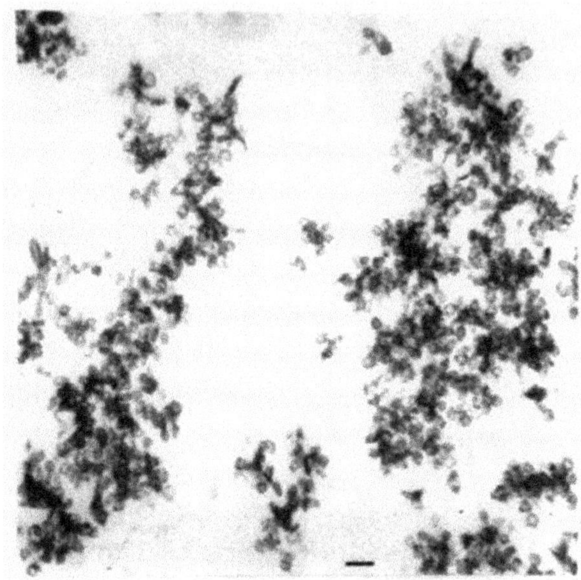

Fig. 7.2 Electronmicrograph of anti-synapsin I IgG-precipitated brain synaptic vesicles. This illustrates the purity of the preparation; it is free of mitochondria, other intracellular organelles, myelin, and plasma membranes. The bar indicates 0.1 μm. Taken from Naito and Ueda (1983)

of highly specific anti-synapsin I IgG (Naito and Ueda 1981), as shown in Fig. 7.2. The immunoprecipitated synaptic vesicle preparation exhibited pure vesicles, free of contamination from other organelles or membranes (Naito and Ueda 1982, 1983). This was the first morphologically pure synaptic vesicle preparation from the brain. Of importance, this synaptic vesicle preparation exhibited ATP-dependent, glutamate-specific uptake under the incubation conditions used (Fig. 7.3; Table 7.1) (Naito and Ueda 1982, 1983). Clear biochemical evidence was provided that synaptic vesicles are capable of accumulating glutamate in a highly specific manner. This demonstration proved key evidence for supporting the concept that glutamate functions as a neurotransmitter and served as a basis for VGLUT research.

Significantly, this evidence was obtained before glutamate had been generally accepted as a neurotransmitter. Three main barriers had prevented acceptance of glutamate's role as a neurotransmitter. The first was teleological. Glutamate occurs as a major biochemical substance, not just in neurons, but in all cells in the entire biological system. As such, it is involved in a variety of important cellular processes, such as energy and nitrogen metabolism, protein and nucleic acid synthesis, and folic acid synthesis, as well as synthesis of the inhibitory neurotransmitter GABA. In contrast, the other transmitters (acetylcholine, norepinephrine, dopamine, serotonin, and GABA) known at that time are synthesized essentially only in neurons and certain endocrine cells, and solely function as a neurotransmitter or hormone. Second, Curtis et al. (1960) showed that glutamate, when iontophoretically applied, fired all the neurons tested. All other neurotransmitters were known to act on certain, but not all, neurons. Based upon this apparent nonspecific effect, it was concluded that glutamate would not serve as a neurotransmitter. Third, evidence was lacking at the time for accumulation of glutamate in synaptic vesicles

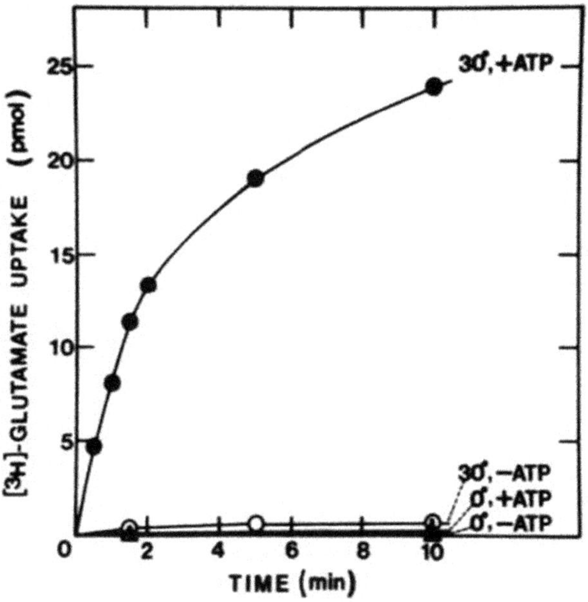

Fig. 7.3 Highly purified brain synaptic vesicles take up glutamate in an ATP-dependent and temperature-dependent manner. Anti-synapsin I IgG-precipitated synaptic vesicles (20 μg) from bovine brain were incubated at 30 or 0 °C with 50 μM [^{3}H]L-glutamate (0.4 Ci/mmol) in the absence or presence of 2 mM ATP in a solution containing 4 mM $MgSO_4$, 5 mM Tris–Cl (pH 7.4), and 320 mM sucrose. This indicates that glutamate is accumulated into synaptic vesicles against the glutamate concentration gradient, namely active transport or uphill transport. Taken from Naito and Ueda (1983)

(Mangan and Whittaker 1966; Rassin 1972; Kontro et al. 1980). Likewise, there was no evidence that synaptic vesicles have the ability to load glutamate (De Belleroche and Bradford 1973), and consequently it was proposed that glutamate is released from the cytosol, rather than from the synaptic vesicle (De Belleroche and Bradford 1977). This failed to support the notion that glutamate functions as a neurotransmitter; well-established neurotransmitters had been found concentrated in synaptic vesicles.

In retrospect, the main reason for the apparent failure of initial investigations to demonstrate enrichment of glutamate in isolated synaptic vesicles seems to have been the inability of synaptic vesicles to maintain accumulated glutamate due to the absence of an electrochemical proton gradient (Burger et al. 1989; Carlson and Ueda 1990). Likewise, the earlier failure to observe glutamate incorporation into isolated synaptic vesicles was due not only to the absence of ATP but also to the presence of a high concentration of chloride in the incubation medium used. Crucial to the clear demonstration of vesicular glutamate uptake was the presence of a low millimolar concentration of chloride as well as of ATP during the uptake, not to mention the purity of synaptic vesicle preparation; high concentrations of chloride were found to inhibit vesicular glutamate

Table 7.1 Substrate specificity for glutamate uptake into highly purified synaptic vesicles

Additions	Uptake	
	cpm	% Control
None (control)	4650±100	100
L-Glutamate	1270±60	27
D-Glutamate	2060±360	44
L-Aspartate	4450±240	96
D-Aspartate	4600±130	99
L-Glutamine	5070±810	109
GABA	3990±40	86

Taken from Naito and Ueda (1983)

Anti-synapsin I IgG-precipitated synaptic vesicles (20 μg) from bovine brain were incubated for 1.5 min with 1 mM [^{3}H]L-glutamate (0.04 Ci/mmol) in the absence or presence of 10 mM glutamate analogues in the same medium as described in Fig. 7.3. Glutamate uptake values represent mean±s.e.m. (n=2 vesicle preparations). Using 50 μM [^{3}H]L-glutamate (0.4 Ci/mmol), the specific activity of glutamate uptake was determined to be 405±44 pmol/min/mg (n=7 preparations)

uptake (Naito and Ueda 1985). This is in great contrast to monoamine uptake into chromaffin granules, which requires much higher chloride concentration, such as 40 mM for maximal uptake (Johnson et al. 1979).

The low concentration chloride stimulation of vesicular glutamate uptake was an accidental discovery. In the initial investigation to characterize synapsin I-containing synaptic vesicles, the chloride concentration was kept minimal in the incubation medium for glutamate uptake, as well as in the vesicle preparation (Naito and Ueda 1983), in order to prevent synapsin I high salt-induced dissociation from vesicles (Ueda and Greengard 1997). A small stimulation by ATP of glutamate uptake in the absence of low millimolar chloride (2-fold as opposed to 40-fold in the presence of a low concentration of chloride) was reported, using crude and partially purified synaptic vesicle preparations (Disbrow et al. 1982). However, the vesicle preparations were not sufficiently pure, so that it was not clear that uptake occurred in synaptic vesicles. Moreover, glutamate specificity was not examined.

The initial clear demonstration of energy-dependent, glutamate-specific uptake into highly purified isolated synaptic vesicles (Naito and Ueda 1983), together with subsequent immunocytochemical (Storm-Mathisen et al. 1983) and biochemical studies (Burger et al. 1989) that glutamate is stored in synaptic vesicles, as well as with the demonstration of Ca^{2+}-dependent vesicular release of glutamate (Nicholls and Sihra 1986), provided strong support for the notion that glutamate plays a role as a neurotransmitter. These lines of evidence led to the subsequent identification of a VGLUT (Takamori et al. 2000; Bellocchio et al. 2000) and demonstration of its control of quantal size (Wojcik et al. 2004) and, together with pharmacological studies (Watkins and Evans 1981; Cotman et al. 1988; Monaghan et al. 1989) and cloning of various glutamate receptors (Hollmann et al. 1989; Masu et al. 1991; Egebjerg et al. 1991; Moriyoshi et al. 1991), proved essential in establishing the neurotransmitter role of glutamate. Thus, synaptic vesicles were demonstrated to be endowed with an energy-dependent molecular machinery able to accumulate glutamate in a highly specific fashion, and this was instrumental in revising the previously held view that the common amino acid glutamate could not function as a neurotransmitter.

7.2 Characteristics of the Vesicular Glutamate Uptake System

As in the case of catecholamine and serotonin uptake into storage granules (Holz 1978; Johnson and Scarpa 1979; Carty et al. 1981; Njus et al. 1986), glutamate uptake into highly purified synaptic vesicles was shown to require ATP (Naito and Ueda 1982, 1983), which is harnessed by v-type proton-pump ATPase to generate an electrochemical proton gradient, the driving force of the uptake (Naito and Ueda 1985; Maycox et al. 1988; Shioi et al. 1989; Fykse et al. 1989; Carlson et al. 1989; Hell et al. 1990; Tabb and Ueda 1991; Tabb et al. 1992). Glutamate uptake was inhibited by the electrochemical proton gradient dissipaters carboxyl cyanide *p*-trifluoromethoxyphenylhydrazone (FCCP) or CCCP, as well as by the v-type proton-pump ATPase inhibitor bafilomycin, but not by mitochondrial proton-pump ATPase inhibitors. It was also inhibited by the H^+/K^+ exchanger nigericin in the presence of K^+. These observations suggested that the vesicular glutamate uptake system consists of v-type proton-pump ATPase and a VGLUT (Ueda 1986; Maycox et al. 1990; Ozkan and Ueda 1998). At the expense of ATP hydrolysis, proton-pump ATPase translocates H^+ across the vesicle membrane from the extravesicular medium to the vesicle interior, generating a membrane potential inside positive. When a membrane-permeable anion such as Cl^- is present in the extravesicular medium, the membrane potential causes anion influx, resulting in formation, for example, of HCl, and decreasing intravesicular pH, accompanied by reduction of the potential difference, namely an increase in ΔpH and a decrease in membrane potential $\Delta\Psi$. The electrochemical proton gradient $\Delta\mu_{H^+}$, defined as a combination of the electrical component $\Delta\Psi$ and the chemical component ΔpH, has been proposed to serve as the driving force for vesicular glutamate uptake. This is in contrast to cellular uptake of glutamate by the plasma membrane transporter, which utilizes the Na^+ gradient as the driving force (for review see ref. Danbolt 2001).

Vesicular glutamate uptake was found surprisingly specific to glutamate, exhibiting no significant interaction with aspartate, GABA, or glutamine (Naito and Ueda 1983, 1985). In essence, the VGLUT failed to recognize glutamate analogs which lack the free α-amino group, free α- and γ-carboxylate groups, and three-carbon skeleton between the two carboxylate groups (Naito and Ueda 1985). The difference

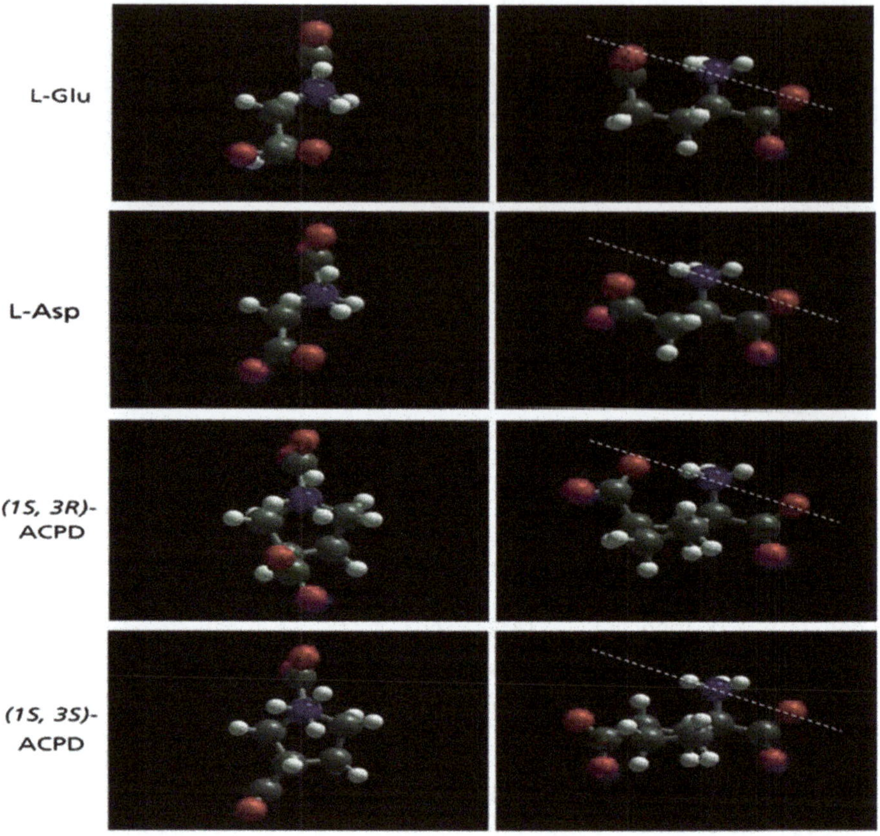

Fig. 7.4 Molecular modeling of glutamate, aspartate, and ACPD. The energy-minimized conformation of glutamate, aspartate, (*1S*, *3R*)-ACPD, and (*1S*, *3S*)-ACPD is shown. When viewed from the side, the negatively charged γ-carboxylate oxygen of glutamate is close to the line connecting its negatively charged α-carboxylate oxygen and positively charged nitrogen. The same applies to the glutamate analog (*1S*, *3R*)-ACPD, but not to the glutamate analogs, aspartate and (*1S*, *3S*)-ACPD. Color codes: *white*, hydrogen; *gray*, carbon; *blue*, nitrogen (positively charged); *red*, negatively charged oxygen of the carboxylate group. Left and right columns show top and side views, respectively. PCMODEL (Serena Software, Bloomington, IN, USA), Ball and Stick, and Mac Molecule (University of Arizona) programs were used for the molecular modeling

in three-dimensional structure between glutamate and its close analog aspartate (which has an amino group, two carboxylate groups, and the two carbon skeleton) is depicted in Fig. 7.4. VGLUT would distinguish the difference in charge distribution between glutamate and aspartate. The charge distribution of glutamate's most stable conformation (the negatively charged oxygen of the α- and γ-carboxylate groups) is similar to that of the glutamate analog (*1S*, *3R*)-ACPD bearing the three carbon skeleton as well as the free α-amino group and the free α- and γ-carboxylate groups, but substantially different from that of aspartate and (*1S*, *3S*)-ACPD. (*1S*, *3R*)-ACPD, but not (*1S*, *3S*)-ACPD, has been shown to serve as a VGLUT substrate.

This high specificity to glutamate was also observed in subsequent studies (Carlson et al. 1989; Tabb and Ueda 1991; Fisher-Bovenkerk et al. 1988; Moriyama and Yamamoto 1995). This contrasts Na^+-dependent cellular glutamate reuptake mediated by a plasma membrane transporter, which does not distinguish between glutamate and aspartate (Logan and Snyder 1972; for review see ref. Danbolt 2001). Moreover, glutamate affinity for the vesicular uptake system was found to be low ($K_m = 1.6$ mM) (Naito and Ueda 1985; Tabb and Ueda 1991), again contrasting with the high affinity for the plasma membrane glutamate transporter ($K_m = 2$–66 μM) (Ueda 1986; Danbolt 2001; Bennett et al. 1974; Kanner and Sharon 1978; Schousboe and Hertz 1981 for review). The low glutamate affinity for the vesicular transporter is compatible with high glutamate concentration in neurons (estimated to be about 10 mM), and the high affinity for the plasma membrane transporter with low concentration in the extracellular milieu (1 μM or less at rest).

Another intriguing finding was that vesicular glutamate uptake was markedly stimulated by a physiologically relevant, low millimolar intracellular concentration of chloride (Naito and Ueda 1985), as shown in Fig. 7.5a, b. However, high concentrations were found to be inhibitory. This inhibition by high chloride concentrations was attenuated by a high concentration of glutamate, suggesting that chloride competes with glutamate for the glutamate binding/transport site. The stimulatory effect of chloride was mimicked by bromide, but not by other halogens or anions (Naito and Ueda 1985; Moriyama and Yamamoto 1995). This interesting, unique property was observed in a number of studies (Fykse et al. 1989; Carlson et al. 1989; Hell et al. 1990; Tabb and Ueda 1991; Tabb et al. 1992; Fisher-Bovenkerk et al. 1988; Moriyama and Yamamoto 1995; Wolosker et al. 1996). These properties of the vesicular uptake system were demonstrated in all vertebrate brains examined (Tabb and Ueda 1991) and in a reconstituted system consisting of solubilized synaptic vesicle proteins and crude soybean phospholipid (which contains phosphatidyl choline, phosphatidyl ethanolamine, phosphatidyl serine, phosphatidyl inositol, triolein, and cholesterol); pure phosphatidyl choline, phosphatidyl ethanolamine, phosphatidyl serine, phosphatidic acid, and phosphatidyl inositol failed to support vesicular uptake (Carlson et al. 1989). These unique features render the vesicular glutamate uptake system distinct from the plasma membrane uptake system (Table 7.2) and led to a model of the vesicular glutamate uptake system (Ueda 1986), as shown in Fig. 7.6.

An electrochemical proton gradient, as opposed to a Na^+ gradient, is harnessed as a driving force to accumulate neurotransmitters into synaptic vesicles; this concept has been extended to GABA and glycine uptake (Fykse et al. 1989; Hell et al. 1990; Kish et al. 1989) as well as to acetylcholine uptake (Anderson et al. 1982; Parsons et al. 1993). Thus, it could be generalized that nonpeptide neurotransmitter uptake into synaptic vesicles occurs at the expense of an electrochemical proton gradient; the relative contribution of its two components (membrane potential and pH gradient) varies, however, depending upon the type of neurotransmitter transported (Hell et al. 1990). The electrochemical proton gradient across the inner mitochondrial membrane, originally conceived of by Mitchell (1961, 1966; Reid et al. 1966) to link electron transport and ATP synthesis, is now recognized as the

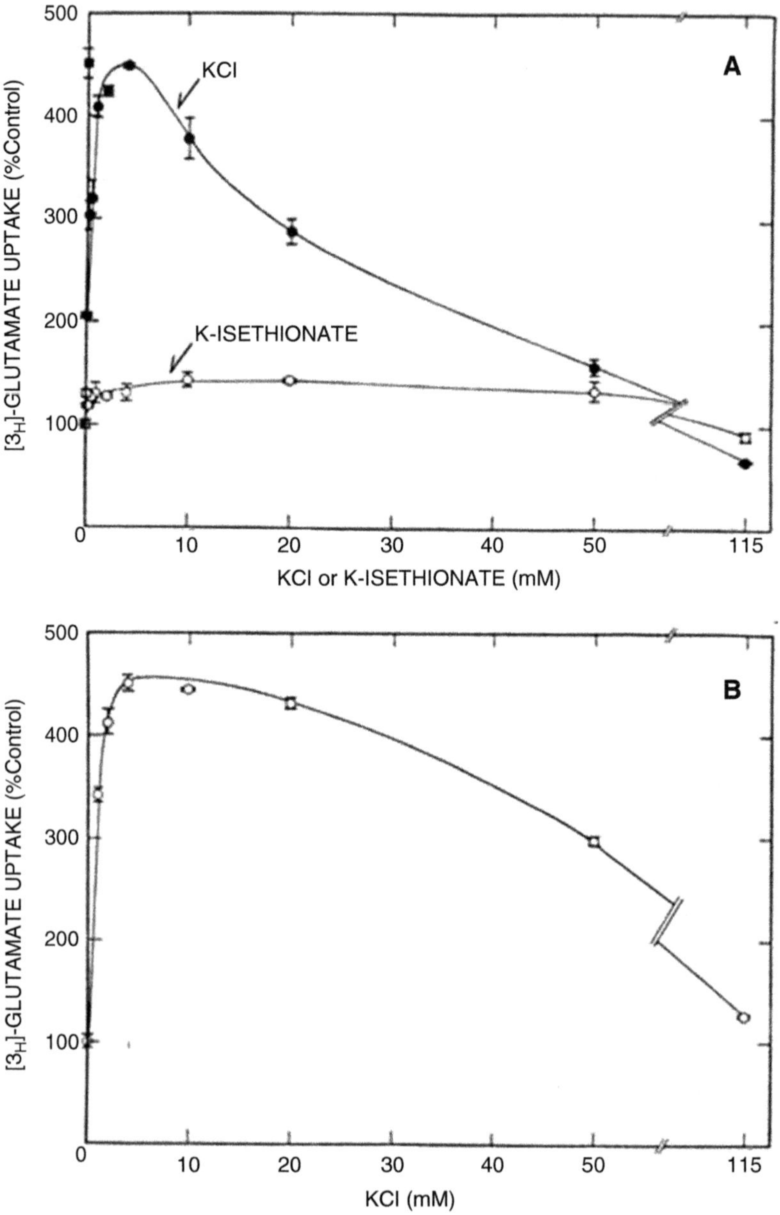

Fig. 7.5 Vesicular glutamate uptake exhibits a biphasic response to chloride. Glutamate uptake into synaptic vesicles (20 μg; purified by sucrose density gradient centrifugation) for 1.5 min at 30 °C was measured with 50 μM (**a**) or 1 mM (**b**) [^{3}H]glutamate in the presence of various concentrations of chloride or isethionate in the same solution as described in Fig. 7.3, except for replacement of 5 mM Tris–Cl and 320 mM sucrose with 5 mM Tris–maleate (pH 7.4) and 250 mM sucrose, respectively. Glutamate uptake is markedly stimulated by low concentrations of chloride, but inhibited by high concentration (**a, b**). The inhibition is attenuated by increasing glutamate concentration (**b**). Taken from Naito and Ueda (1985)

Table 7.2 Comparison of synaptic vesicle and plasma membrane glutamate uptake systems

	Synaptic vesicle membrane	Plasma membrane
Driving force	$\Delta\mu_H{}^+ = \Delta\Psi + \Delta pH$	$Na^+_o > Na^+_i$, $K^+_o > K^+_i$
Stimulator	Low mM Cl^- (1–4 mM)	High mM Cl^- (100 mM)
Affinity for glutamate	Low ($K_m = 1.6$ mM)	High ($K_m = 2$–66 μM)
Interaction with aspartate	No	Yes
Interaction with *t*-ACPD[a]	Yes	No
Glutamate transporter	VGLUT1-3	EAAT1-5

t-ACPD, *trans*-1-amino-1,3-dicarboxycyclopentane; *VGLUT*, vesicular glutamate transporter; *EAAT*, excitatory amino acid transporter (Amara and Fontana 2002)
[a]See ref. Winter and Ueda (1993)

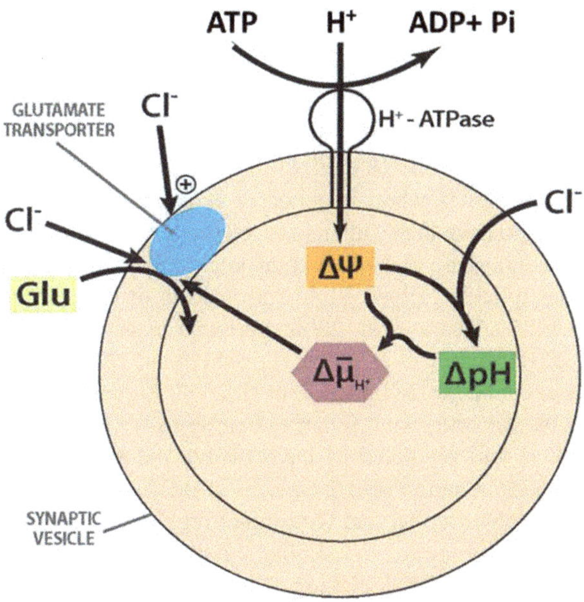

Fig. 7.6 Model of the vesicular glutamate uptake system. The uptake system consists of two functionally different components, v-type proton-pump ATPase and the vesicular glutamate transporter (VGLUT). The H^+-pump ATPase generates a membrane potential at the expense of ATP hydrolysis. In the presence of extravesicular chloride, this induces chloride influx, resulting in HCl formation in the vesicle lumen, thus generating a pH gradient across the vesicle membrane. An electrochemical proton gradient comprised of membrane potential and pH gradient serves as the driving force of vesicular glutamate uptake. Low concentration of chloride could stimulate vesicular glutamate uptake via binding to an allosteric site, as well as via pH gradient formation. High concentrations of chloride could inhibit glutamate uptake by competing with glutamate. $\Delta\mu_H{}^+$, electrochemical proton gradient; $\Delta\Psi$, membrane potential; *Glu*, glutamate

common currency of cellular energy. In mitochondria, it is utilized to generate ATP, whereas in intracellular organelles, such as synaptic vesicles and chromaffin granules, it is harnessed to accumulate transmitters and hormones. Thus, the synaptic vesicular neurotransmitter (NT) uptake system is thought to consist of two functionally different protein components, v-type proton-pump ATPase and a transmitter-specific transporter, in a manner analogous to antibodies (which comprise a common segment and a variable segment):

$$\text{Synaptic vesicular NT uptake system} = \text{v ATPase}\left(\text{common}\right) + \text{NT v transporter}\left(\text{variable}\right)$$

7.3 Vesicular Glutamate Transporter

The first genetic evidence indicating VGLUT is present in glutamatergic neurons was obtained by Fisher-Bovenkerk et al. (1988). Identification of the VGLUT had remained elusive until 2000, when Takamori et al. and Bellocchio et al. (Takamori et al. 2000; Bellocchio et al. 2000) demonstrated that a brain-specific isoform of the sodium-dependent inorganic phosphate transporter (Ni et al. 1994) functioned as VGLUT. This was quickly followed by identification of two additional isoforms, designated VGLUT2 (Aihara et al. 2000; Fremeau et al. 2001; Bai et al. 2001; Takamori et al. 2001; Herzog et al. 2001; Hayashi et al. 2001; Varoqui et al. 2002) and VGLUT3 (Gras et al. 2002; Takamori et al. 2002; Fremeau et al. 2002; Schafer et al. 2002). All the isoforms were found to belong to the SLC 17 family (Reimer and Edwards 2004) and predicted to have twelve membrane-spanning regions. Little variation is seen in amino acid sequences of their greater central domain, but they largely differ in N-terminal and C-terminal regions (Fig. 7.7). As such, their biochemical functional properties are similar (Fremeau et al. 2001, 2002; Bai et al. 2001; Takamori et al. 2001, 2002; Herzog et al. 2001; Hayashi et al. 2001; Varoqui et al. 2002; Gras et al. 2002; Schafer et al. 2002). Their substrate specificity, affinity for glutamate, and chloride stimulation are essentially indistinguishable from those originally found in isolated vesicles (Naito and Ueda 1983, 1985). Detailed comparison of VGLUT1 and two biochemical characteristics has been reported by Kaneko and Fujiyama (2002). Mutational analyses (Juge et al. 2006) and computer modeling (Almqvist et al. 2007) suggest that the active site involves His^{128}, Arg^{184}, and Glu^{191} in the case of VGLUT2 and His^{120} and Arg^{176} in the case of VGLUT1. It is postulated that VGLUT1 His^{120} and Arg^{176} interact with the γ- and α-carboxyl group's negatively charged oxygen of glutamate, respectively, and VGLUT1 Glu^{183} with the positively charged α-amino group of glutamate. Proteomic studies suggest that synaptic vesicles bear 9 and 14 copies per vesicle of VGLUT1 and VGLUT2, respectively (Takamori et al. 2006). These studies also indicate that approximately 80 % of all synaptic vesicles contain VGLUTs 1 and 2, which is compatible with an early observation by Naito and Ueda (1983) suggesting that the majority of brain

```
VGLUT1 1 -------------MEFRQEEFRKLAGRALGKLHRLLEKRQEGAETLELSADGRPVTTQTRDPPVVDCTCFGLPRRYI
VGLUT2 1 -----MESVKQRILAPGKEGLKNFAGKSLGQIYRVLEKKQDTGETIELTEDGKPLEVPERKAPLCDCTCFGLPRRYI
VGLUT3 1 MPFKAFDTFKEKILKPGKEGVKNAVGDSLGILQRKIDGTTEEEDNIELNEEGRPVQTSRPSPPLCDCHCCGLPKRYI

65  IAIMSGLGFCISFGIRCNLGVAIVSMVNNSTTHRGGHVVVQKAQFSWDPETVGLIHGSFFWGYIVTQIPGGFICQKFAANRV
73  IAIMSGLGFCISFGIRCNLGVAIVDMVNNSTTHRGGKVIKEKAKFNWDPETVGMIHGSFFWGYIITQIPGGYIASRLAANRV
78  IAIMSGLGFCISFGIRCNLGVAIVEMVNNSTVYVDGKPEIQTAQFNWDPETVGLIHGSFFWGYIMTQIPGGFISNKFAANRV
                TM1                                             TM2

147 FGFAIVATSTLNMLIPSAARVHYGCVIFVRILQGLVEGVTYPACHGIWSKWAPPLERSRLATTAFCGSYAGAVVAMPLAGVL
155 FGAAILLTSTLNMLIPSAARVHYGCVIFVRILQGLVEGVTYPACHGIWSKWAPPLERSRLATTSFCGSYAGAVIAMPLAGIL
160 FGAAIFLTSTLNMFIPSAARVHYGCVMCVRILQGLVEGVTYPACHGMWSKWAPPLERSRLATTSFCGSYAGAVVAMPLAGVL
            TM3                        TM4                              TM5

229 VQYSGWSSVFYVYGSFGIFWYLFWLLVSYESPALHPSISEEERKYIEDAIGESAKLMNPLTKFSTPWRRFFTSMPVYAIIVA
237 VQYTGWSSVFYVYGSFGMVWYMFWLLVSYESPAKHPTITDEERRYIEESIGESANLLGAMEKFKTPWRKFFTSMPVYAIIVA
242 VQYIGWSSVFYYYGMFGIIWYMFWLLQAYECPAAHPTISNEEKTYIETSIGEGANVVS-LSKFSTPWKRFFTSLPVYAIIVA
                    TM6                                                            TM7

311 NFCRSWTFYLLLISQPAYFEEVFGFEISKVGLVSALPHLVMTIIVPIGGQIADFLRSRRIMSTTNVRKLMNCGGFGMEATLL
319 NFCRSWTFYLLLISQPAYFEEVFGFEISKVGMLSAVPHLVMTIIVPIGGQIADFLRSKQILSTTTVRKIMNCGGFGMEATLL
323 NFCRSWTFYLLLISQPAYFEEVFGFAISKVGLLSAVPHMVMTIVVPIGGQLADYLRSRQILTTTAVRKIMNCGGFGMEATLL
            TM7                             TM8                           TM9

393 LVVGYSHSKGVAISFLVLAVGFSGFAISGFNVNHLDIAPRYASILMGISNGVGTLSGMVCPIIVGAMTKHKTREEWQYVFLI
401 LVVGYSHTRGVAISFLVLAVGFSGFAISGFNVNHLDIAPRYASILMGISNGVGTLSGMVCPIIVGAMTKNKSREEWQYVFLI
405 LVVGFSHTKGVAISFLVLAVGFSGFAISGFNVNHLDIAPRYASILMGISNGVGTLSGMVCPLIVGAMTRHKTREEWQNVFLI
                       TM10                                  TM11                 TM12

475 ASLVHYGGVIFYGVFASGEKQPWAEPEEMSEEKCGFVGHDQLAGSDDSEMEDEAEPPGAPPAPPPSYGATHSTFQPPRPPPP
483 AALVHYGGVIFYAIFASGEKQPWADPEETSEEKCGFIHEDELDEE
487 AALVHYSGVIFYGVFASGEKQEWADPENLSEEKCGIIDQDELAEE
                TM12

557 VRDY
565 VQGE
569 TLDE
```

Fig. 7.7 Deduced amino acid sequence of human VGLUT1, VGLUT2, and VGLUT3. Green highlights and underlines indicate varying and transmembrane regions, respectively. Red letters indicate amino acid residues postulated to be glutamate binding at the active site (Takamori et al. 2002; Fremeau et al. 2002)

synaptic vesicles are glutamate accumulating. These lines of evidence are consistent with the notion that glutamate is the major excitatory neurotransmitter.

VGLUT1 and VGLUT2 have, in general, complementary distributions in the adult brain (Fremeau et al. 2001, 2004a; Herzog et al. 2001; Varoqui et al. 2002; Kaneko and Fujiyama 2002). VGLUT1 predominates in telencephalic regions, including the cerebral cortex and hippocampus, whereas VGLUT2 is primarily expressed in diencephalic and lower brain stem regions. However, in some glutamatergic nerve terminals and certain endocrine cells, VGLUT1 and VGLUT2 are colocalized (Fremeau et al. 2004a, b; Hioki et al. 2003; Morimoto et al. 2003; Hayashi et al. 2003; Bai et al. 2003; Miyazaki et al. 2003; Boulland et al. 2004; Nakamura et al. 2005; Herzog et al. 2006; Persson et al. 2006); some vesicles have both VGLUT1 and VGLUT2 (Herzog et al. 2006). Their coexpression occurs transiently at early developmental stages in multiple glutamatergic neurons (Fremeau et al. 2004b; Miyazaki et al. 2003; Boulland et al. 2004; Nakamura et al. 2005; Herzog et al. 2006); VGLUT2 is expressed first and declines, followed by VGLUT1 expression. A population of synaptic vesicles is endowed with both VGLUT1 and VGLUT2 during early development (Herzog et al. 2006); this could possibly confer greater plasticity to the control of quantal release. Electrophysiological evidence indicates that the copy number of VGLUT1 and VGLUT2 per vesicle can affect quantal size (Wojcik et al. 2004; Daniels et al. 2004; Wilson et al. 2005; Moechars et al. 2006), compatible with the notion that alteration of vesicular glutamate content can change miniature excitatory postsynaptic potential.

In contrast to the more abundant isoforms VGLUT1 and 2, VGLUT3 is expressed by a much smaller number of isolated neuronal populations (Fremeau et al. 2002, 2004a; Gras et al. 2002; Schafer et al. 2002). However, evidence indicates that VGLUT3 is expressed not only in nerve terminals but also in dendrites, a subpopulation of astrocytes, progenitor-like cells, and "nonnervous tissues" (Fremeau et al. 2002, 2004a; Gras et al. 2002; Boulland et al. 2004; Ni and Parpura 2009; Ormel et al. 2012a). Transient expression of VGLUT3 also occurs in particular cells in specific brain regions during the early postnatal period (Boulland et al. 2004; Gras et al. 2005). These observations suggest that the VGLUT isoforms are subject to differential cellular expression and/or intracellular trafficking and involved in different functions. It is interesting to note that VGLUT1 is present in photoreceptor cell terminals (Johnson et al. 2003; Sherry et al. 2003; Fyk-Kolodziej et al. 2004), whereas VGLUT3 is present in auditory inner hair cell terminals and essential for auditory coding (Seal et al. 2008; Ruel et al. 2008); however, VGLUT1 occurs in auditory nerve fiber terminals in the cochlear nucleus (Zhou et al. 2007).

VGLUT1 is mostly localized to synapses distinct from those bearing VGLUT2 and can control the reserve pool size of synaptic vesicles in excitatory nerve terminals (Fremeau et al. 2004b; Todera et al. 2007). VGLUT1 is thought to have a role in regulating the probability of transmitter release (Fremeau et al. 2001, 2004a, b; Varoqui et al. 2002; Weston et al. 2011; Herman et al. 2014) and in synaptic plasticity (Varoqui et al. 2002; Todera et al. 2007; Balschun et al. 2010) including reversal learning (Balschun et al. 2010; Granseth et al. 2015). Studies by Herman et al. (2014) suggest that the probability is affected by the amount of glutamate accumulated in the vesicle, namely the glutamate fill state, not by the VGLUT protein amount.

In contrast, VGLUT2 appears to be present in those synapses with high release probability, which transfer information with high fidelity; these synapses are common in sensory and autonomic nervous systems (Fremeau et al. 2001; Varoqui et al. 2002). Thus, VGLUT2 is expressed in the majority of thalamus neurons (Fremeau et al. 2001, 2004a; Varoqui et al. 2002; Moechars et al. 2006); cerebellar climbing fibers (Fremeau et al. 2001; Hioki et al. 2003); blood pressure-controlling presympathetic neurons (Stornetta et al. 2002), inspiratory augmenting neurons (Stornetta et al. 2003), and chemosensory neurons innervating the main respiratory center (Weston et al. 2004), all in the brainstem; and nociception-mediating spinal cord DRG neurons (Liu et al. 2010; Rogoz et al. 2012; Ma 2014). VGLUT 3 has been found colocalized with vesicular acetylcholine, monoamine, and GABA transporters (Gras et al. 2002, 2005, 2008; Fremeau et al. 2002; Schafer et al. 2002; Boulland et al. 2004; Divito et al. 2015). This has implications for motor control, reward behavior, and neurological and psychiatric disorders (Fremeau et al. 2002; Gras et al. 2008; Divito et al. 2015).

Accumulated evidence now indicates that a subset of each VGLUT is colocalized with a subset of vesicular monoamine, GABA, and acetylcholine transporters (for review see ref. El Mestikawy et al. 2011). In addition, a subset of VGLUT2 is colocalized with a subpopulation of vesicles containing dopamine and norepinephrine and a subset of VGLUT3 with a subpopulation of serotonin-containing vesicles. Thus, VGLUT coexpression levels in nonglutamatergic vesicles could upregulate the vesicle's capacity to accumulate these transmitters, in part via glutamate influx-induced increase in pH gradient, and thereby could modulate GABAergic, cholinergic, and monoaminergic neurotransmission. Conversely, glutamate transmission could be regulated by corelease of GABA, acetylcholine, and monoamine neurotransmitters. These cophenotypes suggest interplay between glutamate transmission and other neurotransmission systems through corelease and have implications for behavioral, neurological, and mental disorders.

VGLUTs are expressed not only in the brain and certain endocrine systems (Hayashi et al. 2001, 2003; Morimoto et al. 2003; Bai et al. 2003) but also in the spinal cord (Persson et al. 2006; Todd et al. 2003; Oliveira et al. 2003; Li et al. 2003; Landry et al. 2004; Wu et al. 2004; Seal et al. 2009) and "nonnervous tissues." VGLUT2 has been shown to occur in pulmonary (sensory) nerve fiber terminals (Brouns et al. 2004) and glucagon-like peptide 1-containing granules in intestinal L cells (Uehara et al. 2006). VGLUT 1 has been observed in intrafusal muscle-spindle afferent endings (Wu et al. 2004) and motor nerve terminals of esophageal striated muscle (Kraus et al. 2004). In the bone, VGLUT1 is expressed in osteoclast cells, responsible for bone reabsorption (Morimoto et al. 2006). VGLUT3 is present in the liver (Gras et al. 2002; Fremeau et al. 2002; Boulland et al. 2004), skeletal muscle fibers, and some motor neuron terminals (Boulland et al. 2004), as well as in a small subset of dorsal root ganglion neurons innervating the spinal cord dorsal horn and mediating mechanical hypersensitivity (Seal et al. 2009). VGLUT3 is also expressed in primary sensory afferents innervating the dental pulp, controlling micro blood circulation (Zerari-Mailly et al.

2012). VGLUT occurrence in other autonomic and sensory nervous systems, including neurons innervating the colorectum (Brumovsky et al. 2011) and bladder (Brumovsky et al. 2013), as well as in the spinal cord, has recently been reviewed extensively by Brumovsky (2013). All these observations indicate VGLUT involvement not only in brain function but in other physiological functions as well.

7.4 Mechanism of Vesicular Glutamate Uptake and Modulation by Cl^- and K^+

The driving force of vesicular glutamate uptake is an electrochemical proton gradient across the synaptic vesicle membrane, analogous to catecholamine uptake into chromaffin granules (Holz 1978; Johnson and Scarpa 1979; Njus et al. 1986; Casey et al. 1977; Phillips and Allison 1978; Johnson et al. 1979; Johnson 1988). This has been demonstrated using the electrogenic proton ionophore FCCP capable of dissipating ΔpH as well as $\Delta\Psi$, the neutral H^+/K^+ exchanger nigericin (and the neutral H^+/Mg^{2+} exchanger A23187) and the electrogenic K^+ ionophore valinomycin. The relative contribution of $\Delta\Psi$ and ΔpH varies with the concentration of permeable anions such as chloride. Evidence supporting this principle has been provided with isolated synaptic vesicles (Naito and Ueda 1985; Maycox et al. 1988; Shioi et al. 1989; Fykse et al. 1989; Tabb and Ueda 1991; Tabb et al. 1992; Moriyama and Yamamoto 1995; Wolosker et al. 1996), reconstituted systems (Maycox et al. 1988; Carlson et al. 1989), cloned VGLUT1 (Takamori et al. 2000; Bellocchio et al. 2000; Schenck et al. 2009), VGLUT2 (Fremeau et al. 2001; Bai et al. 2001; Herzog et al. 2001; Varoqui et al. 2002; Juge et al. 2006), and VGLUT3 (Gras et al. 2002; Takamori et al. 2002; Fremeau et al. 2002). The critical importance of the membrane potential as the driving force has been investigated by several laboratories (Maycox et al. 1988; Shioi et al. 1989; Schenck et al. 2009; Shioi and Ueda 1990; Juge et al. 2010). The membrane potential increases affinity for glutamate (Wolosker et al. 1996; Francis S. Lee, Ph.D. Dissertation, University of Michigan, 1996), suggesting a conformational change in VGLUT. The contribution of the pH gradient has been addressed in a number of studies (Naito and Ueda 1985; Takamori et al. 2000, 2002; Tabb et al. 1992; Wolosker et al. 1996; Fremeau et al. 2001, 2002; Bai et al. 2001; Herzog et al. 2001; Varoqui et al. 2002; Gras et al. 2002; Schenck et al. 2009; Shioi and Ueda 1990).

As to the modulation by chloride, the stimulatory effect of low millimolar chloride is due to an increase in V_{max} with little change in K_m for glutamate (Naito and Ueda 1985; Wolosker et al. 1996). This chloride stimulation was initially postulated to be mediated by a conformational change in the glutamate transporter, as well as by a pH gradient produced by membrane potential-induced chloride influx (Naito and Ueda 1985). Tabb et al. (1992) have provided evidence for the generation of a small pH gradient in the presence of low millimolar chloride and suggested that an intravesicular optimal pH, produced through a small influx of chloride, is responsible for activation of VGLUT. Studies by Hartinger and Jahn (1993) provided the first evidence supporting the notion that chloride stimulation involves its binding to an allosteric site of the glutamate transporter, followed by studies by Moriyama and Yamamoto (1995). Tabb et al. (1992), Wolosker et al. (1996), Schenck et al. (2009), and Juge et al. (2010) showed that activation of vesicular glutamate transport by an extravesicular low chloride concentration can occur in the presence of a membrane potential alone, without a pH gradient.

In elegant experiments using reconstituted liposomes with VGLUT1 and a mitochondrial type of H^+-ATPase, Schenck et al. (2009) demonstrated that intraliposomal chloride together with a membrane potential is sufficient to cause glutamate uptake. This is consistent with an earlier observation by Francis S. Lee (Ph.D. dissertation, University of Michigan, 1996) that chloride-preloaded synaptic vesicles in the presence of ATP and a high concentration of chloride exhibited a twofold increase in glutamate uptake compared to chloride-unloaded intact vesicles, when assayed in the absence of extravesicular chloride. However, in the absence of intraliposomal chloride, low millimolar extraliposomal chloride is required for maximal glutamate uptake. The requirement for intraliposomal chloride was attributed to the charge and osmotic balance maintenance enabled by chloride efflux being loosely coupled to glutamate influx, which would allow continuous glutamate uptake. Similarly, an anti-glutamate/chloride transport mechanism had previously been proposed (Wolosker et al. 1996). It remains to be determined, however, whether the intraliposomal chloride could also activate VGLUT via binding to it on the lumen side.

The stimulation by extraliposomal chloride was attributed to a pH gradient formed by VGLUT-mediated chloride influx. Nigericin (which dissipates the pH gradient) substantially reduced glutamate uptake into reconstituted liposomes with a H^+-pumping ATPase and VGLUT1 (Schenck et al. 2009), as observed with purified brain vesicles (Tabb et al. 1992). It is also noted that there is a small component of low millimolar extraliposomal chloride-stimulated glutamate uptake even in the absence of pH gradient. This could be attributed to the chloride activation via its binding to the extraliposomal side of VGLUT. VGLUT-mediated chloride influx is supported by evidence that VGLUT is capable of chloride transport in the presence of membrane potential, enabling pH gradient formation (Bellocchio et al. 2000; Schenck et al. 2009; Preobraschenski et al. 2014). However, this contrasts studies by Juge et al. (2010), in which a diffusion membrane potential was induced by K^+ in the presence of valinomycin, and chloride flux was measured in the presence of 5 mM glutamate. They showed that VGLUT was unable to mediate chloride flux and suggested that low millimolar extraliposomal chloride, together with a membrane potential alone, is sufficient for maximal vesicular glutamate uptake. Recently, in agreement with studies by Schenck et al. (2009), Preobraschenski et al. (2014) have also demonstrated the intravesicular chloride requirement for vesicular glutamate uptake in the absence of extravesicular chloride, as well as stimulation by low millimolar extravesicular chloride in the absence of intravesicular chloride. Thus, membrane potential can cause vesicular glutamate uptake in the presence of chloride in either the interior or the exterior of the vesicle.

In freshly endocytosed vesicles containing a high concentration of trapped chloride, extravesicular chloride would not be required for maximal glutamate uptake, whereas in chloride-empty vesicles a low physiological concentration of extravesicular chloride would be critical. The degree of glutamate uptake stimulation by extravesicular chloride would presumably vary with the amount of intravesicular chloride: the lower the intravesicular chloride concentration, the higher the glutamate uptake stimulation by low millimolar concentration of extravesicular chloride. This could explain variability in the extent of low millimolar chloride stimulation observed with different vesicle preparations, including those from various species (Tabb and Ueda 1991), possibly reflecting different chloride concentrations present in prepared vesicles. It is likely that, in synaptic vesicles containing low concentrations of chloride, extravesicular chloride, in addition to activating VGLUT via an allosteric binding, substantially contributes to the driving force via generation of a pH gradient induced by chloride influx. The pH gradient-dependent glutamate uptake could be due to proton efflux, possibly coupled with chloride efflux by the same mechanism of Na^+-dependent phosphate transport mediated by VGLUT (cation-dependent anion transport). The chloride efflux in this manner (if the stoichiometry is $2Cl^-/H^+$) could maintain charge and osmotic balance with glutamate uptake. An optimal intravesicular pH brought about by pH gradient formation could be involved in enhancing vesicular glutamate uptake as well.

As to the mechanism underlying the inhibition of vesicular glutamate uptake by high chloride concentration, evidence now supports the notion that chloride competes with glutamate for the transport site. This inhibition is attenuated by increasing glutamate concentration (Naito and Ueda 1985; Wolosker et al. 1996; Schenck et al. 2009). Chloride-dependent inhibition is also observed in the presence of an enhanced, constant membrane potential in the absence of pH gradient (Tabb et al. 1992; Wolosker et al. 1996). Bellocchio et al. (2000) have provided evidence that cloned VGLUT1 is capable of chloride transport and that glutamate competes for anion permeation. This was also demonstrated by Schenck et al. (2009) and Preobraschenski et al. (2014), using reconstituted proteoliposomes containing VGLUT1 and H^+-pump ATPase. Moreover, Schenck et al. (2009) showed that chloride inhibition of glutamate uptake is reduced by increasing glutamate concentration and occurs even under conditions in which the membrane potential is enhanced and the pH gradient is abolished. These lines of evidence together indicate that high chloride concentration inhibits glutamate uptake by competing with glutamate for the anion transport site.

In an effort to address the role of intravesicular H^+ in vesicular glutamate, Goh et al. (2011) found that a physiologically relevant high concentration of extravesicular K^+ stimulates vesicular glutamate uptake via K^+/H^+ exchange. The final level of accumulation, not the initial rate of uptake, is increased. It is proposed that the pH gradient (formed not only by chloride influx but also by glutamate transport) is consumed by a 5-(*N*-ethyl-*N*-isopropyl)amiloride (EIPA)-sensitive cation/H^+ exchanger (potentially NHE6), generating an additional membrane potential when vesicular H^+-pump ATPase is in operation, and that this extra membrane potential is responsible for enhanced glutamate uptake. Preobraschenski et al. (2014) confirmed the K^+/H^+ exchange activity and K^+-stimulated glutamate uptake, as well as the presence of the protein NHE6, in synaptic vesicles. They have also shown, using mitochondrial type

H^+-ATPase-loaded liposomes fused with v-H^+-ATPase-blocked synaptic vesicles, that when liposomes are devoid of chloride, the K^+ stimulation is the highest in the absence of extraliposomal chloride. However, when liposomes are filled with highly concentrated chloride, the K^+-stimulation is minimal or negligible regardless of extraliposomal chloride concentration. The same was observed using reconstituted liposomes with VGLUT1 and H^+-pump ATPase only. VGLUT itself has been proposed to be capable of mediating K^+/H^+ exchange (with high selectivity for K^+ over Na^+). This stimulation is attributed to an increase in membrane potential generated by active H^+-pump ATPase upon K^+/H^+ exchange.

Thus, under physiological conditions [in which the extracellular and intraneuronal chloride concentrations are about 150 mM and 20 mM (Billups and Attwell 2002; Price and Trussell 2006), respectively, and the intraneuronal concentration of glutamate is estimated to be 1–10 mM (Nicholls and Attwell 1990; McMahon and Nicholls 1991)], freshly endocytosed synaptic vesicles would load glutamate, initially driven by a combination of membrane potential and trapped intravesicular chloride (Fig. 7.8a). As lumen chloride decreases, a cytosolic low concentration of chloride would begin to play a role in vesicular glutamate uptake as a VGLUT activator (by directly binding to an allosteric site) as well as a pH gradient cogenerator (Fig. 7.8b, c). The pH gradient thus generated could also serve as a component of the driving force potentially through VGLUT-mediated H^+ efflux coupled with Cl^- efflux (Fig. 7.8b, c). An optimal intravesicular pH (around pH 6.8) could also be produced for maximal function of VGLUT (Fig. 7.8b, c). Cytosolic K^+ could further elevate vesicular glutamate accumulation via H^+/K^+ exchange (mediated by VGLUT and NHE6)-induced enhancement of membrane potential (which occurs with active H^+-pump ATPase) (Fig. 7.8b, c). Based upon the mechanism described above, a time course similar to that shown in Fig. 7.8d would be observed upon sequential additions of ATP, a low concentration of chloride, and a high concentration of potassium ion. In vivo, inhibition by high chloride concentration would not occur, since cytosolic chloride concentration is not high enough, and extravesicular glutamate concentration is sufficiently high to compete with low levels of cytosolic chloride. Price and Trussell have provided evidence that changing the nerve terminal cytosolic chloride concentration does not alter the amount of glutamate in synaptic vesicles in the calyx of Held, a giant glutamatergic nerve ending of the auditory brainstem (Price and Trussell 2006). Although low millimolar chloride is essential in bringing about maximal vesicular uptake in vitro, chloride may not have a modulatory role in vivo. VGLUT could be kept activated at all times with relatively constant low chloride concentration.

7.5 Energy Source for Vesicular Glutamate Uptake

In order to maintain normal synaptic transmission, in particular at the time of rapid neuronal firing when synaptic vesicles are quickly depleted of pre-accumulated glutamate, emptied vesicles demand swift reloading of glutamate. This

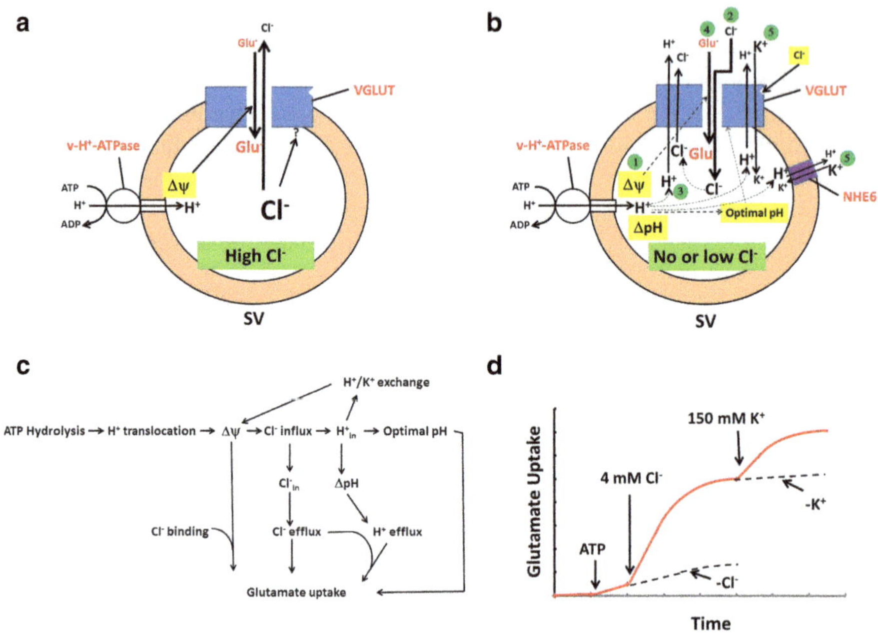

Fig. 7.8 Proposed mechanisms of vesicular glutamate uptake and its modulation by chloride. (**a**) In the presence of a high concentration of intravesicular chloride, a membrane potential generated by H^+-pump ATPase can cause glutamate influx, accumulating glutamate in the vesicle. In order to maintain ionic and osmotic balance, chloride efflux, which would be mediated by VGLUT, must occur; this would allow continuous glutamate uptake. (**b**) In the absence of intravesicular chloride or in the presence of a low concentration of chloride, a low concentration of extravesicular chloride plays a modulatory role by forming a pH gradient and providing an optimal intravesicular pH for VGLUT activity, as well as by activating VGLUT via binding to an allosteric site. VGLUT-mediated proton efflux could occur, possibly coupled with chloride efflux. This would allow ionic and osmotic balance maintenance, which would enable glutamate uptake to continue. Proton efflux could also be coupled with K^+ influx mediated by VGLUT and the cation/proton exchanger NHE6. This would enhance membrane potential when the H^+-pump ATPase is in operation, leading to an increase in glutamate uptake. *ΔΨ*, membrane potential; *Glu*, glutamate; *NHE*, cation/proton exchanger. (**c**) Time sequence of events from ATP hydrolysis to glutamate uptake under conditions described in (**b**). (**d**) Expected time course of vesicular glutamate uptake upon sequential additions of ATP, Cl^- and K^+. This type of time coarse would be observed as a result of sequential additions of ATP, 4 mM Cl^-, and 150 mM K^+ to isolated synaptic vesicles, reflecting the proposed mechanism shown in (**b**, **c**)

demand would be met by a readily available energy supply to fuel the loading, as well as by the availability of glutamate. How is this to be achieved?

Synaptic transmission is well known to vitally rely on glycolysis (Cox and Bachelard 1982), although mitochondria are the major source of cellular ATP production. Under hypoglycemic conditions, abnormal synaptic transmission rapidly develops, yet global cellular ATP levels are hardly changed (Cox et al. 1983; Bachelard et al. 1984). Replacement of glucose with the final glycolysis product pyruvate, which results in production of the majority of cellular ATP by mitochondria,

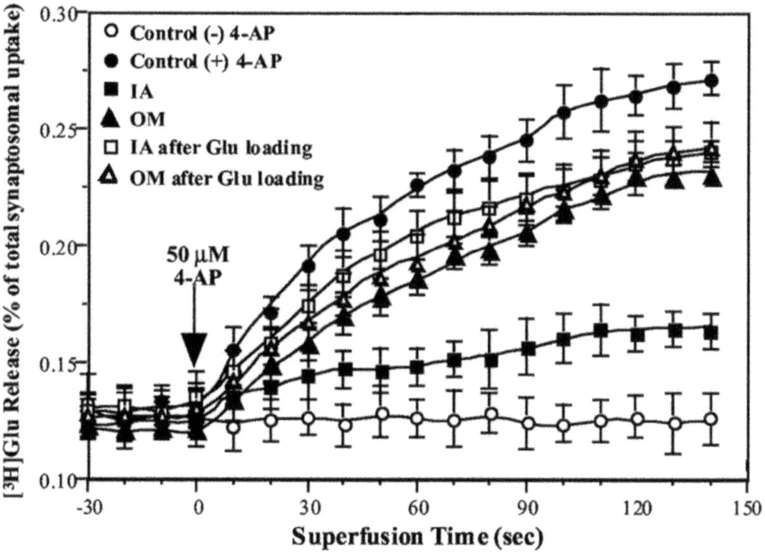

Fig. 7.9 Glycolysis-produced ATP plays a major role in providing energy required for vesicular glutamate loading, prior to its release from synaptosomes. Iodoacetate, an inhibitor of glyceraldehyde 3-phosphate dehydrogenase involved in glycolytic ATP production, reduced the amount of glutamate released in a 4-AP-induced, calcium-dependent manner from synaptosomes, which had been loaded with tritiated glutamate after exposure to iodoacetate, whereas the mitochondrial ATP synthase inhibitor oligomycin had minimal effect. The inhibitory effect of iodoacetate on glutamate is likely to reflect its effect on vesicular glutamate content, as described in the text. *4-AP*, 4-aminopyridine; *Glu*, glutamate; *IA*, iodoacetate; *OM*, oligomycin. Taken from Ikemoto et al. (2003)

does not restore normal synaptic transmission (Cox and Bachelard 1982), even though ATP tissue levels return to normal (Cox et al. 1983). Fleck et al. (1993) showed that hypoglycemic conditions which led to reduction of glutamate release did not significantly alter cellular ATP content. These observations raised the possibility that a glycolytic intermediate or a minor pool of ATP, possibly locally synthesized, might play a critical role in glutamate release and synaptic transmission. Ikemoto et al. (2003) have provided evidence suggesting that glycolytically generated ATP, rather than that produced in mitochondria, plays a major role in providing energy for vesicular glutamate accumulation. The glyceraldehyde-3-phosphate dehydrogenase (GAPDH) inhibitor iodoacetate, which is known to suppress glycolysis, reduced the amount of glutamate released from synaptosomes (pinched-off nerve endings) in a 4-aminopyridine (4-AP)-induced, calcium-dependent manner, whereas the mitochondrial ATP-synthase inhibitor oligomycin had minimal effect (Fig. 7.9). This iodoacetate-induced reduction in glutamate release could reflect its effect on vesicular glutamate content in synaptosomes. These observations are compatible with the finding by Shepherd and Harris (1998) that approximately 50 % of hippocampal CA3/CA1 nerve terminals are devoid of mitochondria; these terminals must rely exclusively on glycolytically generated ATP.

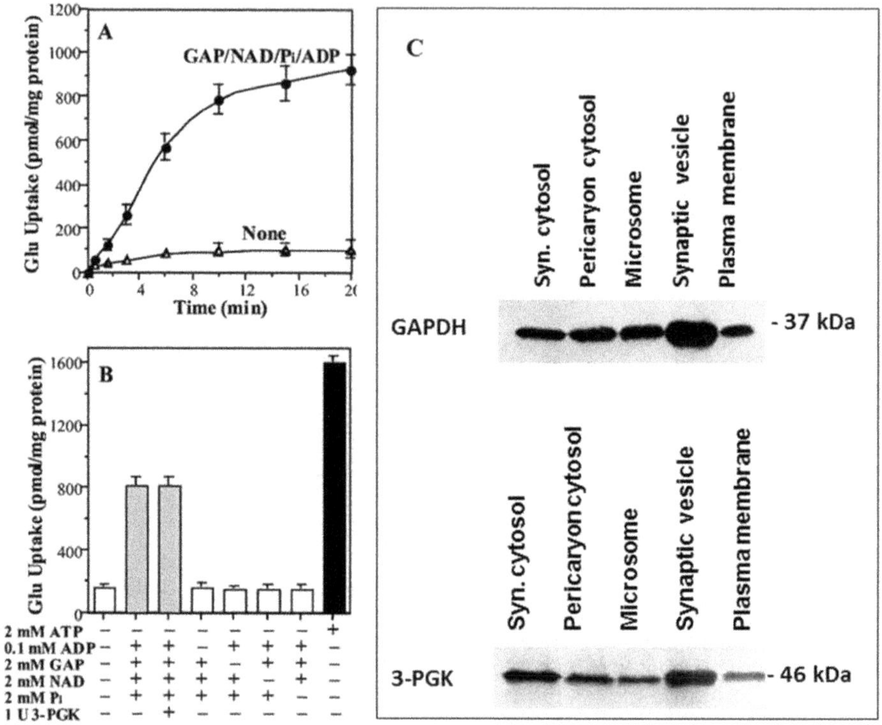

Fig. 7.10 ATP synthesized by synaptic vesicle-bound glycolytic enzymes glyceraldehyde 3-phosphate dehydrogenase and 3-phosphoglycerate kinase is harnessed by the vesicular glutamate uptake system. (**a**) The substrates for the glycolytic ATP-generating enzymes (GAP, NAD, and P_i for GAPDH and ADP for 3-PGK) support glutamate uptake into isolated synaptic vesicles. (**b**) Omission of any of these substrates abolishes vesicular glutamate uptake. (**c**) Western blots of GAPDH and 3-PGK in various subcellular fractions. GAPDH and 3-PGK are most concentrated in the synaptic vesicle fraction of all fractions tested. *GAP*, glyceraldehyde 3-phosphate; *GAPDH*, glyceraldehyde 3-phosphate dehydrogenase; *3-PGK*, 3-phosphoglycerate kinase; P_i, inorganic phosphate. Taken from Ikemoto et al. (2003)

Of importance, in searching for high-energy glycolytic intermediate-modified proteins, Ikemoto et al. (2003) found that synaptic vesicles are endowed with the glycolytic ATP-generating enzyme complex GAPDH/3-phosphoglycerate kinase (3-PGK), which is harnessed to generate energy for immediate uptake of glutamate into synaptic vesicles (Fig. 7.10a, b). Moreover, glutamate uptake into isolated synaptic vesicles was inhibited by iodoacetate. Surprisingly, both GAPDH and 3-PGK were found to be most enriched in the synaptic vesicle fraction, of all subcellular fractions examined (Fig. 7.10c). The association of GAPDH with synaptic vesicles is consistent with evidence obtained by proteomic analyses (Coughenour et al. 2004; Takamori et al. 2006; Burré et al. 2006). Further studies have suggested that a subpopulation of synaptic vesicles bears pyruvate kinase, which is also capable of supporting vesicular glutamate uptake (Ishida et al. 2009). These observations together support the notion that locally synthesized glycolytic ATP is the major energy source harnessed for the efficient loading of glutamate into synaptic vesicles (Ueda and Ikemoto 2007). The role of locally synthesized glycolytic ATP appears not limited to vesicular glutamate uptake and most likely extended to the uptake of other transmitters at other synapses (Ishida et al. 2009). It is feasible that glycolytic ATP synthesized by membrane-bound GAPDH/3-PGK is coupled to all major cation transporters (Ueda and Ikemoto 2007). Moreover, the vital role of local synthesis of glycolytic ATP in fast axonal transport of synaptic vesicles has also been demonstrated recently (Zala et al. 2013).

7.6 Precursor of VGLUT Substrate

How do neurons produce the specific pool of glutamate for immediate use as a neurotransmitter? As the major amino acid, glutamate is involved in a number of metabolic and other cellular functions. In eukaryotes, glutamate is synthesized by three enzymes:

(a) Amino acid aminotransferase

$$\alpha \text{ ketoglutarate} + \text{amino acid} \rightleftharpoons \text{glutamate} + \alpha \text{ keto acid}$$

(b) Glutamate dehydrogenase

$$\alpha \text{ ketoglutarate} + NH_4^+ \underset{(\text{NADPH})}{+\text{NADH}} + H^+ \rightleftharpoons \text{glutamate} + \underset{(\text{NADP}^+)}{\text{NAD}^+} + H_2O$$

(c) Glutaminase

$$\text{glutamine} + H_2O \rightarrow \text{glutamate} + NH_3$$

Glutamine is widely thought to be the principal precursor for synthesis of the neurotransmitter glutamate. Bradford et al. (1978) and Hamberger et al. (1979), using radioactive glutamine and glucose, provided biochemical evidence that glutamine is the major substrate for the neurotransmitter pool of glutamate. This, together with the specific localization of glutamine synthetase in astrocytes (Martinez-Hernandez et al. 1977) and glutaminase enrichment in mitochondria within synaptosomes (Bradford and Ward 1976), as well as with the notion of the glutamate–glutamine cycle (Van den Berg and Garfinkel 1971; Benjamin and Quastel 1972), led to the concept that the glutamate–glutamine cycle plays a central role in supplying a precursor for synthesis of the neurotransmitter glutamate (Cotman et al. 1981; Shank and Aprison 1988; Schousboe et al. 1997). According to this model, astrocytes take up the majority of glutamate released from nerve terminals, followed by conversion to glutamine, which is then transported out to the extracellular space. The released glutamine is taken up by neurons and transformed back to glutamate, largely by mitochondrial glutaminase. Glutamate thus produced and directly released to the cytosol from mitochondria had been thought to be utilized as a neurotransmitter.

This is also supported by studies (Kvamme et al. 1991; Roberg et al. 1995) indicating that phosphate-activated glutaminase is localized to the exterior of the inner mitochondrial membrane. However, other evidence indicates that the active site of this enzyme is on the matrix side of the mitochondrial inner membrane (Shapiro et al. 1985; Aledo et al. 1997; Albrecht et al. 2000; Zieminska et al. 2004; Bak et al. 2008). Moreover, Palaiologos et al. (1988) provided evidence suggesting that glutamine-derived glutamate is transformed by aspartate aminotransferase (AAT) in the mitochondrial matrix to α-ketoglutarate (Dennis et al. 1977), which is then transported out to the cytoplasm, mediated by the dicarboxylic acid exchanger (Bisaccia et al. 1985; Bolli et al. 1989). α-Ketoglutarate exported from mitochondria is then converted to glutamate, which is released in a Ca^{2+}-dependent manner. The resynthesis of glutamate was assumed to be carried out by cytosolic AAT (Palaiologos et al. 1988).

Takeda et al. (2012) have found that isolated synaptic vesicles are capable of synthesizing glutamate from α-ketoglutarate and L-aspartate (as the specific amino donor) for immediate uptake (Fig. 7.11). Vesicle-bound AAT was shown to be responsible for this synthesis. Glutamine was not an effective substrate for glutamate synthesis, consistent with the vesicle's glutaminase deficiency. These observations suggest that local synthesis of the neurotransmitter pool of glutamate could occur at the synaptic vesicle. Of interest, vesicle-bound AAT was found to belong to the mitochondrial type (GOT2), not the cytosolic isozyme (GOT1),

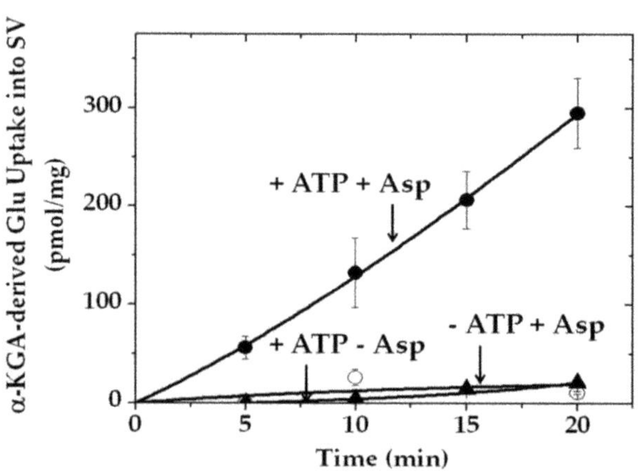

Fig. 7.11 Synaptic vesicle-bound AAT synthesizes the VGLUT substrate glutamate. When isolated synaptic vesicles were incubated with [^{14}C]α-KGA in the presence of aspartate, [^{14}C]glutamate was taken up into vesicles in an ATP-dependent manner. This was not observed in the absence of aspartate, indicating that glutamate is formed from α-KGA by AAT bound to synaptic vesicles and readily transported into vesicles. *α-KGA*, α-ketoglutarate; *Asp*, aspartate; *Glu*, glutamate. With permission from Takeda et al. (2012)

based on its kinetic and immunological properties (Takeda et al. 2012) as well as proteomic analyses (Takamori et al. 2006; Burré et al. 2006). Vesicle-bound AAT has much higher affinity for aspartate than does the cytosolic isozyme. The K_m value of vesicle-bound AAT for aspartate (0.9 mM) is substantially lower than the cellular concentration of aspartate (estimated to be 3.5–8 mM), whereas the K_m value for the cytosolic enzyme (6.7 mM) is higher than or similar to the cytosolic aspartate concentration. This suggests that vesicle-bound AAT would function at close to V_{max} in situ. These findings indicate that α-ketoglutarate can effectively serve as an immediate precursor for synthesis of the neurotransmitter glutamate. Thus, the glutamate synthesis by vesicle-bound AAT could represent an efficient mechanism for vesicular glutamate refilling and, hence, glutamate transmission.

α-Ketoglutarate to be used for synthesis of the neurotransmitter glutamate would derive mainly from nerve terminal mitochondria, where it is produced in large part from glutamine-derived glutamate (Palaiologos et al. 1988), and some possibly from glucose via the TCA cycle. In addition, it could be supplied from astrocytes (Westergaard et al. 1994), where de novo synthesis of α-ketoglutarate occurs via pyruvate carboxylation, coupled with the TCA cycle (Patel 1974; Yu et al. 1983; Shank et al. 1985, 1993; Cesar and Hamprecht 1995; Hertz et al. 2007; Serres et al. 2008; Sonnewald and Rae 2010; Morken et al. 2014; Sonnewald 2014). Synaptosomes have the capacity to take up α-ketoglutarate (Shank and Campbell 1981; Shank and Campbell 1982; Shank and Campbell 1984, Takeda and Ueda, 2016) as well as glutamine (Bradford et al. 1978), Takeda and Ueda, 2016).

Since the catalytic site of mitochondrial glutaminase is likely localized to the matrix side of the mitochondrial inner membrane (Shapiro et al. 1985; Aledo et al. 1997; Albrecht et al. 2000; Zieminska et al. 2004; Bak et al. 2008), the glutamine–glutamate conversion would occur in the matrix, where glutamate, rather than being directly released to the cytoplasm, would largely be transformed to α-ketoglutarate by mitochondrial AAT (Palaiologos et al. 1988) and glutamate dehydrogenase (Tildon et al. 1985; McKenna et al. 1993, 2000; McKenna 2007). A large part of α-ketoglutarate derived from glutamine and a small part of α-ketoglutarate from glucose would be transferred to the cytoplasm, as mentioned above. These lines of evidence taken together suggest that α-ketoglutarate can serve as the major *immediate* precursor for the neurotransmitter pool of glutamate. However, in view of immunocytochemical evidence suggesting that cytosolic AAT is associated with glutamatergic nerve terminals (Altschuler et al. 1981, 1982), the relative contribution of vesicle-bound vs. cytosolic AAT to synthesis of the neurotransmitter pool of glutamate remains to be established.

Although the glutamate–glutamine cycle is thought to play a central role in supplying the precursor of the neurotransmitter glutamate, evidence obtained over the past decade indicates that synthesis of the glutamate responsible for spontaneous synaptic transmission remains less clear (Edwards 2007). Kam and Nicoll (2007) showed that vesicular release of glutamate from neurons in culture devoid of astrocytes occurs even in the absence of glutamine in the culture medium. Masson et al. (2006) observed that genetic deletion of phosphate-activated glutaminase partially reduced, but did not abolish, evoked release, whereas it had little effect on spontaneous release of glutamate.

Moreover, alteration in the major neuronal glutamine transporter SNAT2 does not affect spontaneous glutamatergic transmission or the quantal size (Grewal et al. 2009). These lines of evidence indicate that the glutamate–glutamine cycle may not play an exclusive role in providing the immediate precursor for synthesis of the neurotransmitter pool of glutamate, in particular that of glutamate released spontaneously and responsible for generating the miniature excitatory postsynaptic current. This suggests alternative pathways for such a role. α-Ketoglutarate derived from glucose or some amino acids or odd chain fatty acids (Sonnewald 2014) via the TCA cycle could serve as such a precursor. Glutamate could also be formed from some other amino acids (Sonnewald 2014). Thus, glutamate released by an action potential is likely to derive from glutamine largely via formation of α-ketoglutarate, whereas the precursor of spontaneously released glutamate remains elusive. An effective mechanism for synthesis and vesicular loading of the neurotransmitter glutamate, as well as for ATP synthesis required for rapid vesicular glutamate uptake, is proposed in Fig. 7.12.

7.7 VGLUT Inhibitors

Excessive glutamate transmission is involved in many types of neurological pathologies. Since vesicular glutamate accumulation is an initial pivotal step in glutamate transmission, VGLUT represents a target in controlling glutamate transmission. Hence, considerable efforts have been made to find potent VGLUT inhibitors. Several classes of VGLUT inhibitors have been reported: (a) glutamate analogs such as *trans*-1-amino-1,3-cyclopentanedicarboxylate (*trans*-ACPD) and *trans*-1-amino-1,3-cyclohexanedicarboxylate (*trans*-ACHD) (Winter and Ueda 1993, 2008; Moriyama and Yamamoto 1995); (b) kynurenate analogs such as 7-chloro-kynurenate and 8-hydroxy-kynurenate (Bartlett et al. 1998; Thompson et al. 2005); (c) diazo-dyes bearing disulfonic acid-substituted naphthylamine groups and a biphenyl linker, such as Trypan Blue (Fig. 7.13a) (Roseth et al. 1998; Fonnum et al. 1998) and Evans Blue (Roseth et al. 1995); (d) diazo-dyes bearing sulfonic acid-substituted stilbene and phenol groups, such as Brilliant Yellow (Fig. 7.12a), (Tamura et al. 2014); (e) new monoazo-dyes bearing a disulfonic acid-substituted naphthylamine group and a monophenyl or biphenyl group (Favre-Besse et al. 2014); (f) an inhibitory protein factor referred to as IPF (an *N*-terminal 1–25 amino acid-deleted, 135-kDa fragment of α-fordrin) (Özkan et al. 1997); (g) polyhalogenated fluorescein such as Rose Bengal and its analogs (Ogita et al. 2001; Bole and Ueda 2005; Pietrancosta et al. 2010); and (h) tetrapeptide inhibitors (e.g., D-Gln-L-Ile- D-Glu-L-Try) (Patel et al. 2007). Compounds belonging to classes (a), (b), and (h) are specific to VGLUT, but have relatively low potencies compared to the most potent inhibitors; in addition, they all lack cell membrane permeability. Trypan Blue, IPF, Rose Bengal, and Brilliant Yellow are the most potent inhibitors known (IC_{50} = 20–50 nM). Both Trypan Blue and Brilliant Yellow (Fig. 7.13a) are highly specific to VGLUT (Roseth et al. 1998; Tamura et al. 2014). However, they would not penetrate the cellular membrane, since they are highly charged. The endogenous

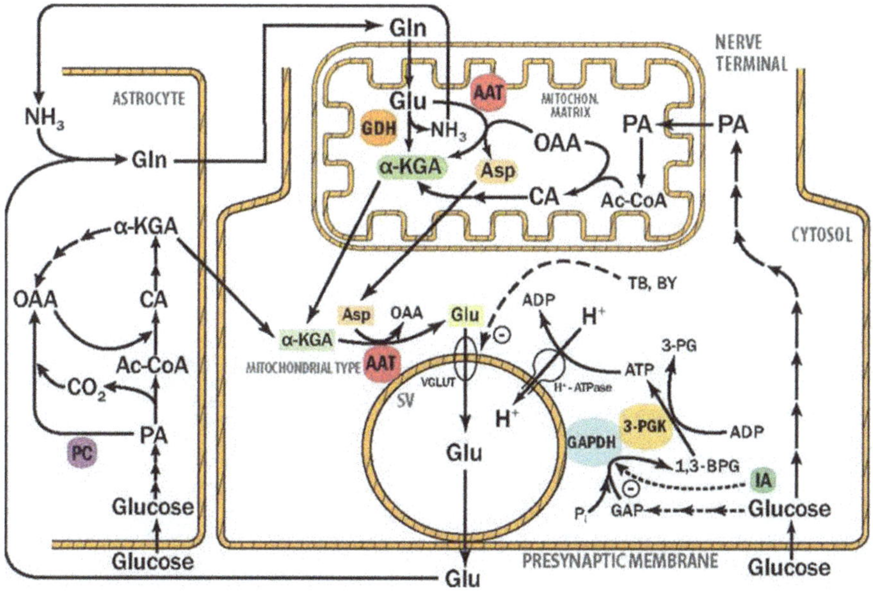

Fig. 7.12 Proposed effective mechanism for accumulating the neurotransmitter glutamate into synaptic vesicles. The VGLUT substrate glutamate is synthesized by vesicle-bound AAT from α-ketoglutarate and aspartate. Energy required for VGLUT's transport function is also generated by the vesicle-bound, glycolytic ATP-generating enzymes, GAPDH and 3-PGK. The majority of α-ketoglutarate would originate from nerve terminal mitochondria, where it is produced from glutamine-derived glutamate by AAT and GDH; this glutamine–glutamate conversion would be carried out largely by glutaminase on the matrix side of the inner membrane. An additional source of α-ketoglutarate would be nerve terminal-linked astrocytes, where it is synthesized de novo via pyruvate carboxylation and the TCA cycle. Another minor source of α-ketoglutarate could be nerve-terminal glucose, which leads to production of α-ketoglutarate via the TCA cycle. α-Ketoglutarate produced in this manner could serve as a substrate for synthesis of the transmitter glutamate responsible for spontaneous firing. This mechanism for supply of the VGLUT substrate does not depend on pre-neuronal activity, in contrast to its supply derived from glutamine, whose synthesis in astrocytes heavily depends on glutamate released upon neuronal activation. Aspartate is supplied from mitochondria in the nerve terminal. *AAT*, aspartate aminotransferase; *Ac-CoA*, acetyl coenzyme A; *α-KGA*, α-ketoglutarate; *Asp*, aspartate; *BY*, Brilliant Yellow; *GAP*, glyceraldehyde-3-phosphate; *GAPDH*, glyceraldehyde-3-phosphate dehydrogenase; *GDH*, glutamate dehydrogenase; *Glu*, glutamate; *Gln*, glutamine; *IA*, iodoacetate; *OAA*, oxaloacetic acid; *PA*, pyruvic acid; *3-PG*, 3-phosphoglycerate; *3-PGK*, 3-phosphoglycerate kinase; P_i, inorganic phosphate, *TB*, Trypan Blue

protein inhibitor IPF, though highly potent, is neither specific to VGLUT nor membrane permeable. Search for potent membrane-permeable VGLUT inhibitors resulted in the discovery of Rose Bengal. Among all known inhibitors, Rose Bengal is unique in that it is not only highly potent but also membrane permeable, leading to reduced glutamate release by inhibiting vesicular glutamate accumulation in synaptosomes (Ogita et al. 2001). However, it proved not highly specific to VGLUT; it exhibited a similar potency for vesicular uptake of GABA and monoamines (Bole and Ueda 2005; Pietrancosta et al. 2010).

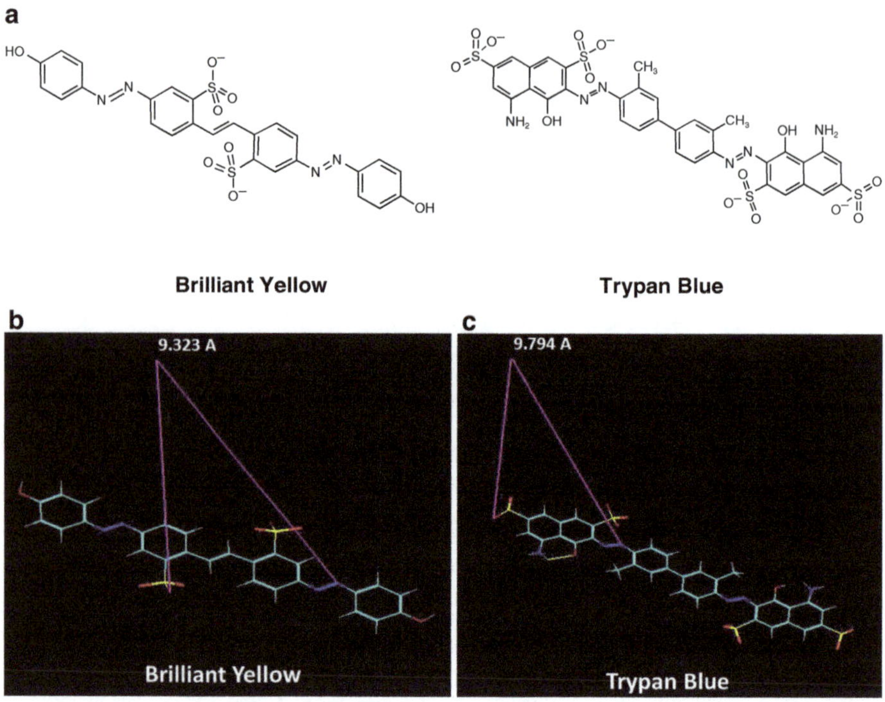

Fig. 7.13 Chemical structure and molecular modeling of Trypan Blue and Brilliant Yellow. (**a**) Chemical structure. (**b**) Molecular modeling of Brilliant Yellow. (**c**) Molecular modeling of Trypan Blue. The distance between the negatively charged oxygen and the azenyl group nitrogen linked to the phenyl group is indicated in purple. PCMODEL V 9.0 (Serena Software, Bloomington, IN, USA) was used for the molecular modeling

Recently, Rose Bengal has been reported to attenuate glutamate synaptic transmission (Neale et al. 2013, 2014). However, it required a much higher concentration than that determined in a glutamate release assay (Ogita et al. 2001). Of interest, Juge et al. (2010) have shown that ketone bodies inhibit VGLUT2 by blocking Cl^- activation, and that acetoacetate, the most potent, is capable of abolishing depolarization-induced glutamate release at millimolar concentrations. However, acetoacetate decreased not only the amplitude but also the frequency of mEPSCs. These ketone bodies were suggested responsible for the benefits of the ketogenic diet in controlling seizures (Juge et al. 2010). Other α-keto acids found in maple syrup urine disorder, such as α-keto-β-methyl-n-methyl valeric acid and α-ketoisocaproic acid, have also been shown to inhibit vesicular glutamate uptake by blocking the Cl^--activating site (Reis et al. 2000). However, these agents appear to also compete with the glutamate transport site. All the VGLUT-inhibitory keto acids reported are endogenous substances, but not potent. The most potent, highly VGLUT-specific inhibitors known today are Trypan Blue (Roseth et al. 1998) and Brilliant Yellow (Tamura et al. 2014), but these are not membrane permeable. Thus, membrane permeable, potent VGLUT-specific inhibitors, which could specifically regulate glutamate release, remain to be developed.

Molecular modeling indicates that one prominent feature shared by Brilliant Yellow and Trypan Blue is the similar distance between the azenyl group nitrogen linked to the phenyl group and the negatively charged oxygen of the sulfonate group: 9.3–9.9 A (Fig. 7.13b, c). Studies by Tamura et al. (2014) suggest that the 9.3 A distance is part of the Brilliant Yellow pharmacophore, which consists of the phenyl group-linked sulfonate group, the azenyl group, and the azenyl group-linked hydroxyphenyl group. Thus, the 9.3–9.9 A distance could represent part of the pharmacophore of VGLUT-specific, highly potent inhibitors.

Brilliant Yellow and Trypan Blue are postulated to block access of glutamate to VGLUT's active site (which are likely to involve VGLUT1 His^{120}, Arg^{176}, and Glu^{183} or VGLUT2 His^{128}, Arg^{184}, and Glu^{191}) by interacting with VGLUT1 His^{120} and Tyr^{187} or VGLUT2 His^{128} and Tyr^{195}; both Brilliant Yellow (Tamura et al. 2014) and Trypan Blue (Roseth et al. 1998) have been shown to compete with glutamate. This could occur in part via (a) the ionic bonding between histidine residue's positively charged nitrogen and the negatively charged oxygen of inhibitor's sulfonate group and (b) the hydrogen bonding between the tyrosine residue's hydroxyl group hydrogen and the inhibitor's azenyl group nitrogen linked to the phenyl group. These considerations suggest that the distance between the positively charged nitrogen of VGLUT1 His^{120} (or VGLUT2 His^{128}) and the hydroxyl group hydrogen of VGLUT1 Tyr^{187} (or VGLUT2 Tyr^{195}) is in the range of 9.3–9.8 A.

7.8 Astrocytic VGLUT

Accumulated evidence suggests that astrocytes, once thought to merely provide structural and trophic support for neurons, can modulate neuronal activity and could play an important role in information processing in the brain. Astrocytes, though electrically silent, have the capability to elevate intracellular Ca^{2+} concentrations through activation of various G-protein-coupled receptor systems (Haydon 2001). Glutamate, exogenously applied or synaptically released, can induce an oscillatory elevation of cytosolic Ca^{2+} in astrocytes, propagating calcium waves, not only within individual cells but also between adjacent astrocytes (Cornell-Bell et al. 1990; Smith 1992; Dani et al. 1992; Porter and McCarthy 1996). Increased cytosolic Ca^{2+} elicits release of glutamate from astrocytes, signaling neighboring neurons (Parpura et al. 1994; Jeftinija et al. 1996; Pasti et al. 1997; Bezzi et al. 1998; Parpura and Haydon 2000; Hassinger et al. 1995). Glutamate released from astrocytes induces neuronal Ca increase via activation of NMDA and non-NMDA receptors, and thereby can modulate synaptic transmission (Parpura et al. 1994; Hassinger et al. 1995; Araque et al. 1998; Newman and Zahs 1998; Jourdain et al. 2007).

In an effort to understand the mechanism of Ca^{2+}-dependent glutamate release from astrocytes, Montana et al. (2004) and Bezzi et al. (2004) provided evidence for the occurrence of VGLUT in astrocytic small vesicles, suggesting that this release originates from VGLUT-bearing vesicles, in agreement with the initial observation by Fremeau et al. (2002). Observations consistent with this, as well those supporting an astrocytic exocytotic mechanism, were also made in a variety of other investigations (Crippa et al. 2006; Stenovec et al. 2007; Platel et al. 2010; Bergersen et al. 2012; Ormel et al. 2012b; Mothet et al. 2005; Marchaland et al. 2008; Zhang et al. 2004). These VGLUT-loaded vesicles are localized in astrocytes adjacent to extra-presynaptic terminals that express NMDA receptors (Jourdain et al. 2007; Montana et al. 2004; Bezzi et al. 2004).

However, VGLUT expression in astrocytes proved controversial. Some other laboratories did not detect VGLUT in astrocytes (Aihara et al. 2000; Franke et al. 2006; Graziano et al. 2008). Transcriptome analysis also failed to support astrocytic expression of VGLUT (Cahoy et al. 2008). Moreover, the ketone body acetoacetate, which was found to downregulate glutamate release from neurons (by interfering with chloride activation of VGLUT), did not inhibit glutamate release from astrocytes (Juge et al. 2010). The nonspecific binding was considered an issue (Fritschy 2008; Li et al. 2013). Tissue fixation can alter the conformation of antigen epitopes, reducing affinity for antibodies or generating nonspecific high background. More recent extensive experiments have failed to provide evidence for VGLUT expression in astrocytes (Li et al. 2013).

Evidence is overwhelming for astrocytic exocytosis involving small vesicles in culture. However, in astrocytes, lysosomes appear to be the major intracellular compartment undergoing exocytosis, which occurs at a much slower rate than does neuronal vesicular exocytosis (Li et al. 2008). Lysosomes are expected to accumulate glutamate and aspartate by sialin, a lysosome marker enzyme, capable of transporting not only sialic acid but also glutamate and aspartate (Miyaji et al. 2008). These lines of evidence suggest that the majority of exocytotically released glutamate could come from lysosomes. The functional importance of astrocytic exocytosis (as judged by physiologically relevant quantitative analyses) in information processing in the brain remains to be determined (Hamilton and Attwell 2010).

7.9 Vesicular Aspartate Uptake

Aspartate has been considered an excitatory neurotransmitter candidate (Curtis and Johnston 1974). The putative neurotransmitter role of this amino acid was supported by demonstration of its release in a calcium-dependent manner, an important criterion for the neurotransmitter candidate (Nadler et al. 1976, 1978). Fleck et al. (1993) also showed calcium-dependent aspartate release in CA1 of the hippocampus, suggesting it is achieved by exocytosis. However the molecular mechanism for vesicular storage of this amino acid has remained elusive.

Miyaji et al. (2008) demonstrated that the lysosomal membrane transporter sialin (SLC17A5) (Mancini et al. 1989) (responsible in vivo for removal of sialic acid produced in lysosomes) is able to transport aspartate into liposomes reconstituted with sialin and proton-pump ATPase. The amino acid sequence of sialin had been determined and compared with VGLUT by Verheijen et al. (1999). Sialin/vesicular aspartate transporter has an arginine residue corresponding to

VGLUT1 Arg^{176} and a glutamate residue corresponding to VGLUT1 Glu^{183}, but lacks a histidine residue corresponding to VGLUT1 His^{120}. These three amino acid residues are thought to be all required for glutamate uptake by VGLUT (Juge et al. 2006; Almqvist et al. 2007). Of interest, aspartate uptake is driven by a membrane potential, while sialic acid transport is driven by a pH gradient (Miyaji et al. 2008; Mancini et al. 1989). Similar to glutamate uptake by VGLUT, aspartate uptake exhibits a biphasic response to extravesicular chloride when ATP is used to generate a membrane potential (Miyaji et al. 2008). However, contrary to VGLUT, sialin/aspartate transporter does not distinguish between aspartate and glutamate (Miyaji et al. 2008). It appears that the sialic acid/aspartate transporter occurs in a subset of hippocampal synaptic vesicle populations, as well as in synaptic-like small vesicles in pinealocytes, but not in the majority of brain synaptic vesicles. However, Herring et al. (2015) have recently provided evidence that synaptically released aspartate is not sufficiently highly concentrated to cause preferential activation of the NMDA receptor, thus ruling out a role of aspartate in fast excitatory synaptic transmission.

Given that sialin occurs in lysosomes, aspartate and glutamate would be accumulated in lysosomes in an ATP-dependent manner. Lysosomes are the major vesicular compartment subject to calcium-dependent exocytosis in astrocytes (Li et al. 2008). However, this astrocytic exocytosis occurs at a much slower rate than that of its neuronal counterpart (Li et al. 2008). Thus, the physiological role of these acidic amino acids released from the lysosomes in synaptic transmission remains elusive.

7.10 VGLUT Pathology

VGLUT is essential for survival of mammals and perhaps other vertebrates. Homozygous knockout (KO) of VGLUT1 or VGLUT2 in mice (which would result in little or no glutamate accumulation in synaptic vesicles) is lethal. VGLUT2-lacking mice die immediately after birth as a result of respiratory failure (Moechars et al. 2006). VGLUT1 KO mice die toward the end of the postnatal third week, most likely due to diminished functioning of multiple organs including the pancreas (Wojcik et al. 2004). A variety of evidence indicates that abnormal vesicular glutamate uptake is involved in various types of neurological and mental disorders. Mentioned here are some of the studies indicating association of certain pathophysiologies with aberrant alterations in vesicular glutamate uptake or VGLUT levels.

7.10.1 Epilepsy

Several lines of evidence suggest that excess glutamate loading in synaptic vesicles is involved in some forms of epilepsy. Vesicular glutamate uptake was shown to be elevated region specifically in the cerebrum of a genetic animal model of human complex partial seizures, in parallel with development of seizure (Lewis et al. 1997). VGLUT1 levels were found to be high in the hippocampus of temporal lobe epilepsy patients (Van der Hel et al. 2009). VGLUT1 in the hippocampus (particularly in the dentate gyrus) was also shown to be overexpressed in a seizure-prone animal model, as well as in animal models with hypoxia-induced and pilocarpin-induced seizures (Kang et al. 2005; Kim et al. 2005, 2007). VGLUT2 is increased in the cortex (but not in the thalamus) of absence-epilepsy animals (Touret et al. 2007), as well as in the hippocampus of an animal model of teratogen-induced seizure (Harrington et al. 2006). These lines of evidence all suggest that brain region-specific vesicular overloading of glutamate and excessive glutamate release underlie certain forms of seizure.

7.10.2 *Schizophrenia*

In schizophrenics, VGLUT1 mRNA is decreased in the hippocampus and dorsal prefrontal cortex (Eastwood and Harrison 2005), whereas VGLUT2 mRNA expression is increased in the inferior temporal gyrus (Uezato et al. 2009). Inta et al. (2012) have shown that heterozygous VGLUT1 KO mice exhibit abnormal behaviors resembling correlates of schizophrenia, such as working- and social-memory impairments and deficits in prepulse inhibition. Recently, VGLUT1 reduction in prefrontal cortex glutamatergic neurons projecting to the subcortical structure nucleus accumbens (a reward circuit component) has been proposed to underlie schizophrenia's cognitive flexibility deficit, based upon impaired reversal learning observed in heterozygous VGLUT1 KO mice (Granseth et al. 2015). These observations suggest that aberrant alterations in VGLUT1-selective vesicular uptake and release in specific brain substructures are involved in some cognitive dysfunctions apparent in schizophrenics.

7.10.3 *Depression*

In both major depression and bipolar depression, VGLUT1 mRNA has been reported to be decreased in the entorhinal cortex (Uezato et al. 2009), while VGLUT1 protein is increased in the lateral amygdala (Varea et al. 2012). In the middle temporal gyrus, VGLUT2 mRNA expression is also reduced in patients with major depression (Uezato et al. 2009). In an animal model of corticosterone-induced depression, VGLUT2 protein has been shown to be increased in the hippocampus but not in the amygdala (Lussier et al. 2013). Tordera et al. (2007) have observed that heterozygous VGLUT1 KO mice exhibit enhanced, depressive-like behavior. These lines of evidence suggest the involvement of abnormal vesicular glutamate uptake in specific brain areas in affective disorders as well.

7.10.4 Alzheimer's Disease

Alzheimer's disease (AD) is a complex neurodegenerative disease with severe cognitive dysfunction, its prevalence increasing among the elderly. In this dementia, VGLUT1 is decreased in the prefrontal dorsal cortex (Kashani et al. 2008). This VGLUT1 reduction is thought to play a role in progression of the cognitive decline. Recent studies suggest that hyperexcitability in the hippocampus contributes to the development of AD. Functional MRI shows that elevated hippocampal activation occurs in individuals at risk for AD.

Miller et al. (2008) proposed that aberrant hippocampal hyperactivity is predictive of the degree and rate of subsequent cognitive decline and conversion to AD. Consistent with this, Bakker et al. (2012) have provided evidence that dampening hippocampal hyperactivity in mild cognitive impairments can result in cognitive improvement. The notion that hippocampal hyperactivation is an early indicator of AD development is also supported by Putcha et al. (2011), demonstrating that hippocampal hyperactivation in mild cognitive impairments is well correlated with degeneration of "AD-signature" regions in the cortex. Compatible with these lines of evidence, Vossel et al. (2013) observed that the absence of epilepsy in patients with mild cognitive impairments and AD retards cognitive decline.

The hallmark pathology of this disease is abundant accumulation of extracellular β-amyloid aggregates and intracellular neurofibrillary tangles containing hyperphosphorylated tau protein. Recently, tau has been shown to mediate β-amyloid-induced cognitive deficits and hyperexcitability in the hippocampus (Roberson et al. 2007, 2011). Tau-induced pathology initially occurs in the entorhinal cortex, followed by the hippocampus, preceding spread to other regions (de Calignon et al. 2012). In a more recent study of a tau mouse model of Alzheimer's disease (TauP301L), Hunsberger et al. (2015) have shown that this mutated, misfolded tau (which coaggregates with endogenous tau) induces increased VGLUT1 expression in the hippocampus. This raises an interesting possibility that VGLUT1 over-expression could be involved in hippocampal hyperexcitability observed near the onset of AD.

7.10.5 Parkinson's Disease

In Parkinson's disease, VGLUT1 and VGLUT2 in the putamen are moderately increased, whereas VGLUT1 is substantially decreased in the prefrontal and temporal cortex; these alterations are thought to contribute to the cognitive impairments associated with the disease (Kashani et al. 2007). In an MPTP-treated monkey model of Parkinsonism, VGLUT1 in corticostriatal nerve terminals was found increased, whereas VGLUT2 in thalamostriatal nerve terminals was not changed (Raju et al. 2008). In MPTP-treated mice, Pflibsen et al. (2015) have recently reported that VGLUT1 is elevated in the dorsolateral striatum as well as in the medial prefrontal cortex, whereas VGLUT2 is diminished in the dorsolateral striatum. Using a 6-hydroxydopamine-treated rat model, Favier et al. (2013) have provided evidence suggesting that reduction of basal ganglia VGLUT2, induced by dopamine depletion, is involved in genesis of the motor deficit associated with the disease, particularly at the advanced stage. Divito et al. (2015) have recently reported interesting findings, suggesting that VGLUT3 can affect neural mechanisms underlying motor deficit and dyskinesia in a mouse model of Parkinson's disease, raising the possibility of a novel therapeutic target.

7.10.6 Pain

In morbidity associated with inflammation, nerve injury, and trauma, mechanical pain hypersensation plays a major role. Using global VGLUT3 knockout mice, Seal et al. (2009) provided evidence suggesting that this is mediated by a small subset population of sensory neurons in the spinal cord dorsal root ganglia containing VGLUT3. Lou et al. (2013), however, did not observe major changes in acute and chronic mechanical pain in VGLUT3-depleted mice; these mice were produced through knockout of a transcription factor, which controls development of unmyelinated sensory neurons including pain-related nociceptors as well as VGLUT3 expression. Experiments with heterozygous VGLUT2 KO mice suggest that VGLUT2 plays a role in neuropathy-induced allodynia and hypersensitivity (Leo et al. 2009). Using conditional VGLUT2 KO mice, Liu et al. (2010) have provided evidence indicating that VGLUT2 in nociceptor-bearing neurons in dorsal root ganglia mediates a variety of pain sensations, including mechanical, thermal, capsaicin-evoked, inflammatory, and neuropathic. Recent evidence suggests that VGLUT2 in the thalamus, periaqueductal gray, and amygdala is involved in neuropathic pain induction (Wang et al. 2015). The involvement of VGLUT2 and VGLUT3 in nociception and pain has been extensively reviewed by Brumovsky (2013).

7.11 Summary and Concluding Remarks

Vesicular glutamate uptake is now recognized as a pivotal step in glutamate synaptic transmission, allowing glutamate to enter the neurotransmitter pathway away from its metabolic pathway. This process requires ATP and a low millimolar concentration of extravesicular chloride or high intravesicular chloride and is highly specific to glutamate. The vesicular uptake system is composed of a vesicular glutamate transporter, VGLUT, and v-type proton-pump ATPase, which generates an electrochemical proton gradient, the driving force for active transport into intracellular organelles. Energy harnessed by v-type proton-pump ATPase is likely to be largely supplied by a vesicle-bound, glycolytic ATP-synthesizing enzyme complex, GAPDH/3-phosphoglycerate kinase. The VGLUT substrate glutamate is postulated to be synthesized by vesicle-bound AAT from α-ketoglutarate at the expense of aspartate. The majority of α-ketoglutarate would derive from mitochondria where astrocyte-derived glutamine is converted to glutamate then to α-ketoglutarate. The presence of both ATP- and glutamate-generating enzymes in the synaptic vesicle would represent an efficient mechanism for vesicular glutamate loading.

VGLUT has three isoforms. Their biochemical properties are similar, but in general they have complementary distributions not only in nervous tissue but also in nonnervous tissue, serving various physiological functions, from learning and memory formation to visual and auditory sensing as well as respiratory control and mechanical sensation. As such, aberrant VGLUT expression or vesicular glutamate uptake is involved in various types of neurological pathophysiologies.

Under physiological conditions, freshly endocytosed synaptic vesicles would take up glutamate at the expense of a high concentration of lumen chloride as well as of a membrane potential generated by H^+-pump ATPase. As the intravesicular chloride concentration decreases, a cytosolic low concentration of chloride (in the presence of a membrane potential) would begin to play a significant role in glutamate uptake by forming a pH gradient as well as activating VGLUT upon allosteric binding. An optimal intravesicular pH produced could also contribute to Cl^--enhanced glutamate uptake. Moreover, cytosolic K^+ could elevate vesicular glutamate accumulation through K^+/H^+ exchange.

In view of VGLUT involvement in a number of neurological pathophysiologies, drug development targeting VGLUT should be explored. Several types of inhibitors which act on VGLUT were discovered. The most potent and specific are Trypan Blue and Brilliant Yellow, with IC_{50} in the 20–50 nM range. However, these agents are not membrane permeable. Moreover, although they are specific to VGLUT, their selectivity among VGLUT isoforms remains to be determined.

7.11.1 Anticipated Developments

(a) Membrane-permeable VGLUT isoform-specific inhibitors and activators, which could serve as prototype therapeutic agents for treatment of certain pathological conditions involving aberrant hyper- or hypoglutamate neurotransmission.
(b) Membrane-permeable VGLUT isoform-specific agents useful for noninvasive imaging of VGLUT isoform-specific glutamatergic nerve endings in various brain regions and nonnervous tissue.
(c) Determination of the three-dimensional structure of VGLUTs and ultimately elucidation of the glutamate transport mechanism at the atomic level.

Acknowledgement I thank Dr. Minor J Coon, Dr. Stephen K. Fisher, Dr. John T. Hackett, Dr. Francis S. Lee, and Yasuko Ueda for critical reading and comments, and the computer consultant Douglass Smith and Yasuko Ueda for making model figure illustrations.

References

Aihara Y, Mashima H, Onda H, Hisano S, Kasuya H, Hori T, Yamada S, Tomura H, Yamada Y, Inoue I, Kojima I, Takeda J (2000) Molecular cloning of a novel brain-type Na^+-dependent inorganic phosphate cotransporter. J Neurochem 74:2622–2625

Albrecht J, Dolinska M, Hilgier W, Lipkowski AW, Nowacki J (2000) Modulation of glutamine uptake and phosphate-activated glutaminase activity in rat brain mitochondria by amino acids and their synthetic analogues. Neurochem Int 36:341–347

Aledo JC, de Pedro E, Gomez-Fabre PM, de Castro IN, Marquez J (1997) Submitochondrial localization and membrane topography of Ehrlich ascitic tumour cell glutaminase. Biochim Biophys Acta 1323:173–184

Almqvist J, Huang Y, Laaksonen A, Wang DN, Hovmöller S (2007) Docking and homology modeling explain inhibition of the human vesicular glutamate transporters. Protein Sci 16:1819–1829

Altschuler RA, Neises GR, Harmison GG, Wenthold RJ, Fex J (1981) Immunocytochemical localization of aspartate aminotransferase immunoreactivity in cochlear nucleus of the guinea pig. Proc Natl Acad Sci U S A 78:6553–6657

Altschuler RA, Mosinger JL, Harmison GG, Parakkal MH, Wenthold RJ (1982) Aspartate aminotransferase-like immunoreactivity as a marker for aspartate/glutamate in guinea pig photoreceptors. Nature 298:657–659

Amara SG, Fontana ACK (2002) Excitatory amino acid transporters: keeping up with glutamate. Neurochem Int 41:313–318

Anderson DC, King SC, Parsons SM (1982) Proton gradient linkage to active uptake of [^{3}H]acetylcholine by Torpedo electric organ synaptic vesicles. Biochemistry 21:3037–3043

Araque A, Parpura V, Sanzgiri RP, Haydon PG (1998) Glutamate-dependent astrocyte modulation of synaptic transmission between cultured hippocampal neurons. Eur J Neurosci 10:2129–2142

Bachelard HS, Cox DW, Drower J (1984) Sensitivity of guinea-pig hippocampal granule cell field potentials to hexoses in vitro: an effect on cell excitability? J Physiol 352:91–102

Bai L, Xu H, Collins JF, Ghishan FK (2001) Molecular and functional analysis of a novel neuronal vesicular glutamate transporter. J Biol Chem 276:36764–36769

Bai L, Zhang X, Ghishan FK (2003) Characterization of vesicular glutamate transporter in pancreatic alpha- and beta- cells and its regulation by glucose. Am J Physiol Gastrointest Liver Physiol 284:G808–G814

Bak LK, Zieminnska E, Waagepetersen HS, Schousboe A, Albrecht J (2008) Metabolism of [U-^{13}C] glutamine and [U-^{13}C]glutamate in isolated rat brain mitochondria suggests functional phosphate-activated glutaminase activity in matrix. Neurochem Res 33:273–278

Bakker A, Krauss GL, Albert MS, Speck CL, Jones LR, Stark CE, Yassa MA, Bassett SS, Shelton AL, Gallagher M (2012) Reduction of hippocampal hyperactivity improves cognition in amnestic mild cognitive impairment. Neuron 74:467–474

Balschun D, Moechars D, Callaerts-Vegh Z, Vermaercke B, Van Acker N, Andries L, D'Hooge R (2010) Vesicular glutamate transporter VGLUT1 has a role in hippocampal long-term potentiation and spatial reversal learning. Cereb Cortex 20:684–693

Bartlett RD, Esslinger CS, Thompson CM, Bridges RJ (1998) Substituted quinolines as inhibitors of L-glutamate transport into synaptic vesicles. Neuropharmacology 37:839–846

Bellocchio EE, Reimer RJ, Fremeau RTJ, Edwards RH (2000) Uptake of glutamate into synaptic vesicles by an inorganic phosphate transporter. Science 289:957–960

Benjamin AM, Quastel JH (1972) Locations of amino acids in brain slices from the rat. Biochem J 128:631–646

Bennett JP Jr, Mulder AH, Snyder SH (1974) Neurochemical correlates of synaptically active amino acids. Life Sci 15:1045–1056

Bergersen LH, Morland C, Ormel L, Rinholm JE, Larsson M, Wold JF, Røe AT, Stranna A, Santello M, Bouvier D, Ottersen OP, Volterra A, Gundersen V (2012) Immunogold detection of L-glutamate and D-serine in small synaptic-like microvesicles in adult hippocampal astrocytes. Cereb Cortex 22:1690–1697

Bezzi P, Carmignoto G, Pasti L, Vesce S, Rossi D, Rizzini BL, Pozzan T, Volterra A (1998) Prostaglandins stimulate calcium-dependent glutamate release in astrocytes. Nature 391:281–285

Bezzi P, Gundersen V, Galbete JL, Seifert G, Steinhauser C, Pilati E, Volterra A (2004) Astrocytes contain a vesicular compartment that is competent for regulated exocytosis of glutamate. Nat Neurosci 7:613–620

Billups D, Attwell D (2002) Control of intracellular chloride concentration and GABA response polarity in rat retinal ON-bipolar cells. J Physiol 545:183–198

Bisaccia F, Indiveri C, Palmieri F (1985) Purification of reconstitutively active alpha-oxoglutarate carrier from pig heart mitochondria. Biochim Biophys Acta 810:362–369

Bloom FE, Ueda T, Battenberg E, Greengard P (1979) Immunocytochemical localization in synapses, of protein I, an endogenous substrate for protein kinases in mammalian brain. Proc Natl Acad Sci U S A 76:5982–5986

Bole DG, Ueda T (2005) Inhibition of vesicular uptake by Rose Bengal-related compounds: structure activity relationship. Neurochem Res 30:363–369

Bolli R, Nałecz KA, Azzi A (1989) Monocarboxylate and alpha-ketoglutarate carriers from bovine heart mitochondria. Purification by affinity chromatography on immobilized 2-cyano-4-hydroxycinnamate. J Biol Chem 264:18024–18030

Boulland JL, Qureshi T, Seal RP, Rafiki A, Gundersen V, Bergersen LH, Fremeau RT Jr, Edwards RH, Storm-Mathisen J, Chaudhry FA (2004) Expression of the vesicular glutamate transporters during development indicates the widespread corelease of multiple neurotransmitters. J Comp Neurol 480:264–280

Bradford HF, Ward HK (1976) On glutaminase activity in mammalian synaptosomes. Brain Res 110:115–125

Bradford HF, Ward HK, Thomas AJ (1978) Glutamine—a major substrate for nerve endings. J Neurochem 30:1453–1459

Brouns I, Pintelon I, Van Genechten J, De Proost I, Timmermans JP, Adriaensen D (2004) Vesicular glutamate transporter 2 is expressed in different nerve fibre populations that selectively contact pulmonary neuroepithelial bodies. Histochem Cell Biol 121:1–12

Brumovsky PR (2013) VGLUTs in peripheral neurons and the spinal cord: time for review. ISRN Neurol 829753

Brumovsky PR, Robinson DR, La JH, Seroogy KB, Lundgren KH, Albers KM, Kiyatkin ME, Seal RP, Edwards RH, Watanabe M, Hökfelt T, Gebhart GF (2011) Expression of vesicular glutamate transporters type 1 and 2 in sensory and autonomic neurons innervating the mouse colorectum. J Comp Neurol 519:3346–3366

Brumovsky PR, Seal RP, Lundgren KH, Seroogy KB, Watanabe M, Gebhart GF (2013) Expression of vesicular glutamate transporters in sensory and autonomic neurons innervating the mouse bladder. J Urol 189:2342–2349

Burger PM, Mehl E, Cameron PL, Maycox PR, Baumert M, Lottspeich F, De Camilli P, Jahn R (1989) Synaptic vesicles immunoisolated from rat cerebral cortex contain high levels of glutamate. Neuron 3:715–720

Burré J, Beckhaus T, Schägger H, Corvey C, Hofmann S, Karas M, Zimmermann H, Volknandt W (2006) Analysis of the synaptic vesicle proteome using three gel-based protein separation techniques. Proteomics 6:6250–6262

Cahoy JD, Emery B, Kaushal A, Foo LC, Zamanian JL, Christopherson KS, Xing Y, Lubischer JL, Krieg PA, Krupenko SA, Thompson WJ, Barres BA (2008) A transcriptome database for astrocytes, neurons, and oligodendrocytes: a new resource for understanding brain development and function. J Neurosci 28:264–278

Carlson MD, Ueda T (1990) Accumulated glutamate levels in the synaptic vesicle are not maintained in the absence of active transport. Neurosci Lett 110:325–330

Carlson MD, Kish PE, Ueda T (1989) Characterization of the solubilized and reconstituted ATP-dependent vesicular glutamate uptake system. J Biol Chem 264:7369–7376

Carty SE, Johnson RG, Scarpa A (1981) Serotonin transport in isolated platelet granules. J Biol Chem 256:11244–11250

Casey RP, Njus D, Radda GK, Sehr DA (1977) Active proton uptake by chromaffin granules: observation by amine distribution and phosphorus-31 nuclear magnetic resonance techniques. Biochemistry 16:972–977

Cesar M, Hamprecht B (1995) Immunocytochemical examination of neural rat and mouse primary cultures using monoclonal antibodies raised against pyruvate carboxylase. J Neurochem 64:2312–2318

Chaudhry FA, Edwards RH, Fonnum F (2008) Vesicular neurotransmitter transporters as targets for endogenous and exogenous toxic substances. Annu Rev Pharmacol Toxicol 48:277–301

Cornell-Bell AH, Finkbeiner S, Cooper MS, Smith SJ (1990) Glutamate induces calcium waves in cultured astrocytes: long-range glial signaling. Science 247:470–473

Cotman CW, Foster AC, Lanthorn TH (1981) An overview of glutamate as a neurotransmitter. In: Di Chiara G, Gessa GL (eds) Glutamate as a Neurotransmitter. Raven, New York, pp 1–27

Cotman CW, Monaghan DT, Ganong AH (1988) Excitatory amino acid neurotransmission: NMDA receptors and Hebb-type synaptic plasticity. Annu Rev Neurosci 11:61–80

Coughenour HD, Spaulding RS, Thompson CM (2004) The synaptic vesicle proteome: a comparative study in membrane protein identification. Proteomics 4:3141–3155

Cox DW, Bachelard HS (1982) Attenuation of evoked field potentials from dentate granule cells by low glucose, pyruvate + malate, and sodium fluoride. Brain Res 239:527–534

Cox DW, Morris PG, Feeney J, Bachelard HS (1983) ^{31}P n. m. r. studies on cerebral energy metabolism under conditions of hypoglycaemia and hypoxia in vitro. Biochem J 212:365–370

Crippa D, Schenk U, Francolini M, Rosa P, Verderio C, Zonta M, Pozzan T, Matteoli M, Carmignoto G (2006) Synaptobrevin2-expressing vesicles in rat astrocytes: insights into molecular characterization, dynamics and exocytosis. J Physiol 570:567–582

Curtis DR, Johnston GA (1974) Amino acid transmitters in the mammalian central nervous system. Ergebn Physiol 69:97–188

Curtis DR, Phillis JW, Watkins JC (1960) The chemical excitation of spinal neurones by certain acidic amino acids. J Physiol 150:656–682

Danbolt NC (2001) Glutamate uptake. Prog Neurobiol 65:1–105

Dani JW, Chernjavsky A, Smith SJ (1992) Neuronal activity triggers calcium waves in hippocampal astrocyte networks. Neuron 8:429–440

Daniels RW, Collins CA, Gelfand MV, Dant J, Brooks ES, Krantz DE, DiAntonio A (2004) Increased expression of the Drosophila vesicular glutamate transporter leads to excess glutamate release and a compensatory decrease in quantal content. J Neurosci 24:10466–10474

De Belleroche JS, Bradford HF (1973) Amino acids in synaptic vesicles from mammalian cerebral cortex: a reappraisal. J Neurochem 21:441–451

De Belleroche JS, Bradford HF (1977) On the site of origin of transmitter amino acids released by depolarization of nerve terminals in vitro. J Neurochem 29:335–343

de Calignon A, Polydoro M, Suárez-Calvet M, William C, Adamowicz DH, Kopeikina KJ, Pitstick R, Sahara N, Ashe KH, Carlson GA, Spires-Jones TL, Hyman BT (2012) Propagation of tau pathology in a model of early Alzheimer's disease. Neuron 73:685–697

De Camilli P, Harris SM Jr, Huttner WB, Greengard P (1983) Synapsin I (Protein I), a nerve terminal-specific phosphoprotein. II. Its specific association with synaptic vesicles demonstrated by immunocytochemistry in agarose-embedded synaptosomes. J Cell Biol 96:1355–1373

Dennis SC, Lai JC, Clark JB (1977) Comparative studies on glutamate metabolism in synaptic and non-synaptic rat brain mitochondria. Biochem J 164:727–736

Disbrow JK, Gershten MJ, Ruth JA (1982) Uptake of L-[^{3}H] glutamic acid by crude and purified synaptic vesicles from rat brain. Biochem Biophys Res Commun 108:1221–1227

Divito CB, Steece-Collier K, Case DT, Williams SP, Stancati JA, Zhi L, Rubio ME, Sortwell CE, Collier TJ, Sulzer D, Edwards RH, Zhang H, Seal RP (2015) Loss of VGLUT3 produces circadian-dependent hyperdopaminergia and ameliorates motor dysfunction and L-dopa-mediated dyskinesias in a model of Parkinson's disease. J Neurosci 35:14983–14999

Eastwood SL, Harrison PJ (2005) Decreased expression of vesicular glutamate transporter 1 and complexin II mRNAs in schizophrenia: further evidence for a synaptic pathology affecting glutamate neurons. Schizophr Res 73:159–172

Edwards RH (2007) The neurotransmitter cycle and quantal size. Neuron 55:835–858
Egebjerg J, Bettler B, Hermans-Borgmeyer I, Heinemann S (1991) Cloning of a cDNA for a glutamate receptor subunit activated by kainate but not AMPA. Nature 351:745–774
El Mestikawy S, Wallén-Mackenzie A, Fortin GM, Descarries L, Trudeau LE (2011) From glutamate co-release to vesicular synergy: vesicular glutamate transporters. Nat Rev Neurosci 12:204–216
Erickson JD, De Gois S, Varoqui H, Schafer MKH, Weihe E (2006) Activity-dependent regulation of vesicular glutamate and GABA transporters: a means to scale quantal size. Neurochem Int 48:643–649
Favier M, Carcenac C, Drui G, Boulet S, El Mestikawy S, Savasta M (2013) High-frequency stimulation of the subthalamic nucleus modifies the expression of vesicular glutamate transporters in basal ganglia in a rat model of Parkinson's disease. BMC Neurosci 14:152
Favre-Besse FC, Poirel O, Bersot T, Kim-Grellier E, Daumas S, El Mestikawy S, Acher FC, Pietrancosta N (2014) Design, synthesis and biological evaluation of small-azo-dyes as potent Vesicular Glutamate Transporters inhibitors. Eur J Med Chem 78:236–247
Fisher-Bovenkerk C, Kish PE, Ueda T (1988) ATP-dependent glutamate uptake into synaptic vesicles from cerebellar mutant mice. J Neurochem 51:1054–1059
Fleck MW, Henze DA, Barrionuevo G, Palmer AM (1993) Aspartate and glutamate mediate excitatory synaptic transmission in area CA1 of the hippocampus. J Neurosci 13:3944–3955
Fonnum F, Fykse EM, Roseth S (1998) Uptake of glutamate into synaptic vesicles. Prog Brain Res 116:87–101
Franke H, Grummich B, Härtig W, Grosche J, Regenthal R, Edwards RH, Illes P, Krügel U (2006) Changes in purinergic signaling after cerebral injury—involvement of glutamatergic mechanisms? Int J Dev Neurosci 24:123–132
Fremeau RT Jr, Troyer MD, Pahner I, Nygaard GO, Tran CH, Reimer RJ, Bellocchio EE, Fortin D, Storm-Mathisen J, Edwards RH (2001) The expression of vesicular glutamate transporters defines two classes of excitatory synapse. Neuron 31:247–260
Fremeau RT Jr, Burman J, Qureshi T, Tran CH, Proctor J, Johnson J, Zhang H, Sulzer D, Copenhagen DR, Storm-Mathisen J, Reimer RJ, Chaudhry FA, Edwards RH (2002) The identification of vesicular glutamate transporter 3 suggests novel modes of signaling by glutamate. Proc Natl Acad Sci U S A 99:14488–14493
Fremeau RT Jr, Voglmaier S, Seal RP, Edwards RH (2004a) VGLUTs define subsets of excitatory neurons and suggest novel roles for glutamate. Trends Neurosci 27:98–103
Fremeau RT Jr, Kam K, Qureshi T, Johnson J, Copenhagen DR, Storm-Mathisen J, Chaudhry FA, Nicoll RA, Edwards RH (2004b) Vesicular glutamate transporters 1 and 2 target to functionally distinct synaptic release sites. Science 304:1815–1819
Fritschy JM (2008) Is my antibody-staining specific? How to deal with pitfalls of immunohistochemistry. Eur J Neurosci 28:2365–2370
Fyk-Kolodziej B, Dzhagaryan A, Qin P, Pourcho RG (2004) Immunocytochemical localization of three vesicular glutamate transporters in the cat retina. J Comp Neurol 475:518–530
Fykse EM, Christensen H, Fonnum F (1989) Comparison of the properties of γ-aminobutyric acid and L-glutamate uptake into synaptic vesicles isolated from rat brain. J Neurochem 52:946–951
Goh GY, Huang H, Ullman J, Borre L, Hnasko TS, Trussell LO, Edwards RH (2011) Presynaptic regulation of quantal size: K^+/H^+ exchange stimulates vesicular glutamate transport. Nat Neurosci 14:1285–1292
Granseth B, Andersson FK, Lindström SH (2015) The initial stage of reversal learning is impaired in mice hemizygous for the vesicular glutamate transporter (VGluT1). Genes Brain Behav 14:477–485
Gras C, Herzog E, Bellenchi GC, Bernard V, Ravassard P, Pohl M, Gasnier B, Giros B, El Mestikawy S (2002) A third vesicular glutamate transporter expressed by cholinergic and serotoninergic neurons. J Neurosci 22:5442–5451
Gras C, Vinatier J, Amilhon B, Guerci A, Christov C, Ravassard P, Giros B, El Mestikawy S (2005) Developmentally regulated expression of VGLUT3 during early post-natal life. Neuropharmacology 49:901–911

Gras C, Amilhon B, Lepicard EM, Poirel O, Vinatier J, Herbin M, Dumas S, Tzavara ET, Wade MR, Nomikos GG, Hanoun N, Saurini F, Kemel ML, Gasnier B, Giros B, El Mestikawy S (2008) The vesicular glutamate transporter VGLUT3 synergizes striatal acetylcholine tone. Nat Neurosci 11:292–300

Graziano A, Liu XB, Murray KD, Jones EG (2008) Vesicular glutamate transporters define two sets of glutamatergic afferents to the somatosensory thalamus and two thalamocortical projections in the mouse. J Comp Neurol 507:1258–1276

Greengard P (1976) Possible role for cyclic nucleotides and phosphorylated membrane proteins in postsynaptic actions of neurotransmitters. Nature 260:101–110

Grewal S, Defamie N, Zhang X, De Gois S, Shawki A, Mackenzie B, Chen C, Varoqui H, Erickson JD (2009) SNAT2 amino acid transporter is regulated by amino acids of the SLC6 gamma-aminobutyric acid transporter subfamily in neocortical neurons and may play no role in delivering glutamine for glutamatergic transmission. J Biol Chem 284:11224–11236

Hackett J, Ueda T (2015) Glutamate release. Neurochem Res 40:2443–2460

Hamberger AC, Chiang GH, Nylén ES, Scheff SW, Cotman CW (1979) Glutamate as a CNS transmitter. I. Evaluation of glucose and glutamine as precursors for the synthesis of preferentially released glutamate. Brain Res 168:513–530

Hamilton NB, Attwell D (2010) Do astrocytes really exocytose neurotransmitters? Nat Rev Neurosci 11:227–238

Harrington EP, Möddel G, Najm IM, Baraban SC (2006) Altered glutamate receptor—transporter expression and spontaneous seizures in rats exposed to methylazoxymethanol in utero. Epilepsia 48:158–168

Hartinger J, Jahn R (1993) An anion binding site that regulates the glutamate transporter of synaptic vesicles. J Biol Chem 268:23122–23127

Hassinger TD, Atkinson PB, Strecker GJ, Whalen LR, Dudek FE, Kossel AH, Kater SB (1995) Evidence for glutamate-mediated activation of hippocampal neurons by glial calcium waves. J Neurobiol 28:159–170

Hayashi M, Otsuka M, Morimoto R, Hirota S, Yatsushiro S, Takeda J, Yamamoto A, Moriyama Y (2001) Differentiation-associated Na^+-dependent inorganic phosphate cotransporter (DNPI) is a vesicular glutamate transporter in endocrine glutamatergic systems. J Biol Chem 276:43400–43406

Hayashi M, Yamada H, Uehara S, Morimoto R, Muroyama A, Yatsushiro S, Takeda J, Yamamoto A, Moriyama Y (2003) Secretory granule-mediated co-secretion of L-glutamate and glucagon triggers glutamatergic signal transmission in islets of Langerhans. J Biol Chem 278:1966–1974

Haydon PG (2001) Glia: listening and talking to the synapse. Nat Rev Neurosci 2:185–193

Hell JW, Maycox PR, Jahn R (1990) Energy dependence and functional reconstitution of the gamma-aminobutyric acid carrier from synaptic vesicles. J Biol Chem 265:2111–2117

Herman MA, Ackermann F, Trimbuch T, Rosenmund C (2014) Vesicular glutamate transporter expression level affects synaptic vesicle release probability at hippocampal synapses in culture. J Neurosci 34:11781–11791

Herring BE, Silm K, Edwards RH, Nicoll RA (2015) Is aspartate an excitatory neurotransmitter? J Neurosci 35:10168–10171

Hertz L, Peng L, Dienel GA (2007) Energy metabolism in astrocytes: high rate of oxidative metabolism and spatiotemporal dependence on glycolysis/glycogenolysis. J Cereb Blood Flow Metab 27:219–249

Herzog E, Bellenchi GC, Gras C, Bernard V, Ravassard P, Bedet C, Gasnier B, Giros B, El Mestikawy S (2001) The existence of a second vesicular glutamate transporter specifies subpopulations of glutamatergic neurons. J Neurosci 21:RC181

Herzog E, Takamori S, Jahn R, Brose N, Wojcik SM (2006) Synaptic and vesicular co-localization of the glutamate transporters VGLUT1 and VGLUT2 in the mouse hippocampus. J Neurochem 99:1011–1018

Hioki H, Fujiyama F, Taki K, Tomioka R, Furuta T, Tamamaki N, Kaneko T (2003) Differential distribution of vesicular glutamate transporters in the rat cerebellar cortex. Neuroscience 117:1–6

Hollmann M, O'Shea-Greenfield A, Rogers SW, Heinemann S (1989) Cloning by functional expression of a member of the glutamate receptor family. Nature 342:643–648

Holz RW (1978) Evidence that catecholamine transport into chromaffin vesicles is coupled to vesicle membrane potential. Proc Natl Acad Sci U S A 75:5190–5194

Hunsberger HC, Rudy CC, Batten SR, Gerhardt GA, Reed MN (2015) P301L tau expression affects glutamate release and clearance in the hippocampal trisynaptic pathway. J Neurochem 132:169–182

Ikemoto A, Bole DG, Ueda T (2003) Glycolysis and glutamate accumulation into synaptic vesicles. Role of glyceraldehyde phosphate dehydrogenase and 3-phosphoglycerate kinase. J Biol Chem 278:5929–5940

Inta D, Vogt M, Perreau-Lenz S, Schneider M, Pfeiffer N, Wojcik S, Spanagel R, Gass P (2012) Sensorimotor gating, working and social memory deficits in mice with reduced expression of the vesicular glutamate transporter VGLUT1. Behav Brain Res 228:328–332

Ishida A, Noda Y, Ueda T (2009) Synaptic vesicle-bound pyruvate kinase can support vesicular glutamate uptake. Neurochem Res 34:807–818

Jeftinija SD, Jeftinija KV, Stefanovic G, Liu F (1996) Neuroligand-evoked calcium-dependent release of excitatory amino acids from cultured astrocytes. J Neurochem 66:676–684

Johnson RG Jr (1988) Accumulation of biological amines into chromaffin granules: a model for hormone and neurotransmitter transport. Physiol Rev 68:232–307

Johnson RG, Scarpa A (1979) Protonmotive force and catecholamine transport in isolated chromaffin granules. J Biol Chem 254:3750–3760

Johnson RG, Pfister D, Carty SE, Scarpa A (1979) Biological amine transport in chromaffin ghosts. J Biol Chem 254:10963–10972

Johnson J, Tian N, Caywood MS, Reimer RJ, Edwards RH, Copenhagen DR (2003) Vesicular neurotransmitter transporter expression in developing postnatal rodent retina: GABA and glycine precede glutamate. J Neurosci 23:518–529

Jourdain P, Bergersen LH, Bhaukaurally K, Bezzi P, Santello M, Domercq M, Matute C, Tonello F, Gundersen V, Volterra A (2007) Glutamate exocytosis from astrocytes controls synaptic strength. Nat Neurosci 10:331–339

Juge N, Yoshida Y, Yatsushiro S, Omote H, Moriyama Y (2006) Vesicular glutamate transporter contains two independent transport machineries. J Biol Chem 281:39499–39506

Juge N, Gray JA, Omote H, Miyaji T, Inoue T, Hara C, Uneyama H, Edwards RH, Nicoll RA, Moriyama Y (2010) Metabolic control of vesicular glutamate transport and release. Neuron 68:99–112

Kam K, Nicoll R (2007) Excitatory synaptic transmission persists independently of the glutamate–glutamine cycle. J Neurosci 27:9192–9200

Kaneko T, Fujiyama F (2002) Complementary distribution of vesicular glutamate transporters in the central nervous system. Neurosci Res 42:243–250

Kang T-C, Kim D-S, Kwak S-E, Kim J-E, Kim DW, Kang JH, Won MH, Kwon O-S, Choi S-Y (2005) Valproic acid reduces enhanced vesicular glutamate transporter immunoreactivities in the dentate gyrus of the seizure prone gerbil. Neuropharmacology 49:912–921

Kanner BI, Sharon I (1978) Active transport of L-glutamate by membrane vesicles isolated from rat brain. Biochemistry 17:3949–3953

Kashani A, Betancur C, Giros B, Hirsch E, Mestikawy S (2007) Altered expression of vesicular glutamate transporters VGLUT1 and VGLUT2 in Parkinson disease. Neurobiol Aging 28:568–578

Kashani A, Lepicard E, Poirel O, Videau C, David JP, Fallet-Bianco C, Simon A, Delacourte A, Giros B, Epelbaum J, Betancur C, El Mestikawy S (2008) Loss of VGLUT1 and VGLUT2 in the prefrontal cortex is correlated with cognitive decline in Alzheimer disease. Neurobiol Aging 29:1619–1630

Kim DS, Kwak SE, Kim JE, Won MH, Choi HC, Song HK, Kwon OS, Kim YI, Choi SY, Kang TC (2005) Bilateral enhancement of excitation via up-regulation of vesicular glutamate transporter subtype 1, not subtype 2, immunoreactivity in the unilateral hypoxic epilepsy model. Brain Res 1055:122–130

Kim JE, Kim DS, Kwak SE, Choi HC, Song HK, Choi SY, Kwon OS, Kim YI, Kang TC (2007) Anti-glutamatergic effect of riluzole: comparison with valproic acid. Neuroscience 147:136–145

Kish PE, Fischer-Bovenkerk C, Ueda T (1989) Active transport of gamma-aminobutyric acid and glycine into synaptic vesicles. Proc Natl Acad Sci U S A 86:3877–3881

Kontro P, Marnela KM, Oja SS (1980) Free amino acids in the synaptosome and synaptic vesicle fractions of different bovine brain areas. Brain Res 184:129–141

Kraus T, Neuhuber WL, Raab M (2004) Vesicular glutamate transporter 1 immunoreactivity in motor endplates of striated esophageal but not skeletal muscles in the mouse. Neurosci Lett 360:53–56

Kvamme E, Torgner IA, Roberg B (1991) Evidence indicating that pig renal phosphate-activated glutaminase has a functionally predominant external localization in the inner mitochondrial membrane. J Biol Chem 266:13185–13192

Landry M, Bouali-Benazzouz R, El Mestikawy S, Ravassard P, Nagy F (2004) Expression of vesicular glutamate transporters in rat lumbar spinal cord, with a note on dorsal root ganglia. J Comp Neurol 468:380–394

Leo S, Moechars D, Callaerts-Vegh Z, D'Hooge R, Meert T (2009) Impairment of VGLUT2 but not VGLUT1 signaling reduces neuropathy-induced hypersensitivity. Eur J Pain 13:1008–1017

Lewis SM, Lee FS, Todorova M, Seyfried TN, Ueda T (1997) Synaptic vesicle glutamate uptake in epileptic (EL) mice. Neurochem Int 31:581–585

Li JL, Fujiyama F, Kaneko T, Mizuno N (2003) Expression of vesicular glutamate transporters, VGluT1 and VGluT2, in axon terminals of nociceptive primary afferent fibers in the superficial layers of the medullary and spinal dorsal horns of the rat. J Comp Neurol 457:236–249

Li D, Ropert N, Koulakoff A, Giaume C, Oheim M (2008) Lysosomes are the major vesicular compartment undergoing Ca^{2+}-regulated exocytosis from cortical astrocytes. J Neurosci 28:7648–7658

Li D, Hérault K, Silm K, Evrard A, Wojcik S, Oheim M, Herzog E, Ropert N (2013) Lack of evidence for vesicular glutamate transporter expression in mouse astrocytes. J Neurosci 33:4434–4455

Liu Y, Abdel Samad O, Zhang L, Duan B, Tong Q, Lopes C, Ji RR, Lowell BB, Ma Q (2010) VGLUT2-dependent glutamate release from nociceptors is required to sense pain and suppress itch. Neuron 68:543–556

Logan WJ, Snyder SH (1972) High affinity uptake systems for glycine, glutamic and aspartic acids in synaptosomes of rat central nervous tissues. Brain Res 42:413–431

Lou S, Duan B, Vong L, Lowell BB, Ma Q (2013) Runx1 controls terminal morphology and mechanosensitivity of VGLUT3-expressing C-mechanoreceptors. J Neurosci 33:870–882

Lussier AL, Romay-Tallón R, Caruncho HJ, Kalynchuk LE (2013) Altered GABAergic and glutamatergic activity within the rat hippocampus and amygdala in rats subjected to repeated corticosterone administration but not restraint stress. Neuroscience 231:38–48

Ma Q (2014) Itch modulation by VGLUT2-dependent glutamate release from somatic sensory neurons. In: Carstens E, Akiyama T (eds) Itch: mechanisms and treatment. CRC Press, Boca Raton, Chapter 21, Frontiers in Neuroscience

Mancini GM, de Jonge HR, Galjaard H, Verheijen FW (1989) Characterization of a proton-driven carrier for sialic acid in the lysosomal membrane. Evidence for a group-specific transport system for acidic monosaccharides. J Biol Chem 264:15247–15254

Mangan JL, Whittaker VP (1966) The distribution of free amino acids in subcellular fractions of guinea-pig brain. Biochem J 98:128–137

Marchaland J, Calì C, Voglmaier SM, Li H, Regazzi R, Edwards RH, Bezzi P (2008) Fast subplasma membrane Ca^{2}-transients control exoendocytosis of synaptic-like microvesicles in astrocytes. J Neurosci 28:9122–9132

Martinez-Hernandez A, Bell KP, Norenberg MD (1977) Glutamine synthetase: glial localization in brain. Science 195:1356–1358

Masson J, Darmon M, Conjard A, Chuhma N, Ropert N, Thoby-Brisson M, Foutz AS, Parrot S, Miller GM, Jorisch R, Polan J, Hamon M, Hen R, Rayport S (2006) Mice lacking brain/kidney phosphate-activated glutaminase have impaired glutamatergic synaptic transmission, altered breathing, disorganized goal-directed behavior and die shortly after birth. J Neurosci 26:4660–4671

Masu M, Tanabe Y, Tsuchida K, Shigemoto R, Nakanishi S (1991) Sequence and expression of a metabotropic glutamate receptor. Nature 349:760–765

Maycox PR, Deckwerth T, Hell JW, Jahn R (1988) Glutamate uptake by brain synaptic vesicles. Energy dependence of transport and functional reconstitution in proteoliposomes. J Biol Chem 263:15423–15428
Maycox PR, Hell JW, Jahn R (1990) Amino acid neurotransmission: spotlight on synaptic vesicles. Trends Neurosci 13:83–87
McKenna MC (2007) The glutamate–glutamine cycle Is not stoichiometric: fates of glutamate in brain. J Neurosci Res 85:3347–3358
McKenna MC, Tildon JT, Stevenson JH, Boatright R, Huang S (1993) Regulation of energy metabolism in synaptic terminals and cultured rat brain astrocytes: differences revealed using aminooxyacetate. Dev Neurosci 15:320–329
McKenna MC, Stevenson JH, Huan X, Hopkins IB (2000) Differential distribution of the enzymes glutamate dehydrogenase and aspartate aminotransferase in cortical synaptic mitochondria contributes to metabolic compartmentation in cortical synaptic terminals. Neurochem Int 37:229–241
McMahon HT, Nicholls DG (1991) The bioenergetics of neurotransmitter release. Biochim Biophys Acta 1059:243–264
Miller SL, Fenstermacher E, Bates J, Blacker D, Sperling RA, Dickerson BC (2008) Hippocampal activation in adults with mild cognitive impairment predicts subsequent cognitive decline. J Neurol Neurosurg Psychiatry 79:630–635
Mitchell P (1961) Coupling of phosphorylation to electron and hydrogen transfer by a chemi-osmotic type of mechanism. Nature 191:144–148
Mitchell P (1966) Chemiosmotic coupling in oxidative and photosynthetic phosphorylation. Biol Rev Camb Philos Soc 41:445–502
Miyaji T, Echigo N, Hiasa M, Senoh S, Omote H, Moriyama Y (2008) Identification of a vesicular aspartate transporter. Proc Natl Acad Sci U S A 105:11720–11724
Miyazaki T, Fukaya M, Shimizu H, Watanabe M (2003) Subtype switching of vesicular glutamate transporters at parallel fibre-Purkinje cell synapses in developing mouse cerebellum. Eur J Neurosci 17:2563–2572
Moechars D, Weston MC, Leo S, Callaerts-Vegh Z, Goris I, Daneels G, Buist A, Cik M, van der Spek P, Kass S, Meert T, D'Hooge R, Rosenmund C, Hampson RM (2006) Vesicular glutamate transporter VGLUT2 expression levels control quantal size and neuropathic pain. J Neurosci 26:12055–12066
Monaghan DT, Bridges RJ, Cotman CW (1989) The excitatory amino acid receptors: their classes, pharmacology, and distinct properties in the function of the central nervous system. Annu Rev Pharmacol Toxicol 29:365–402
Montana V, Ni Y, Sunjara V, Hua X, Parpura V (2004) Vesicular glutamate transporter-dependent glutamate release from astrocytes. J Neurosci 24:2633–2642
Morimoto R, Hayashi M, Yatsushiro S, Otsuka M, Yamamoto A, Moriyama Y (2003) Co-expression of vesicular glutamate transporters (VGLUT1 and VGLUT2) and their association with synaptic-like microvesicles in rat pinealocytes. J Neurochem 84:382–391
Morimoto R, Uehara S, Yatsushiro S, Juge N, Hua Z, Senoh S, Echigo N, Hayashi M, Mizoguchi T, Ninomiya T, Udagawa N, Omote H, Yamamoto A, Edwards RH, Moriyama Y (2006) Secretion of L-glutamate from osteoclasts through transcytosis. EMBO J 25:4175–4186
Moriyama Y, Yamamoto A (1995) Vesicular L-glutamate transporter in microvesicles from bovine pineal glands. Driving force, mechanism of chloride anion activation, and substrate specificity. J. Biol Chem 270:22314–22320
Moriyoshi K, Masu M, Ishii T, Shigemoto R, Mizuno N, Nakanishi S (1991) Molecular cloning and characterization of the rat NMDA receptor. Nature 354:31–37
Morken TS, Brekke E, Håberg A, Widerøe M, Brubakk AM, Sonnewald U (2014) Neuron-astrocyte interactions, pyruvate carboxylation and the pentose phosphate pathway in the neonatal rat brain. Neurochem Res 39:556–569
Mothet JP, Pollegioni L, Ouanounou G, Martineau M, Fossier P, Baux G (2005) Glutamate receptor activation triggers a calcium-dependent and SNARE protein-dependent release of the gliotransmitter D-serine. Proc Natl Acad Sci U S A 102:5606–5611

Nadler JV, Vaca KW, White WF, Lynch GS, Cotman CW (1976) Aspartate and glutamate as possible transmitters of excitatory hippocampal afferents. Nature 260:538–540

Nadler JV, White WF, Vaca KW, Perry BW, Cotman CW (1978) Biochemical correlates of transmission mediated by glutamate and aspartate. J Neurochem 31:147–155

Naito S, Ueda T (1981) Affinity-purified anti-protein I antibody. Specific inhibitor of phosphorylation of protein I, a synaptic protein. J Biol Chem 256:10657–10663

Naito S, Ueda T (1982) ATP-dependent glutamate uptake into Protein I-associated synaptic vesicles. Soc Neurosci Abst 8:878

Naito S, Ueda T (1983) Adenosine triphosphate-dependent uptake of glutamate into protein I-associated synaptic vesicles. J Biol Chem 258:696–699

Naito S, Ueda T (1985) Characterization of glutamate uptake into synaptic vesicles. J Neurochem 44:99–109

Nakamura K, Hioki H, Fujiyama F, Kaneko T (2005) Postnatal changes of vesicular glutamate transporter (VGluT)1 and VGluT2 immunoreactivities and their colocalization in the mouse forebrain. J Comp Neurol 492:263–288

Neale SA, Copeland CS, Uebele VN, Thomson FJ, Salt TE (2013) Modulation of hippocampal synaptic transmission by the kynurenine pathway member xanthurenic acid and other VGLUT inhibitors. Neuropsychopharmacology 38:1060–1067

Neale SA, Copeland CS, Salt TE (2014) Effect of VGLUT inhibitors on glutamatergic synaptic transmission in the rodent hippocampus and prefrontal cortex. Neurochem Int 73:159–165

Newman EA, Zahs KR (1998) Modulation of neuronal activity by glial cells in the retina. J Neurosci 18:4022–4028

Ni Y, Parpura V (2009) Dual regulation of Ca^{2+}-dependent glutamate release from astrocytes: vesicular glutamate transporters and cytosolic glutamate levels. Glia 57:1296–1305

Ni B, Rosteck PR Jr, Nadi NS, Paul SM (1994) Cloning and expression of a cDNA encoding a brain-specific Na^{+}-dependent inorganic phosphate cotransporter. Proc Natl Acad Sci U S A 91:5607–5611

Nicholls D, Attwell D (1990) The release and uptake of excitatory amino acids. Trends Pharmacol Sci 11:462–468

Nicholls DG, Sihra TS (1986) Synaptosomes possess an exocytotic pool of glutamate. Nature 321:772–773

Njus D, Kelley PM, Harnadek GJ (1986) Bioenergetics of secretary vesicles. Biochim Biophys Acta 853:237–265

Ogita K, Hirata K, Bole DG, Yoshida S, Tamura Y, Leckenby AM, Ueda T (2001) Inhibition of vesicular glutamate storage and exocytotic release by Rose Bengal. J Neurochem 77:34–42

Oliveira AL, Hydling F, Olsson E, Shi T, Edwards RH, Fujiyama F, Kaneko T, Hökfelt T, Cullheim S, Meister B (2003) Cellular localization of three vesicular glutamate transporter mRNAs and proteins in rat spinal cord and dorsal root ganglia. Synapse 50:117–129

Omote H, Miyaji T, Juge N, Moriyama Y (2011) Vesicular neurotransmitter transporter: bioenergetics and regulation of glutamate transport. Biochemistry 50:5558–5565

Ormel L, Stensrud MJ, Chaudhry FA, Gundersen V (2012a) A distinct set of synaptic-like microvesicles in atroglial cells contain VGLUT3. Glia 60:1289–1300

Ormel L, Stensrud MJ, Bergersen LH, Gundersen V (2012b) VGLUT1 is localized in astrocytic processes in several brain regions. Glia 60:229–238

Ozkan ED, Ueda T (1998) Glutamate transport and storage in synaptic vesicles. Jpn J Pharmacol 77:1–10

Özkan ED, Lee FS, Ueda T (1997) A protein factor that inhibits ATP-dependent glutamate and γ-aminobutyric acid accumulation into synaptic vesicles: purification and initial characterization. Proc Natl Acad Sci U S A 94:4137–4142

Palaiologos G, Hertz L, Schousboe A (1988) Evidence that aspartate aminotransferase activity and ketodicarboxylate carrier function are essential for biosynthesis of transmitter glutamate. J Neurochem 51:317–320

Parpura V, Haydon PG (2000) Physiological astrocytic calcium levels stimulate glutamate release to modulate adjacent neurons. Proc Natl Acad Sci U S A 97:8629–8634

Parpura V, Basarsky TA, Liu F, Jeftinija K, Jeftinija S, Haydon PG (1994) Glutamate-mediated astrocyte-neuron signalling. Nature 369:744–747
Parsons SM, Prior C, Marshall IG (1993) Acetylcholine transport, storage, and release. Int Rev Neurobiol 35:279–390
Pasti L, Volterra A, Pozzan T, Carmignoto G (1997) Intracellular calcium oscillations in astrocytes: a highly plastic, bidirectional form of communication between neurons and astrocytes in situ. J Neurosci 17:7817–7830
Patel MS (1974) The relative significance of CO_2-fixing enzymes in the metabolism of rat brain. J Neurochem 22:717–724
Patel SA, Nagy JO, Bolstad ED, Gerdes JM, Thompson CM (2007) Tetrapeptide inhibitors of the glutamate vesicular transporter (VGLUT). Bioorg Med Chem Lett 17:5125–5128
Persson S, Boulland JL, Aspling M, Larsson M, Fremeau RT Jr, Edwards RH, Storm-Mathisen J, Chaudhry FA, Broman J (2006) Distribution of vesicular glutamate transporters 1 and 2 in the rat spinal cord, with a note on the spinocervical tract. J Comp Neurol 497:683–701
Pflibsen L, Stang KA, Sconce MD, Wilson VB, Hood RL, Meshul CK, Mitchell SH (2015) Executive function deficits and glutamatergic protein alterations in a progressive 1-methyl-4-phenyl-1,2,3,6-tetrahydropyridine mouse model of Parkinson's disease. J Neurosci Res 93:1849–1864
Phillips JH, Allison YP (1978) Proton translocation of the bovine chromaffin-granule membrane. Biochem J 170:661–672
Pietrancosta N, Kessler A, Favre-Besse FC, Triballeau N, Quentin T, El Mestikawy S, Acher FC (2010) Rose Bengal analogs and vesicular glutamate transporters (VGLUTs). Bioorg Med Chem 18:6922–6933
Platel JC, Dave KA, Gordon V, Lacar B, Rubio ME, Bordey A (2010) NMDA receptors activated by subventricular zone astrocytic glutamate are critical for neuroblast survival prior to entering a synaptic network. Neuron 65:859–872
Porter JT, McCarthy KD (1996) Hippocampal astrocytes *in situ* respond to glutamate released from synaptic terminals. J Neurosci 16:5073–5081
Preobraschenski J, Zander JF, Suzuki T, Ahnert-Hilger G, Jahn R (2014) Vesicular glutamate transporters use flexible anion and cation binding sites for efficient accumulation of neurotransmitter. Neuron 84:1287–1301
Price GD, Trussell LO (2006) Estimate of the chloride concentration in a central glutamatergic terminal: a gramicidin perforated-patch study on the calyx of Held. J Neurosci 26:11432–11436
Putcha D, Brickhouse M, O'Keefe K, Sullivan C, Rentz D, Marshall G, Dickerson B, Sperling R (2011) Hippocampal hyperactivation associated with cortical thinning in Alzheimer's disease signature regions in non-demented elderly adults. J Neurosci 31:17680–17688
Raju DV, Ahern TH, Shah DJ, Wright TM, Standaert DG, Hall RA, Smith Y (2008) Differential synaptic plasticity of the corticostriatal and thalamostriatal systems in an MPTP-treated monkey model of parkinsonism. Eur J Neurosci 27:1647–1658
Rassin DK (1972) Amino acids as putative transmitters: failure to bind to synaptic vesicles of guinea pig cerebral cortex. J Neurochem 19:139–148
Reid RA, Moyle J, Mitchell P (1966) Synthesis of adenosine triphosphate by a protonmotive force in rat liver mitochondria. Nature 212:257–258
Reimer RJ, Edwards RH (2004) Organic anion transport is the primary function of the SLC17/type I phosphate transporter family. Pflugers Arch 447:629–635
Reis M, Farage M, Wolosker H (2000) Chloride-dependent inhibition of vesicular glutamate uptake by alpha-keto acids accumulated in maple syrup urine disease. Biochim Biophys Acta 1475:114–118
Roberg B, Torgner IA, Kvamme E (1995) The orientation of phosphate activated glutaminase in the inner mitochondrial membrane of synaptic and non-synaptic rat brain mitochondria. Neurochem Int 27:367–376

Roberson ED, Scearce-Levie K, Palop JJ, Yan F, Cheng IH, Wu T, Gerstein H, Yu GQ, Mucke L (2007) Reducing endogenous tau ameliorates amyloid beta-induced deficits in an Alzheimer's disease mouse model. Science 316:750–754

Roberson ED, Halabisky B, Yoo JW, Yao J, Chin J, Yan F, Wu T, Hamto P, Devidze N, Yu GQ, Palop JJ, Noebels JL, Mucke L (2011) Amyloid-β/Fyn-induced synaptic, network, and cognitive impairments depend on tau levels in multiple mouse models of Alzheimer's disease. J Neurosci 31:700–711

Rogoz K, Lagerström MC, Dufour S, Kullander K (2012) VGLUT2-dependent glutamatergic transmission in primary afferents is required for intact nociception in both acute and persistent pain modalities. Pain 153:1525–1536

Roseth S, Fykse EM, Fonnum F (1995) Uptake of L-glutamate into rat brain synaptic vesicles: effect of inhibitors that bind specifically to the glutamate transporter. J Neurochem 65:96–103

Roseth S, Fykse EM, Fonnum F (1998) Uptake of L-glutamate into synaptic vesicles: competitive inhibition by dyes with biphenyl and amino and sulphonic acid substituted naphthyl groups. Biochem Pharmacol 56:1243–1249

Ruel J, Emery S, Nouvian R, Bersot T, Amilhon B, Van Rybroek JM, Rebillard G, Lenoir M, Eybalin M, Delprat B, Sivakumaran TA, Giros B, El Mestikawy S, Moser T, Smith RJ, Lesperance MM, Puel JL (2008) Impairment of SLC17A8 encoding vesicular glutamate transporter-3, VGLUT3, underlies nonsyndromic deafness DFNA25 and inner hair cell dysfunction in null mice. Am J Hum Genet 83:278–292

Schafer MK-H, Varoqui H, Defamie N, Weihe E, Erickson JD (2002) Molecular cloning and functional identification of mouse vesicular glutamate transporter 3 and its expression in subsets of novel excitatory neurons. J Biol Chem 277:50734–50748

Schenck S, Wojcik SM, Brose N, Takamori S (2009) A chloride conductance in VGLUT1 underlies maximal glutamate loading into synaptic vesicles. Nat Neurosci 12:156–162

Schousboe A, Hertz L (1981) Role of astroglial cells in glutamate homeostasis. In: Di Chiara G, Gessa GL (eds) Glutamate as a neurotransmitter. Raven Press, New York, pp 103–113

Schousboe A, Westergaard N, Waagepetersen H, Larsson OM, Bakken IJ, Sonnewald U (1997) Trafficking between glia and neurons of TCA cycle intermediates and related metabolites. Glia 21:99–105

Seal RP, Akil O, Yi E, Weber CM, Grant L, Yoo J, Clause A, Kandler K, Noebels JL, Glowatzki E, Lusting LR, Edwards RH (2008) Sensorineural deafness and seizures in mice lacking vesicular glutamate transporter 3. Neuron 57:263–275

Seal RP, Wang X, Guan Y, Raja SN, Woodbury CJ, Basbaum AI, Edwards RH (2009) Injury-induced mechanical hypersensitivity requires C-low threshold mechanoreceptors. Nature 462:651–655

Serres S, Raffard G, Franconi JM, Merle M (2008) Close coupling between astrocytic and neuronal metabolisms to fulfill anaplerotic and energy needs in the rat brain. J Cereb Blood Flow Metab 28:712–724

Shank RP, Aprison MH (1988) Glutamate as a neurotransmitter. In: Kvamme E (ed) Glutamine and Glutamate in Mammals, vol II. CRC Press, Boca Raton, pp 3–20

Shank RP, Campbell GL (1981) Avid Na^+-dependent high-affinity uptake of alpha-ketoglutarate by nerve terminal enriched material from mouse cerebellum. Life Sci 28:843–850

Shank RP, Campbell GL (1982) Glutamine and alpha-ketoglutarate uptake and metabolism by nerve terminals enriched material from mouse cerebellum. Neurochem Res 7:601–616

Shank RP, Campbell GL (1984) Alpha-ketoglutarate and malate uptake and metabolism by synaptosomes: further evidence for an astrocyte-to-neuron metabolic shuttle. J Neurochem 42:1153–1161

Shank RP, Bennett GS, Freytag SO, Campbell GL (1985) Pyruvate carboxylase: astrocyte-specific enzyme implicated in the replenishment of amino acid neurotransmitter pools. Brain Res 329:364–367

Shank RP, Leo GC, Zielke HR (1993) Cerebral metabolic compartmentation as revealed by nuclear magnetic resonance analysis of D-[1-^{13}C]glucose metabolism. J Neurochem 61:315–323

Shapiro RA, Haser WG, Curthoys NP (1985) The orientation of phosphate-dependent glutaminase on the inner membrane of rat renal mitochondria. Arch Biochem Biophys 243:1–7

Shepherd GMG, Harris KM (1998) Three-dimensional structure and composition of CA3-CA1 axons in rat hippocampal slices: implications for presynaptic connectivity and compartmentalization. J Neurosci 18:8300–8310

Sherry DM, Wang MM, Bates J, Frishman LJ (2003) Expression of vesicular glutamate transporter 1 in the retina reveals temporal ordering in development of rod vs. cone and ON vs. OFF circuits. J Comp Neurol 465:480–498

Shioi J, Ueda T (1990) Artificially imposed electrical potentials drive L-glutamate uptake into synaptic vesicles of bovine cerebral cortex. Biochem J 267:63–68

Shioi J, Naito S, Ueda T (1989) Glutamate uptake into synaptic vesicles of bovine cerebral cortex and electrochemical potential difference of proton across the membrane. Biochem J 258:499–504

Smith SJ (1992) Do astrocytes process neural information? Prog Brain Res 94:119–136

Sonnewald U (2014) Glutamate synthesis has to be matched by its degradation—where do all the carbons go? J Neurochem 131:399–406

Sonnewald U, Rae C (2010) Pyruvate carboxylation in different model systems studied by ^{13}C MRS. Neurochem Res 35:1916–1921

Stenovec M, Kreft M, Grilc S, Potokar M, Kreft ME, Pangrsic T, Zorec R (2007) Ca^{2+}-dependent mobility of vesicles capturing anti-VGLUT1 antibodies. Exp Cell Res 313:3809–3818

Storm-Mathisen J, Leknes AK, Bore AT, Vaaland JL, Edminson P, Haug E-M S, Ottersen OP (1983) First visualization of glutamate and GABA in neurons by immunocytochemistry. Nature 301:517–520

Stornetta RL, Sevigny CP, Guyenet PG (2002) Vesicular glutamate transporter DNPI/VGLUT2 mRNA is present in C1 and several other groups of brainstem catecholaminergic neurons. J Comp Neurol 444:191–206

Stornetta RL, Sevigny CP, Guyenet PG (2003) Inspiratory augmenting bulbospinal neurons express both glutamatergic and enkephalinergic phenotypes. J Comp Neurol 455:113–124

Südhof TC (1990) The structure of the human synapsin I gene and protein. J Biol Chem 265:7849–7852

Südhof TC, Czernik AJ, Kao HT, Takei K, Johnston PA, Horiuchi A, Kanazir SD, Wagner MA, Perin MS, De Camilli P, Greengard P (1989) Synapsins: mosaics of shared and individual domains in a family of synaptic vesicle phosphoproteins. Science 245:1474–1480

Tabb JS, Ueda T (1991) Phylogenetic studies on the synaptic vesicle glutamate transport system. J Neurosci 11:1822–1828

Tabb JS, Kish PE, Van Dyke R, Ueda T (1992) Glutamate transport into synaptic vesicles. Roles of membrane potential, pH gradient, and intravesicular pH. J Biol Chem 267:15412–15418

Takamori S, Rhee JS, Rosenmund C, Jahn R (2000) Identification of a vesicular glutamate transporter that defines a glutamatergic phenotype in neurons. Nature 407:189–194

Takamori S, Rhee JS, Rosenmund C, Jahn R (2001) Identification of differentiation-associated brain-specific phosphate transporter as a second vesicular glutamate transporter (VGLUT2). J Neurosci 21:RC182

Takamori S, Malherbe P, Broger C, Jahn R (2002) Molecular cloning and functional characterization of human vesicular glutamate transporter 3. EMBO Rep 3:798–803

Takamori S, Holt M, Stenius K, Lemke EA, Grønborg M, Riedel D, Urlaub H, Schenck S, Brügger B, Ringler P, Müller SA, Rammner B, Gräter F, Hub JS, De Groot BL, Mieskes G, Moriyama Y, Klingauf J, Grubmüller H, Heuser J, Wieland F, Jahn R (2006) Molecular Anatomy of a Trafficking Organelle. Cell 127:831–846

Takeda K, Ueda T (2016) Effective mechanism for synthesis of neurotransmitter glutamate and its loading into synaptic vesicles. Neurochem Res. doi:10.1007/s11064-016-2037-3

Takeda K, Ishida A, Takahashi K, Ueda T (2012) Synaptic vesicles are capable of synthesizing the VGLUT substrate glutamate from α-ketoglutarate for vesicular loading. J Neurochem 121:184–196

Tamura Y, Ogita K, Ueda T (2014) A new VGLUT-specific potent inhibitor: pharmacophore of Brilliant Yellow. Neurochem Res 39:117–128

Thompson CM, Davis E, Carrigan CN, Cox HD, Bridges RJ, Gerdes JM (2005) Inhibitors of the glutamate vesicular transporter (VGLUT). Curr Med Chem 12:2041–2056

Tildon JT, Roeder LM, Stevenson JH (1985) Substrate oxidation by isolated rat brain mitochondria and synaptosomes. J Neurosci Res 14:207–215

Todd AJ, Hughes DI, Polgár E, Nagy GG, Mackie M, Ottersen OP, Maxwell DJ (2003) The expression of vesicular glutamate transporters VGLUT1 and VGLUT2 in neurochemically defined axonal populations in the rat spinal cord with emphasis on the dorsal horn. Eur J Neurosci 17:13–27

Todera R, Totterdell S, Wojcik S, Brose N, Elizalde N, Lasheras B, Rio J (2007) Enhanced anxiety, depressive-like behaviour and impaired recognition memory in mice with reduced expression of the vesicular glutamate transporter 1 (VGLUT1). Eur J Neurosci 25:281–290

Touret M, Parrot S, Denoroy L, Belin M-F, Didier-Bazes M (2007) Glutamatergic alterations in the cortex of genetic absence epilepsy rats. BMC Neurosci 8:69–76

Ueda T (1986) Glutamate transport in the synaptic vesicle. In: Roberts PJ, Storm-Mathisen J, Bradford H (eds) Excitatory amino acids. Macmillan, London, pp 173–195

Ueda T, Greengard P (1997) Adenosine 3′:5′-monophosphate-regulated phosphoprotein system of neuronal membranes. I. Solubilization, purification, and some properties of an endogenous phosphoprotein. J Biol Chem 252:5155–5163

Ueda T, Ikemoto A (2007) Cytoplasmic glycolytic enzymes. Synaptic vesicle-associated glycolytic ATP-generating enzymes: coupling to neurotransmitter accumulation. In: Gibson G, Dienel G (eds) Handbook of neurochemistry and molecular neurobiology, 3rd edn, Brain energetics, cellular and molecular integration. Springer, Heidelberg, pp 241–259

Ueda T, Greengard P, Berzins K, Cohen RS, Blomberg F, Grab DJ, Siekevitz P (1979) Subcellular distribution in cerebral cortex of two proteins phosphorylated by a cAMP-dependent protein kinase. J Cell Biol 83:308–319

Uehara S, Jung SK, Morimoto R, Arioka S, Miyaji T, Juge N, Hiasa M, Shimizu K, Ishimura A, Otsuka M, Yamamoto A, Maechler P, Moriyama Y (2006) Vesicular storage and secretion of L-glutamate from glucagon-like peptide 1-secreting clonal intestinal L cells. J Neurochem 96:550–560

Uezato A, Meador-Woodruff JH, McCullumsmith RE (2009) Vesicular glutamate transporter mRNA expression in the medial temporal lobe in major depressive disorder, bipolar disorder, and schizophrenia. Bipolar Disord 11:711–725

Van den Berg CJ, Garfinkel D (1971) A simulation study of brain compartments. Metabolism of glutamate and related substances in mouse brain. Biochem J 123:211–218

Van der Hel WS, Verlinde SA, Meijer DH, de Wit M, Rensen MG, van Gassen KL, van Rijen PC, van Veelen CW, de Graan PN (2009) Hippocampal distribution of vesicular glutamate transporter 1 in patients with temporal lobe epilepsy. Epilepsia 50:1717–1728

Varea E, Guirado R, Gilabert-Juan J, Martí U, Castillo-Gomez E, Blasco-Ibáñez JM, Crespo C, Nacher J (2012) Expression of PSA-NCAM and synaptic proteins in the amygdala of psychiatric disorder patients. J Psychiatr Res 46:189–197

Varoqui H, Schafer MK-H, Zhu H, Weihe E, Erickson JD (2002) Identification of the differentiation-associated Na^+/Pi transporter as a novel vesicular glutamate transporter expressed in a distinct set of glutamatergic synapses. J Neurosci 22:142–155

Verheijen FW, Verbeek E, Aula N, Beerens CE, Havelaar AC, Joosse M, Peltonen L, Aula P, Galjaard H, van der Spek PJ, Mancini GM (1999) A new gene, encoding an anion transporter, is mutated in sialic acid storage diseases. Nat Genet 23:462–465

Vossel KA, Beagle AJ, Rabinovici GD, Shu H, Lee SE, Naasan G, Hegde M, Cornes SB, Henry ML, Nelson AB, Seeley WW, Geschwind MD, Gorno-Tempini ML, Shih T, Kirsch HE, Garcia PA, Miller BL, Mucke L (2013) Seizures and epileptiform activity in the early stages of Alzheimer disease. JAMA Neurol 70:1158–1166

Wang ZT, Yu G, Wang HS, Yi SP, Su RB, Gong ZH (2015) Changes in VGLUT2 expression and function in pain-related supraspinal regions correlate with the pathogenesis of neuropathic pain in a mouse spared nerve injury model. Brain Res 1624:515–524

Watkins JC, Evans RH (1981) Excitatory amino acid transmitters. Annu. Rev Pharmacol Toxicol 21:165–204

Westergaard N, Sonnewald U, Schousboe A (1994) Release of α-ketoglutarate, malate and succinate from cultured astrocytes: possible role in amino acid neurotransmitter homeostasis. Neurosci Lett 176:105–109

Weston MC, Stornetta RL, Guyenet PG (2004) Glutamatergic neuronal projections from the marginal layer of the rostral ventral medulla to the respiratory centers in rats. J Comp Neurol 473:73–85

Weston MC, Nehring RB, Wojcik SM, Rosenmund C (2011) Interplay between VGLUT isoforms and endophilin A1 regulates neurotransmitter release and short-term plasticity. Neuron 69:1147–1159

Wilson NR, Kang J, Hueske EV, Leung T, Varoqui H, Murnick JG, Erickson JD, Liu G (2005) Presynaptic regulation of quantal size by the vesicular glutamate transporter VGLUT1. J Neurosci 25:6221–6234

Winter HC, Ueda T (1993) Glutamate uptake system in the presynaptic vesicle: glutamic acid analogs as inhibitors and alternative substrates. Neurochem Res 18:79–85

Winter HC, Ueda T (2008) The glutamate uptake system in presynaptic vesicles: further characterization of structural requirements for inhibitors and substrates. Neurochem Res 33:223–231

Wojcik SM, Rhee CJS, Herzog E, Sigler A, Jahn R, Takamori S, Brose N, Rosenmund C (2004) An essential role for vesicular glutamate transporter 1 (VGLUT1) in postnatal development and control of quantal size. Proc Natl Acad Sci U S A 101:7158–7163

Wolosker H, de Souza DO, de Meis L (1996) Regulation of glutamate transport into synaptic vesicles by chloride and proton gradient. J Biol Chem 271:11726–11731

Wu SX, Koshimizu Y, Feng YP, Okamoto K, Fujiyama F, Hioki H, Li YQ, Kaneko T, Mizuno N (2004) Vesicular glutamate transporter immunoreactivity in the central and peripheral endings of muscle-spindle afferents. Brain Res 1011:247–251

Yu AC, Drejer J, Hertz L, Schousboe A (1983) Pyruvate carboxylase activity in primary cultures of astrocytes and neurons. J Neurochem 41:1484–1487

Zala D, Hinckelmann MV, Yu H, Lyra da Cunha MM, Liot G, Cordelières FP, Marco S, Saudou F (2013) Vesicular glycolysis provides on-board energy for fast axonal transport. Cell 152:479–491

Zerari-Mailly F, Braud A, Davido N, Touré B, Azérad J, Boucher Y (2012) Glutamate control of pulpal blood flow in the incisor dental pulp of the rat. Eur J Oral Sci 120:402–407

Zhang Q, Pangrsic T, Kreft M, Krzan M, Li N, Sul JY, Halassa M, Van Bockstaele E, Zorec R, Haydon PG (2004) Fusion-related release of glutamate from astrocytes. J Biol Chem 279:12724–12733

Zhou J, Nannapaneni N, Shore S (2007) Vesicular glutamate transporters 1 and 2 are differentially associated with auditory nerve and spinal trigeminal inputs to the cochlear nucleus. J Comp Neurol 500:777–787

Zieminska E, Hilgier W, Waagepetersen HS, Hertz L, Sonnewald U, Schousboe A, Albrecht J (2004) Analysis of glutamine accumulation in rat brain mitochondria in the presence of a glutamine uptake inhibitor, histidine, reveals glutamine pools with a distinct access to deamidation. Neurochem Res 29:2121–2123

Chapter 8
The Glutamine Transporters and Their Role in the Glutamate/GABA–Glutamine Cycle

Renata Leke and Arne Schousboe

Abstract Glutamine is a key amino acid in the CNS, playing an important role in the glutamate/GABA–glutamine cycle (GGC). In the GGC, glutamine is transferred from astrocytes to neurons, where it will replenish the inhibitory and excitatory neurotransmitter pools. Different transporters participate in this neural communication, i.e., the transporters responsible for glutamine efflux from astrocytes and influx into the neurons, such as the members of the SNAT, LAT, y^+LAT, and ASC families of transporters. The SNAT family consists of the transporter isoforms SNAT3 and SNAT5 that are related to efflux from the astrocytic compartment, and SNAT1 and SNAT2 that are associated with glutamine uptake into the neuronal compartment. The isoforms SNAT7 and SNAT8 do not have their role completely understood, but they likely also participate in the GGC. The isoforms LAT2 and y^+LAT2 facilitate the exchange of neutral amino acids and cationic amino acids (y^+LAT2 isoform) and have been associated with glutamine efflux from astrocytes. ASCT2 is a Na^+-dependent antiporter, the participation of which in the GGC also remains to be better characterized. All these isoforms are tightly regulated by transcriptional and translational mechanisms, which are induced by several determinants such as amino acid deprivation, hormones, pH, and the activity of different signaling pathways. Dysfunctional glutamine transporter activity has been associated with the pathophysiological mechanisms of certain neurologic diseases, such as Hepatic Encephalopathy and Manganism. However, there might also be other neuropathological conditions associated with an altered GGC, in which glutamine transporters are dysfunctional. Hence, it appears to be of critical importance that the physiological and pathological aspects of glutamine transporters are thoroughly investigated.

Keywords Glutamine transporters • Glutamate/GABA–glutamine cycle • SNAT • LAT • y^+LAT • ASC

R. Leke (✉)
Department of Clinical Neuroscience and Rehabilitation, Institute of Neuroscience and Physiology, Sahlgrenska Academy at the University of Gothenburg, Göteborg 40530, Sweden
e-mail: renata.leke@neuro.gu.se; renataleke@hotmail.com

A. Schousboe
Department of Drug Design and Pharmacology, Faculty of Health and Medical Sciences, University of Copenhagen, Copenhagen 2100, Denmark
e-mail: arne.schousboe@sund.ku.dk

A. Schousboe, U. Sonnewald (eds.), *The Glutamate/GABA-Glutamine Cycle*, Advances in Neurobiology 13, DOI 10.1007/978-3-319-45096-4_8

Abbreviations

AAR	Amino acid response
AARE	AAR elements
APC	Amino acid- polyamine- and organocations family
BCH	2-Aminobicyclo-(2-2-1)-hepatane-2-caboxylic acid
CATs	Cationic amino acid transporters
CNS	Central nervous system
DHT	Dihydrotestosterone
EAAT	Excitatory amino acid transporters
EGFR	Epidermal growth factor receptor
eIF2α	Eukaryotic initiation factor 2α
ERK	Extracellular regulated kinase
FXR	Farnesoid X receptor
GABA	Gamma-aminobutyric acid
GAT	Gamma-aminobutyric acid transporter
GGC	Glutamate/GABA–glutamine cycle
GS	Glutamine synthetase
HATs	Heteromeric amino acid transporters
HE	Hepatic encephalopathy
IPSC	Inhibitory postsynaptic potential
LATs	L-type amino acid transporters
L-DOPA	Dihydroxyphenylalanine
MAP2	Anti-micro-tubule-associated protein-2
MAPK	Mitogen-activated protein kinase
MeAIB	α-Methylamino-iso-butyric acid
Mn	Manganese
PAG	Phosphate-activated glutaminase
PI3K	Phosphatidylinositol 3-kinase
PKB	Protein kinase B
PMA	Phorbol-12-myristate-13-acetate
SGK	Glucocorticoid inducible kinase
SNAT	Solute neutral amino acid transporters
TCA	Tricarboxylic acid
TMB	Transmembrane domain
UTR	3′-untranslated region
vGAT	Vesicular gamma-aminobutyric acid transporter

8.1 Introduction

Glutamine is one of the most important amino acids in the human body, participating in several pathways and functions, such as ammonia detoxification and the production of urea in the liver, acid-base buffering in the kidneys, and as nitrogen donor for other amino acids, purines, pyrimidines, and amino sugars (Bode 2001; McGivan and Bungard 2007). It also serves as an energy substrate since its carbon skeleton gets access to the tricarboxylic acid cycle (TCA), as described for the intestinal epithelia as well as the liver and brain (Ziegler et al. 2000; Verrey et al. 2009; Pochini et al. 2014; Schousboe et al. 2014a). Its vital role is underlined by its abundancy in the human body, where concentrations range from 2.0 to 20 mM intracellularly and 0.2 to 0.8 mM in the extracellular compartment (Bode 2001; Newsholme et al. 2003). In the central nervous system (CNS) specifically, extracellular glutamine concentrations are very low ranging from 0.13 to 0.5 mM, being the upper concentration similar to that found in the CSF in rodents and humans (Erecińska and Silver 1990; Aasly et al. 1997; Xu et al. 1998; White et al. 2004). The intracellular glutamine concentrations vary in the range 5–10 mM and can even reach 20 mM in cultured astrocytes and neurons when exposed to media containing 2 mM glutamine (Patel and Hunt 1985; Brookes 1992). This important amino acid participates in several pathways, including scavenging of ammonia, as a nitrogen donor, maintenance of redox state by glutathione, and importantly, in the glutamate/gamma-aminobutyric acid–glutamine cycle (GGC) (Bröer and Brookes 2001; Bak et al. 2006). In this cycle, the transfer of glutamine from astrocytes to neurons is a key point for replenishment of the amino acid neurotransmitter pools in the neuronal compartment, and its proper function and regulation rely upon specific transporters (Bak et al. 2006; Schousboe et al. 2013).

For that reason, the glutamine transporters are of fundamental importance in sustaining the GGC properly and they have been extensively investigated in several studies. As it will be discussed in the present chapter, glutamine carriers are known as solute neutral amino acid transporters (SNAT), system L, system y^+L, and ASC, and the isoforms of these carriers are expressed and regulated differently, depending upon the cellular compartment in which they are expressed as well as the intra and extracellular environments. Other transporters expressed in the brain have affinity for glutamine, as the isoforms B0AT2 and arpAT, but they are not involved in the GGC and therefore will not be addressed in this chapter (Pochini et al. 2014; Fotiadis et al. 2013).

Considering the importance of the GGC in the maintenance and optimal function of the excitatory and inhibitory neurotransmitter systems, it is not surprising that disturbances in the cycle have been related to different neurologic diseases. However, many of these diseases have been linked mainly to defects in the functioning of the high affinity transporters for glutamate and gamma-aminobutyric acid (GABA) as well as to derangements in the activity of glutamine synthetase (GS). However, alterations in the activity and/or expression of glutamine transporters are likely to play a significant role in the etiology of such diseases as well as the neurological symptoms associated with hepatic encephalopathy. The present chapter will provide a detailed account of the expression and regulation of the different glutamine transporters as well as a discussion of their importance for GGC in brain diseases as well as under physiological conditions.

8.2 Glutamine Transporters

Studies by Oxender and Christensen (1963) were the first to describe the existence of two different transporters called system L and A that were able to transfer glutamine between cellular compartments. The nomenclature, system L and A refers to their preference to carry the amino acids leucine and alanine, respectively (Collarini and Oxender 1987). Later, it was described that system L likely does not contribute significantly to glutamine transport across the plasma membrane in most cells, since this transporter is Na^+ independent and works as an exchanger rather than performing net uptake (Chaudhry et al. 2002a). However, system L activity has been described as part of the machinery for glutamine exchange in the CNS, which will be discussed in more detail later. System A plays an important role in glutamine transport, but also has affinity for small aliphatic amino acids. These transporters are Na^+ dependent and therefore can work against cellular concentration gradients in a concentrative fashion. Subsequently, another Na^+-dependent glutamine transport system was discovered and this was originally classified as being constituted by two isoforms Nm and Nb, found in muscle and brain tissue, respectively (Tamarappoo et al. 1997). Later, a carrier with similar functional characteristics and affinity for glutamine and to a lesser extent histidine and asparagine was found in hepatocytes. This transporter was described as being the main, if not the solely, carrier of glutamine in this tissue (Kilberg et al. 1980).

System N and A have since been thoroughly studied and different isoforms for these systems have been discovered and characterized. Both are of great importance for the transfer of glutamine into different cellular compartments, and therefore participating in the communication between neurons and astrocytes. Other transporters that have affinity for glutamine were also described in the CNS, as system L, system y^+L, and ASC, but their importance for regulating glutamine concentrations in the CNS and participation at the GGC is controversial.

System N and A isoforms, which are now denominated solute neutral amino acid transporters (SNAT), have been originally classified based on their substrate affinity, ion dependence, tolerance to Na^+ substitution by Li^+, kinetics and regulatory properties, and pH sensitivity (Bode 2001). However, in 1999 the first isoform of system N (SNAT3) was cloned, which led to the molecular characterization and better understating of its function and classification (Chaudhry et al. 1999). The cloning strategy for SNAT3 was based on its homology to the vesicular inhibitory amino acid transporter (VIAAT) family, i.e., the vesicular gamma-aminobutyric acid transporter (vGAT). SNAT3 shares the characteristics of vGAT such as cotransport of Na^+ and the exchange of lumenal protons by cytoplasmatic transmitters/amino acids. However, SNAT3 was found to be expressed in the cellular plasma membrane in contrast to vGAT, that is present in the vesicular membranes (Chaudhry et al. 1999; Gu et al. 2000). Subsequently, SNAT1 was cloned also facilitated by its sequence similarity to vGAT (Varoqui et al. 2000), followed by the other isoforms SNAT2, SNAT4, SNAT5, SNAT7 and most recently SNAT8 and based on the previously described functional characteristics, these isoforms were classified into a new family of proteins (Sugawara 2000; Yao et al. 2000; Varoqui et al. 2000; Nakanishi et al. 2001a; Hägglund et al. 2011, 2015).

Table 8.1 Glutamine transporters

Gene name	Protein name	Alias	System	Mechanism of transport	Substrate affinity
SLC38A1	SNAT1	SAT1, ATA1, SA2, NAT2	System A	Na^+-Glutamine cotransport (electrogenic)	A, S, C, M, H, Q, N
SLC38A2	SNAT2	SAT2, ATA2, SA1	System A	Na^+-Glutamine cotransport (electrogenic)	A, N, C, Q, G, H, M, P, S
SLC38A3	SNAT3	SN1, NAT	System N	Na^+-Glutamine/H^+ antiport (electroneutral)	Q, H, N
SLC38A4	SNAT4	ATA3, NAT3, SAT3	System A	Na^+-Glutamine cotransport (electrogenic)	A, N, C, G, S, T
SLC38A5	SNAT5	SN2	System N	Na^+-Glutamine/H^+ antiport (electroneutral)	G, N, A, S, Q, H
SLC38A7	SNAT7	–	System N		Q, H, A, S, N
SLC38A8	SNAT8	–	System A	Na^+-Glutamine/H^+ antiport (electroneutral)	Q, N, A, H, S
SLC7A5	LAT1	–	System L	Glutamine/amino acid antiport	Large neutral and aromatic amino acids (W,F, L, H, Q, T)
SLC7A8	LAT2	–	System L	Glutamine/amino acid antiport	Neutral amino acids
SLC7A7	y+LAT1	–	System y+L	Na^+-Glutamine/cationic amino acid antiport	Cationic and large neutral amino acids
SLC7A6	y+LAT2	–	System y+L	Na^+-Glutamine/cationic amino acid antiport	Cationic and large neutral amino acids
SLC1A5	ASCT2	–	System ASC	Glutamine/amino acid antiport	Neutral amino acids

Description of the different glutamine transporters and their gene family, mechanism of transport, and substrate affinity. The different nomenclatures used for the glutamine transporters are shown in the Alias column. *A* alanine, *C* cysteine, *F* phenylalanine, *G* glycine, *H* histidine, *L* leucine, *M* methionine, *N* Asparagine, *P* proline, *Q* glutamine, *S* serine, *T* threonine, *W* tryptophan

The glutamine transporters belong to the solute carriers 38 (SLC38) gene family and thus the superfamily of amino acid and auxin permeases (AAAP). The SLC38 family is composed of 11 members described in the human genome, 7 of which have been characterized in more detail (Table 8.1). The vesicular GABA transporter, which was the starting point for cloning the glutamine transporter, is part of the SLC32 gene family (Mackenzie and Erickson 2004). The SLC36, constituted by the proton-coupled amino acid transporters located in lysosomes and plasma membranes is also part of the AAAP superfamily (Boll et al. 2004).

The glutamine transporter isoforms have been given different nomenclature through the years, due to the simultaneous studies focusing on characterizing these proteins. However, with the advent of molecular characterization with the information obtained from the functional studies, the isoforms for glutamine transporters belonging to System N or A were accordingly identified and classified, and currently their nomenclature is based on the chronologic order in which they were discovered (Mackenzie and Erickson 2004). As seen in Table 8.1, which shows the current and older nomenclature used for these proteins, System A is constituted by SNAT1, SNAT2, SNAT4, and recently SNAT8, while System N includes SNAT3, SNAT5, and SNAT7. Their particular characteristics, focusing on those functionally related to the GGC, will be discussed in more detail in the following sections.

Regarding the protein structure of SCL38 members, their predicted topology suggests a protein with an intracellular N terminus, 11 transmembrane helices, and an extracellular C terminus (Pochini et al. 2014; Mackenzie and Erickson 2004). Studies that employed the crystallographic structure of hydantoin transporter Mhp1 as a template, transporter belonging to the APC family and sharing homology based on profile alignment and secondary structure prediction, predicted that the transmembrane domain (TMD) 1 contributes to a conserved Na^+ binding site (Bröer et al. 2009). In addition, mutations in the asparagine 76 of SNAT3, which is located in the TMD1 abolished all substrate-induced currents (Bröer et al. 2009). Mutation of N82A of SNAT2, which is localized at TMD1, also leads to complete depletion of the transporter currents and amino acid uptake (Zhang et al. 2008).

As previously mentioned, system L was identified due to its preference for leucine as substrate, but it also has affinity for other neutral amino acids such as glutamine. Contrary to system A, system L is characterized by being Na^+ independent and functioning as an antiporter of intracellular and extracellular amino acids, having affinity for branched chain and bulky neutral amino acids (Kanai et al. 1998; Fotiadis et al. 2013). The system L isoforms started to be identified molecularly in 1998 and were subsequently classified as part of the SLC7 gene family (Kanai et al. 1998; Mastroberardino et al. 1998; Rajan et al. 2000). This family of proteins is formed by two distinct subfamilies, which are named cationic amino acid transporters (CATs) and light or catalytic subunits (L-type amino acid transporters—LATs), of the heteromeric amino acid transporters (HATs) (Fotiadis et al. 2013). Both CATs and LATs are part of the large amino acid-polyamine- and organocations-(APC) family. With respect to LATs, they are known as glycoprotein associated transporters, due to the covalently association with a heavy subunit (Verrey et al. 2000). The heavy subunit can be the glycoprotein named 4F2hc, which is important for routing the transporters to the plasma membrane, whereas the light subunit is necessary for the transport activity (Palacín and Kanai 2004; Fotiadis et al. 2013). At the moment, six LATs have been identified as heterodimers with 4F2hc, i.e., LAT1 (SLC7A5), LAT2 (SLC7A8), y^+LAT1 (SLC7A7), y^+LAT2 (SCL7A6), Asc-1 (SLC7A10), and xCT (SLC7A11), all belonging to the SLC7 family. It is important to mention that the isoforms LAT1 and LAT2 are part of System L, while y^+LAT1 and y^+LAT2 constitute System y^+LAT (Table 8.1). The latter, however, represents a different transport system, mediating the exchange of cationic amino acids and neutral amino acids in the presence of sodium (Bröer et al. 2000a; Kanai et al. 2000). From all members mentioned earlier, only those related to GGC will be discussed in more detail in the following sections.

Regarding the protein structure for the LAT proteins, as described for LAT1, a 12 TMD was predicted that is linked with the heavy subunit 4F2hc through a conserved disulfide bridge. They exhibit an intracellular terminus and are not glycosylated (Fotiadis et al. 2013). The heavy subunit is described as type II membrane N-glycoproteins, which have an intracellular N-terminal and a bulky extracellular C-terminal (Fort et al. 2007).

System ASC was originally named on the basis of its affinity for the substrates alanine, serine, and cysteine (ASC), to differentiate it from System A activity (Christensen et al. 1967). This family of transporters is characterized by the affinity for neutral amino acids and as a transport system being Na^+ dependent. The first transporter gene was isolated in 1996 from mouse testis and from human choriocarcinoma (Kekuda et al. 1996; Utsunomiya-Tate et al. 1996), and subsequently classified as part of the gene family SLC1. This gene family encompasses the five high affinity glutamate transporters (SLC1A1, SLC1A2, SLC1A3, SLC1A6, and SLC1A7) and two neutral amino acid transporters, ASCT1 (SLC1A4) and ASCT2 (SLC1A5). Among the neutral amino acid transporters, ASCT2 has affinity for glutamine, whereas ASCT1 does not transport glutamine (Shafqat et al. 1993; Arriza et al. 1993). The structure for these proteins as determined subsequent to the cloning revealed a protein with 6N-terminal domains followed by a C-terminal domain. Although the crystal structure for these transporters is unknown, it is predicted that the C terminal domain is the one that has the binding site for the amino acid to be transported (Albers et al. 2012).

8.2.1 Glutamine Transporters: Isoforms, Function, and Localization

8.2.1.1 System N, Isoforms SNAT3, SNAT5, and SNAT7

System N isoforms are characterized by a narrower substrate specificity, with affinity for glutamine, asparagine, and histidine (Chaudhry et al. 1999; Fei et al. 2000; Nakanishi et al. 2001a). These carriers function as a Na^+-glutamine symporter and H^+ antiporter, a characteristic that distinguishes them from the isoforms belonging to system A (Chaudhry et al. 1999, 2001; Bröer et al. 2002). The carriers that are members of this system also have the particular property of high tolerance to the substitution of Na^+ by Li^+ for the amino acid cotransport (Chaudhry et al. 1999; Fei et al. 2000; Nakanishi et al. 2001a, b). This feature has for a long time been used to distinguish system N from system A but later studies demonstrated that SNAT1, SNAT2, and SNAT4 also tolerate Li^+ (Albers et al. 2001; Chaudhry et al. 2002b; Gu et al. 2001a). However, more recent studies have been inconclusive regarding the tolerance of system A transporters to Li^+. As discussed by Mackenzie and coworkers, it is likely that the affinity for Li^+ decreases upon membrane depolarization to a higher extent than that observed for Na^+ (Mackenzie and Erickson 2004; Mackenzie et al. 2003). Therefore, this criterion should be carefully taken into account when evaluating the isoforms of glutamine transporters with respect to their functionality (Chaudhry et al. 2002a; Gu et al. 2001a; Hägglund et al. 2011). System N also has the property of marked inhibition by low pH environments, with impairment of influx transport as external pH is lowered (Kilberg et al. 1980; Gu et al. 2001a).

When the first isoform of the glutamine transporter SNAT3 was cloned and characterized, it was revealed that it resembled the proteins that belong to System N, due to its properties regarding Na^+-dependent transport, affinity to glutamine, histidine, and asparagine and the fact that it is found highly expressed in hepatocytes (Chaudhry et al. 1999; Gu et al. 2000). The SNAT3 gene encodes a protein of 503 amino acids and exhibits a K_m for glutamine of 2.4 mM at pH 7.0 (Chaudhry et al. 1999; Bröer et al. 2002). Its mRNA was found expressed predominantly in liver, but also in brain, kidney, and heart (Chaudhry et al. 1999; Gu et al. 2000). In the brain, specifically, SNAT3 is found in the neocortex, cerebellum, olfactory bulb, and brain stem. At the cellular level, this isoform is abundant in astrocytes, while it is absent from neurons and oligodendrocytes (Boulland et al. 2002, 2003). With respect to transport activity, it is currently agreed that this isoform mediates the exchange of 1 glutamine molecule together with 1 Na^+ for the cotransport of 1 H^+, and therefore it acts as an electroneutral transporter (Bröer et al. 2002; Chaudhry et al. 2001). It is important to mention that ionic inward currents were observed when the SNAT3 transporter was expressed in *Xenopus oocytes*, a finding leading to confusion about the stoichiometry of this carrier. However, these currents turned out to be a consequence of uncoupled H^+ fluxes through the transporter, which were not related to the mechanisms of the amino acid transport per se (Chaudhry et al. 2001; Schneider et al. 2007). This uncoupled proton flux might be of importance for restoration of the intracellular pH thereby securing the correct amino acid transporter activity in between cellular compartments (Chaudhry et al. 2001). In addition, it was found that other cations also elicit these currents, in a fashion dependent on or independent of substrate, but its relevance is still unknown and further investigations are needed to elucidate its physiological importance (Hamdani et al. 2012; Schneider et al. 2007).

Subsequently, SNAT5 and SNAT7 were also shown to be part of system N due to their molecular and functional characteristics (Nakanishi et al. 2001a; Hägglund et al. 2011). The SNAT5 gene encodes a protein of 471 amino acids and has a sequence homology to SNAT3 of about 63 %, for the isoforms described in rats (Nakanishi et al. 2001a). Although SNAT5 shares genetic similarity with SNAT3, it differs with respect to substrate specificity, since it accepts a broader spectrum of amino acids such as glycine, asparagine, alanine, serine, glutamine, and histidine (Nakanishi et al. 2001a). With regard to glutamine, SNAT5 has a K_m of 4.1 mM, as demonstrated in the murine isoform of this transporter. It is found expressed in different organs, including brain, lung, intestines, and spleen (Nakanishi et al. 2001a), and it is described as an electroneutral transporter, i.e., analogous to its family counterpart SNAT3. In the CNS, SNAT5 is ubiquitously expressed, with higher levels in the neocortex, hippocampus, and striatum and at the cellular level, this transporter is found in astrocytes (Cubelos et al. 2005).

The isoform SNAT7 was more recently characterized as a member of system N, due to its characteristic of substrate preference toward glutamine and histidine and due to its insensitivity toward MeAIB (α-methylamino-iso-butyric acid), which is the hallmark inhibitor of the isoforms belonging to system A (Christensen et al. 1965; Hägglund et al. 2011). In addition to the amino acids mentioned earlier, SNAT7 also has affinity for other amino acids, such as alanine, serine, and asparagine (Hägglund et al. 2011). This transporter is Na^+ dependent as also described for the other members of the system N family. However, contrary to SNAT3 and SNAT5, this carrier does not tolerate Na^+ substitution by Li^+. This supports the concept that this parameter, i.e., substitution of Na^+ to Li^+, may not alone be adequate to characterize the carrier isoforms (Hägglund et al. 2011). Its gene encodes a protein of 461 amino acids and it is markedly expressed in the liver and brain. In the CNS it is ubiquitously expressed, with highest abundance in hypothalamus, basolateral amygdala, ventral tegmental area, and locus coeruleus (Hägglund et al. 2011). At the cellular level, SNAT7 is present in layers in the cortex and at the dentate gyrus and pyramidal cell layers in the hippocampus, zones that are rich in neurons. In this context, it was demonstrated that this isoform is predominantly present in the neuronal compartment, while it is absent from astrocytes (Hägglund et al. 2011).

8.2.1.2 System A, Isoforms SNAT1, SNAT2, SNAT4, and SNAT8

System A is constituted by the amino acid carriers that are Na^+-dependent electrogenic transporters. Contrary to the isoforms belonging to system N, System A transporters do not exhibit electroneutral transport, as they transport Na^+ together with the neutral amino acid without H^+ exchange (Oxender and Christensen 1963; Collarini and Oxender 1987). Therefore, System A works in a concentrative fashion against transmembrane concentration gradients, as demonstrated by its susceptibility to cell depolarization (Chaudhry et al. 2002b). With respect to the substrates accepted by System A, these include a broader range of aliphatic amino acids, such as alanine, asparagine, cysteine, glutamine, glycine, methionine, and serine (Mackenzie and Erickson 2004), i.e., differently from system N that has a restricted substrate preference.

The isoforms of system A are characterized by the competitive inhibition by a nonmetabolizable substrate analog MeAIB (Christensen et al. 1965; Christensen 1990). Such glutamine transport inhibition has been demonstrated in oocytes expressing SNAT1 (Varoqui et al. 2000), in cultured neurons (Su et al. 1997) and in rat brain in vivo (Kanamori and Ross 2006). Therefore, this analogous substrate has been used as a functional marker to distinguish the isoforms that constitute system A from those that belong to system N (Christensen et al. 1965). However, the use of MeAIB has its limitations, since it has been demonstrated that at lower concentrations it does not inhibit amino acid transport, resulting in confusion with respect to subfamily classification (Gu et al. 2001a). Therefore, the combination of functional and molecular characteristics constitutes the safer way of classifying these amino acid transporter isoforms. In analogy to system N, system A is influenced by pH, having its activity decreased at low extracellular pH, due to the competitive inhibition of H^+ to Na^+ transporter activation (Albers et al. 2001; Chaudhry et al. 2002b).

The isoform SNAT1, the gene of which encodes a protein of 486 amino acids, has glutamine as the preferred substrate (K_m of 230–300 μM), but it also transports alanine, cysteine, and asparagine (Varoqui et al. 2000; Chaudhry et al. 2002b; Albers et al. 2001). In addition, it has affinity for histidine, methionine, and serine, but since the concentration of these amino acids is very low in the cerebral spinal fluid, in comparison of that to glutamine, it is likely that under physiological conditions they are not transported by SNAT1 (Albers et al. 2001; Varoqui et al. 2000; Chaudhry et al. 2002b; Mackenzie et al. 2003). This isoform is mainly expressed in brain, but it has also been described at lower levels in heart and colon tissues (Varoqui et al. 2000; Chaudhry et al. 2002b). Other studies also verified SNAT1 in retina and placenta (Gu et al. 2001b; Wang et al. 2000). In the CNS, SNAT1 is found widespread with greater expression in the cerebellum and thalamus, followed by brainstem, cerebral cortex, hippocampus, and striatum (Varoqui et al. 2000; Melone et al. 2004). In terms of its cellular localization, it was demonstrated that this glutamine transporter isoform is mainly related to the neuronal compartment, but also described in ependymal cells lining the ventricles as well as in astrocytes (Varoqui et al. 2000; Melone et al. 2004; Mackenzie et al. 2003). Regarding its functional activity, the SNAT1 isoform has been well characterized as a unidirectional transporter that couples substrate with Na^+ in a 1:1 stoichiometry ratio (Albers et al. 2001; Mackenzie et al. 2003). As mentioned previously, the isoforms of system A exhibit sensitivity toward low pH, a characteristic shared by SNAT1, but interestingly it also exhibited greater glutamine uptake at pH 8 when compared to more acid environments (Varoqui et al. 2000). The effect of H^+ apparently results from reducing the affinity to Na^+, subsequently affecting the transporter activity, likely as a result of an allosteric effect. This is in agreement with the fact that when studying the transporter functionality it was verified that Na^+ binding is voltage dependent and occurs prior to the substrate biding and subsequent transport across the membrane (Mackenzie et al. 2003).

The isoform SNAT2, the gene of which encodes a protein with 506 amino acids, shares a very similar spectrum of the amino acids for which SNAT1 has affinity, albeit SNAT2 has a lower affinity for glutamine (K_m 1.65 mM) and higher capacity to transport alanine in comparison to SNAT1 (Reimer et al. 2000; Yao et al. 2000). In addition, SNAT2 demonstrated affinity toward proline, while this amino acid is almost not recognized by SNAT1 (Yao et al. 2000; Mackenzie et al. 2003). This carrier has a widespread expression, being found in all tissues in which its expression was tested for (Yao et al. 2000; Reimer et al. 2000). By Northern blot analysis, SNAT2 mRNA was found in heart, adrenal, skeletal muscle, stomach, lung, brain, spinal cord, kidneys, and adipose tissues, with lower expression in the spleen (Yao et al. 2000). With respect to its cellular localization in the CNS, SNAT2 protein is detected in neuron-rich cultures, while absent in astrocyte-rich cultures (Yao et al. 2000). However, a study employing in situ hybridization demonstrated mRNA expression for this transporter in certain glial cells and cells that are part of the blood–brain barrier (Reimer et al. 2000; Takanaga et al. 2002). Its functional characteristics are analogous to those described for SNAT1, i.e., electrogenic with a stoichiometry 1:1 substrate to Na^+ (Yao et al. 2000; Chaudhry et al. 2002b) and SNAT2 is inhibited by low pH by the same mechanisms as those described for its counterpart SNAT1 (Chaudhry et al. 2002b).

A third transporter SNAT4, a protein with 546 amino acids, is predominantly expressed in liver and bladder (Hatanaka et al. 2001; Sugawara et al. 2000). It is a carrier of short neutral amino acids and has low affinity for glutamine (Hatanaka et al. 2001; Gu et al. 2001a). Its mechanism of substrate transport is the same as that for the other members of this subfamily of transporters (Hatanaka et al. 2001). Since this isoform is not expressed in the CNS, its functional characteristics and role will not be discussed further.

Another isoform was recently characterized and named SNAT8. This carrier demonstrated higher preference for glutamine, alanine, arginine, histidine, and aspartate, using a Na^+-dependent cotransport mechanism of action (Hägglund et al. 2015). When analyzed in the mouse brain, SNAT8 was found expressed in the gray matter, highly expressed in thalamus, hypothalamus, amygdale, and pons, while its expression was lower in hippocampus, cerebellum, and layers of the cortex (Hägglund et al. 2015). This recently discovered isoform is found exclusively expressed in neurons, as demonstrated by in situ hybridization combined with immunohistochemistry (Hägglund et al. 2015). This expression pattern is different from that of SNAT1 and SNAT2, isoforms being present in both neurons and astrocytes. The mechanisms of transport appear similar to those described for system A isoforms, using a Na^+-dependent substrate transport and also showing intolerance to Na^+ substitution by Li^+, as demonstrated in studies using oocytes overexpressing SNAT8 (Hägglund et al. 2015). Interestingly, when flux studies employing MeAIB were performed it was observed that positive currents are generated for this transporter, contrary to the other members for this subfamily. The reason why this isoform specifically is not inhibited by MeAIB is not clear, and due to the transporter substrate profile and mechanism of transport SNAT8 was classified as part of the System A subfamily (Hägglund et al. 2015).

8.2.1.3 System L, Isoforms LAT1 and LAT2

System L is constituted by the carriers LAT1 and LAT2 that facilitate amino acid exchange in a Na^+-independent manner, with affinity toward large neutral amino acids, with a 1:1 stoichiometry for the amino acids (Kanai et al. 1998; Rossier et al. 1999). The light subunit LAT1 gene was the first one among the HATs to be cloned and characterized (Kanai et al. 1998; Mastroberardino et al. 1998). It was isolated from both rats and humans, and the only protein that has been characterized until now consists of 507 amino acids (Kanai et al. 1998; Mastroberardino et al. 1998). The isoform LAT1 is a heterodimer (4F2hc/LAT1) that exerts the exchange of large neutral amino acids in a Na^+- and pH-independent manner, demonstrating high affinity for tryptophan, phenylalanine, leucine, and histidine as substrates (Del Amo et al. 2008). This isoform has lower affinity for glutamine and threonine, and it does not recognize alanine, proline, and charged amino acids (Mastroberardino et al. 1998). In addition, the LAT1 transporter is selectively inhibited by 2-aminobicyclo-(2-2-1)-hepatane-2-caboxylic acid (BCH), which is a nonmetabolizable compound (Meier et al. 2002; Kim et al. 2008). LAT1 is an obligatory exchanger; however, intracellular amino acid concentrations control its activity due to a high intracellular apparent affinity, in comparison to a low extracellular apparent affinity. Therefore, when in combination with other transporter systems, such as System A, it can participate in regulating the amino acid equilibrium between intra and extracellular compartments (Verrey et al. 2004).

LAT1 mRNA is found expressed in the placenta, brain, spleen, testis, colon, blood–brain barrier (Kanai et al. 1998; Wittmann et al. 2015). In the CNS, LAT1 is expressed in the micro-vessels, being responsible for the transport of dihydroxyphenylalanine (L-DOPA) across the blood–brain barrier (Kageyama et al. 2000). In addition, it has been ascribed an important role for the maintenance of large neutral amino acid and neurotransmitter amino acid homeostasis in the inner blood retinal barrier (Tomi et al. 2005). Although the LAT1 isoform has been described in cultured astrocytes and neurons (Deitmer et al. 2003), its significance for the GGC is likely minimal or absent, due to its low affinity toward glutamine as a substrate.

The second isoform LAT2 was isolated in 1999 from both human and rat tissues and consists of a protein with 535 amino acids and 12 putative transmembrane domains (Pineda et al. 1999; Segawa et al. 1999; Rossier et al. 1999). This isoform is also characterized as an exchanger of amino acids, in a Na^+-independent fashion. Contrary to its counterpart, LAT2 is sensitive to more acidic environments (Christensen 1990). In addition, LAT2 has a broad substrate preference, preferring all neutral amino acids, including the small ones alanine, glycine, and serine (Pineda et al. 1999; Bröer and Brookes 2001). Importantly, LAT2 exhibits much higher affinity for glutamine when compared to LAT1 (Rajan et al. 2000; Rossier et al. 1999; Segawa et al. 1999). This isoform, like LAT1, is inhibited by the substrate analog BCH (Kim et al. 2008). LAT2 mRNA was found expressed in kidney, placenta, brain, liver, spleen, skeletal muscle, heart, intestine, and lung (Pineda et al. 1999). As for LAT1, LAT2 is a heterodimer with 4F2hc, which is the heavy subunit responsible for anchoring the transporter in the cytosolic membrane (Rossier et al. 1999). Regarding its expression in neural cells, LAT2 is like LAT1 been described in astrocyte cultures (Nagaraja and Brookes 1996; Deitmer et al. 2003) but also in neurons and in the choroid plexus (Núñez et al. 2014).

8.2.1.4 System y^+L, Isoform y^+LAT2

The isoform y^+LAT2 also belonging to the SLC7 gene family was characterized in 1999 by Torrents and colleagues, from the human cDNA clone number KIAA0245 described in 1996 by Nagase and colleagues (1996). The human protein has 515 amino acids (Pfeiffer et al. 1999) and is predicted to exhibit 12 transmembrane domains (Torrents et al. 1998). In analogy to system L isoforms, y^+LAT2 forms a heterodimer with the heavy chain of 4F2hc via a disulfide linkage, which is related to its coexpression and activity (Kanai et al. 2000; Pfeiffer et al. 1999; Torrents et al. 1998). mRNA for this isoform was found expressed in many tissues in the mouse, such as heart, kidney, small intestine, and parotid but its highest expression was in the brain (Bröer et al. 2000a). It was also demonstrated in the rat brain

(Deitmer et al. 2003). With respect to the cellular localization of y^+LAT2, the isoform has been described in astrocytes, as demonstrated in primary cultures, albeit, the gene expression of y^+LAT2 was less abundant when compared to its counterparts LAT1 and LAT2 (Deitmer et al. 2003). In the in vitro system, y^+LAT2 was also found in cultured neurons (Bröer and Brookes 2001; Bröer et al. 2000a). The mRNA expression of y^+LAT2 has also been verified in the cortex of rats (Zielińska et al. 2011). With respect to the mechanism of transport of this isoform, it works as an exchanger of cationic amino acids and neutral amino acids, the latter being coupled to Na^+. This leads to an electroneutral transport, since the cationic amino acid compensates the charge of the ion flux that accompanies the neutral amino acid (Bröer et al. 2000a).

8.2.2 System ASC, Isoform ASCT2

The isoform ASCT2 was cloned in two independent studies, which isolated the cDNA for this transporter from both murine testis and human placenta (Kekuda et al. 1996; Utsunomiya-Tate et al. 1996). The cDNA for the human homolog encodes a polypeptide containing 541 amino acids that is described to have 10 putative transmembrane domains (Kekuda et al. 1996). ASCT2 is a transporter obligatory working as a Na^+-dependent antiporter of neutral amino acids. Differently from system N, ASCT2 does not tolerate Na^+ substitution for Li^+, and in contrast to system A and L, the substrate analogs MeAIB and BCH do not inhibit this carrier (Pingitore et al. 2013). The ASCT2 transport stoichiometry is described as 1:1 extracellular Na^+-glutamine by intracellular amino acid, but its electrical characteristics of transport are not completely clear, being described as electroneutral or electrogenic (Bröer et al. 2000b). However, recently it was described that the human ASCT2 transporter exerts the inward flux of one glutamine with one Na^+ molecule, leading to an electrogenic transport albeit the authors do not discard the possibility that more than one Na^+ ion might be involved in the transporter activity (Scalise et al. 2014). In fact, Zander and colleagues verified that the rat isoform of ASCT2 exhibited the Na^+/amino acid stoichiometry of at least 2:1, with one sodium ion binding to the amino acid–free apo form of the transporter (Zander et al. 2013).

In this sense, ASCT2 is strongly dependent on Na^+ concentrations at both intracellular and extracellular compartments, but the mechanism of action for its activation is not completely clear (Bröer et al. 2000b; Scalise et al. 2014). It has been described that Na^+ binds to the ASCT2 isoform before the amino acid association, which would favor the catalysis due to structural conformation changes in the transporter (Bröer et al. 2000b). Another study, however, pointed to Na^+ binding to the transporter after the amino acid association is complete, resulting in the complex formation and subsequent amino acids exchange (Grewer and Grabsch 2004). Recently, the effect of intracellular Na^+ was verified as part of a regulatory mechanism, likely promoting the coupling of the transporter subunits or stabilization of the protein during the transition state (Scalise et al. 2014). It is also important to mention that this transporter exhibits substrate-induced anion conductance, which was also described for its family counterparts, the high affinity glutamate transporters (Bröer et al. 2000b). However, the physiological role of this mechanism is still not completely understood.

The preferred substrates for ASCT2 are alanine, serine, and cysteine, giving rise to its name, but also for other neutral amino acids (Utsunomiya-Tate et al. 1996; Bröer et al. 2000b). It was demonstrated that the ASCT2 transporter has asymmetric specificity for certain amino acids, being alanine, cysteine, valine, and methionine only inwardly transported, while glutamine, serine, asparagine, and threonine were transported bidirectionally (Pingitore et al. 2013). The K_m values for glutamine were of 0.097 and 1.8 mM, when measured on the external and internal sides of the proteoliposome in which the transporter was expressed for the functional studies (Pingitore et al. 2013). Regarding its expression, the mRNA for ASCT2 transporter has been described widespread throughout almost all tissues analyzed. Interestingly, earlier studies did not detect its presence in the brain (Utsunomiya-Tate et al. 1996) but subsequent studies proved that this isoform is present in the CNS (Bröer et al. 1999; Gliddon et al. 2009). In the brain, this transporter isoform was found widespread in areas such as cortex, hippocampus, striatum, and cerebellum, and at the cellular level it was described in the neuronal compartment (Gliddon et al. 2009).

8.3 Glutamine Transporters and the Glutamate/GABA–Glutamine Cycle

Transfer of glutamine from astrocytes to neurons is of fundamental importance for maintaining the proper function of glutamatergic neurotransmission (Schousboe et al. 2013; Nissen-Meyer and Chaudhry 2013). This is highlighted by the fact that during neurotransmission glutamate is taken up by astrocytes via the efficient high affinity glutamate transporters of the excitatory amino acid transporter family (EAAT), which terminate the excitatory stimulation (see Schousboe et al. 2013). Hence, neurons rely upon glutamate recycling in the astrocytic compartment for the replenishment of the glutamate neurotransmitter pool, since neurons are unable to perform de novo synthesis of glutamate due to the absence of the anaplerotic enzyme pyruvate carboxylase (Yu et al. 1983; Schousboe et al. 1997; Hertz et al. 1999). In the astrocytes glutamine can be produced either from glutamate taken up from the synaptic cleft or from glutamate synthesized de novo via anaplerosis and conversion of α-ketoglutarate by transamination (Schousboe et al. 2013). The synthesis of glutamine from these glutamate sources depends on the enzyme glutamine synthetase (GS) which is exclusively expressed in astrocytes (Norenberg and Martinez-Hernandez 1979). Subsequently, glutamine is released from the astrocytic compartment and taken up by neurons, where it will give rise to the synthesis of glutamate by the action of phosphate-activated glutaminase (PAG) and serve mainly as a precursor for the neurotransmitter pools of either glutamate or GABA. In the GABAergic synapses, the majority of the inhibitory neurotransmitter is recycled in neurons, but part of the released GABA is taken up by surrounding astrocytes, by the activity of GAT, and therefore lost from the transmitter pool (Schousboe et al. 2013). Therefore, the neurotransmitter portion that is consumed by the other compartment is also compensated by the transfer of glutamine to replenish the neurotransmitter pool of GABA (Schousboe 2003; Patel et al. 2001; Liang et al. 2006). Hence, the glutamine transporters are of critical importance in sustaining the appropriate transfer of this neutral amino acid between the neural compartments. From the astrocytic compartment, glutamine needs to be released to the extracellular space and subsequently be taken up by carriers in the neuronal compartment. Although the exact mechanisms of transport are not completely elucidated, the transporters that have been mainly related to the action of glutamine efflux from glia are SNAT3, SNAT5, LAT2, y+LAT2, and ASCT2 (see Fig. 8.1).

The isoform SNAT3, as previously mentioned, is found confined to the astroglial compartment, specifically at the processes ensheathing glutamatergic and GABAergic synapses, and has been considered one of the major transporters involved in glutamine release from astrocytes (Chaudhry et al. 1999; Boulland et al. 2002). It has been reported that SNAT3 immunostaining was more closely associated with the GABAergic than the glutamatergic synapses, suggesting that this isoform could be linked to the transfer of glutamine from astrocytes to neurons for the synthesis of GABA (Boulland et al. 2002). This is in agreement with the fact that this transporter exhibits the ability of performing flux reversal (Chaudhry et al. 2002a). It has been described that SNAT3 activity is influenced by the intracellular Na^+ concentration and membrane potential, since cells that undergo progressive depolarization showed a switch from glutamine uptake to release. Also, the direction of flux is influenced by the glutamine concentration, efflux being favored at concentrations bellow 400 μM (Bröer et al. 2002; Chaudhry et al. 1999). The importance of SNAT3 for glutamine efflux was demonstrated in cultured astrocytes, since in the absence of extracellular amino acids this transporter was the predominant mediator, but a part was also mediated by a system independent of Na^+ (Deitmer et al. 2003). Interestingly, the presence of extracellular amino acids changed the profile of transport, glutamine efflux being greatly stimulated, indicating the participation of amino acid exchangers for this type of transport. In this sense, LAT2 seems to participate actively in this process, since histidine, leucine, and alanine stimulated glutamine efflux (Deitmer et al. 2003). In addition, the Na^+-dependent transporter isoform ASCT2 exerted an important function for glutamine release in the in vitro environment. Although this transporter mRNA was clearly demonstrated in cultured astrocytes, in previous studies ASCT2 levels were found very low in the murine brain tissue (Deitmer et al. 2003). Low mRNA levels were also observed for the transporter y^+LAT2 (Deitmer et al. 2003). However, a subsequent study described the presence of ASCT2 in astrocyte cultures and cerebral cortex homogenates at the protein level. In addition, it was demonstrated by immunolabeling that this transporter localizes at the dendrites of neurons, and it is absent from the astrocytic compartment (Gliddon et al. 2009). Therefore, ASCT2 might not participate in glutamine efflux from astrocytes, since in the adult brain this isoform is not expressed in this cellular compartment. However, it is abundant in neurons, as will be further discussed.

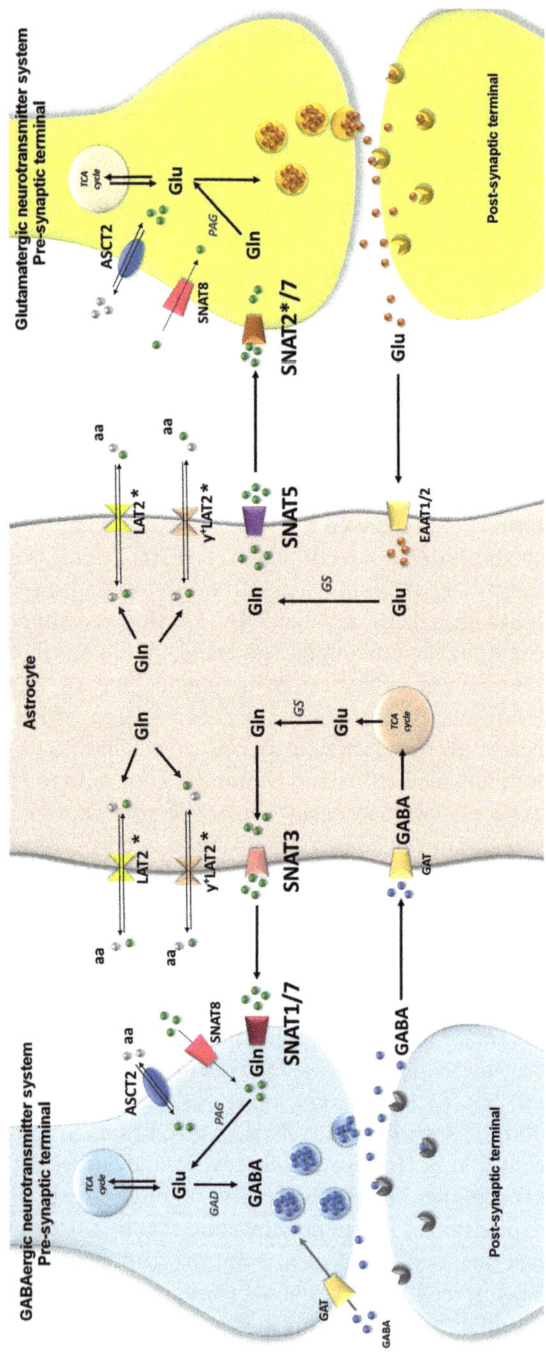
Glutamatergic neurotransmitter system
Pre-synaptic terminal
TCA cycle
Glu
ASCT2
SNAT8
PAG
Gln
SNAT2*/7
Glu
Post-synaptic terminal
Astrocyte
aa
LAT2 *
y⁺LAT2 *
Gln
SNAT5
SNAT3
GS
Glu
EAAT1/2
GABA
GAT
SNAT1/7
GABAergic neurotransmitter system
Pre-synaptic terminal
GAD
GABA
Post-synaptic terminal

Fig. 8.1 Glutamine transporters and the glutamate/GABA–glutamine cycle: Schematic representation of the glutamine transporters involved in the glutamate/GABA–glutamine cycle (GGC). In the glutamatergic synapse, the released glutamate is taken up into astrocytes by the high affinity glutamate transporters (EAAT1/2) and glutamate (Glu) is subsequently amidated by glutamine synthetase (GS), giving rise to glutamine (Gln). Thereafter, glutamine is released to the extracellular space by the action of SNAT5, and subsequently taken up by neurons by the transporter SNAT2. In the neuronal compartment glutamine is deamidated by the activity of phosphate-activated glutaminase (PAG), giving rise to glutamate. In the GABAergic neurotransmitter system, part of GABA released as neurotransmitter is taken up via gamma-aminobutyric acid transporters (GAT) into astrocytes where it will be converted to succinate, which is metabolized in the TCA cycle to α-ketoglutarate, which can be transaminated to glutamate. In the GS catalyzed reaction, glutamate is amidated to form glutamine which is released to the extracellular space by the action of SNAT3, and subsequently taken up by SNAT1 into the neuronal compartment. In the GABAergic neuron, glutamine is deamidated by PAG to form glutamate, which is decarboxylated to GABA by the action of glutamate decarboxylase (GAD) hereby replenishing the GABAergic neurotransmitter pool. Glutamate may in both the glutamatergic and the GABAergic neurotransmitter system additionally serve as an energy substrate and be metabolized in the TCA cycle. The SNAT7 isoform is also described as being responsible for transport glutamine into the neuronal compartment and has been reported to be expressed in both glutamatergic and GABAergic synapses. Its family counterpart SNAT8 might also participate in the same functions. The transporters LAT2 and y^+LAT2 have been related to glutamine efflux from the astrocytic compartment, although their cellular localization remains to be determined. The transporter ASCT2 is expressed in the neuronal compartment in vivo and might also be part of the mechanisms of glutamine uptake. Note that the transporters with an asterisk do not have their cellular localization clarified (LAT2 and y^+LAT2) or are not found at the nerve terminals (SNAT2). *aa* amino acid, *EAAT1/2* excitatory amino acid transporters 1 and 2, *GABA* gamma-aminobutyric acid, *GAD* glutamate decarboxylase, *Glu* glutamate, *Gln* glutamine, *GS* glutamine synthetase, *PAG* phosphate-activated glutaminase, *TCA* tricarboxylic acid

It is important to note that the study performed by Deitmer and coauthors (2003) specifically analyzed the efflux of glutamine from glial cells, while other studies have investigated the profile of glutamine uptake, which was very different with respect to the glutamine transporter isoforms involved in this process (Heckel et al. 2003). Uptake in cultured astrocytes was mediated by, in descending order from the highest to lowest percentage of participation, LAT2, ASCT2, y^+LAT2, and SNAT3 (Heckel et al. 2003). However, the importance of SNAT3 in cultured astrocytes was found different in previous studies, which described a significant participation of this transporter isoform (Nagaraja and Brookes 1996; Su et al. 1997). The discrepancies in these results might be a consequence of the cell culture preparation and experimental protocols employed for these studies. Furthermore, the importance of SNAT3 in relation to the glutamate/GABA–glutamine cycle is highlighted by the observation that exogenous glutamate mediates a decrease in the K_m for this isoform in astrocyte cultures (Bröer et al. 2004).

It may be noted that the observations described earlier represent results from in vitro experiments, and to which degree they are translated into an in vivo condition is unknown. Recently, the importance of SNAT3 has been demonstrated in a mutant mouse model deficient in this SNAT isoform (Chan et al. 2016). SNAT3 deficient mice had a short life span (18–20 days old) and exhibited smaller body size, lower body weight, and ataxia when compared to the wild-type mice (Chan et al. 2016). At the CNS level, the SNAT3 deficient mice showed higher brain glutamine levels, while glutamate and GABA were found reduced, demonstrating that this transporter is fundamental for sustaining the GGC for the replenishment of the neurotransmitter pools (Chan et al. 2016). In addition, an increase in SNAT1 and LAT1 expression was observed in these animals, likely as a compensatory mechanism due to the higher concentrations of glutamine (Chan et al. 2016).

With respect to the isoform SNAT5, its role for glutamine efflux is less well understood, but it has also been regarded as a transporter responsible for glutamine efflux from astrocytes. It was shown that its distribution paralleled that of the vesicular glutamate transporter vGLUT1 in the forebrain of mice (Cubelos et al. 2005). Therefore, SNAT5 has been related to the glutamatergic synapses, while SNAT3 correlated more closely with the distribution of GABAergic synapses (Cubelos et al. 2005; Rodríguez et al. 2014). In this regard, SNAT5 demonstrates partial overlap with SNAT3 according to its distribution within the brain, but also exhibits complementary distribution depending on the brain region. SNAT5 is enriched in the hippocampus and striatum, where SNAT3 has a lower expression, whereas it has low immunoreactivity in the diencephalon, brainstem, and cerebellum that are brain areas with high expression of SNAT3 (Cubelos et al. 2005; Hamdani et al. 2012). In addition, it has been described that SNAT5 is related to the modulation of glutamatergic neurotransmission by the release of glycine. This carrier isoform has under certain conditions the property of releasing glycine, which serves as a cotransmitter at NMDA receptors thereby influencing synaptic plasticity in the CNS (Hamdani et al. 2012).

Once at the extracellular space, glutamine needs to be taken up and subsequently internalized in the neuronal compartment in order to complete the GGC (see Fig. 8.1). The transporters that have been shown to be the main facilitators of this process are SNAT1, SNAT2, and SNAT7 (Chaudhry et al. 2002b; Blot et al. 2009; Hägglund et al. 2011). The isoform SNAT1 is considered one of the main transporters responsible for glutamine uptake in the neuronal compartment due to its characteristics of concentrative, Na^+-dependent transport. This isoform was found in GABAergic and glutamatergic neurons, but it is also expressed in the astrocytic compartment (Mackenzie et al. 2003; Melone et al. 2004; Varoqui et al. 2000; Solbu et al. 2010). In the adult murine brain, SNAT1 has been found in the neuronal cell body region as well as neuronal processes including axons at or near synaptic terminals (Weiss et al. 2003; Solbu et al. 2010; Brown and Mathews 2010). With respect to function, it appears that SNAT1 is mostly related to the GABAergic synapses, since it was demonstrated in brain regions that are enriched in GABAergic neurons in proximity to vGAT. In addition, the pattern of labeling for SNAT1 in GABAergic neurons resembles that for glutamate decarboxylase (GAD) responsible for the synthesis of GABA from the precursor glutamate (Solbu et al. 2010). The pattern of distribution of SNAT1 and its relation to the vesicular GABA transporter could be an indication of its relation with glutamine uptake as a prerequisite to replenish the GABA neurotransmitter pool (Solbu et al. 2010; Varoqui et al. 2000). This is in agreement with the observation that glutamine substantially contributes to the vesicular and cytoplasmatic pools of GABA in cultures of cerebral cortical neurons (Waagepetersen et al. 2001), and that inhibition of glutamine uptake by MeAIB reduces inhibitory synaptic transmission in rat hippocampal CA1 pyramidal neurons, indicating that synaptic vesicle content of GABA was reduced (Fricke et al. 2007). In line with this, it has been shown that MeAIB blocked the increase of inhibitory postsynaptic potential (mIPSC) amplitude following depolarization of hippocampal slices (Brown and Mathews 2010). These studies underline the importance of system A activity, in this case most likely the SNAT1 transporter, for the replenishment of vesicular GABA.

SNAT2, on the other hand, being ubiquitously expressed, is localized mainly in glutamatergic neurons, apparently in a manner complementary to the expression of SNAT1 (González-González et al. 2005; Jenstad et al. 2009; Solbu et al. 2010). As an example, Jenstad and coauthors (2009) described that SNAT2 immunolabeling is barely detectable in interneurons, while pyramidal neurons and hilar mossy cells, all having glutamatergic phenotypes, are clearly SNAT2 positive. In the cerebral structures of neocortex and hippocampus, high levels of SNAT1 expression are described in dispersed GABAergic interneurons, while adjacent pyramidal neurons lack immunoreactivity (Jenstad et al. 2009). Regarding the cellular localization of SNAT2 in the neuronal compartment, it has been verified that this isoform is present at the somatodendritic region of glutamatergic neurons.

Previously the participation of SNAT2 in the GGC has been questioned, since its cellular localization apart from the synaptic terminal is not in accordance with the concept of a fast and efficient glutamate neurotransmitter recycling system, and it has been suggested that SNAT2 participates in taking up glutamine for other reasons than replenishing the neurotransmitter pool (Melone et al. 2006; Grewal et al. 2009). However, Jenstad and coauthors (2009) showed synthesis of glutamate at the expense of glutamine in the nerve endings of electrically stimulated cells from the perforant path, although they also failed to demonstrate SNAT2 expression at the nerve terminals of glutamatergic neurons (Jenstad et al. 2009). In addition, MeAIB led to depletion of glutamate, and therefore the authors suggested that SNAT2 at the dendrites might be supplying glutamatergic nerve terminals with glutamine, but the participation of other transporters in this process was not excluded. Moreover, SNAT2 has been related to glutamate retrograde signaling, which modulates neurotransmission and therefore synaptic plasticity (Jenstad et al. 2009).

The isoform SNAT7, which is part of system N, is widely expressed in both glutamatergic and GABAergic neurons in the mouse brain, while absent from astrocytes. At the cellular level, this isoform is distributed in the neuronal soma and axon, and studies of colocalization failed to demonstrate its presence at the neuronal terminals (Hägglund et al. 2011). It is possible that SNAT7 could also play a role in the GGC by acting in the uptake and recycling of glutamate/GABA (Hägglund et al. 2011), but further investigation is necessary to better understand the importance of this SNAT isoform in this context. As its family counterpart SNAT7, SNAT8 has not yet its role very well defined in the GGC. It is exclusively expressed in the neuronal compartment in the cell bodies and axons of both excitatory and inhibitory neurons (Hägglund et al. 2015). Interestingly, SNAT7 and SNAT8 were found colocalized in the same neuronal compartment and subcellular region, being apart by 40 nm or less of each other (Hägglund et al. 2015). The significance of this transporter interaction is not clear at the moment. While it has been suggested that both transporters might be important for internalization of glutamine as an energy source and participation in general biosynthesis, their function and interaction in the GGC remain to be elucidated (Hägglund et al. 2011, 2015).

Recently, another transporter has been functionally characterized in hippocampal neurons, and contrary of those described for the SNAT family of transporters, this carrier exhibits a Ca^{2+}-dependent glutamine transport activity and it is inhibited by P type voltage-gated Ca^{2+} channel blockers. Moreover, it saturates at 200 μM and has a higher affinity for glutamine (K_m 30 ± 4 μM) when compared to SNAT 1 and 2 neuronal isoforms (app 1.5–0.4 mM range). This recent discovery brings a new participant for the GGC, which contributes with the concept of a fast and concentrative transporter for the replenishment of the neurotransmitter pools of glutamate and GABA (Erickson 2016).

As previously mentioned, the isoform ASCT2, which was earlier considered important for astrocyte glutamine efflux, is absent from this cellular compartment, but largely expressed in neurons, as verified by immunohistochemistry in the adult mouse brain. This transporter isoform is found in neuronal areas that are anti-microtubule-associated protein-2 positive (MAP2), suggesting that its localization may be restricted to the dendrites, while absent in the cell bodies. In addition, ASCT2-mediated low affinity Na^+-dependent uptake of D-serine and L-glutamine was observed in synaptosomal preparations. These findings indicate that ASCT2 may be an important neuronal neutral amino acid transporter (Gliddon et al. 2009).

Therefore, as exemplified in Fig. 8.1, the transporters that have been related to glutamine uptake in the astrocytic compartment are SNAT3 and SNAT5 mostly expressed in GABAergic and glutamatergic synapses, respectively. While at the neuronal compartment, the isoforms SNAT1, SNAT2, SNAT7, and likely SNAT8 are related to glutamine uptake. SNAT1 and SNAT2 are found in the inhibitory and excitatory neurons, respectively, whereas SNAT7 and SNAT8 are expressed at both neuronal systems. It is important to mention that SNAT1 and SNAT7 are found expressed close to the synaptic cleft (Armano et al. 2002; Hägglund et al. 2011), which is in agreement with the concept of the glutamate/GABA–glutamine cycle but they are also to a considerable extent present at the soma regions of the neurons. SNAT2 is rarely found at the axonal terminals, but is abundant at the somato-dendritic regions of the neuronal compartment (Jenstad et al. 2009). SNAT8 is also localized at the soma and axonal region of the neuronal compartment, albeit its importance for the GGC is still unknown. The role of ASCT2 at the neuronal compartment and the activity of system LAT and y^+LAT remains to be better elucidated.

8.4 Regulation of Glutamine Transporter Isoforms

Glutamine transporters are regulated by different mechanisms at translational as well as transcriptional levels. In the following section, this will be described for those isoforms which play important roles in the GGC, i.e., those that belong to the SNAT, ASC, and LAT family of proteins. As previously mentioned, the SNAT family of transporters is considered fundamental for the transfer of glutamine between the neuronal and astrocytic compartments. Therefore, their function is regulated by several mechanisms that modulate the expression of the transporters as well as their function. In line with this, the first mechanism for modulating the transporter activity is affecting their trafficking across the cytoplasmic membrane, which will alter the availability of the transporter at the plasma membrane and thereby increase or decrease the influx/efflux of glutamine between the intracellular and extracellular spaces. This mechanism has been well characterized for SNAT2, which is present in both plasma and internal membrane fractions from rat adipose and muscle tissues (Hyde et al. 2001, 2002), and the intracellular pool of this isoform functions as a reservoir of the transporter that can quickly respond to a higher demand. In accordance with this, it has been described that increased recruitment of SNAT2 from the cytoplasm to the plasma membrane was associated with an acute stimulus, namely, the absence of amino acids in the media (Ling et al. 2001). Moreover, in a chronic phase, amino acid starvation led to an increase in gene expression, enhancing de novo synthesis of SNAT2. This phenomenon resulted in a larger transporter activity when compared to the acute phase (Ling et al. 2001). In addition, the stability of the SNAT2 protein appears to participate in the regulation of the transporter function. This is shown by the finding that HeLa cells expressing the CMV-driven SNAT2 construct exhibited more stable SNAT2 protein during amino acid restriction. Furthermore, L6 myotubes treated with the protein synthesis inhibitor cycloheximide exhibited decreased system A activity when these cells were incubated in a medium containing amino acids, in contrast to what was observed after incubation in an amino acid free medium, suggesting that SNAT2 is stabilized in an amino acid-dependent manner (Hyde et al. 2007). This mechanism seems to be related to a domain located at the N terminus of the transporter (Hyde et al. 2007). Besides these adaptive regulatory mechanisms, SNAT2 expression is stimulated by hormones, such as insulin, growth factors, as well as by cell stress (Guma et al. 1988; McDowell et al. 1998; Hundal et al. 1994). It has been shown that insulin increases System A transport in L6 cells by stimulating the exocytosis of SNAT2 carriers from an endosomal compartment, an event that is related to the activation of phosphatidylinositol 3-kinase (PI3K) and protein kinase B (PKB) (Hyde et al. 2002). Another parameter that influences SNAT2 availability/activity is its rate of degradation that has been described to be regulated by polyubiquitination by Nedd4-2, a specific ubiquitin ligase with HECT domain, leading to endocytosis of the transporter and subsequent proteoasomal degradation (Hatanaka et al. 2006).

The SNAT1 isoform has also been shown to be regulated at the protein level, as demonstrated by an increased transporter activity in synaptosomes and enhanced inhibitory synaptic strength as a result of an increase in metabolic demand. This may be very important for modulation of GABAergic transmission (Brown and Mathews 2010).

For those isoforms that are responsible for glutamine efflux, posttranslational regulation has been described for SNAT3 (Karinch and Lin 2002; Nissen-Meyer and Chaudhry 2013). In analogy to SNAT2, it was demonstrated in vitro and in vivo that SNAT3 is upregulated in hepatocytes by dietary restriction and serum deprivation (Gu 2005). Moreover, acute treatment of hepatocytes with insulin stimulates the recruitment of SNAT3 to the plasma membrane resulting in an increase in substrate uptake. On the other hand, chronic stimulation by this hormone led to down-regulation of the expression, a mechanism involving PI3K (Gu 2005). In addition, SNAT3 exhibits different putative phosphorylation sites that are involved in down-regulation of the activity of the transporter. In keeping with this, activation of PKC by phorbol-12-myristate-13-acetate (PMA) leads to down-regulation of SNAT3 in *Xenopus laevis* oocytes (Balkrishna et al. 2010). However, this study showed that the reduced activity of SNAT3 was not a consequence of direct phosphorylation of the transporter, but rather the result of trafficking and internalization of the transporter in a caveolin-dependent manner (Balkrishna et al. 2010). In addition, other studies explored the effect of phosphorylation and demonstrated that SNAT3 is directly phosphorylated by the protein kinases PKCα, PKCγ, and PKCδ (Sidoryk-Wegrzynowicz et al. 2011; Nissen-Meyer et al. 2011). More specifically, PKC phosphorylates SNAT3 at one single serine residue at the N-terminal of the transporter, leading to sequestration of the transporter to intracellular reservoirs, which consequently regulates the transporter activity. In addition, prolonged activation of PKC results in the degradation of the SNAT3 transporter (Nissen-Meyer et al. 2011). In analogy to SNAT2, SNAT3 is regulated by degradation in a ubiquitin ligase Nedd4-2 manner (Boehmer et al. 2003). This appears to be regulated by the pathways involving serum and glucocorticoid inducible kinase (SGK) and PKB, which reverted the Nedd4-2 induced degradation of SNAT3 (Boehmer et al. 2003).

An alternative form of regulation of SNAT isoforms involves changes in pH. As mentioned earlier, SNAT proteins are sensitive to changes in the pH, having their activity decreased in more acidic environments, while at more alkaline pH they exhibit higher activity.

The SNAT isoforms are also tightly regulated at the transcriptional level, which consequently influences the availability of the transporters. SNAT2 transcriptional regulation has been extensively studied and several factors affect its expression, such as hypertonic conditions, amino acid deprivation, and the availability of hormones (Ling et al. 2001; Palii et al. 2004; Franchi-Gazzola et al. 2004; Kashiwagi et al. 2009). Moreover, osmotic shock which is stressful to cells has been demonstrated to result in increased mRNA levels of SNAT2 (Franchi-Gazzola et al. 2004; Kashiwagi et al. 2009). Activation of mitogen-activated protein kinase (MAPK) cascades, such as extracellular regulated kinase (ERK) is possible intracellular signaling pathways involved in this process (Matsuda et al. 1995). The intracellular signaling pathway involved in the stimulation of mRNA synthesis seems to depend upon the cell type, since other studies with L6 cells demonstrated that the up-regulation of the transcription of SNAT2 transporter did not involve PI3K, ERK, p38, and JNK pathways (Kashiwagi et al. 2009), while in CHO-K1 cells, the effect of hyperosmotic conditions was partially blocked by inhibition of p38 (Lopez-Fontanals 2003).

Amino acid deprivation has also been described to increase mRNA levels for SNAT2 and subsequently protein levels (Kashiwagi et al. 2009; Lopez-Fontanals 2003). This was verified when inhibitors of mRNA synthesis suppressed the increase in the transporter expression, demonstrating that the increase in the transporter was related to de novo synthesis of the transporter (Kashiwagi et al. 2009). It is known that amino acid deprivation involves the activation of a pathway called amino acid response (AAR), leading to the phosphorylation of the translation initiation factor eIF2α (eukaryotic initiation factor 2α), which consequently promotes increased translation of a selected number of mRNAs, including activating transcription factor 4 (ATF4) (Harding et al. 2000). ATF4, in turn, increases the transcription from a subset of target genes, including those related to AAR elements (AARE) that mediate the enhanced transcription (Kilberg et al. 2005). In relation to this, the AARE site has been identified in the SCL38A2 gene, which is located at intron 1 (Palii et al. 2004). This region together with a CAAT box is responsible for mediating the amino acid response as an enhancer, while another site, called Purine rich box acts as an inhibitor of gene transcription (Palii et al. 2004). In addition, it has been described that multiple transcription factors bind to the SNAT2 AARE and thereby contribute to the amino acid-dependent regulation, being a key role demonstrated for ATF4, likely a component that triggers the mechanism for the SNAT2 activation (Palii et al. 2006). In addition, other signaling cascades that appear to be involved in the increased transcription of SNAT2 under amino acid deprivation are ERK and JNK (Kashiwagi et al. 2009; Lopez-Fontanals 2003). Insulin has also an effect on stimulating the transcription of SNAT2, being the signaling pathway of this event related to P38 in L6 muscle cells (Kashiwagi et al. 2009).

Increases in mRNA expression have also been demonstrated for the System N isoform SNAT3 during metabolic acidosis in the cortex of kidney of acidotic rats. However, the mechanisms related to this increase in translation for SNAT3 are not completely clear but they might involve the pH-response element present in the 3′-untranslated region (UTR), and/or mRNA stabilization (Karinch and Lin 2002). Also, glucocorticoids, directly or indirectly regulate the expression of the SNAT3 transporter during chronic metabolic acidosis (Karinch et al. 2007).

It has also been described that glutamine itself can stimulate ASCT2 transporter expression in human hepatoma cells by indirectly inducing the expression of Farnesoid X receptor (FXR), which binds to an IR-1 sequence on the ASCT2 promoter region (Bungard and McGivan 2005). This leads to an increase in expression of ASCT2 mRNA and protein thereby increasing glutamine uptake, as observed in HepG2 cells (Bungard and McGivan 2005). Moreover, the signaling pathways SGK1 and SGK3 modulate ASCT2 by increasing the transporter abundance in the plasma membrane (Palmada et al. 2005). In addition, a role for the EGF signaling pathway has also been suggested. Insulin like growth factor (IGF) has an effect on stimulating the transporter activity, involving the PI3K signaling pathway (Karlsen and Serck-Hanssen 2002). Furthermore, ASCT2 is modulated by pH, its activity being increased in more acidic environments (Bröer et al. 1999).

With respect to the system L isoform LAT2, it was shown that its activity and expression were increased in skeletal muscle fibers by the hormone dihydrotestosterone (DHT). This regulation is likely mediated by epidermal growth factor receptor (EGFR) involving the activation of ERK1/2 module of the MAPK pathway (Hamdi and Mutungi 2011).

It is important to mention, however, that the regulatory mechanisms described earlier were studied in different cell types and experimental models, and for the different stimuli, the response and pathway involved in these processes vary according to the cell type studied. Therefore, the mechanisms that regulate glutamine transporters in neurons and astrocytes may be identical or similar to those that have been mentioned here but it should be emphasized that neural cells might exhibit distinct regulation involving other mechanisms.

8.5 Glutamine Transporters in Neurologic Diseases

Altered function of the GGC has been linked to the pathophysiological mechanisms of different neurological disorders, such as epilepsy, Alzheimer's disease, and stroke (Walton and Dodd 2007; Coulter and Eid 2012; Jeitner et al. 2015). In the case of epilepsy, some studies have described a decrease in the high affinity glutamate transporters expression in patients with temporal lobe epilepsy and decreased expression and activity of GS (Mathern et al. 1999; Proper et al. 2002). Patients with Alzheimer's disease also present disturbances in GS, which has been shown to be reduced at the vascular end-feet of astrocytes in the cortical brain region (Robinson 2001), as well as the function of astrocytic glutamate transporters (Li et al. 1997).

The involvement of GS in neurological disorders is further elaborated on in another chapter by Jayakumar and Norenberg (2016).

It is clear that defects in key steps of the GGC, as glutamate uptake by astrocytes and its conversion to glutamine by GS will have a significant impact on the performance of the cycle thereby affecting neurotransmitter systems. Nevertheless, the role of glutamine transporters in neurological diseases is sometimes overlooked. Although as discussed earlier, the glutamine transporter family consists of members having a great variety of properties, some isoforms share similar functional characteristics and have a widespread expression in the brain tissue. It should therefore be expected that they play important roles in the mechanisms underlying a variety of neurologic disorders.

In this regard disturbances in different glutamine transporter isoforms have been described during conditions of hyperammonemia, such as that associated with hepatic encephalopathy (HE). Increases in the ammonia concentration in the blood and CNS are a hallmark of HE, a neurologic disease that occurs due to the reduced ability of the liver to synthesize urea for the clearance of circulating blood ammonia (Albrecht and Jones 1999; Butterworth 2002). As a consequence, increased ammonia levels reach the CNS, which lacks a metabolizing system for ammonia disposal as urea. In the brain the main system responsible for ammonia detoxification is constituted by the astrocyte-specific enzyme GS catalyzing the ATP-dependent synthesis of glutamine from glutamate and ammonia. As a result, glutamine synthesis, exclusively occurring in astrocytes, is increased during this condition and since most histopathological findings of HE are found in this cell type, HE has been designated as a gliopathy (Rama Rao and Norenberg 2014; Norenberg 1998). Extracellular glutamine can be transferred from the CNS to the periphery. However, this event together with the detoxification via GS is not enough to prevent the pathological consequences, which lead to several disturbances in various neurotransmitter systems and brain energy metabolism (Butterworth 2000; Leke et al. 2011a, b; Schousboe et al. 2014b).

As mentioned earlier, HE leads to altered GS activity as well as glutamate and GABA transporter activity (Butterworth 2000; Suarez et al. 2002) and importantly an altered expression of glutamine transporters. It has been demonstrated that acute HE leads to a diminished gene expression for SNAT5, although other studies described unaltered expression at both mRNA and protein levels (Desjardins et al. 2012). Decreases in the astrocytic transporter would not be in agreement with the pathophysiological findings of acute HE, i.e., increased levels of glutamine at the extracellular space (Swain et al. 1992). On the contrary, in animals with chronic HE resulting from bile duct ligation, SNAT5 mRNA levels were found increased, which could be an adaptive response due to the increased intracellular concentrations of glutamine. However, the protein levels remained unchanged during this condition (Leke et al. 2015). The other astrocytic isoform SNAT3 has been described to be augmented in the brain tissue of rats with acute HE due to liver failure induced by thioacetamide (Zielińska et al. 2014). In addition, an increase in the brain expression and activity of y^{+}LAT2 has been documented in rats acutely exposed to hyperammonemia (Zielińska et al. 2011). These authors suggest that the increased y^{+}LAT2 function leads to enhanced efflux of arginine in exchange of extracellular glutamine, events that are linked to an altered NO/cGMP pathway (Zielińska et al. 2011). Although the authors described that the decrease in y⁺LAT2 is related to the defects in astrocytes (Skowrońska et al. 2012), it is still needed to elucidate in which cellular compartment y⁺LAT2 is expressed in vivo.

Manganism is another neurologic disease that alters the function of GGC. It is characterized by accumulation of manganese (Mn) due to food intoxication or occupational exposure, leading to CNS toxicity, which culminates in a neurological brain disorder characterized by symptoms of idiopathic Parkinson's disease (Sidoryk-Wegrzynowicz and Aschner 2013). Intoxication with Mn is linked to dysfunction of different steps of GGC, such as down-regulation of GS and decreased expression of the glutamate transporters GLT-1 and GLAST (Erikson et al. 2008). A decrease in the expression of the GABA transporter GAT1 has also been shown in cerebral tissue of rats, together with increases in the extracellular GABA concentration in striatum of Mn-exposed rats (Brouillet et al. 1993). In this regard and in analogy with HE, astrocytes are the main target for Mn intoxication (Aschner et al. 1992). As a result, decreased levels for SNAT3, SNAT2, and LAT2 at mRNA and protein levels in astrocyte cultures have been demonstrated (Sidoryk-Węgrzynowicz et al. 2009). Interestingly, it was shown that SNAT3 protein during Mn exposure is completely degraded, a process that could be a result of increased activity of the ubiquitin-mediated proteolytic system (Sidoryk-Wegrzynowicz et al. 2010).

It is likely that other neurological conditions also may lead to defects in glutamine transporter expression and activity. However, the role of glutamine transporters in the pathological mechanisms of different neuropathologies, in which the dysfunction of the excitatory and/or inhibitory neurotransmitter systems and likely GGC is acknowledged, is often disregarded. Hence, for better understanding of the role of glutamine transporters in physiological and pathological conditions it appears necessary to elucidate in greater detail their role and participation in the GGC.

8.6 Concluding Remarks

It is clear from the presentation earlier that glutamine transport in the brain is extremely complex being mediated by an extraordinarily large number of different isoforms of transporters belonging to several transporter families. This complexity extends to the regulatory mechanisms and altogether this indicates that this transport capacity is of remarkable importance for brain function. Since the amino acid neurotransmitter systems account for a significant fraction of all neurotransmission and the transfer of glutamine between astrocytes and neurons is obligatory for its function, it is not surprising that the transporters responsible for this are numerous and highly regulated. Extensive research in this area will unquestionably lead to a better understanding of the mechanism of action of these transporters providing new therapeutic avenues for treatment of neurological disorders in which malfunction of the glutamine transporters is involved.

References

Aasly J, Gårseth M, Sonnewald U, Zwart JA, White LR, Unsgård G (1997) Cerebrospinal fluid lactate and glutamine are reduced in multiple sclerosis. Acta Neurol Scand 95:9–12

Albers A, Bröer A, Wagner CA, Setiawan I, Lang PA, Kranz EU, Lang F, Bröer S (2001) Na^+ transport by the neural glutamine transporter ATA1. Pflugers Arch 443:92–101

Albers T, Marsiglia W, Thomas T, Gameiro A, Grewer C (2012) Defining substrate and blocker activity of alanine-serine-cysteine transporter 2 (ASCT2) ligands with novel serine analogs. Mol Pharmacol 81:356–365

Albrecht J, Jones EAA (1999) Hepatic encephalopathy: molecular mechanisms underlying the clinical syndrome. J Neurol Sci 170:138–146

Armano S, Coco S, Bacci A, Pravettoni E, Schenk U, Verderio C, Varoqui H, Erickson JD, Matteoli M (2002) Localization and functional relevance of system a neutral amino acid transporters in cultured hippocampal neurons. J Biol Chem 277:10467–10473

Arriza JL, Kavanaugh MP, Fairman WA, Wu YN, Murdoch GH, North RA, Amara SG (1993) Cloning and expression of a human neutral amino acid transporter with structural similarity to the glutamate transporter gene family. J Biol Chem 268:15329–15332

Aschner M, Gannon M, Kimelberg HK (1992) Manganese uptake and efflux in cultured rat astrocytes. J Neurochem 58:730–735

Bak LK, Schousboe A, Waagepetersen HS (2006) The glutamate/GABA-glutamine cycle: aspects of transport, neurotransmitter homeostasis and ammonia transfer. J Neurochem 98:641–653

Balkrishna S, Broer A, Kingsland A, Broer S (2010) Rapid downregulation of the rat glutamine transporter SNAT3 by a caveolin-dependent trafficking mechanism in Xenopus laevis oocytes. Am J Physiol Cell Physiol 299:C1047–C1057

Blot A, Billups D, Bjørkmo M, Quazi AZ, Uwechue NM, Chaudhry FA, Billups B (2009) Functional expression of two system A glutamine transporter isoforms in rat auditory brainstem neurons. Neuroscience 164:998–1008

Bode BP (2001) Recent molecular advances in mammalian glutamine transport. J Nutr 131:2475s–2485s

Boehmer C, Okur F, Setiawan I, Bröer S, Lang F (2003) Properties and regulation of glutamine transporter SN1 by protein kinases SGK and PKB. Biochem Biophys Res Commun 306:156–162

Boll M, Daniel H, Gasnier B (2004) The SLC36 family: proton-coupled transporters for the absorption of selected amino acids from extracellular and intracellular proteolysis. Pflugers Arch 447:776–779

Boulland J-LL, Osen KK, Levy LM, Danbolt NC, Edwards RH, Storm-Mathisen J, Chaudhry FA (2002) Cell-specific expression of the glutamine transporter SN1 suggests differences in dependence on the glutamine cycle. Eur J Neurosci 15:1615–1631

Boulland J-L, Rafiki A, Levy LM, Storm-Mathisen J, Chaudhry FA (2003) Highly differential expression of SN1, a bidirectional glutamine transporter, in astroglia and endothelium in the developing rat brain. Glia 41:260–275

Bröer S, Brookes N (2001) Transfer of glutamine between astrocytes and neurons. J Neurochem 77:705–719

Bröer A, Brookes N, Ganapathy V, Dimmer KS, Wagner CA, Lang F, Bröer S (1999) The astroglial ASCT2 amino acid transporter as a mediator of glutamine efflux. J Neurochem 73:2184–2194

Bröer A, Wagner CA, Lang F, Bröer S (2000a) The heterodimeric amino acid transporter 4F2hc/y+LAT2 mediates arginine efflux in exchange with glutamine. Biochem J 349:787–795

Bröer A, Wagner C, Lang F, Bröer S (2000b) Neutral amino acid transporter ASCT2 displays substrate-induced Na^+ exchange and a substrate-gated anion conductance. Biochem J 346:705–710

Bröer A, Albers A, Setiawan I, Edwards RH, Chaudhry FA, Wagner CA, Bröer S (2002) Regulation of the glutamine transporter SN1 by extracellular pH and intracellular sodium ions. J Physiol 539:3–14

Bröer A, Deitmer JW, Bröer S (2004) Astroglial glutamine transport by system N is upregulated by glutamate. Glia 48:298–310

Bröer S, Schneider H-P, Bröer A, Deitmer JW (2009) Mutation of asparagine 76 in the center of glutamine transporter SNAT3 modulates substrate-induced conductances and Na^+ binding. J Biol Chem 284:25823–25831

Brookes N (1992) Effect of intracellular glutamine on the uptake of large neutral amino acids in astrocytes: concentrative Na (+)-independent transport exhibits metastability. J Neurochem 59:227–235

Brouillet EP, Shinobu L, McGarvey U, Hochberg F, Beal MF (1993) Manganese injection into the rat striatum produces excitotoxic lesions by impairing energy metabolism. Exp Neurol 120:89–94

Brown MN, Mathews GC (2010) Activity- and age-dependent modulation of GABAergic neurotransmission by system A-mediated glutamine uptake. J Neurochem 114:909–920

Bungard CI, McGivan JD (2005) Identification of the promoter elements involved in the stimulation of ASCT2 expression by glutamine availability in HepG2 cells and the probable involvement of FXR/RXR dimers. Arch Biochem Biophys 443:53–59

Butterworth RF (2000) Hepatic encephalopathy: a neuropsychiatric disorder involving multiple neurotransmitter systems. Curr Opin Neurol 13:721–727

Butterworth RF (2002) Pathophysiology of hepatic encephalopathy: a new look at ammonia. Metab Brain Dis 17:221–227

Chan K, Busque SM, Sailer M, Stoeger C, Bröer S, Daniel H, Rubio-Aliaga I, Wagner CA (2016) Loss of function mutation of the Slc38a3 glutamine transporter reveals its critical role for amino acid metabolism in the liver, brain, and kidney. Pflügers Arch Eur J Physiol 468:213–227

Chaudhry FA, Reimer RJ, Krizaj D, Barber D, Storm-Mathisen J, Copenhagen DR, Edwards RH (1999) Molecular analysis of system N suggests novel physiological roles in nitrogen metabolism and synaptic transmission. Cell 99:769–780

Chaudhry FA, Krizaj D, Larsson P, Reimer RJ, Wreden C, Storm-Mathisen J, Copenhagen D, Kavanaugh M, Edwards RH (2001) Coupled and uncoupled proton movement by amino acid transport system N. EMBO J 20:7041–7051

Chaudhry FA, Reimer RJ, Edwards RH (2002a) The glutamine commute: take the N line and transfer to the A. J Cell Biol 157:349–355

Chaudhry FA, Schmitz D, Reimer RJ, Larsson P, Gray AT, Nicoll R, Kavanaugh M, Edwards RH (2002b) Glutamine uptake by neurons: interaction of protons with system a transporters. J Neurosci 22:62–72

Christensen HN (1990) Role of amino acid transport and countertransport in nutrition and metabolism. Physiol Rev 70:43–77

Christensen HN, Oxender DL, Liang M, Vatz KA (1965) The use of N-methylation to direct route of mediated transport of amino acids. J Biol Chem 240:3609–3616

Christensen HN, Liang M, Archer EG (1967) A distinct Na^+-requiring transport system for alanine, serine, cysteine, and similar amino acids. J Biol Chem 242:5237–5246

Collarini EJ, Oxender DL (1987) Mechanisms of transport of amino acids across membranes. Annu Rev Nutr 7:75–90

Coulter DA, Eid T (2012) Astrocytic regulation of glutamate homeostasis in epilepsy. Glia 60:1215–1226

Cubelos B, González-González IM, Giménez C, Zafra F (2005) Amino acid transporter SNAT5 localizes to glial cells in the rat brain. Glia 49:230–244

Deitmer JW, Bröer A, Bröer S (2003) Glutamine efflux from astrocytes is mediated by multiple pathways. J Neurochem 87:127–135

Del Amo EM, Urtti A, Yliperttula M (2008) Pharmacokinetic role of L-type amino acid transporters LAT1 and LAT2. Eur J Pharm Sci 35:161–174

Desjardins P, Du T, Jiang W, Peng L, Butterworth RF (2012) Pathogenesis of hepatic encephalopathy and brain edema in acute liver failure: role of glutamine redefined. Neurochem Int 60:690–696

Erecińska M, Silver IA (1990) Metabolism and role of glutamate in mammalian brain. Prog Neurobiol 35:245–296

Erickson J (2016) Functional identification of an activity-induced, Ca^{2+}-dependent glutamine transporter in hippocampal neurons. Trans Am Soc Neurochem 47:68

Erikson KM, Dorman DC, Lash LH, Aschner M (2008) Duration of airborne-manganese exposure in rhesus monkeys is associated with brain regional changes in biomarkers of neurotoxicity. Neurotoxicology 29:377–385

Fei YJ, Sugawara M, Nakanishi T, Huang W, Wang H, Prasad PD, Leibach FH, Ganapathy V (2000) Primary structure genomic organization, and functional and electrogenic characteristics of human system N 1, a Na^+- and H^+ -coupled glutamine transporter. J Biol Chem 275:23707–23717

Fort J, De La Ballina LR, Burghardt HE, Ferrer-Costa C, Turnay J, Ferrer-Orta C, Usón I et al (2007) The structure of human 4F2hc ectodomain provides a model for homodimerization and electrostatic interaction with plasma membrane. J Biol Chem 282:31444–31452

Fotiadis D, Kanai Y, Palacín M (2013) The SLC3 and SLC7 families of amino acid transporters. Mol Aspects Med 34:139–158

Franchi-Gazzola R, Gaccioli F, Bevilacqua E, Visigalli R, Dall'Asta V, Sala R, Varoqui H, Erickson JD, Gazzola GC, Bussolati O (2004) The synthesis of SNAT2 transporters is required for the hypertonic stimulation of system A transport activity. Biochim Biophys Acta Biomembr 1667:157–166

Fricke MN, Jones-Davis DM, Mathews GC (2007) Glutamine uptake by system A transporters maintains neurotransmitter GABA synthesis and inhibitory synaptic transmission. J Neurochem 102:1895–1904

Gliddon CM, Shao Z, LeMaistre JL, Anderson CM (2009) Cellular distribution of the neutral amino acid transporter subtype ASCT2 in mouse brain. J Neurochem 108:372–383

González-González IM, Cubelos B, Giménez C, Zafra F (2005) Immunohistochemical localization of the amino acid transporter SNAT2 in the rat brain. Neuroscience 130:61–73

Grewal S, Defamie N, Zhang X, de Gois S, Shawki A, Mackenzie B, Chen C, Varoqui H, Erickson JD (2009) SNAT2 amino acid transporter is regulated by amino acids of the SLC6 γ-aminobutyric acid transporter subfamily in neocortical neurons and may play no role in delivering glutamine for glutamatergic transmission. J Biol Chem 284:11224–11236

Grewer C, Grabsch E (2004) New inhibitors for the neutral amino acid transporter ASCT2 reveal its Na^+-dependent anion leak. J Physiol 557:747–759

Gu S (2005) Differential regulation of amino acid transporter SNAT3 by insulin in hepatocytes. J Biol Chem 280:26055–26062

Gu S, Roderick HL, Camacho P, Jiang JX (2000) Identification and characterization of an amino acid transporter expressed differentially in liver. Proc Natl Acad Sci U S A 97:3230–3235

Gu S, Adan-Rice D, Leach RJ, Jiang JX (2001a) A novel human amino acid transporter, hNAT3: cDNA cloning, chromosomal mapping, genomic structure, expression, and functional characterization. Genomics 74:262–272

Gu S, Roderick HL, Camacho P, Jiang JX (2001b) Characterization of an N-system amino acid transporter expressed in retina and its involvement in glutamine transport. J Biol Chem 276:24137–24144

Guma A, Testar X, Palacin M, Zorzano A (1988) Insulin-stimulated a-(methyl) aminoisobutyric acid uptake skeletal muscle protein synthesis. Biochem J 253:625–629

Hägglund MG, Sreedharan S, Nilsson VCO, Shaik JH, Almkvist IM, Bäcklin S, Wrange Ö, Fredriksson R (2011) Identification of SLC38A7 (SNAT7) protein as a glutamine transporter expressed in neurons. J Biol Chem 286:20500–20511

Hägglund MG, Hellsten SV, Bagchi S, Philippot G, Löfqvist E, Nilsson VCO, Almkvist I et al (2015) Transport of l-glutamine, l-alanine, l-arginine and l-histidine by the neuron-specific Slc38a8 (SNAT8) in CNS. J Mol Biol 427:1495–1512

Hamdani EH, Gudbrandsen M, Bjørkmo M, Chaudhry FA (2012) The system N transporter SN2 doubles as a transmitter precursor furnisher and a potential regulator of NMDA receptors. Glia 60:1671–1683

Hamdi MM, Mutungi G (2011) Dihydrotestosterone stimulates amino acid uptake and the expression of LAT2 in mouse skeletal muscle fibres through an ERK1/2-dependent mechanism. J Physiol 589:3623–3640

Harding HP, Novoa I, Zhang Y, Zeng H, Wek R, Schapira M, Ron D (2000) Regulated translation initiation controls stress-induced gene expression in mammalian cells. Mol Cell 6:1099–1108

Hatanaka T, Huang W, Ling R, Prasad PD, Sugawara M, Leibach FH, Ganapathy V (2001) Evidence for the transport of neutral as well as cationic amino acids by ATA3, a novel and liver-specific subtype of amino acid transport system A. Biochim Biophys Acta Biomembr 1510:10–17
Hatanaka T, Hatanaka Y, Setou M (2006) Regulation of amino acid transporter ATA2 by ubiquitin ligase Nedd4-2. J Biol Chem 281:35922–35930
Heckel T, Bröer A, Wiesinger H, Lang F, Bröer S (2003) Asymmetry of glutamine transporters in cultured neural cells. Neurochem Int 43:289–298
Hertz L, Dringen R, Schousboe A, Robinson SR (1999) Astrocytes: glutamate producers for neurons. J Neurosci Res 428:417–428
Hundal HS, Bilan PJ, Tsakiridis T, Marette A, Klip A (1994) Structural disruption of the trans-Golgi network does not interfere with the acute stimulation of glucose and amino acid uptake by insulin-like growth factor I in muscle cells. Biochem J 297:289–295
Hyde R, Christie GR, Litherland GJ, Hajduch E, Taylor PM, Hundal HS (2001) Subcellular localization and adaptive up-regulation of the system A (SAT2) amino acid transporter in skeletal-muscle cells and adipocytes. Biochem J 355:563–568
Hyde R, Peyrollier K, Hundal HS (2002) Insulin promotes the cell surface recruitment of the SAT2/ATA2 system A amino acid transporter from an endosomal compartment in skeletal muscle cells. J Biol Chem 277:13628–13634
Hyde R, Cwiklinski EL, MacAulay K, Taylor PM, Hundal HS (2007) Distinct sensor pathways in the hierarchical control of SNAT2, a putative amino acid transceptor, by amino acid availability. J Biol Chem 282:19788–19798
Jayakumar AR, Norenberg MD (2016). Glutamine synthetase: role in neurological disorders. Adv Neurobiol 13:327–350 (in press)
Jeitner T, Battaile K, Cooper AL (2015) Critical evaluation of the changes in glutamine synthetase activity in models of cerebral stroke. Neurochem Res 40:1–13
Jenstad M, Quazi AZ, Zilberter M, Haglerød C, Berghuis P, Saddique N, Goiny M, Buntup D, Davanger S, Haug FM, Barnes CA, McNaughton BL, Ottersen OP, Storm-Mathisen J, Harkany T, Chaudhry FA (2009) System A transporter SAT2 mediates replenishment of dendritic glutamate pools controlling retrograde signaling by glutamate. Cereb Cortex 19:1092–1106
Kageyama T, Nakamura M, Matsuo A, Yamasaki Y, Takakura Y, Hashida M, Kanai Y, Naito M, Tsuruo T, Minato N, Shimohama S (2000) The 4F2hc/ LAT1 complex transports L-DOPA across the blood–brain barrier. Brain Res 879:115–121
Kanai Y, Segawa H, Miyamoto KI, Uchino H, Takeda E, Endou H (1998) Expression cloning and characterization of a transporter for large neutral amino acids activated by the heavy chain of 4F2 antigen (CD98). J Biol Chem 273:23629–23632
Kanai Y, Fukasawa Y, Cha SH, Segawa H, Chairoungdua A, Kim DK, Matsuo H et al (2000) Transport properties of a system y+L neutral and basic amino acid transporter. Insights into the mechanisms of substrate recognition. J Biol Chem 275:20787–20793
Kanamori K, Ross BD (2006) Kinetics of glial glutamine efflux and the mechanism of neuronal uptake studied *in vivo* in mildly hyperammonemic rat brain. J Neurochem 99:1103–1113
Karinch A, Lin C (2002) Regulation of expression of the SN1 transporter during renal adaptation to chronic metabolic acidosis in rats. Am J Physiol Renal Physiol 283:1011–1019
Karinch AM, Lin CM, Meng Q, Pan M, Souba WW (2007) Glucocorticoids have a role in renal cortical expression of the SNAT3 glutamine transporter during chronic metabolic acidosis. Am J Physiol Renal Physiol 292:F448–F455
Karlsen TV, Serck-Hanssen G (2002) Acute stimulation by IGF-I of amino acid transport system A in chromaffin cells depends on PI3 kinase activation and the electrochemical gradient of Na^+. Ann NY Acad Sci 971:573–575
Kashiwagi H, Yamazaki K, Takekuma Y, Ganapathy V, Sugawara M (2009) Regulatory mechanisms of SNAT2, an amino acid transporter, in L6 rat skeletal muscle cells by insulin, osmotic shock and amino acid deprivation. Amino Acids 36:219–230
Kekuda R, Prasad PD, Fei YJ, Torres-Zamorano V, Sinha S, Yang-Feng TL, Leibach FH, Ganapathy V (1996) Cloning of the sodium-dependent, broad-scope, neutral amino acid transporter Bo from a human placental choriocarcinoma cell line. J Biol Chem 271:18657–18661

Kilberg MS, Handlogten ME, Christensen HN (1980) Characteristics of an amino acid transport system in rat liver for glutamine, asparagine, histidine, and closely related analogs. J Biol Chem 255:4011–4019

Kilberg MS, Pan XY, Chen H, Leung-Pineda V, Pan Y (2005) Nutritional control of gene expression: how mammalian cells respond to amino acid limitation. Annu Rev Nutr 25:59–85

Kim CS, Cho SH, Chun HS, Lee SY, Ndou HE (2008) BCH, an inhibitor of system L amino acid transporters, induces apoptosis in cancer cells. Biol Pharm Bull 31:1096–1100

Leke R, Bak L, Iversen P, Sørensen M, Keiding S, Vilstrup H, Ott P, Portela LV, Schousboe A, Waagepetersen HS (2011a) Synthesis of neurotransmitter GABA via the neuronal tricarboxylic acid cycle is elevated in rats with liver cirrhosis consistent with a high GABAergic tone in chronic hepatic encephalopathy. J Neurochem 117:824–832

Leke R, Bak LK, Anker M, Melø TM, Sørensen M, Keiding S, Vilstrup H et al (2011b) Detoxification of ammonia in mouse cortical GABAergic cell cultures increases neuronal oxidative metabolism and reveals an emerging role for release of glucose-derived alanine. Neurotox Res 19:496–510

Leke R, Escobar TDC, Rama Rao KV, Silveira TR, Norenberg MD, Schousboe A (2015) Expression of glutamine transporter isoforms in cerebral cortex of rats with chronic hepatic encephalopathy. Neurochem Int 39:605–611

Li S, Mallory M, Alford M, Tanaka S, Masliah E (1997) Glutamate transporter alterations in Alzheimer disease are possibly associated with abnormal APP expression. J Neuropathol Exp Neurol 56:901–911

Liang S-L, Carlson GC, Coulter DA (2006) Dynamic regulation of synaptic GABA release by the glutamate-glutamine cycle in hippocampal area CA1. J Neurosci 26:8537–85348

Ling R, Bridges CC, Sugawara M, Fujita T, Leibach FH, Prasad PD, Ganapathy V (2001) Involvement of transporter recruitment as well as gene expression in the substrate-induced adaptive regulation of amino acid transport system A. Biochim Biophys Acta 1512:15–21

Lopez-Fontanals M (2003) The osmoregulatory and the amino acid-regulated responses of system A are mediated by different signal transduction pathways. J Gen Physiol 122:5–16

Mackenzie B, Erickson JD (2004) Sodium-coupled neutral amino acid (system N/A) transporters of the SLC38 gene family. Pflugers Arch 447:784–795

Mackenzie B, Schäfer MKH, Erickson JD, Hediger MA, Weihe E, Varoqui H (2003) Functional properties and cellular distribution of the system A glutamine transporter SNAT1 support specialized roles in central neurons. J Biol Chem 278:23720–23730

Mastroberardino L, Spindler B, Pfeiffer R, Skelly PJ, Loffing J, Shoemaker CB, Verrey F (1998) Amino-acid transport by heterodimers of 4F2hc/CD98 and members of a permease family. Nature 395:288–291

Mathern GW, Mendoza D, Lozada A, Pretorius JK, Dehnes Y, Danbolt NC, Nelson N, Leite JP, Chimelli L, Born DE, Sakamoto AC, Assirati JA, Fried I, Peacock WJ, Ojemann GA, Adelson PD (1999) Hippocampal GABA and glutamate transporter immunoreactivity in patients with temporal lobe epilepsy. Neurology 52:453–472

Matsuda S, Kawasaki H, Moriguchi T, Gotoh Y, Nishida E (1995) Activation of protein kinase cascades by osmotic shock. J Biol Chem 270:12781–12786

McDowell HE, Eyers PA, Hundal HS (1998) Regulation of system A amino acid transport in L6 rat skeletal muscle cells by insulin, chemical and hyperthermic stress. FEBS Lett 441:15–19

McGivan JD, Bungard CI (2007) The transport of glutamine into mammalian cells. Front Biosci 12:874–882

Meier C, Ristic Z, Klauser S, Verrey F (2002) Activation of system L heterodimeric amino acid exchangers by intracellular substrates. EMBO J 21:580–589

Melone M, Quagliano F, Barbaresi P, Varoqui H, Erickson JD, Conti F (2004) Localization of the glutamine transporter SNAT1 in rat cerebral cortex and neighboring structures, with a note on its localization in human cortex. Cereb Cortex 14:562–574

Melone M, Varoqui H, Erickson JD, Conti F (2006) Localization of the Na^+-coupled neutral amino acid transporter 2 in the cerebral cortex. Neuroscience 140:281–292

Nagaraja TN, Brookes N (1996) Glutamine transport in mouse cerebral astrocytes. J Neurochem 66:1665–1674

Nagase T, Seki N, Ishikawa K, Tanaka A, Nomura N (1996) Prediction of the coding sequences of unidentified human genes. V. The coding sequences of 40 new genes (KIAA0161-KIAA0200) deduced by analysis of cDNA clones from human cell line KG-1. DNA Res 3:17–24

Nakanishi T, Kekuda R, Fei YJ, Hatanaka T, Sugawara M, Martindale RG, Leibach FH, Prasad PD, Ganapathy V (2001a) Cloning and functional characterization of a new subtype of the amino acid transport system N. Am J Physiol Cell Physiol 281:C1757–C1768

Nakanishi T, Sugawara M, Huang W, Martindale RG, Leibach FH, Ganapathy ME, Prasad PD, Ganapathy V (2001b) Structure, function, and tissue expression pattern of human SN2, a subtype of the amino acid transport system N. Biochem Biophys Res Commun 281:1343–1348

Newsholme P, Procopio J, Lima MM, Pithon-Curi TC, Curi R (2003) Glutamine and glutamate—their central role in cell metabolism and function. Cell Biochem Funct 21:1–9

Nissen-Meyer LSH, Chaudhry FA (2013) Protein kinase C phosphorylates the system N glutamine transporter SN1 (Slc38a3) and regulates its membrane trafficking and degradation. Front Endocrinol (Lausanne) 4:138

Nissen-Meyer LSH, Popescu MC, Hamdani EH, Chaudhry FA (2011) Protein kinase C-mediated phosphorylation of a single serine residue on the rat glial glutamine transporter SN1 governs its membrane trafficking. J Neurosci 31:6565–6575

Norenberg MD (1998) Astroglial dysfunction in hepatic encephalopathy. Metab Brain Dis 13:319–335

Norenberg M, Martinez-Hernandez A (1979) Fine structural localization of glutamine synthetase in astrocytes of rat brain. Brain Res 161:303–310

Núñez B, Martínez de Mena R, Obregon MJ, Font-Llitjós M, Nunes V, Palacín M, Dumitrescu AM, Morte B, Bernal J (2014) Cerebral cortex hyperthyroidism of newborn Mct8-deficient mice transiently suppressed by Lat2 inactivation. PLoS One 9:e96915

Oxender DL, Christensen HN (1963) Distinct mediating systems for the transport of neutral amino acids by the Ehrlich cell. J Biol Chem 238:3686–3699

Palacín M, Kanai Y (2004) The ancillary proteins of HATs: SLC3 family of amino acid transporters. Pflugers Arch Eur J Physiol 447:490–494

Palii SS, Chen H, Kilberg MS (2004) Transcriptional control of the human sodium-coupled neutral amino acid transporter system A gene by amino acid availability is mediated by an intronic element. J Biol Chem 279:3463–3471

Palii SS, Thiaville MM, Pan Y-X, Zhong C, Kilberg MS (2006) Characterization of the amino acid response element within the human sodium-coupled neutral amino acid transporter 2 (SNAT2) system A transporter gene. Biochem J 395:517–527

Palmada M, Speil A, Jeyaraj S, Böhmer C, Lang F (2005) The serine/threonine kinases SGK1, 3 and PKB stimulate the amino acid transporter ASCT2. Biochem Biophys Res Commun 331:272–277

Patel AJ, Hunt A (1985) Concentration of free amino acids in primary cultures of neurones and astrocytes. J Neurochem 44:1816–1821

Patel AB, Rothman DL, Cline GW, Behar KL (2001) Glutamine is the major precursor for GABA synthesis in rat neocortex *in vivo* following acute GABA-transaminase inhibition. Brain Res 919:207–220

Pfeiffer R, Rossier G, Spindler B, Meier C, Kühn L, Verrey F (1999) Amino acid transport of y+L-type by heterodimers of 4F2hc/CD98 and members of the glycoprotein-associated amino acid transporter family. EMBO J 18:49–57

Pineda M, Fernández E, Torrents D, Estévez R, López C, Camps M, Lloberas J, Zorzano A, Palacín M (1999) Identification of a membrane protein, LAT-2, that co-expresses with 4F2 heavy chain, an L-type amino acid transport activity with broad specificity for small and large zwitterionic amino acids. J Biol Chem 274:19738–19744

Pingitore P, Pochini L, Scalise M, Galluccio M, Hedfalk K, Indiveri C (2013) Large scale production of the active human ASCT2 (SLC1A5) transporter in Pichia pastoris—functional and kinetic asymmetry revealed in proteoliposomes. Biochim Biophys Acta 1828:2238–2246

Pochini L, Scalise M, Galluccio M, Indiveri C (2014) Membrane transporters for the special amino acid glutamine: structure/function relationships and relevance to human health. Front Chem 2:61

Proper EA, Hoogland G, Kappen SM, Jansen GH, Rensen MG, Schrama LH, van Veelen CW, van Rijen PC, van Nieuwenhuizen O, Gispen WH, de Graan PN (2002) Distribution of glutamate transporters in the hippocampus of patients with pharmaco-resistant temporal lobe epilepsy. Brain 125:32–43

Rajan DP, Kekuda R, Huang W, Devoe LD, Leibach FH, Prasad PD, Ganapathy V (2000) Cloning and functional characterization of a Na(+)-independent, broad-specific neutral amino acid transporter from mammalian intestine. Biochim Biophys Acta 1463:6–14

Rama Rao KV, Norenberg MD (2014) Glutamine in the pathogenesis of hepatic encephalopathy: the Trojan horse hypothesis revisited. Neurochem Res 39:593–598

Reimer RJ, Chaudhry FA, Gray AT, Edwards RH (2000) Amino acid transport system A resembles system N in sequence but differs in mechanism. Proc Natl Acad Sci U S A 97:7715–7720

Robinson S (2001) Changes in the cellular distribution of glutamine synthetase in Alzheimer's disease. J Neurosci Res 980:972–980

Rodríguez A, Ortega A, Berumen LC, García-Alcocer MG, Giménez C, Zafra F (2014) Expression of the system N transporter (SNAT5/SN2) during development indicates its plausible role in glutamatergic neurotransmission. Neurochem In. 73:166–71

Rossier G, Meier C, Bauch C, Summa V, Sordat B, Verrey F, Kühn LC (1999) LAT2, a new basolateral 4F2hc/CD98-associated amino acid transporter of kidney and intestine. J Biol Chem 274:34948–34954

Scalise M, Pochini L, Panni S, Pingitore P, Hedfalk K, Indiveri C (2014) Transport mechanism and regulatory properties of the human amino acid transporter ASCT2 (SLC1A5). Amino Acids 46:2463–2475

Schneider HP, Bröer S, Bröer A, Deitmer JW (2007) Heterologous expression of the glutamine transporter SNAT3 in *Xenopus oocytes* is associated with four modes of uncoupled transport. J Biol Chem 282:3788–3798

Schousboe A (2003) Role of astrocytes in the maintenance and modulation of glutamatergic and GABAergic neurotransmission. Neurochem Res 28:347–352

Schousboe A, Westergaard N, Waagepetersen HS, Larsson OM, Bakken IJ, Sonnewald U (1997) Trafficking between glia and neurons of TCA cycle intermediates and related metabolites. Glia 21:99–105

Schousboe A, Bak LK, Waagepetersen HS (2013) Astrocytic control of biosynthesis and turnover of the neurotransmitters glutamate and GABA. Front Endocrinol (Lausanne) 4:102. doi:10.3389/fendo.2013.00102

Schousboe A, Scafidi S, Bak LK, Waagepetersen HS, McKenna MC (2014a) Glutamate metabolism in the brain focusing on astrocytes. Adv Neurobiol 11:13–30

Schousboe A, Waagepetersen HS, Leke R, Bak LK (2014b) Effects of hyperammonemia on brain energy metabolism: controversial findings *in vivo* and *in vitro*. Metab Brain Dis 29:913–917

Segawa H, Fukasawa Y, Miyamoto K, Takeda E, Endou H, Kanai Y (1999) Identification and functional characterization of a Na+-independent neutral amino acid transporter with broad substrate selectivity. J Biol Chem 274:19745–19751

Shafqat S, Tamarappoo BK, Kilberg MS, Puranam RS, McNamara JO, Guadaño-Ferraz A, Fremeau RT (1993) Cloning and expression of a novel Na(+)-dependent neutral amino acid transporter structurally related to mammalian Na+/glutamate cotransporters. J Biol Chem 268:15351–15355

Sidoryk-Wegrzynowicz M, Aschner M (2013) Role of astrocytes in manganese mediated neurotoxicity. BMC Pharmacol Toxicol 14:23

Sidoryk-Węgrzynowicz M, Lee E, Albrecht J, Aschner M (2009) Manganese disrupts astrocyte glutamine transporter expression and function. J Neurochem 110:822–830

Sidoryk-Wegrzynowicz M, Lee E-S, Ni M, Aschner M (2010) Manganese-induced downregulation of astroglial glutamine transporter SNAT3 involves ubiquitin-mediated proteolytic system. Glia 58:1905–1912

Sidoryk-Wegrzynowicz M, Lee E, Mingwei N, Aschner M (2011) Disruption of astrocytic glutamine turnover by manganese is mediated by the protein kinase C pathway. Glia 59:1732–1743

Skowrońska M, Zielińska M, Albrecht J, Fręśko I (2012) Upregulation of the heteromeric y+LAT2 transporter contributes to ammonia-induced increase of arginine uptake in rat cerebral cortical astrocytes. Neurochem Int 61:531–535

Solbu TT, Bjørkmo M, Berghuis P, Harkany T, Chaudhry FA (2010) SAT1, a glutamine transporter, is preferentially expressed in GABAergic neurons. Front Neuroanat 4:1

Su TZ, Campbell GW, Oxender DL (1997) Glutamine transport in cerebellar granule cells in culture. Brain Res 757:69–78

Suarez I, Bodega G, Fernandez B (2002) Glutamine synthetase in brain: effect of ammonia. Neurochem Int 41:123–142

Sugawara M (2000) Cloning of an amino acid transporter with functional characteristics and tissue expression pattern identical to that of system A. J Biol Chem 275:16473–16477

Sugawara M, Nakanishi T, Fei YJ, Martindale RG, Ganapathy ME, Leibach FH, Ganapathy V (2000) Structure and function of ATA3, a new subtype of amino acid transport system A, primarily expressed in the liver and skeletal muscle. Biochim Biophys Acta Biomembr 1509:7–13

Swain M, Butterworth RF, Blei AT (1992) Ammonia and related amino-acids in the pathogenesis of brain edema in acute ischemic liver-failure in rats. Hepatology 15:449–453

Takanaga H, Tokuda N, Ohtsuki S, Hosoya K-I, Terasaki T (2002) ATA2 is predominantly expressed as system A at the blood-brain barrier and acts as brain-to-blood efflux transport for L-proline. Mol Pharmacol 61:1289–1296

Tamarappoo BK, Raizada MK, Kilberg MS (1997) Identification of a system N-like Na(+)-dependent glutamine transport activity in rat brain neurons. J Neurochem 68:954–960

Tomi M, Mori M, Tachikawa M, Katayama K, Terasaki T, Hosoya K (2005) L-type amino acid transporter 1-mediated L-leucine transport at the inner blood-retinal barrier. Invest Ophthalmol Vis Sci 46:2522–2530

Torrents D, Pineda M, Ferna E, Lloberas J, Shi Y, Zorzano A, Palacı M (1998) Identification and characterization of a membrane protein (y+L amino acid transporter-1) that associates with 4F2hc to encode the amino acid transport activity y+L. J Biol Chem 273:32437–32445

Utsunomiya-Tate N, Endou H, Kanai Y (1996) Cloning and functional characterization of a system ASC-like Na^+-dependent neutral amino acid transporter. J Biol Chem 271:14883–14890

Varoqui H, Zhu H, Yao D, Ming H, Erickson JD (2000) Cloning and functional identification of a neuronal glutamine transporter. J Biol Chem 275:4049–4054

Verrey F, Meier C, Rossier G, Kühn LC (2000) Glycoprotein-associated amino acid exchangers: broadening the range of transport specificity. Pflugers Arch Eur J Physiol 440:503–512

Verrey F, Closs EI, Wagner CA, Palacin M, Endou H, Kanai Y (2004) CATs and HATs: the SLC7 family of amino acid transporters. Pflugers Arch Eur J Physiol 447:532–542

Verrey F, Singer D, Ramadan T, Vuille-Dit-Bille RN, Mariotta L, Camargo SMR (2009) Kidney amino acid transport. Pflugers Arch Eur J Physiol 458:53–60

Waagepetersen HS, Sonnewald U, Gegelashvili G, Larsson OM, Schousboe A (2001) Metabolic distinction between vesicular and cytosolic GABA in cultured GABAergic neurons using 13C magnetic resonance spectroscopy. J Neurosci Res 63:347–355

Walton HS, Dodd PR (2007) Glutamate-glutamine cycling in Alzheimer's disease. Neurochem Int 50:1052–1066

Wang H, Huang W, Sugawara M, Devoe LD, Leibach FH, Prasad PD, Ganapathy V (2000) Cloning and functional expression of ATA1, a subtype of amino acid transporter A, from human placenta. Biochem Biophys Res Commun 273:1175–1179

Weiss MD, Derazi S, Rossignol C, Varoqui H, Erickson JD, Kilberg MS, Anderson KJ (2003) Ontogeny of the neutral amino acid transporter SAT1/ATA1 in rat brain. Dev Brain Res 143:151–159

White LR, Gårseth M, Aasly J, Sonnewald U (2004) Cerebrospinal fluid from patients with dementia contains increased amounts of an unknown factor. J Neurosci Res 78:297–301

Wittmann G, Mohácsik P, Balkhi MY, Gereben B, Lechan RM (2015) Endotoxin-induced inflammation down-regulates l-type amino acid transporter 1 (LAT1) expression at the blood–brain barrier of male rats and mice. Fluids Barriers CNS 12:21

Xu GY, McAdoo DJ, Hughes MG, Robak G, de Castro R (1998) Considerations in the determination by microdialysis of resting extracellular amino acid concentrations and release upon spinal cord injury. Neuroscience 86:1011–1021

Yao D, Mackenzie B, Ming H, Varoqui H, Zhu H, Hediger MA, Erickson JD (2000) A novel system A isoform mediating Na^+/neutral amino acid cotransport. J Biol Chem 275:22790–22797

Yu AC, Drejer J, Hertz L, Schousboe A (1983) Pyruvate carboxylase activity in primary cultures of astrocytes and neurons. J Neurochem 41:1484–1487

Zander CB, Albers T, Grewer C (2013) Voltage-dependent processes in the electroneutral amino acid exchanger ASCT2. J Gen Physiol 141:659–672

Zhang Z, Gameiro A, Grewer C (2008) Highly conserved asparagine 82 controls the interaction of Na^+ with the sodium-coupled neutral amino acid transporter SNAT2. J Biol Chem 283:12284–12292

Ziegler TR, Bazargan N, Leader LM, Martindale RG (2000) Glutamine and the gastrointestinal tract. Curr Opin Clin Nutr Metab Care 3:355–362

Zielińska M, Ruszkiewicz J, Hilgier W, Fresko I, Albrecht J (2011) Hyperammonemia increases the expression and activity of the glutamine/arginine transporter y+LAT2 in rat cerebral cortex: implications for the nitric oxide/cGMP pathway. Neurochem Int 58:190–195

Zielińska M, Popek M, Albrecht J (2014) Roles of changes in active glutamine transport in brain edema development during hepatic encephalopathy: an emerging concept. Neurochem Res 39:599–604

Chapter 9
Glutamine Metabolism in Gliomas

Monika Szeliga and Jan Albrecht

Abstract By histological, morphological criteria, and malignancy, brain tumors are classified by WHO into grades I (most benign) to IV (highly malignant), and gliomas are the most frequently occurring class throughout the grades. Similar to peripheral tumors, the growth of glia-derived tumor cells largely depends on glutamine (Gln), which is vividly taken up by the cells, using mostly ASCT2 and SN1 as Gln carriers. Tumor growth-promoting effects of Gln are associated with its phosphate-activated glutaminase (GA) (specifically KGA)-mediated degradation to glutamate (Glu) and/or with its entry to the energy- and intermediate metabolite-generating pathways related to the tricarboxylic acid cycle. However, a subclass of liver-type GA are absent in glioma cells, a circumstance which allows phenotype manipulations upon their transfection to the cells. Gln-derived Glu plays a major role in promoting tumor proliferation and invasion. Glu is relatively inefficiently recycled to Gln and readily leaves the cells by exchange with the extracellular pool of the glutathione (GSH) precursor Cys mediated by xc- transporter. This results in (a) cell invasion-fostering interaction of Glu with ionotropic Glu receptors in the surrounding tissue, (b) intracellular accumulation of GSH which increases tumor resistance to radio- and chemotherapy.

Keywords Gliomas • Glutamine • Glutamate • Glutathione • Cell proliferation • Tumor invasion • Glutaminase

9.1 Introduction and Outlines

Neoplastic transformation of a tissue involves complex genetic and metabolic remodeling of the cells, wherein no single event can be considered as either decisive or dispensable in the whole process. This chapter aims at elucidating the specific roles of Gln in the biology of gliomas, the most frequently occurring class of brain tumors. Gln and its metabolites contribute outstandingly to the different aspects of

M. Szeliga (✉) • J. Albrecht
Department of Neurotoxicology, Mossakowski Medical Research Centre, Polish Academy of Sciences, Pawińskiego St. 5, 02-106, Warsaw, Poland
e-mail: mszeliga@imdik.pan.pl

A. Schousboe, U. Sonnewald (eds.), *The Glutamate/GABA-Glutamine Cycle*, Advances in Neurobiology 13, DOI 10.1007/978-3-319-45096-4_9

the biology of tumors residing in, or stemming from, the central nervous system (CNS) and the peripheral tissues alike. In so far, most of the data describing the role and fate of Gln have been derived from studies on peripheral tumors. Indeed, interest in the subject commenced with the seminal observation that growth of the archetypical human epithelial cancer cell line, the HeLa cell, is better promoted by the presence of Gln than glucose in the growth medium (Reitzer et al. 1979). The Gln dependence of the growth of glia-derived tumor cells has been later confirmed in a rat C6 glioma cell line (Dolińska et al. 2001; Martin et al. 1998; Portais et al. 1996).

Studies on different cell lines or solid tumors have confirmed the role of Gln not only as a major energy fuel, but not less so as a precursor of metabolites critically involved in tumor progression. It is by now clear that all the processes, which contribute to the tumor growth-promoting effects of Gln are associated with its glutaminase-mediated degradation to Glu and/or with its entry into the energy- and intermediate metabolite-generating pathways related to the tricarboxylic acid cycle. Subsequent sections of the chapter will focus on the pertinent changes of Gln metabolism in gliomas which promote their proliferation and invasiveness. Wherever possible (in so far a few instances only), relation between distinct traits of Gln metabolism and tumor grade will be emphasized.

To position the high-grade gliomas, further referred to as glioblastomas (GB), among the diversified world of brain tumors, the chapter commences with a brief account on brain tumor classification, epidemiology, and treatment (Sect. 9.2). The next section (Sect. 9.3) characterizes Gln uptake to the tumor cells, accounting for the diverse involvement of distinct Gln transporters and presents data on the relative Gln accumulation within the tumor tissue vs. surrounding nontransformed brain tissue. Section 9.4 will be devoted to the Gln-degrading enzymes, phosphate-activated glutaminases (GAs). Diverse status and functions of GA isoforms will be emphasized. Similar to other neoplastic and metabolically active tissues, the kidney-type GA contained in glioma cells is upregulated and bears the brunt for increased Gln degradation to Glu. KGA appears as a plausible therapeutic target. By contrast, the liver-type GA which in nontransformed cells is hypothesized to be involved in the regulation of transcription (Olalla et al. 2002), is absent in glioma cells, a circumstance which allows phenotype manipulations upon their transfection to the cells. Section 9.5 will account for the role of Gln-derived Glu in promoting tumor proliferation and invasion. The tumor-promoting role of the xc- transporter, which catalyzes the exchange of intracellular Glu for the extracellular pool of the GSH precursor Cys, will be emphasized. Finally, results of magnetic resonance spectroscopic analysis of Gln fluxes via the tricarboxylic acid cycle and their role in glioma metabolism will be described in Sect. 9.6. The fates of Gln described in the chapter are summarized in Fig. 9.1.

9.2 Characteristics of Glial Tumors: Classification, Prognosis, and Treatment

Primary brain tumors comprise a heterogeneous group of neoplasms arising from the tissues of the brain or its surroundings. The world incidence rate for all primary brain tumors ranges from 8.63 to 13.56 per 100,000 per year (de Robles et al. 2015).

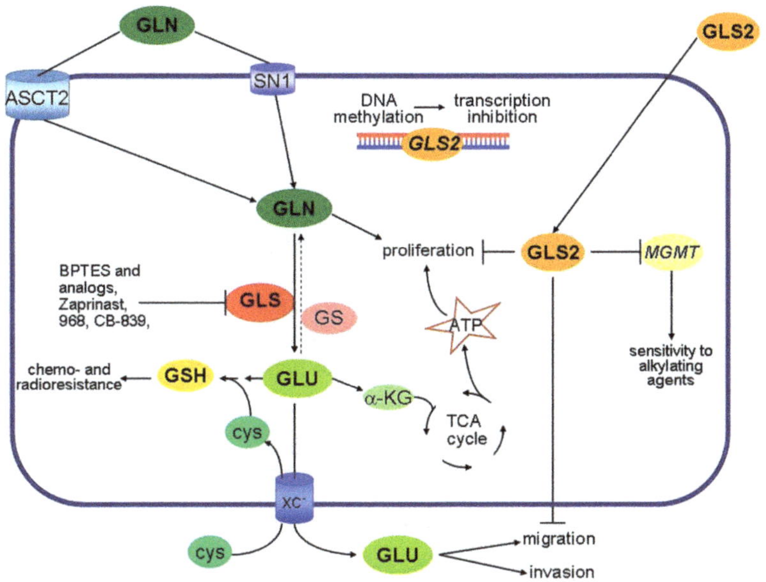

Fig. 9.1 A scheme summarizing metabolic fate and biological roles of Gln in brain tumors. Gln is taken up by ASCT2 and SN1, the two Gln carriers which abound in brain tumor cells as compared to the surrounding normal brain tissue. Gln is primarily degraded by the upregulated kidney-type glutaminase (GLS). Gln serves as a precursor for the synthesis of nucleic acids and proteins and in this way promotes cell proliferation. Excess of Gln-derived Glu is released from the cells via the xc- transporter exchanging Glu for Cys, with two key consequences: (1) increased extracellular Glu which promotes cell invasion and (2) increased intracellular GSH synthesis, responsible for increased chemo- and radioresistance. Part of Glu enters the GGC replenishing TCA cycle constituents (anaplerosis) and giving rise to other important metabolic substrates and ATP, used among other purposes in cell proliferation. Only a relatively minor proportion of Glu is recycled to Gln, because of low GS activity. Malignant brain tumors contain very low amounts of the liver-type glutaminase (GLS2), most likely due to excessive DNA methylation in the promoter region of *GLS2*. Introduction of GLS2 into the cells elicits a chain of as yet not fully described events that inhibit cell proliferation, migration, and invasion. Experimental studies point to the utility of GLS inhibitors (BPTES and analogs, zaprinast, 968, CB-839) in curbing malignant properties of the cells

The World Health Organization (WHO) classification of tumors of the central nervous system (CNS) divides the tumors by histological criteria, and based on morphological features of anaplasia assigns them a malignancy grade from I (benign) to IV (highly malignant) (Louis et al. 2007).

Gliomas arising from any type of glial cells (i.e., astrocytes, oligodendrocytes, ependymocytes) are the most common primary brain tumors accounting for more than 70 % of all neoplasms of the CNS (Ohgaki and Kleihues 2009). Pilocytic astrocytoma (WHO I) develops in children and young adults and is usually cured by resection with low risk of tumor recurrence (Fernandez et al. 2003; Tibbetts et al. 2009). Diffuse astrocytoma (WHO II) is a slow-growing tumor with a tendency to recur after surgery which is often associated with progression to high-grade gliomas. Median survival time of patients with diffuse astrocytoma ranges from 5 to 8 years

(Lind-Landström et al. 2012). Oligodendroglioma (WHO II) has better prognosis with median survival time from 10 to 17 years (Scheie et al. 2011). Mixed oligoastrocytoma (WHO II) has an intermediate prognosis compared to the other low-grade gliomas with median survival time 6 years (Scheie et al. 2011). Anaplastic astrocytoma (WHO grade III) is a rapidly growing tumor with median survival time 2–3 years (Riemenschneider and Reifenberger 2009). Glioblastoma (GB) (WHO IV) is the most frequent of all gliomas (Ohgaki and Kleihues 2009). Median survival time of patients with newly diagnosed GB is less than 10 months (Weller et al. 2013a). Grade III and IV tumors are referred to as high-grade gliomas. Standard treatment of these tumors, which consists of surgery followed by radiation plus concomitant and adjuvant chemotherapy with an alkylating agent, temozolomide (TMZ), results in only marginal prolongation of patient's survival.

Histological examination is still the basic method used in diagnosis of brain tumors. However, a growing body of evidence points to a significant role of some molecular markers as diagnostic, prognostic, or predictive factors (Hegi et al. 2006; Weller et al. 2013b). Specifically, *IDH1* and *IDH2* mutations, 1p and 19q codeletions, and *MGMT* promoter methylation are used in diagnosis of some types of brain tumors (Weller et al. 2013b). Mutations in *IDH1* and *IDH2* genes encoding isocitrate dehydrogenase 1 and 2, respectively, occur mainly in low-grade gliomas. Patients with mutated *IDH1* or *IDH2* have significantly longer overall survival than patients with wild-type *IDH* of the same histological grade (Cohen et al. 2013). Codeletion of 1p and 19q is strongly associated with oligodendroglial histology and tumors carrying this aberration are more sensitive to radio- and chemotherapy (Weller et al. 2013b). Hypermethylation of *MGMT* promoter leads to silencing of the gene coding for a DNA repair protein, O-6-methylguanine-DNA methyltransferase (MGMT), promoting resistance of tumor cells to alkylating agents. Reduced level of MGMT protein results in diminished DNA repair activity which in turn is associated with prolonged overall survival in GB patients treated with alkylating agents (Weller et al. 2013b).

The detailed description of classification, prognosis, and treatment of distinct brain tumors goes far beyond the scope of this chapter. The readers interested in this issue are encouraged to study comprehensive reports cited in this section.

9.3 Gln Uptake and Accumulation in Brain Tumor Cells

Studies on peripheral tumor cell lines (Collins et al. 1998; Wasa et al. 1996) and tumor spheroids (Pawlik et al. 2000) have equivocally pointed to increased Gln uptake as a determinant of rapid growth of transformed cells, more specifically of the increased DNA and protein synthesis in these cells (Wasa et al. 1996). The studies have emphasized the role of the neutral amino acid transporting system ASC in the upregulation of Gln transport. The ASC system variant, ASCT2 has been implicated as the predominant Gln transporting moiety in a C6 rat glioma cell line (Dolińska et al. 2003). Interestingly, diminished proliferation of C6 cells grown in the absence of Gln paralleled the reduction in the activity of system ASC-mediated Gln uptake (Dolińska et al. 2001). While ASCT2 is likewise overexpressed in all classes of native human gliomas, the most malignant phenotypes (grades III/IV) are distinguished from the more benign tumors by abundance of mRNA coding for the glia-specific N system transporter SN1 (SNAT3) (Sidoryk et al. 2004).

Since Gln synthesis in cerebral tumors is comparatively very low (see Sect. 9.6), their Gln content is thought to mainly reflect its import from the surrounding brain tissue and/or from blood. Proton magnetic resonance spectroscopy (^{1}H MRS) carried out in high-grade glioma patients has consistently revealed elevated Gln content in the affected tissue (Li et al. 2015, and references therein). Elevation of Gln content is more pronounced and more consistently found in primary human gliomas than in brain metastases of peripheral tumors (Fan et al. 2004). Moreover, Gln content has been reported to be higher in malignant recurrences of primary tumors than in the primary tumors themselves (Lehnhardt et al. 2005). Increased Gln concentrations have also been detected in the contralateral normal-appearing white matter of GB patients suggesting that elevated Gln may serve as a marker of early neoplastic infiltration of an apparently healthy tissue (Kallenberg et al. 2009). Combined gene expression and ^{1}H NMR metabolomic analysis of nine different glioblastoma cell lines revealed their considerable heterogeneity with regard to their amino acid content and expression of pertinent transporter molecules differing in their content of different amino acids (Cuperlovic-Culf et al. 2012). The heterogeneity encompassed the Gln status: of the four distinct classes of the cell lines, only two showed elevated Gln content, which coincided with simultaneous overexpression in both, of two Gln-transporting moieties—the N system transporter SNAT1 and the L system transporter LAT1 (Cuperlovic-Culf et al. 2012). In a more general perspective, it will be of interest to see whether and in what degree, the existence of four metabolically distinct types of glioma cells in culture correspond to the four subtypes of glioblastoma identified in clinical material, recently defined by Verhaak and colleagues as proneural, neural, classical, and mesenchymal, respectively (Verhaak et al. 2010).

9.4 Diverse Roles and Status of Phosphate-Activated Glutaminase (GA) Isoforms

Apart from increased Gln uptake, a wide range of cancer cells present also elevated glutaminolysis. Phosphate-activated glutaminase (GA, EC 3.5.1.2), the initial enzyme of the glutaminolysis, metabolizes Gln to Glu and ammonia. In humans, there are two genes coding for GA isoforms: the *GLS* gene encodes kidney-type isoforms—KGA and GAC (Curthoys and Watford 1995; Elgadi et al. 1999), while the *GLS2* gene encodes liver-type isoforms—LGA and GAB (Aledo et al. 2000; de la Rosa et al. 2009). GA isoforms present distinct tissue specificity, kinetic properties, and cell localization (Aoki et al. 1991; Campos et al. 2003; Campos-Sandoval et al. 2007; Castell et al. 2004; Curthoys and Watford 1995; Elgadi et al. 1999; Olalla et al. 2002).

Impaired expression and activity of GA isoforms is a hallmark of different tumors and neoplastic cell lines (Szeliga and Obara-Michlewska 2009). *GLS* is ubiquitously overexpressed in neoplastic tissues of different origin (Huang et al. 2014; Pan et al. 2015; Pérez-Gómez et al. 2005; Turner and McGivan 2003; Yu et al. 2015) and silencing of this gene significantly reduced proliferation rate of neoplastic cells in vitro (Lobo et al. 2000; Meng et al. 2014). The expression of *GLS2* in neoplastic tissues is more diverse and ranges from its lack (Hu et al. 2010; Liu et al. 2014; Yu et al. 2015; Zhang et al. 2013) to abundance (Xiang et al. 2013). Tumor regression was obtained by both ectopic expression of *GLS2* (Hu et al. 2010; Liu et al. 2014; Zhang et al. 2013) or its silencing (Lee et al. 2014), depending on the cell type and initial status of the *GLS2*. Recent findings indicate that GA isoforms may be regulated at several levels by oncogenes or tumor suppressor genes. *GLS* expression has been shown to be indirectly upregulated by the oncogene MYC via suppression of miRNA23a/b expression in lymphoma and prostate cancer cells (Gao et al. 2009). Moreover, the other mediators of oncogenesis, Rho GTPases, have been shown to activate the NFkB transcriptional program which, in turn, elevates the activity of GAC isoform in different cancer cells (Wang et al. 2010). By contrast, the transcription of *GLS2* is regulated by tumor suppressor p53 (Hu et al. 2010; Suzuki et al. 2010). Of note, GLS2 mediates the role of p53 in antioxidant defense and energy metabolism (Hu et al. 2010; Suzuki et al. 2010).

High levels of the *GLS* isoforms and only traces of transcripts arising from the *GLS2* were found in GB tissues (Szeliga et al. 2005) and in several human GB cell lines (Cheng et al. 2011; Szeliga et al. 2009). Very recent data strongly suggest that downregulation of *GLS2* in GB may be caused by DNA methylation found in the promoter region and in the first intron of this gene (Szeliga et al. 2016). Increasing evidence suggests that GA isoforms play opposing role in GB. Overexpression of *GLS2* in GB cells reduced their proliferation and migration ability (Szeliga et al. 2009) and sensitized them to the alkylating agents, most likely through diminishing the expression of *MGMT* (Szeliga et al. 2012). On the other hand, significant decrease in GB cell proliferation was also observed after silencing of *GLS* (Martín-Rufián et al. 2014; Szeliga et al. 2014). Inhibition of *GLS* diminished GSH-dependent antioxidant capacity and triggered apoptosis mediated by mitochondrial dysfunction. Similar tendency was observed in GB cells after ectopic overexpression of GAB (Martín-Rufián et al. 2014).

Inhibition of Glu production in glioma cells offers therapeutic perspectives. A growing body of evidence shows that GLS silencing reduces proliferation of cancer cells of different origin. Thus, pharmacological inhibition of GLS seems to be a potential therapeutic approach for treating cancer. Research conducted in last decade identified some GLS-selective inhibitors increasing the possibility to block GLS with minimal off-target effects. Of these, the most deeply described is an allosteric GLS inhibitor, bis-2-(5-phenylacetamido-1,3,4-thiadiazol-2-yl)ethyl sulfide (BPTES). BPTES binds to an allosteric pocket of GLS triggering conformational changes leading to inactivation of the enzyme (DeLaBarre et al. 2011; Thangavelu et al. 2012). Antitumor activity of BPTES has been demonstrated in different models (Emadi et al. 2014; Thangavelu et al. 2012), including gliomas

(Seltzer et al. 2010). Of note, in case of glioma and leukemia, IDH mutant cells were more sensitive to BPTES as compared to the wild-type counterparts (Emadi et al. 2014; Seltzer et al. 2010). An increasing number of other GLS inhibitors, such as: analogs of BPTES, compound 968 (bromo-benzophenanthridinone), CB-839, Zaprinast, have been shown to diminish growth of some cancer cells in vitro and in vivo (Elhammali et al. 2014; Gross et al. 2014; Katt et al. 2012; Shukla et al. 2012; Wang et al. 2010). Insofar, however, none of them has been tested in gliomas in the clinical setting.

While GLS has emerged as a critical enzyme in tumorigenesis, the role of GLS2 is still elusive. It is possible that GAB isoform suppresses malignant phenotype of GB cells through its potential ability to modulate transcription (Szeliga et al. 2009). Of note, based on nuclear localization, existence of consensus motifs in the structure and identification of novel interacting proteins, GAB has been proposed to be a multifunctional protein (Márquez et al. 2006). The molecular mechanism underlying the mode of action of distinct GLS2 proteins in different cancer models in vivo and in vitro is a subject of ongoing research in different laboratories (Liu et al. 2014; Martín-Rufián et al. 2014; Szeliga et al. 2009, 2012, 2014; Xiang et al. 2013; Zhang et al. 2013).

9.5 Increased Glu Accumulation in Glioma Cells: Mechanisms and Implication for Tumor Progression and Epileptic Seizures

Excessive intra- and/or extracellular Glu accumulation has been consistently documented in rodent glioma cell lines and glioma implants (Buckingham et al. 2011; Lyons et al. 2007; Takano et al. 2001; Ye and Sontheimer 1999; Ye et al. 1999), and in human brain tumors and peritumoral tissue (Cuperlovic-Culf et al. 2012; Liubinas et al. 2014; Yuen et al. 2012). Increased Glu content and release in glial tumor cells has been ascribed to three factors: (1) decreased intracellular conversion of Glu to Gln, due to the decrease of GS activity (Rosati et al. 2009, but see Sect. 9.6 for contradictory views); (2) decreased GLU reuptake associated with downregulation, misplacement, and/or impaired RNA splicing of the astrocytic Glu transporters EAAT-2 (GLT-1) (Münch et al. 2001; Ye et al. 1999) and/or EAAT1 (Cuperlovic-Culf et al. 2012); (3) increased Glu release in exchange for cystine, due to upregulation of the Glu/Cys antiporter xc- (Ye et al. 1999). Coincident changes in the expression of EAAT-2 and xc-, originally reported for glioma cell lines, were recently documented in human brain tumor tissue (Yuen et al. 2012). Increasing body of evidence suggests that increased peritumoral accumulation of Glu promotes excitotoxicity by interacting with ionotropic and/or metabotropic Glu receptors in adjacent neurons (de Groot et al. 2008; Rzeski et al. 2001), and subsequently, tumor proliferation and invasion (reviewed by de Groot and Sontheimer 2011). One of the important clinical complications of the overexposure of brain tissue to the adjacent glioma-derived Glu is epileptic seizures (Buckingham et al. 2011; Rosati et al. 2009; Yuen et al. 2012). Glutamatergic hyperexcitability underlying the seizures was suppressed by application of the xc- inhibitor, sulfasalazine (SAS), underscoring the role of this transporter in their pathogenesis (Yuen et al. 2012).

One other advantage of increased xc- activity for tumor growth stems from the fact that it couples increased GLU release to increased uptake of Cys, an essential precursor of GSH. GSH is a key cellular antioxidant which is present in high amounts in GBs (Li et al. 2015; Santandreu et al. 2008), ensuring defense of GB against oxidative stress, and their resistance against radio- and chemotherapy (Rocha et al. 2015, and references therein). Indeed, application of the xc- inhibitor SAS (see earlier) decreased GSH content and simultaneously slowed down tumor growth in a xenograft animal model of human glioma (Chung et al. 2005).

There are data showing that metabolic coupling of glioma cells to peritumoral astrocytes protects neurons against excitotoxic effects of glioma-derived Glu. Glioma growth rate in glioma/astrocyte cocultures (AGC) was slowed down with increasing astroglia/glioma ratio in these cultures, and growth media from cocultures with higher glioma/astrocyte ratio produced a stronger calcium signal in cultured neurons and was more toxic to these cells (Yao et al. 2014). In an experimental glioma model in rats, proliferation of glioma cells and neuronal demise were attenuated upon in situ upregulation of the astrocytic glutamate transporter GLT-1 (EAAT-2) following administration an antibiotic, thiamphenicol (Sattler et al. 2013).

9.6 Gln Fluxes via the TCA Cycle and Their Implications for Glioma Metabolism

Similar to other neoplastic cells, glioma cells support their energy demand mostly by anaerobic glycolysis subsiding to the classic pattern of Warburg effect (Warburg et al. 1927; Warburg 1956). Glioma cells also exhibit aerobic glycolysis (Marin-Valencia et al. 2012), which is facilitated by relatively high expression of hexokinase 2 isoform (Wolf et al. 2011). Nonetheless, the strong requirement of cultured glioma cells for Gln (Dolińska et al. 2001; Martin et al. 1998; Portais et al. 1996) suggested that Gln is likewise actively metabolized in these cells. However, evidence from ^{13}C NMR studies is controversial: there is no good match between the results of cell culture studies and those carried out in the in vivo setting.

Comparative analysis of ^{13}C glucose utilization in C6 glioma cells grown in the presence or absence of excess of unlabelled Gln demonstrated that while Gln is actively metabolized, it is not an energy substrate for the cells, but rather serves as a carbon donor to replenish the tricarboxylic acid cycle (Portais et al. 1996). In a study in which ^{13}C Gln metabolism was followed in the presence or absence of glucose (^{13}C labeled or unlabelled), glioblastoma cells performing anaerobic glycolysis (SF 1898 cell line), metabolized ^{13}C Gln at a rate high enough to promote efficient generation of NADPH to support fatty acid synthesis, and anaplerosis (restoration of oxaloacetate), at the same time facilitating the use of glucose carbon for synthesis of amino acids, nucleotides, and nucleosides (DeBerardinis et al. 2007). By contrast, a very recent study demonstrated that human GB harbored in mice and human patient GBs alike, preferentially oxidized ^{13}C acetate (Mashimo et al. 2014) and less so ^{13}C glucose, but oxidation of ^{13}C Gln was virtually absent in these cells. While further studies are needed to definitely unravel the contribution of Gln to various intracellular pathways in glioma cells, at present one can safely exclude its significant role as direct energy fuel.

There is also disagreement in the literature as to whether and in what degree glioma cells are capable of glutamine synthesis. No conversion of ^{13}C glucose to Gln was detected in cultured C6 glioma cells (Portais et al. 1993) or in C6 glioma-bearing rat model (Bouzier et al. 1999). On the other hand, analysis of ^{13}C glucose metabolism in human brain GBs in situ showed that a discernable proportion of glucose-derived carbon becomes incorporated to Gln (Maher et al. 2012). Incompatibility of literature data extends to the enrichment of the content and pathophysiological role of glutamine synthetase (GS). Proteomic analysis of a representative group of human brain tumors revealed GS enrichment in astrocytomas of all grades as compared to oligodendrogliomas (Zhuang et al. 2011). In one other study, however, a lowered GS content is considered as a factor predisposing the patients to epileptic bursts associated with tumor progression (Rosati et al. 2009), most likely in relation to excessive Glu accumulation (see also Sect. 9.5). It may thus be speculated that the GS status in GBs is at least partly stage dependent. In support of this notion, in glioma cell lines, GS expression and activity increase along with cell differentiation and decreased proliferation rates (Davies and Vernadakis 1986), and a recent study disclosed a negative growth regulatory role of GS, related to upregulation of N-cadherin synthesis and subsequent strengthening of cell–cell contact (Yin et al. 2013). Irrespective of the controversies, it appears that glia-derived neoplastic cells convert relatively less glutamate to Gln than nontransformed astrocytes, rendering their intracellular glutamine cycling relatively inefficient, which causes their metabolism and growth to rely in a great degree on the delivery of extracellular Gln.

References

Aledo JC, Gómez-Fabre PM, Olalla L, Márquez J (2000) Identification of two human glutaminase loci and tissue-specific expression of the two related genes. Mamm Genome 11(12):1107–1110

Aoki C, Kaneko T, Starr A, Pickel VM (1991) Identification of mitochondrial and non-mitochondrial glutaminase within select neurons and glia of rat forebrain by electron microscopic immunocytochemistry. J Neurosci Res 28(4):531–548

Bouzier AK, Quesson B, Valeins H, Canioni P, Merle M (1999) [1-(13)C] glucose metabolism in the tumoral and nontumoral cerebral tissue of a glioma-bearing rat. J Neurochem 72(6):2445–2455

Buckingham SC, Campbell SL, Haas BR, Montana V, Robel S, Ogunrinu T, Sontheimer H (2011) Glutamate release by primary brain tumors induces epileptic activity. Nat Med 17(10):1269–1274. doi:10.1038/nm.2453

Campos JA, Aledo JC, Segura JA, Alonso FJ, Gómez-Fabre PM, Núñez de Castro I, Márquez J (2003) Expression of recombinant human L-glutaminase in *Escherichia coli*: polyclonal antibodies production and immunological analysis of mouse tissues. Biochim Biophys Acta 1648(1–2):17–23

Campos-Sandoval JA, López de la Oliva AR, Lobo C, Segura JA, Matés JM, Alonso FJ, Márquez J (2007) Expression of functional human glutaminase in baculovirus system: affinity purification, kinetic and molecular characterization. Int J Biochem Cell Biol 39(4):765–773

Castell L, Vance C, Abbott R, Marquez J, Eggleton P (2004) Granule localization of glutaminase in human neutrophils and the consequence of glutamine utilization for neutrophil activity. J Biol Chem 279(14):13305–13310

Cheng T, Sudderth J, Yang C, Mullen AR, Jin ES, Matés JM, DeBerardinis RJ (2011) Pyruvate carboxylase is required for glutamine-independent growth of tumor cells. Proc Natl Acad Sci U S A 108(21):8674–8679. doi:10.1073/pnas.1016627108

Chung WJ, Lyons SA, Nelson GM, Hamza H, Gladson CL, Gillespie GY, Sontheimer H (2005) Inhibition of cystine uptake disrupts the growth of primary brain tumors. J Neurosci 25(31):7101–7110

Cohen AL, Holmen SL, Colman H (2013) IDH1 and IDH2 mutations in gliomas. Curr Neurol Neurosci Rep 13(5):345. doi:10.1007/s11910-013-0345-4

Collins CL, Wasa M, Souba WW, Abcouwer SF (1998) Determinants of glutamine dependence and utilization by normal and tumor-derived breast cell lines. J Cell Physiol 176(1):166–178

Cuperlovic-Culf M, Ferguson D, Culf A, Morin P Jr, Touaibia M (2012) 1H NMR metabolomics analysis of glioblastoma subtypes: correlation between metabolomics and gene expression characteristics. J Biol Chem 287(24):20164–20175. doi:10.1074/jbc.M111.337196

Curthoys NP, Watford M (1995) Regulation of glutaminase activity and glutamine metabolism. Annu Rev Nutr 15:133–159

Davies DL, Vernadakis A (1986) Responses in astrocytic C6 glioma cells to ethanol and dibutyryl cyclic AMP. Brain Res 389(1–2):253–260

de Groot J, Sontheimer H (2011) Glutamate and the biology of gliomas. Glia 59(8):1181–1189. doi:10.1002/glia.21113

de Groot JF, Piao Y, Lu L, Fuller GN, Yung WK (2008) Knockdown of GluR1 expression by RNA interference inhibits glioma proliferation. J Neurooncol 88(2):121–133. doi:10.1007/s11060-008-9552-2

de la Rosa V, Campos-Sandoval JA, Martín-Rufián M, Cardona C, Matés JM, Segura JA, Alonso FJ, Márquez J (2009) A novel glutaminase isoform in mammalian tissues. Neurochem Int 55(1–3):76–84. doi:10.1016/j.neuint.2009.02.021

de Robles P, Fiest KM, Frolkis AD, Pringsheim T, Atta C, St Germaine-Smith C, Day L, Lam D, Jette N (2015) The worldwide incidence and prevalence of primary brain tumors: a systematic review and meta-analysis. Neuro Oncol 17(6):776–783. doi:10.1093/neuonc/nou283

DeBerardinis RJ, Mancuso A, Daikhin E, Nissim I, Yudkoff M, Wehrli S, Thompson CB (2007) Beyond aerobic glycolysis: transformed cells can engage in glutamine metabolism that exceeds the requirement for protein and nucleotide synthesis. Proc Natl Acad Sci U S A 104(49):19345–19350

DeLaBarre B, Gross S, Fang C, Gao Y, Jha A, Jiang F, Song JJ, Wei W, Hurov JB (2011) Full-length human glutaminase in complex with an allosteric inhibitor. Biochemistry 50(50):10764–10770. doi:10.1021/bi201613d

Dolińska M, Dybel A, Zielińska M, Zabłocka B, Bużańska L, Albrecht J (2001) Glutamine transport in C6 glioma cells: substrate specificity and modulation in a glutamine-deprived medium. J Neurosci Res 66(5):959–966

Dolińska M, Dybel A, Zabłocka B, Albrecht J (2003) Glutamine transport in C6 glioma cells shows ASCT2 system characteristics. Neurochem Int 43(4–5):501–507

Elgadi KM, Meguid RA, Qian M, Souba WW, Abcouwer SF (1999) Cloning and analysis of unique human glutaminase isoforms generated by tissue-specific alternative splicing. Physiol Genomics 1(2):51–62

Elhammali A, Ippolito JE, Collins L, Crowley J, Marasa J, Piwnica-Worms D (2014) A high-throughput fluorimetric assay for 2-hydroxyglutarate identifies Zaprinast as a glutaminase inhibitor. Cancer Discov 4(7):828–839. doi:10.1158/2159-8290.CD-13-0572

Emadi A, Jun SA, Tsukamoto T, Fathi AT, Minden MD, Dang CV (2014) Inhibition of glutaminase selectively suppresses the growth of primary acute myeloid leukemia cells with IDH mutations. Exp Hematol 42(4):247–251. doi:10.1016/j.exphem.2013.12.001

Fan G, Sun B, Wu Z, Guo Q, Guo Y (2004) In vivo single-voxel proton MR spectroscopy in the differentiation of high-grade gliomas and solitary metastases. Clin Radiol 59(1):77–85

Fernandez C, Figarella-Branger D, Girard N, Bouvier-Labit C, Gouvernet J, Paz Paredes A, Lena G (2003) Pilocytic astrocytomas in children: prognostic factors—a retrospective study of 80 cases. Neurosurgery 53(3):544–553; discussion 554–555

Gao P, Tchernyshyov I, Chang TC, Lee YS, Kita K, Ochi T, Zeller KI, De Marzo AM, Van Eyk JE, Mendell JT, Dang CV (2009) c-Myc suppression of miR-23a/b enhances mitochondrial glutaminase expression and glutamine metabolism. Nature 458(7239):762–765. doi:10.1038/nature07823

Gross MI, Demo SD, Dennison JB, Chen L, Chernov-Rogan T, Goyal B, Janes JR, Laidig GJ, Lewis ER, Li J, Mackinnon AL, Parlati F, Rodriguez ML, Shwonek PJ, Sjogren EB, Stanton TF, Wang T, Yang J, Zhao F, Bennett MK (2014) Antitumor activity of the glutaminase inhibitor CB-839 in triple-negative breast cancer. Mol Cancer Ther 13(4):890–901. doi:10.1158/1535-7163.MCT-13-0870

Hegi ME, Murat A, Lambiv WL, Stupp R (2006) Brain tumors: molecular biology and targeted therapies. Ann Oncol 17(Suppl 10):x191–x197

Hu W, Zhang C, Wu R, Sun Y, Levine A, Feng Z (2010) Glutaminase 2, a novel p53 target gene regulating energy metabolism and antioxidant function. Proc Natl Acad Sci U S A 107(16):7455–7460. doi:10.1073/pnas.1001006107

Huang F, Zhang Q, Ma H, Lv Q, Zhang T (2014) Expression of glutaminase is upregulated in colorectal cancer and of clinical significance. Int J Clin Exp Pathol 7(3):1093–1100

Kallenberg K, Bock HC, Helms G, Jung K, Wrede A, Buhk JH, Giese A, Frahm J, Strik H, Dechent P, Knauth M (2009) Untreated glioblastoma multiforme: increased myo-inositol and glutamine levels in the contralateral cerebral hemisphere at proton MR spectroscopy. Radiology 253(3):805–812. doi:10.1148/radiol.2533071654

Katt WP, Ramachandran S, Erickson JW, Cerione RA (2012) Dibenzophenanthridines as inhibitors of glutaminase C and cancer cell proliferation. Mol Cancer Ther 11(6):1269–1278. doi:10.1158/1535-7163.MCT-11-0942

Lee YZ, Yang CW, Chang HY, Hsu HY, Chen IS, Chang HS, Lee CH, Lee JC, Kumar CR, Qiu YQ, Chao YS, Lee SJ (2014) Discovery of selective inhibitors of Glutaminase-2, which inhibit mTORC1, activate autophagy and inhibit proliferation in cancer cells. Oncotarget 5(15):6087–6101

Lehnhardt FG, Bock C, Röhn G, Ernestus RI, Hoehn M (2005) Metabolic differences between primary and recurrent human brain tumors: a 1H NMR spectroscopic investigation. NMR Biomed 18(6):371–382

Li Y, Larson P, Chen AP, Lupo JM, Ozhinsky E, Kelley D, Chang SM, Nelson SJ (2015) Short-echo three-dimensional H-1 MR spectroscopic imaging of patients with glioma at 7 tesla for characterization of differences in metabolite levels. J Magn Reson Imaging 41(5):1332–1341. doi:10.1002/jmri.24672

Lind-Landström T, Habberstad AH, Sundstrøm S, Torp SH (2012) Prognostic value of histological features in diffuse astrocytomas WHO grade II. Int J Clin Exp Pathol 5(2):152–158

Liu J, Zhang C, Lin M, Zhu W, Liang Y, Hong X, Zhao Y, Young KH, Hu W, Feng Z (2014) Glutaminase 2 negatively regulates the PI3K/AKT signaling and shows tumor suppression activity in human hepatocellular carcinoma. Oncotarget 5(9):2635–2647

Liubinas SV, Drummond KJ, Desmond PM, Bjorksten A, Morokoff AP, Kaye AH, O'Brien TJ, Moffat BA (2014) Glutamate quantification in patients with supratentorial gliomas using chemical shift imaging. NMR Biomed 27(5):570–577. doi:10.1002/nbm.3095

Lobo C, Ruiz-Bellido MA, Aledo JC, Márquez J, Núñez De Castro I, Alonso FJ (2000) Inhibition of glutaminase expression by antisense mRNA decreases growth and tumourigenicity of tumour cells. Biochem J 348(Pt 2):257–261

Louis DN, Ohgaki H, Wiestler OD, Cavenee WK, Burger PC, Jouvet A, Scheithauer BW, Kleihues P (2007) The 2007 WHO classification of tumours of the central nervous system. Acta Neuropathol 114(2):97–109. Epub 2007 Jul 6. Review. Erratum in: Acta Neuropathol 114(5):547

Lyons SA, Chung WJ, Weaver AK, Ogunrinu T, Sontheimer H (2007) Autocrine glutamate signaling promotes glioma cell invasion. Cancer Res 67(19):9463–9471

Maher EA, Marin-Valencia I, Bachoo RM, Mashimo T, Raisanen J, Hatanpaa KJ, Jindal A, Jeffrey FM, Choi C, Madden C, Mathews D, Pascual JM, Mickey BE, Malloy CR, DeBerardinis RJ (2012) Metabolism of [U-13 C]glucose in human brain tumors in vivo. NMR Biomed 25(11):1234–1244. doi:10.1002/nbm.2794

Marin-Valencia I, Yang C, Mashimo T, Cho S, Baek H, Yang XL, Rajagopalan KN, Maddie M, Vemireddy V, Zhao Z, Cai L, Good L, Tu BP, Hatanpaa KJ, Mickey BE, Matés JM, Pascual JM, Maher EA, Malloy CR, Deberardinis RJ, Bachoo RM (2012) Analysis of tumor metabolism reveals mitochondrial glucose oxidation in genetically diverse human glioblastomas in the mouse brain in vivo. Cell Metab 15(6):827–837

Márquez J, de la Oliva AR, Matés JM, Segura JA, Alonso FJ (2006) Glutaminase: a multifaceted protein not only involved in generating glutamate. Neurochem Int 48(6–7):465–471

Martin M, Beauvoit B, Voisin PJ, Canioni P, Guerin B, Rigoulet M (1998) Energetic and morphological plasticity of C6 glioma cells grown on 3-D support; effect of transient glutamine deprivation. J Bioenerg Biomembr 30(6):565–578

Martín-Rufián M, Nascimento-Gomes R, Higuero A, Crisma AR, Campos-Sandoval JA, Gómez-García MC, Cardona C, Cheng T, Lobo C, Segura JA, Alonso FJ, Szeliga M, Albrecht J, Curi R, Márquez J, Colquhoun A, Deberardinis RJ, Matés JM (2014) Both GLS silencing and GLS2 overexpression synergize with oxidative stress against proliferation of glioma cells. J Mol Med (Berl) 92(3):277–290. doi:10.1007/s00109-013-1105-2

Mashimo T, Pichumani K, Vemireddy V, Hatanpaa KJ, Singh DK, Sirasanagandla S, Nannepaga S, Piccirillo SG, Kovacs Z, Foong C, Huang Z, Barnett S, Mickey BE, DeBerardinis RJ, Tu BP, Maher EA, Bachoo RM (2014) Acetate is a bioenergetic substrate for human glioblastoma and brain metastases. Cell 159(7):1603–1614. doi:10.1016/j.cell.2

Meng G, Xia M, Xu C, Yuan D, Schnurr M, Wei J (2014) Multifunctional antitumor molecule 5′-triphosphate siRNA combining glutaminase silencing and RIG-I activation. Int J Cancer 134(8):1958–1971. doi:10.1002/ijc.28416

Münch C, Penndorf A, Schwalenstöcker B, Troost D, Ludolph AC, Ince P, Meyer T (2001) Impaired RNA splicing of 5′-regulatory sequences of the astroglial glutamate transporter EAAT2 in human astrocytoma. J Neurol Neurosurg Psychiatry 71(5):675–678

Ohgaki H, Kleihues P (2009) Genetic alterations and signaling pathways in the evolution of gliomas. Cancer Sci 100(12):2235–2241. doi:10.1111/j.1349-7006.2009.01308.x

Olalla L, Gutierrez A, Campos JA, Khan ZU, Alonso FJ, Segura JA, Marquez J, Aledo JC (2002) Nuclear localization of L-type glutaminase in mammalian brain. J Biol Chem 277(41) :38939–38944

Pan T, Gao L, Wu G, Shen G, Xie S, Wen H, Yang J, Zhou Y, Tu Z, Qian W (2015) Elevated expression of glutaminase confers glucose utilization via glutaminolysis in prostate cancer. Biochem Biophys Res Commun 456(1):452–458. doi:10.1016/j.bbrc.2014.11.105

Pawlik TM, Souba W, Sweeney TJ, Bode BP (2000) Amino acid uptake and regulation in multicellular hepatoma spheroids. J Surg Res 91(1):15–25

Pérez-Gómez C, Campos-Sandoval JA, Alonso FJ, Segura JA, Manzanares E, Ruiz-Sánchez P, González ME, Márquez J, Matés JM (2005) Co-expression of glutaminase K and L isoenzymes in human tumour cells. Biochem J 386(Pt 3):535–542

Portais J, Martin M, Canioni P, Merle M (1993) Glutathione, but not glutamine, is detected in 13C-NMR spectra of perchloric acid extracts from C6 glioma cells. FEBS Lett 327(3):301–306

Portais JC, Voisin P, Merle M, Canioni P (1996) Glucose and glutamine metabolism in C6 glioma cells studied by carbon 13 NMR. Biochimie 78(3):155–164

Reitzer LJ, Wice BM, Kennell D (1979) Evidence that glutamine, not sugar, is the major energy source for cultured HeLa cells. J Biol Chem 254(8):2669–2676

Riemenschneider MJ, Reifenberger G (2009) Astrocytic tumors. Recent results. Cancer Res 171:3–24. doi:10.1007/978-3-540-31206-2_1

Rocha CR, Garcia CC, Vieira DB, Quinet A, de Andrade-Lima LC, Munford V, Belizário JE, Menck CF (2015) Glutathione depletion sensitizes cisplatin- and temozolomide-resistant glioma cells in vitro and in vivo. Cell Death Dis 6:e1727. doi:10.1038/cddis.2015.101

Rosati A, Marconi S, Pollo B, Tomassini A, Lovato L, Maderna E, Maier K, Schwartz A, Rizzuto N, Padovani A, Bonetti B (2009) Epilepsy in glioblastoma multiforme: correlation with glutamine synthetase levels. J Neurooncol 93(3):319–324. doi:10.1007/s11060-008-9794-z

Rzeski W, Turski L, Ikonomidou C (2001) Glutamate antagonists limit tumor growth. Proc Natl Acad Sci U S A 98(11):6372–6377

Santandreu FM, Brell M, Gene AH, Guevara R, Oliver J, Couce ME, Roca P (2008) Differences in mitochondrial function and antioxidant systems between regions of human glioma. Cell Physiol Biochem 22(5–6):757–768. doi:10.1159/000185559

Sattler R, Tyler B, Hoover B, Coddington LT, Recinos V, Hwang L, Brem H, Rothstein JD (2013) Increased expression of glutamate transporter GLT-1 in peritumoral tissue associated with prolonged survival and decreases in tumor growth in a rat model of experimental malignant glioma. J Neurosurg 119(4):878–886. doi:10.3171/2013.6.JNS122319

Scheie D, Meling TR, Cvancarova M, Skullerud K, Mørk S, Lote K, Eide TJ, Helseth E, Beiske K (2011) Prognostic variables in oligodendroglial tumors: a single-institution study of 95 cases. Neuro Oncol 13(11):1225–1233. doi:10.1093/neuonc/nor114

Seltzer MJ, Bennett BD, Joshi AD, Gao P, Thomas AG, Ferraris DV, Tsukamoto T, Rojas CJ, Slusher BS, Rabinowitz JD, Dang CV, Riggins GJ (2010) Inhibition of glutaminase preferentially slows growth of glioma cells with mutant IDH1. Cancer Res 70(22):8981–8987. doi:10.1158/0008-5472.CAN-10-1666

Shukla K, Ferraris DV, Thomas AG, Stathis M, Duvall B, Delahanty G, Alt J, Rais R, Rojas C, Gao P, Xiang Y, Dang CV, Slusher BS, Tsukamoto T (2012) Design, synthesis, and pharmacological evaluation of bis-2-(5-phenylacetamido-1,2,4-thiadiazol-2-yl)ethyl sulfide 3 (BPTES) analogs as glutaminase inhibitors. J Med Chem 55(23):10551–10563. doi:10.1021/jm301191p

Sidoryk M, Matyja E, Dybel A, Zielińska M, Bogucki J, Jaskólski DJ, Liberski PP, Kowalczyk P, Albrecht J (2004) Increased expression of a glutamine transporter SNAT3 is a marker of malignant gliomas. Neuroreport 15(4):575–578

Suzuki S, Tanaka T, Poyurovsky MV, Nagano H, Mayama T, Ohkubo S, Lokshin M, Hosokawa H, Nakayama T, Suzuki Y, Sugano S, Sato E, Nagao T, Yokote K, Tatsuno I, Prives C (2010) Phosphate-activated glutaminase (GLS2), a p53-inducible regulator of glutamine metabolism and reactive oxygen species. Proc Natl Acad Sci U S A 107(16):7461–7466. doi:10.1073/pnas.1002459107

Szeliga M, Obara-Michlewska M (2009) Glutamine in neoplastic cells: focus on the expression and roles of glutaminases. Neurochem Int 55(1–3):71–75. doi:10.1016/j.neuint.2009.01.008

Szeliga M, Sidoryk M, Matyja E, Kowalczyk P, Albrecht J (2005) Lack of expression of the liver-type glutaminase (LGA) mRNA in human malignant gliomas. Neurosci Lett 374(3):171–173

Szeliga M, Obara-Michlewska M, Matyja E, Łazarczyk M, Lobo C, Hilgier W, Alonso FJ, Márquez J, Albrecht J (2009) Transfection with liver-type glutaminase cDNA alters gene expression and reduces survival, migration and proliferation of T98G glioma cells. Glia 57(9):1014–1023. doi:10.1002/glia.20825

Szeliga M, Zgrzywa A, Obara-Michlewska M, Albrecht J (2012) Transfection of a human glioblastoma cell line with liver-type glutaminase (LGA) down-regulates the expression of DNA-repair gene MGMT and sensitizes the cells to alkylating agents. J Neurochem 123(3):428–436. doi:10.1111/j.1471-4159.2012.07917.x

Szeliga M, Bogacińska-Karaś M, Różycka A, Hilgier W, Marquez J, Albrecht J (2014) Silencing of GLS and overexpression of GLS2 genes cooperate in decreasing the proliferation and viability of glioblastoma cells. Tumour Biol 35(3):1855–1862. doi:10.1007/s13277-013-1247-4

Szeliga M, Bogacińska-Karaś M, Kuźmicz K, Rola R, Albrecht J (2016) Downregulation of GLS2 in glioblastoma cells is related to DNA hypermethylation but not to the p53 status. Mol Carcinog 55(9):1309–1316. doi:10.1002/mc.22372

Takano T, Lin JH, Arcuino G, Gao Q, Yang J, Nedergaard M (2001) Glutamate release promotes growth of malignant gliomas. Nature Med 7(9):1010–1015

Thangavelu K, Pan CQ, Karlberg T, Balaji G, Uttamchandani M, Suresh V, Schüler H, Low BC, Sivaraman J (2012) Structural basis for the allosteric inhibitory mechanism of human kidney-type glutaminase (KGA) and its regulation by Raf-Mek-Erk signaling in cancer cell metabolism. Proc Natl Acad Sci U S A 109(20):7705–7710. doi:10.1073/pnas.1116573109

Tibbetts KM, Emnett RJ, Gao F, Perry A, Gutmann DH, Leonard JR (2009) Histopathologic predictors of pilocytic astrocytoma event-free survival. Acta Neuropathol 117(6):657–665. doi:10.1007/s00401-009-0506-3

Turner A, McGivan JD (2003) Glutaminase isoform expression in cell lines derived from human colorectal adenomas and carcinomas. Biochem J 370(Pt 2):403–408

Verhaak RG, Hoadley KA, Purdom E, Wang V, Qi Y, Wilkerson MD, Miller CR, Ding L, Golub T, Mesirov JP, Alexe G, Lawrence M, O'Kelly M, Tamayo P, Weir BA, Gabriel S, Winckler W, Gupta S, Jakkula L, Feiler HS, Hodgson JG, James CD, Sarkaria JN, Brennan C, Kahn A, Spellman PT, Wilson RK, Speed TP, Gray JW, Meyerson M, Getz G, Perou CM, Hayes DN, Cancer Genome Atlas Research Network (2010) Integrated genomic analysis identifies clinically relevant subtypes of glioblastoma characterized by abnormalities in PDGFRA, IDH1, EGFR, and NF1. Cancer Cell 17(1):98–110. doi:10.1016/j.ccr.2009.12.020

Wang JB, Erickson JW, Fuji R, Ramachandran S, Gao P, Dinavahi R, Wilson KF, Ambrosio AL, Dias SM, Dang CV, Cerione RA (2010) Targeting mitochondrial glutaminase activity inhibits oncogenic transformation. Cancer Cell 18(3):207–219. doi:10.1016/j.ccr.2010.08.009. Erratum in: Cancer Cell. 2010 18(4):397

Warburg O (1956) On the origin of cancer cells. Science 123(3191):309–314

Warburg O, Wind F, Negelein E (1927) The metabolism of tumors in the body. J Gen Physiol 8:519–530

Wasa M, Bode BP, Abcouwer SF, Collins CL, Tanabe KK, Souba WW (1996) Glutamine as a regulator of DNA and protein biosynthesis in human solid tumor cell lines. Ann Surg 224(2):189–197

Weller M, Cloughesy T, Perry JR, Wick W (2013a) Standards of care for treatment of recurrent glioblastoma—are we there yet? Neuro Oncol 15(1):4–27. doi:10.1093/neuonc/nos273

Weller M, Pfister SM, Wick W, Hegi ME, Reifenberger G, Stupp R (2013b) Molecular neuro-oncology in clinical practice: a new horizon. Lancet Oncol 14(9):e370–e379. doi:10.1016/S1470-2045(13)70168-2

Wolf A, Agnihotri S, Micallef J, Mukherjee J, Sabha N, Cairns R, Hawkins C, Guha A (2011) Hexokinase 2 is a key mediator of aerobic glycolysis and promotes tumor growth in human glioblastoma multiforme. J Exp Med 208(2):313–326. doi:10.1084/jem.20101470

Xiang L, Xie G, Liu C, Zhou J, Chen J, Yu S, Li J, Pang X, Shi H, Liang H (2013) Knock-down of glutaminase 2 expression decreases glutathione, NADH, and sensitizes cervical cancer to ionizing radiation. Biochim Biophys Acta 1833(12):2996–3005. doi:10.1016/j.bbamcr.2013.08.003

Yao PS, Kang DZ, Lin RY, Ye B, Wang W, Ye ZC (2014) Glutamate/glutamine metabolism coupling between astrocytes and glioma cells: neuroprotection and inhibition of glioma growth. Biochem Biophys Res Commun 450(1):295–299. doi:10.1016/j.bbrc.2014.05.120

Ye Z, Sontheimer H (1999) Glioma cells release excitotoxic concentrations of glutamate. Cancer Res 59(17):4383–4391

Ye ZC, Rothstein JD, Sontheimer H (1999) Compromised glutamate transport in human glioma cells: reduction-mislocalization of sodium-dependent glutamate transporters and enhanced activity of cystine-glutamate exchange. J Neurosci 19(24):10767–10777

Yin Y, Sun W, Xiang J, Deng L, Zhang B, Xie P, Qiao W, Zou J, Liu C (2013) Glutamine synthetase functions as a negative growth regulator in glioma. J Neurooncol 114(1):59–69. doi:10.1007/s11060-013-1168-5

Yu D, Shi X, Meng G, Chen J, Yan C, Jiang Y, Wei J, Ding Y (2015) Kidney-type glutaminase (GLS1) is a biomarker for pathologic diagnosis and prognosis of hepatocellular carcinoma. Oncotarget 6(10):7619–7631

Yuen TI, Morokoff AP, Bjorksten A, D'Abaco G, Paradiso L, Finch S, Wong D, Reid CA, Powell KL, Drummond KJ, Rosenthal MA, Kaye AH, O'Brien TJ (2012) Glutamate is associated with a higher risk of seizures in patients with gliomas. Neurology 79(9):883–889. doi:10.1212/WNL.0b013e318266fa89

Zhang J, Wang C, Chen M, Cao J, Zhong Y, Chen L, Shen HM, Xia D (2013) Epigenetic silencing of glutaminase 2 in human liver and colon cancers. BMC Cancer 13:601. doi:10.1186/1471-2407-13-601

Zhuang Z, Qi M, Li J, Okamoto H, Xu DS, Iyer RR, Lu J, Yang C, Weil RJ, Vortmeyer A, Lonser RR (2011) Proteomic identification of glutamine synthetase as a differential marker for oligodendrogliomas and astrocytomas. J Neurosurg 115(4):789–795. doi:10.3171/2011.5.JNS11451

Chapter 10
Oligodendrocytes: Development, Physiology and Glucose Metabolism

Ana I. Amaral, Joana M. Tavares, Ursula Sonnewald, and Mark R.N. Kotter

Abstract The glutamate–glutamine cycle is an outstanding example of how essential neuronal–glial interactions are for brain function. For several decades, this and other metabolic cycles in the brain have only included neurons and astrocytes but not oligodendrocytes, the myelinating cells of the central nervous system (CNS). Recent data revealed that oligodendrocytes are highly metabolically active cells in the brain and, therefore, should not be ignored. Using ^{13}C-labelled glucose in combination with nuclear magnetic resonance spectroscopy (MRS) and/or mass spectrometry (MS) it is possible to characterize metabolic functions in primary oligodendrocyte cultures. Mature rat oligodendrocytes avidly metabolize glucose in the cytosol and pyruvate derived from glucose in mitochondria. Moreover, they seem to have the ability of performing anaplerosis from pyruvate, which might enable them to synthesize metabolites de novo and transfer them to neighbouring cells. All these original findings highlight the importance of investigating oligodendrocyte metabolism separately from that of astrocytes and neurons to be able to discern the roles played by the individual partners. This is of particular importance in the white matter where the number of oligodendrocytes is considerable. The present book chapter provides some background on oligodendrocyte biology and physiology and summarizes the not very extensive information published on glucose metabolism in oligodendrocytes.

Keywords Pyruvate carboxylation • Glucose • Glutamate • Lactate • Citrate • Glutamine–glutamate cycle

---

A.I. Amaral • J.M. Tavares • M.R.N. Kotter
Anne McLaren Laboratory, Wellcome Trust Medical Research Council Cambridge Stem Cell Institute and Department of Clinical Neurosciences, University of Cambridge, Cambridge, UK
e-mail: aa705@cam.ac.uk; jt603@cam.ac.uk; mrk25@cam.ac.uk

U. Sonnewald (✉)
Department of Neuroscience, Faculty of Medicine, Norwegian University of Science and Technology, Trondheim 7491, Norway

Department of Drug Design and Pharmacology, Faculty of Health and Medical Sciences, University of Copenhagen, Copenhagen 2100, Denmark
e-mail: Ursula.sonnewald@ntnu.no

A. Schousboe, U. Sonnewald (eds.), *The Glutamate/GABA-Glutamine Cycle*, Advances in Neurobiology 13, DOI 10.1007/978-3-319-45096-4_10

Abbreviations

ATP	Adenosine triphosphate
CNP	20,30-cyclicnucleotide 30-phosphodiesterase
CNS	Central nervous system
Fructose-6P	Fructose-6-phosphate
GA3P	Glyceraldehyde-3-phosphate
GABA	γ-Aminobutyric acid
MAG	Myelin antigen glycoprotein
MBP	Myelin basic protein
MOG	Myelin oligodendrocyte glycoprotein
MRS	Magnetic resonance spectroscopy
MS	Multiple sclerosis
OLs	Oligodendrocytes
OPCs	Oligodendrocyte precursor cells
PC	Pyruvate carboxylase
PDGFR-α	Platelet-derived growth factor receptor alpha
PEPCK	Phosphoenolpyruvate carboxy kinase
PK	Pyruvate kinase
PLP	Proteolipid protein
PPP	Pentose phosphate pathway
T3	Triiodothyronine/thyroid hormone 3
TBI	Traumatic brain injury
TCA	Tricarboxylic acid

10.1 Introduction

Oligodendrocytes make up a large proportion of the cells in the central nervous system (CNS). Although oligodendrocytes are vulnerable to hypoxic/hypoglycaemic conditions (Lyons and Kettenmann 1998; Yan and Rivkees 2006), their metabolic properties, in particular, their glucose metabolism, have not been investigated in depth (Amaral et al. 2013). In contrast, the metabolic interactions revolving around glucose between neurons and astrocytes have received considerable attention since their discovery in the 1970s (van den Berg and Garfinkel 1971). Specifically, the shuttling of glutamine–glutamate–γ-Aminobutyric acid (GABA) between astrocytes and neurons is thought to be fundamentally important for neuronal function and is the focus of other chapters in this book. Glucose is primarily metabolized to pyruvate via glycolysis in the cytosol. Stepwise conversion of a single glucose molecule into pyruvate generates two molecules of ATP. This reaction is not oxygen dependent. Glucose metabolism can also take an alternative route via a biosynthetic pathway termed phosphate pentose pathway (PPP). This complex detour bypasses several steps of glycolysis without consuming or producing ATP. In the first, oxidative phase of the PPP,

NADP+ is converted into NADPH. NADPH acts as a reducing agent that may participate in lipid and steroid synthesis or in the production of glutathione and thioredoxin, which are involved in the cell's defence mechanism against oxidative stress. In the second phase of the PPP, 5-carbon sugars are non-oxidatively synthesized. The PPP joins the glycolytic pathway at the level of glyceraldehyde-3-phosphate (GA3P) and fructose-6-phosphate (fructose-6P). Fructose-6P is subsequently converted into pyruvate, which constitutes the endpoint of both glycolysis and the PPP.

In the presence of oxygen, the pyruvate produced by glycolysis or by the PPP can be metabolized in the mitochondrial tricarboxylic acid cycle (TCA) to further produce ATP via coupling to the mitochondrial electron transport chain. Alternatively, pyruvate can be (reversibly) converted into lactate in the cytosol, which results in the production of NAD+ from NADH. Net synthesis of TCA cycle intermediates and related compounds, including glutamate and glutamine, depend on the entry of pyruvate via anaplerotic replenishment of intermediates in the TCA cycle. In the brain, this is mediated by pyruvate carboxylase (PC) (Patel 1974). PC is not expressed in neurons, but it is present in astrocytes (Yu et al. 1983; Shank et al. 1985; Cesar and Hamprecht 1995) [for review see Sonnewald and Rae (2010) and Sonnewald (2014)]. Consequently, neurons depend on astrocytes as an external source of glutamine for the production of neurotransmitters. Conversion of pyruvate by PC generates a 'new' molecule of oxaloacetate. Oxaloacetate may subsequently condense with acetyl CoA to synthesize the TCA cycle intermediate citrate which, after several steps, is converted to α-ketoglutarate from, which, glutamate can be formed by transamination or amination. In a subsequent step, glutamine synthetase, which is known to be expressed in astrocytes (Norenberg and Martinez-Hernandez 1979), is able to convert glutamate into glutamine (Fig. 10.1).

In the present chapter, we summarize the available data on how oligodendrocyte lineage cells may contribute to the metabolic interactions in the CNS. We propose that intercellular shuttling of metabolites occurs between all three major cell groups of the CNS (neurons, astrocytes and oligodendrocytes), instead of being restricted to closed-loop interactions between astrocytes and neurons (Fig. 10.1).

10.2 Oligodendrocyte Precursor Cells (OPCs): Functions and Metabolic Role in the Brain

10.2.1 Origin and Development of Oligodendrocyte Precursors

OPCs are small unipolar or bipolar cells and can be identified by the expression of the NG2 proteoglycan and platelet-derived growth factor receptor alpha (PDGFR-α), as well as the more general oligodendroglial lineage markers such as SOX10 and OLIG2 (Richardson et al. 2011) (Fig. 10.2). OPCs are scattered randomly throughout the CNS, occurring in both white matter and grey matter; they make up approximately 5–8 % of the cell population in the CNS (Levine et al. 2001).

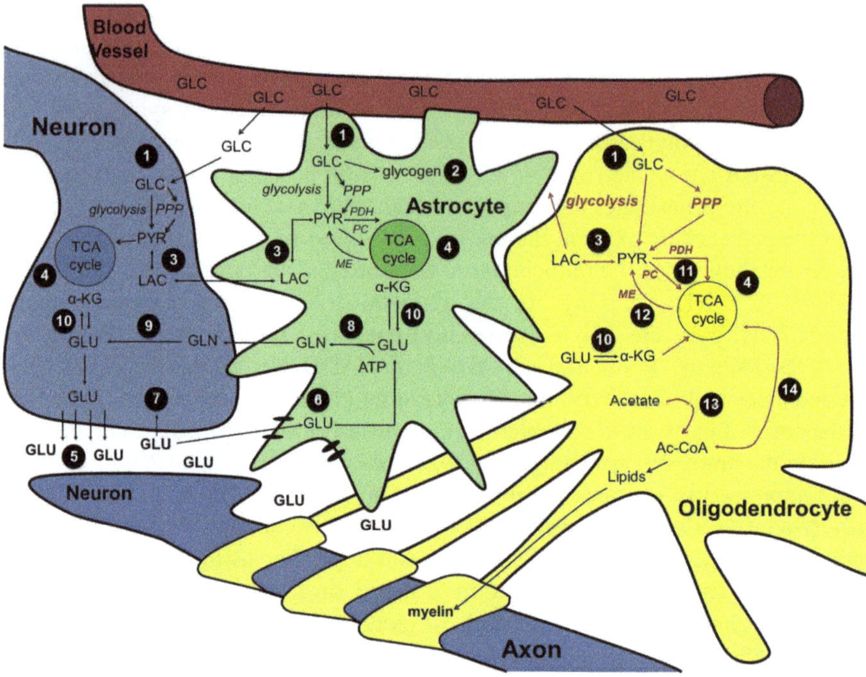

Fig. 10.1 Schematic overview of metabolic interactions between neurons–astrocytes–oligodendrocytes. Glucose (GLC) from the blood is taken up by neurons, astrocytes and oligodendrocytes and can be metabolized via glycolysis and the pentose phosphate pathway (PPP) (*1*), both giving rise to pyruvate (PYR) formation. In astrocytes, GLC can also be stored in the form of glycogen (*2*). PYR can be reduced to lactate (LAC) (*3*) which can be released and taken up by cells with lower lactate concentration or be converted into acetyl CoA (Ac CoA) and subsequently oxidized in the tricarboxylic acid (TCA) cycle (*4*). After synaptic release of glutamate (GLU) by neurons (*5*), astrocytes are responsible for the most of its uptake via specific high-affinity glutamate transporters (GLAST and GLT-1) (*6*) to prevent neuronal excitotoxicity, although some pre-synaptic re-uptake can also occur (*7*). GLU taken up by astrocytes can be converted to glutamine (GLN) (*8*) which can be transferred back to neurons where it is transformed into GLU (*9*), making it available again for neurotransmission and, in this way, closing the GLU–GLN cycle. The close association between GLU, GLN and TCA cycle metabolism is indicated in the three cell compartments: GLU can be additionally converted into α-ketoglutarate (α-KG) and be subsequently oxidized (*10*). The pathways investigated in the present study are highlighted in red in the oligodendrocyte: after glucose is metabolized either via glycolysis only or also via the PPP, the resulting pyruvate produced can be carboxylated via pyruvate carboxylase (PC) or enter the TCA cycle after being converted to acetyl CoA via PDH (*11*). Pyruvate can be further completely oxidized if it is decarboxylated via malic enzyme (ME) (*12*), which also seems to be present in this cell type. Oligodendrocytes can also metabolize acetate into acetyl CoA (*13*) that can be then incorporated into lipids or oxidized in the TCA cycle (*14*)

Despite the widespread distribution of OPCs in the adult CNS, they originate from the ventral (Sonic Hedgehog (Shh)-dependent) and dorsal (Shh-independent) regions during embryonic development (Richardson et al. 1997; Rowitch 2004; Cai et al. 2005; Vallstedt et al. 2005). It was estimated that dorsally derived OPCs

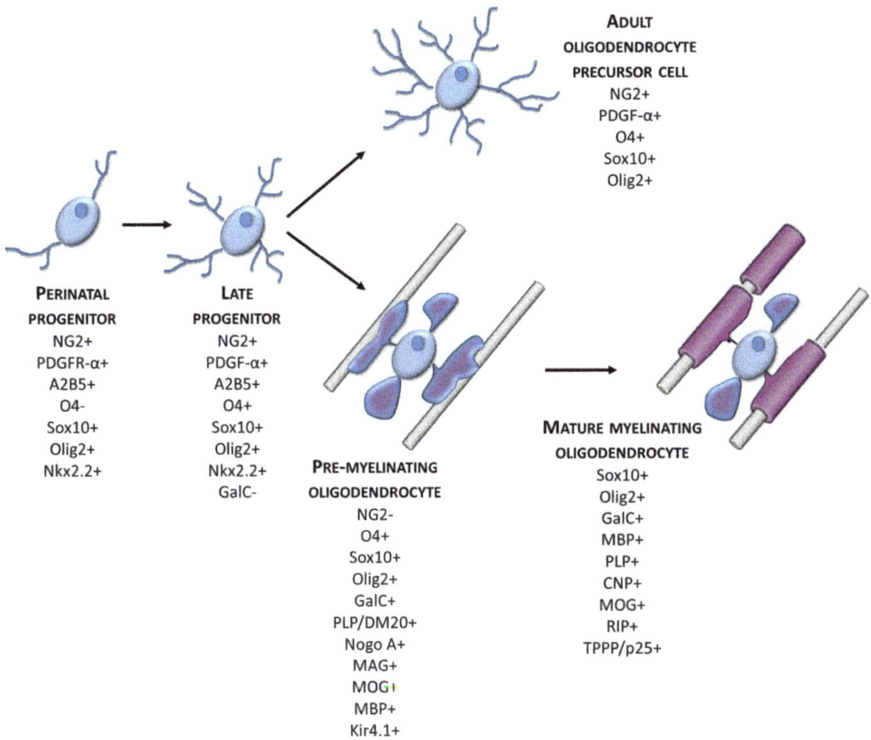

Fig. 10.2 Transcription factors and markers expressed by oligodendrocytes during differentiation. Oligodendrocyte progenitors can be identified by the lineage markers Sox10 and Olig2, in addition to the early stage-specific markers NG2, PDGFR-α and A2B5. Late stage progenitors start expressing O4, which remains being expressed throughout subsequent stages. As progenitors differentiate, they become more branched and might establish contact with axons and start myelinating. Pre-myelinating oligodendrocytes thus distinctively express lipid markers such as Galactoceramides (Galc) and myelin proteins including PLP, MBP, MAG and MOG that continue to be expressed in mature myelinating oligodendrocytes

constitute between 20 and 30 % of the total OPC number seen in the spinal cord of wild-type animals (Vallstedt et al. 2005) whereas the majority of adult oligodendrocytes in the spinal cord seems to be ventrally derived (Mitew et al. 2014). Nevertheless, it is still unclear which cellular source drives developmental myelination and which is the contribution of the different populations to different myelinated tracts in the CNS.

During the early postnatal period, OPCs migrate into the surrounding grey and white matter of the developing CNS. A proportion of OPCs engage with axons and differentiate into myelinating cells, whilst others are reserved as immature cells (adult progenitors) in the parenchyma (Fancy et al. 2011). While OPCs with distinct embryonic origins exhibit a preference to myelinate particular axonal tracts, their electrical properties seem to be comparable (Tripathi et al. 2011).

OPCs exhibit stem cell-like features as they can give rise to daughter cells, including astrocytes and oligodendrocytes, as first described by Raff and colleagues (Raff et al. 1983; Ffrench-Constant and Raff 1986; Kondo and Raff 2000). More recently, their unexpected ability to form cells that are usually associated with PNS development, namely Schwann cells, was reported (Zawadzka et al. 2010). This multipotent capacity, also confirmed by the generation of astrocytes in vivo, in Rivers et al. (2008), Tatsumi et al. (2008), and Guo et al. (2010), consolidated the view of OPCs as an internal source of cells for repair within the CNS. The potential of these cells for CNS remyelination (repair of damaged myelin) has been recently reviewed (Crawford et al. 2014).

In addition to their multipotency, OPCs are also capable of self-renewal (Rivers et al. 2008; Psachoulia et al. 2009; Kang et al. 2010). However, whether these cells have the unlimited potential of self-renewal required of a stem cell or whether their proliferation potential is limited akin to that of a precursor cell still remains to be determined (Crang et al. 1998). More recently, it has been proposed that they can also undergo asymmetrical cellular division and, as a consequence, a growing field investigates OPCs as a source for glioma formation (Sugiarto et al. 2011; Lewis and Petritsch 2013; Galvao et al. 2014).

10.2.2 OPC Migration, Differentiation and Myelination

Before myelination takes place, multiple signalling molecules and mitogens guide OPCs throughout the CNS to myelinating sites. These include the mitogenic factors PDGF and FGF-2 that stimulate cell motility; a variety of extracellular matrix molecules (e.g. fibronectin, N-cadherin, laminin) and cell adhesion molecules (e.g. polysialylated neuronal cell adhesion molecule (PSA-NCAM)) on the surface of CNS cells providing a dynamic environment of signals and cues that help guide the OPCs, as well as chemoattractant and repellent cues (see Mitew et al. 2014 for a comprehensive review).

Once OPCs have populated the neuraxis, another set of signalling pathways balancing inhibitors and effectors drive differentiation. Triiodothyronine/thyroid hormone 3 (T3) signalling in conjunction with the withdrawal of mitogenic cues regulates a crucial step for differentiation and myelin development (Barres et al. 1994). IGF-1 signalling is another positive inducer of myelination, promoting OPC differentiation and expression of myelin genes (Carson et al. 1993; Ye et al. 1995; Goddard et al. 1999) (See Mitew et al. 2014 for a full list of related references). Differentiation into post-mitotic, premyelinating oligodendrocytes generally proceeds in a caudal-to-rostral gradient in the brain but rostrocaudally in the spinal cord (Brody et al. 1987) and entails profound changes in both cellular behaviour and gene expression. The dynamic nature of oligodendrocytes being able to extend processes, sense axons and myelinate multiple axons simultaneously requires a complex and dynamic system of signalling mechanisms converging on the oligodendrocyte to promote and coordinate each action. Although these extracellular signals have various roles in the myelination process, they can be grossly categorized into two groups: those that modulate oligodendrocyte process extension and initial axonal contact, and factors that dictate subsequent myelin thickening (Mitew et al. 2014).

To form myelin, OPCs exit the cell cycle and begin to express a subset of myelin-associated proteins (Campagnoni and Macklin 1988) such as proteolipid protein (PLP), myelin basic protein (MBP) and myelin oligodendrocyte glycoprotein (MOG), among others, typical markers of mature OL (Duchala et al. 1995) (Fig. 10.2). As part of OPC differentiation, OPCs extend processes to nearby axons; recognizing and adhering to axonal targets the processes flatten and become ensheathing. Myelin components are then synthesized and transported to the appropriate location within the sheath, axons are wrapped and the myelin membrane is compacted such that the cytoplasmic myelin leaflets are practically fused and almost devoid of cytoplasm (Snaidero et al. 2014). It is estimated that each oligodendrocyte can produce up to 40 myelin segments in multiple axons (Miron et al. 2011).

In addition to the role of signalling pathways operating during development and the early post-natal period to regulate OPC differentiation, also neurons regulate myelination via cAMP (Trajkovic et al. 2006), BDNF and neuregulin (Lundgaard et al. 2013) or synaptic activity (Wake et al. 2011; Hines et al. 2015; Mensch et al. 2015) to oligodendroglial glutamate receptors (Karadottir and Attwell 2007). However, it is still unclear in which conditions axons drive the myelination process since cultured oligodendrocytes can also myelinate synthetic nanofibres (Lee et al. 2012a). In this model, only nanofibres of >400 nm were myelinated, suggesting a size-dependent threshold for myelination. Neuronal activity seems to trigger OPC differentiation and myelination of an active neural circuit, a concept called adaptive myelination (Gibson et al. 2014), which is important, for example, to the acquisition of improved motor functions. Moreover, neuronal signalling is also being investigated in the context of myelin repair (Gautier et al. 2015), to understand how adult OPCs migrate to areas of damaged myelin and remyelinate demyelinated axons.

In vitro studies of oligodendrocyte biology were mostly based on primary rodent oligodendrocyte-enriched cultures, which are normally purified from mixed glia cultures (McCarthy and de Vellis 1980). This approach results in high yields of relatively pure (>93 %) cultures (Amaral et al. 2016). The morphological and transcriptional events that occur in vivo are recapitulated when O4+, A2B5+, O1−, CNP (20,30-cyclicnucleotide 30-phosphodiesterase)- and MBP-rodent OPCs are cultured in Sato's medium (McCarthy and de Vellis 1980). These culture conditions enable the formation of a monolayer of mature oligodendrocytes with complex-branched processes and membrane sheets, which express late stage markers, including MBP, CNP and MAG (Figs. 10.2 and 10.3). Although a significant proportion of cells (approximately 60 %) reach a mature oligodendrocyte stage, the cultures also include late stage progenitors.

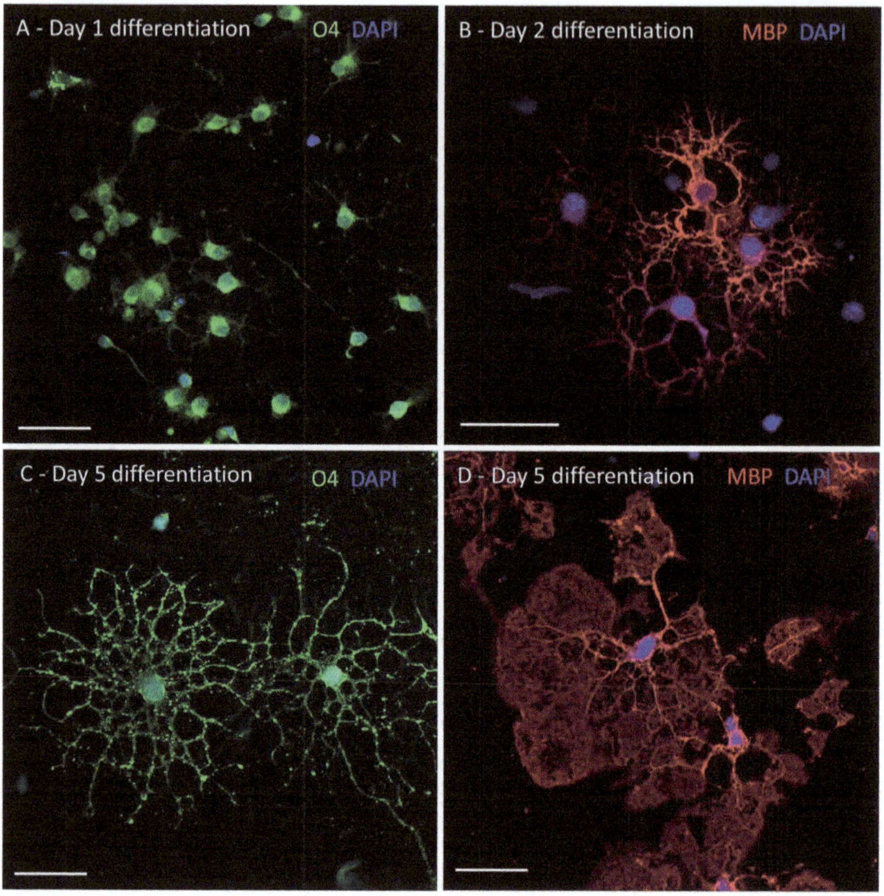

Fig. 10.3 Primary cultures of rat oligodendrocytes at different stages of differentiation. At day 1 in culture, oligodendrocyte lineage cells (O4-positive) present a simple and mostly bipolar morphology with only a few branches (**a**). At day 2 in culture, the morphology is more complex and branched, with some cells already expressing MBP (**b**). At day 5 of differentiation, cells are highly ramified and present intricate morphology and complex branched structure, as indicated by O4 staining (**c**). Above 60 % of the cells in culture at 5 days of differentiation express the mature marker myelin basic protein (MBP) (**d**)

10.2.3 Oligodendrocyte Physiology and Function

Recent data indicate that the oligodendrocyte population is heterogeneous and can differ in location, morphology, activity and function. They have a highly ramified morphology and extend processes to synapses (Bergles et al. 2000) and nodes of Ranvier (Huang et al. 2005) suggesting that they may also monitor or modulate neuronal activity. Indeed, a subpopulation expresses voltage-gated sodium and potassium channels, generates action potentials when depolarized and senses the environment by receiving excitatory and inhibitory synaptic input from axons (Karadottir et al. 2008). This suggests that the oligodendrocyte lineage cells may process information as well as act as a reservoir of new oligodendrocytes: It is possible that subpopulations exist that are specialized to perform these different tasks (Crawford et al. 2014).

Nevertheless, the main function of oligodendrocytes is to make myelin. Myelin accounts for approximately 70 % of the dry weight of the mammalian CNS (Siegel and Albers 2006) and is critical in maintaining electrical impulse conduction and maximizing its velocity. Myelin sheaths are arranged in segments separated by nodes of Ranvier, where sodium channels are clustered in high density in the axon membrane so that they can produce action potentials (Waxman and Ritchie 1993). Myelin covers and masks the internodal parts of the axon, which contain fewer sodium channels and a higher density of potassium channels, which tend to oppose the generation of action potentials (Waxman and Ritchie 1993). Myelination results in axons with high resistance and low capacitance increasing the speed of electrical conduction and improving its metabolic efficiency, thereby optimizing space and energy utilization (Waxman and Ritchie 1993). In addition, myelin sheaths play an important role in maintaining axonal integrity (Nave 2010a, b). Although deletion of single myelin genes such as PLP (Griffiths et al. 1998) or CNP (Lappe-Siefke et al. 2003) does not lead to overt structural changes of myelin sheaths, animals thus affected display enhanced chronic axonal degeneration. Therefore, myelin sheaths not only provide physiological and structural but also trophic support for axons, promoting their viability and integrity (Nave and Trapp 2008; Nave 2010a, b). These interactions may involve metabolic interactions between OL and axons via the transfer of lactate (Funfschilling et al. 2012) and, eventually, other molecules released by OLs. The disruption of such metabolic interactions might contribute to the pathology of neurodegenerative disorders such as Amyotrophic Lateral Sclerosis (Lee et al. 2012b). Therefore, a new niche of investigation has emerged on the metabolic role of OLs in the brain (Amaral et al. 2013; Hirrlinger and Nave 2014).

10.2.4 Oligodendrocyte Metabolism

Glucose is the main cerebral energy substrate. Brain glucose metabolism exhibits large regional variations at rest (consider, for example, the difference between grey and white matter) and is profoundly affected by brain activation (Siegel and Albers 2006), indicating a strong link between energy metabolism and brain function. Glucose oxidation has been measured in neurons, astrocytes and oligodendrocytes from the developing brain and these measurements showed that oligodendrocytes oxidized twice as much glucose as astrocytes in the TCA cycle and had only slightly less oxidation than neurons (Edmond et al. 1987; Amaral et al. 2016). Moreover, glucose and lactate have been identified as important regulators of oligodendrocyte development and myelination (Rinholm et al. 2011). Recent findings indicate that axons and oligodendrocytes are coupled with respect to their metabolism, with oligodendrocytes providing aerobic glycolysis products to axons under conditions that cause energy deprivation (Funfschilling et al. 2012).

The recently recognized trophic and metabolic roles of myelin provide an explanation for the progressive atrophy of demyelinated axons. Conversely, they also suggest that promoting remyelination could reinstate this supportive function and prevent the neurodegenerative component of demyelinating disease (Franklin and Ffrench-Constant 2008; Crawford et al. 2014).

10.2.4.1 Glucose and Lactate Metabolism in Oligodendrocytes

A few studies have investigated glucose metabolism in oligodendrocyte lineage cells (Sanchez-Abarca et al. 2001; Rinholm et al. 2011; Amaral et al. 2016). Our recent findings indicate that glucose metabolism is very active in oligodendrocyte lineage cells, in particular, their glucose consumption rate is comparable to that of astrocytes (Amaral et al. 2016). Moreover, based on the ratios of lactate release to glucose consumption for both types of cultures, it has been reported that oligodendrocytes use a larger proportion of glucose for mitochondrial metabolism than astrocytes (Sanchez-Abarca et al. 2001; Amaral et al. 2016) although they also release substantial amounts of lactate (Amaral et al. 2016). In fact, lactate is a metabolic intermediate thought to mediate oligodendroglial trophic support to axons as proposed recently by Funfschilling et al. (2012) and Lee et al. (2012b).

Lactate is also used as substrate by oligodendrocytes. It supports myelination and oligodendrocyte development in the absence of sufficient glucose levels (Rinholm et al. 2011) and is metabolized for lipid synthesis to a larger extent than in astrocytes and neurons (Sanchez-Abarca et al. 2001). On the other hand, extracellular lactate also appears to function as a metabolic activator in oligodendrocytes (Amaral et al. 2016). Investigating glucose metabolism with isotope labelling indicated that both glucose consumption and the proportion of glucose-derived pyruvate metabolized in the mitochondria were increased in the presence of exogenous lactate (Amaral et al. 2016). A net production of lactate was observed indicating that glycolysis in oligodendrocytes is not inhibited by extracellular lactate (Amaral et al. 2016).

The Pentose Phosphate Pathway (PPP)

The PPP is a glucose shunt, which is thought to be active in neurons and astrocytes (Almeida et al. 2002; Garcia-Nogales et al. 2003; Amaral et al. 2010; Bolanos and Almeida 2010; Brekke et al. 2012). Using ^{14}C tracing techniques, Edmond et al. (1987) showed active PPP in oligodendrocyte lineage cells, whereas Sanchez-Abarca et al. (2001) reported PPP activity in immature OPC cultures as being two-fold higher than in astrocytes and four-fold higher than in neurons. We recently measured the contribution of the PPP to the synthesis of glycolytic intermediates and glutamate synthesis based on ^{13}C-tracing techniques. This demonstrated that mature oligodendrocytes use approximately 10–15 % of glucose in the PPP compared to glycolysis as observed in the labelling of the glycolytic intermediates PEP and 3PG. These values are in the range of previously published data for cortical astrocyte cultures in a metabolic modelling study also using ^{13}C-labelled substrates (Amaral et al. 2011a, b), which contrasts with the report by Sanchez-Abarca et al. (2001). A potential explanation for the differences is that our cultures contained more mature oligodendrocytes than those used by Sanchez-Abarca et al. (2001), which may be associated with lower PPP activity, presenting a rate closer to the one observed in astrocytes (Amaral et al. 2011a). On the basis of ^{14}C tracing experiments, (Sykes et al. 1986) reported that the PPP produces more CO_2 than the TCA cycle in primary oligodendrocyte lineage cells due to its close link to de novo synthesis of fatty acids and cholesterol (Sykes et al. 1986). Since it was not feasible to measure its contribution to lipid synthesis, it is therefore possible that our approach underestimated the total use of glucose via the PPP. Furthermore, our study showed for the first time that pyruvate generated from glucose via the PPP contributes to the synthesis of acetyl CoA for oxidation and generation of metabolites in the TCA cycle in oligodendrocytes, contributing to approximately 10 % of the glutamate produced (Amaral et al. 2016). Estimation of the PPP activity on the basis of glutamate isotopomers indicated that the PPP accounted for approximately 6 % of glucose metabolism in cortical neurons and approximately 4 % in cerebellar neurons (Brekke et al. 2012). Similar to what was reported in neurons (Brekke et al. 2012), we found that oligodendrocytes incorporate ^{13}C label in glutamate produced by [1,2-^{13}C]glucose metabolism via the PPP.

Despite significant PPP activity in oligodendrocytes, it remains largely unknown whether the PPP plays a role during pathological events affecting oligodendrocyte function. Interestingly, increased activity of the enzyme transaldolase (which is part of the non-oxidative branch of the PPP and is involved in lipid and nucleotide synthesis) has been reported in oligodendrocytes in brains of Multiple Sclerosis (MS) patients compared to healthy controls (Banki et al. 1994). Banki and colleagues suggested that the PPP might be involved in the protection of myelin sheaths from oxidative stress and that transaldolase appears to be a target to break immunological tolerance and kill oligodendrocytes in MS. Altered PPP activity has also been reported in a study on traumatic brain injury (TBI) patients although the authors did not establish any correlation with oligodendrocyte function (Dusick et al. 2007). It is likely that part of PPP alterations observed in the study by Dusick et al. (2007) could be linked to extensive demyelination and remyelination that is known to occur in TBI patients (Armstrong et al. 2015).

Mitochondrial Metabolism of Pyruvate Generated from Glucose

Using different forms of ^{13}C labelled glucose our recent study also demonstrated that mature oligodendrocytes exhibit a high rate of mitochondrial metabolism (Amaral et al. 2016). These findings echo the results from previous studies by Sanchez-Abarca et al. (2001). Functional mitochondria seem to be particularly important when OPCs differentiate (Schoenfeld et al. 2010; Ziabreva et al. 2010). It is likely that OPC differentiation, which involves a complex expansion of cell morphology (Fig. 10.3), requires high levels of glucose metabolism to meet the increased demands of ATP, citrate and amino acids required for the synthesis of cellular components. In contrast, it was proposed that myelinating oligodendrocytes are not dependent on mitochondrial activity in vivo (Funfschilling et al. 2012). The most likely explanation for the discrepancy between the high mitochondrial demands of the cultures used in Amaral et al. (2016) and the findings in vivo is the significant presence of pre-myelinating (MBP-negative) OPCs that have not yet reached fully mature stages in vitro. Our data suggest a prominent role of oligodendrocyte oxidative metabolism at the late stages of OPC differentiation, including the pre-myelinating and early myelinating stages.

Pyruvate Carboxylation

An important question with respect to mitochondrial metabolism in oligodendrocytes is whether they can replenish TCA cycle intermediates via anaplerosis. It is well established that neurons depend on astrocytes for replenishing their TCA cycle intermediates. As neurons cannot carboxylate pyruvate, external TCA cycle intermediates are required for the synthesis of amino acid neurotransmitters (McKenna et al. 2012). We therefore investigated whether and to which extent oligodendrocytes are self-sufficient with respect to the production of anaplerotic substrates. Incubation of cells with [1,2-^{13}C]glucose and subsequent mass spectrometry analysis of cell extracts indicated that pyruvate carboxylation indeed takes place in oligodendrocytes as shown by the production of [2,3-^{13}C]glutamate. This was further confirmed by label incorporation in citrate in the medium of cells incubated with [1-^{13}C]lactate, which is only possible via pyruvate carboxylation (Amaral et al. 2016). Comparable label incorporation from [1-^{13}C]lactate was found in citrate in the medium of astrocyte cultures. This suggests that lactate is metabolized in a similar way in the TCA cycle of both cell types. It must be noted that the overall contribution of pyruvate carboxylation to oligodendrocyte and astrocyte metabolism is underestimated in experiments using [1-^{13}C]lactate because the [1-^{13}C]pyruvate from [1-^{13}C]lactate competes with unlabelled pyruvate from glucose, which was also present in the incubation medium.

Pyruvate can be carboxylated to oxaloacetate by PC or to malate and NADP+ by malic enzyme. Whether PC or malic enzyme is responsible for pyruvate carboxylation in oligodendrocytes remains to be established. Murin et al. (2009) reported PC expression in cultured oligodendroglia. Whether oligodendrocytes express malic enzyme remains unknown. In neurons and astrocytes malic enzyme only works in the direction of pyruvate production (McKenna et al. 1995, 2000). Moreover, PC has been shown to be the most important anaplerotic enzyme in the brain (Patel 1974). Irrespective of which enzyme is responsible for pyruvate carboxylation, the detection of carboxylation (and thus anaplerosis) has consequences for oligodendrocyte metabolism. If oligodendrocytes (similarly to neurons) were not capable of anaplerosis they would depend on the provision of glutamine by astrocytes, which have a net production of glutamine via pyruvate carboxylation (Gamberino et al. 1997; Waagepetersen et al. 2001).

An alternative route to replenish the TCA cycle in oligodendrocytes is to use aspartate, liberated from *N*-acetyl aspartate (NAA), which in turn is supplied by neurons (Moffett et al. 2007). However, aspartate released by NAA hydrolysis in oligodendrocytes could potentially be sent back to neurons, thus avoiding the depletion of anaplerotic substrates in neurons (NAA synthesis in neurons is dependent on glutamine entry from astrocytes). Our recent study demonstrated that oligodendrocytes are capable of anaplerosis, which suggests that they are potentially independent of astrocytic pyruvate carboxylation. Whether the level of anaplerosis in oligodendrocytes is sufficient to meet their entire requirements is not known at present. It is also unclear whether aspartate is shuttled back to neurons or whether it is metabolized in oligodendrocytes. Both possibilities have previously been suggested but evidence is lacking (Baslow and Guilfoyle 2006; Moffett et al. 2007).

10.2.4.2 Acetate Metabolism

Acetyl CoA is an essential molecule in the TCA cycle. Most acetyl CoA derives from pyruvate via PDH; however, oligodendrocytes are also known to express the enzyme aspartoacylase, which catalyses the hydrolysis of NAA into aspartate and acetate (Moffett et al. 2011). NAA-derived acetate significantly contributes to myelin lipid synthesis in the CNS (Chakraborty et al. 2001) and is also thought to support oxidative metabolism during myelination (Francis et al. 2012). Furthermore, oligodendrocytes express acetyl CoA synthetase-1 (catalyses the synthesis of acetyl coenzyme A from acetate and coenzyme A), especially during postnatal brain development, supporting the contribution of acetate to lipid synthesis in these cells (Ariyannur et al. 2010). Currently, it is thought that astrocytes can convert acetate into acetyl CoA whereas neurons are not able to do so (Muir et al. 1986; Sonnewald et al. 1993). Consequently, acetate has been used extensively to assess astrocyte metabolism in the context of astrocytic-neuronal interactions in vivo (e.g. Melo et al. 2005; Morken et al. 2014; Nilsen et al. 2014), in neurospheres (Sa Santos et al. 2011) and in culture (Sonnewald et al. 1993). However, these studies did not consider potential metabolic contributions from oligodendrocytes. Our recent study demonstrated that mature oligodendrocyte cultures are able to convert acetate into acetyl CoA and oxidize it in the mitochondria as shown by the incorporation of ^{13}C label from [1,2-^{13}C]acetate into the TCA cycle intermediates malate and citrate and the amino acids glutamate and glutamine. Labelling from [1,2-^{13}C]acetate was not as pronounced as labelling from [1,2-^{13}C]glucose but nevertheless significant and comparable to that observed in astrocytes (Amaral et al. 2016). Surprisingly, we found that alanine was also labelled from [1,2-^{13}C]acetate, and that alanine M+1 enrichment was detected in oligodendrocytes incubated with [1,2-^{13}C]glucose. Both isotopologues of alanine could not have been produced without the participation of the TCA cycle and malic enzyme or pyruvate kinase (PK) and phosphoenolpyruvate carboxy kinase (PEPCK) (Cruz et al. 1998). This indicates that pyruvate recycling, a catabolic pathway (Cerdan et al. 1990; Kunnecke et al. 1993; Haberg et al. 1998; Olstad et al. 2007; Amaral et al. 2011b), is not only active in astrocytes and neurons but also in oligodendrocytes.

10.3 Conclusion

It is now clear that oligodendrocytes have an important metabolic role in the brain. Metabolic support to axons is essential and its dysfunction might be implicated in many neurological disorders, including ischemia (Bakiri et al. 2008), ALS (Lee et al. 2012a, b) and psychiatric disorders (Nave and Ehrenreich 2014), in addition to the classical myelin disorders (Fancy et al. 2010). The observation of pyruvate carboxylation activity in oligodendrocytes suggests a possible role in a glutamate–glutamine cycle (Fig. 10.1) with axons, taking into account the studies reporting glutamate release from axons targeting NMDA receptors on oligodendrocytes (Karadottir and Attwell 2007). A glutamate–glutamine cycle might also play a role in OPC differentiation and myelination in the sequence of studies showing that neuronal signalling promotes OPC differentiation in certain conditions (Wake et al. 2011; Gibson et al. 2014; Gautier et al. 2015). Further studies are required to address these particular questions.

Moreover, the stem cell features of adult OPCs suggest that metabolic remodelling is likely to occur during differentiation as in other stem cell populations. Such metabolic changes might be determinant for the correct migration, proliferation and differentiation of OPCs during development or during adulthood in contexts of remyelination. It is likely that the demand for building blocks required to expand cell processes and cytoskeleton of a bipolar cell into a complex ramified cell extending myelin sheaths is high and requires specialized energetic demand. It is also likely that damage induced by reactive oxygen species (Lassmann 2010) and glutamate in MS and ischemic injury might affect oligodendrocyte metabolic functions and contribute to an incorrect differentiation/remyelination.

Finally, the possibility of obtaining human oligodendrocytes derived from pluripotent stem cells, e.g. (Stacpoole et al. 2013; Douvaras and Fossati 2015; Gorris et al. 2015) is creating the unique opportunity of investigating metabolism in human cells, including in models of disease. The metabolic characterization of human oligodendrocytes will enable to compare rodent and human cells and investigate metabolic abnormalities linked to myelin disorders in patient-derived cells.

In conclusion, oligodendrocyte metabolism is a growing field of research and future studies will further elucidate oligodendrocyte metabolic function, both in immature and mature cells and also in vivo, to shed new light into the mechanisms of OPC differentiation and neuronal–glial interactions in physiological and disease contexts.

Acknowledgements The preparation of this book chapter was supported by funding from the UK Multiple Sclerosis Society.

References

Almeida A, Delgado-Esteban M, Bolanos JP, Medina JM (2002) Oxygen and glucose deprivation induces mitochondrial dysfunction and oxidative stress in neurones but not in astrocytes in primary culture. J Neurochem 81(2):207–217

Amaral AI, Teixeira AP, Martens S, Bernal V, Sousa MFQ, Alves PM (2010) Metabolic alterations induced by ischemia in primary cultures of astrocytes: merging 13C NMR spectroscopy and metabolic flux analysis. J Neurochem 113(3):735–748

Amaral AI, Teixeira AP, Haakonsen BI, Sonnewald U, Alves AM (2011a) A comprehensive metabolic profile of cultured astrocytes using isotopic transient metabolic flux analysis and 13-C-labeled glucose. Front Neuroenerg 3(5)

Amaral AI, Teixeira AP, Sonnewald U, Alves PM (2011b) Estimation of intracellular fluxes in cerebellar neurons after hypoglycemia: importance of the pyruvate recycling pathway and glutamine oxidation. J Neurosci Res 89(5):700–710

Amaral AI, Meisingset TW, Kotter MR, Sonnewald U (2013) Metabolic aspects of neuron-oligodendrocyte-astrocyte interactions. Front Endocrinol (Lausanne) 4:54

Amaral AI, Hadera MG, Tavares JM, Kotter MR, Sonnewald U (2016) Characterization of glucose-related metabolic pathways in differentiated rat oligodendrocyte lineage cells. Glia 64(1):21–34

Ariyannur PS, Moffett JR, Madhavarao CN, Arun P, Vishnu N, Jacobowitz DM, Hallows WC, Denu JM, Namboodiri AM (2010) Nuclear-cytoplasmic localization of acetyl coenzyme a synthetase-1 in the rat brain. J Comp Neurol 518(15):2952–2977

Armstrong RC, Mierzwa AJ, Sullivan GM, Sanchez MA (2015) Myelin and oligodendrocyte lineage cells in white matter pathology and plasticity after traumatic brain injury. Neuropharmacology. doi:10.1016/j.neuropharm.2015.04.029. [Epub ahead of print]

Bakiri Y, Hamilton NB, Karadottir R, Attwell D (2008) Testing NMDA receptor block as a therapeutic strategy for reducing ischaemic damage to CNS white matter. Glia 56(2):233–240

Banki K, Colombo E, Sia F, Halladay D, Mattson DH, Tatum AH, Massa PT, Phillips PE, Perl A (1994) Oligodendrocyte-specific expression and autoantigenicity of transaldolase in multiple sclerosis. J Exp Med 180(5):1649–1663

Barres BA, Lazar MA, Raff MC (1994) A novel role for thyroid hormone, glucocorticoids and retinoic acid in timing oligodendrocyte development. Development 120(5):1097–1108

Baslow MH, Guilfoyle DN (2006) Functions of N-acetylaspartate and N-acetylaspartyl-glutamate in brain—evidence of a role in maintenance of higher brain integrative activities of information processing and cognition. In: Moffett JR, Tieman SB, Weinberger DR, Coyle JT, Namboodiri AMA (eds) N-acetylaspartate: a unique neuronal molecule in the central nervous system, vol 576. Springer, Bethesda, MD, pp 95–112

Bergles DE, Roberts JD, Somogyi P, Jahr CE (2000) Glutamatergic synapses on oligodendrocyte precursor cells in the hippocampus. Nature 405(6783):187–191

Bolanos JP, Almeida A (2010) The pentose-phosphate pathway in neuronal survival against nitrosative stress. IUBMB Life 62(1):14–18

Brekke EM, Walls AB, Schousboe A, Waagepetersen HS, Sonnewald U (2012) Quantitative importance of the pentose phosphate pathway determined by incorporation of 13C from [2-13C]- and [3-13C]glucose into TCA cycle intermediates and neurotransmitter amino acids in functionally intact neurons. J Cereb Blood Flow Metab 32(9):1788–1799

Brody BA, Kinney HC, Kloman AS, Gilles FH (1987) Sequence of central nervous system myelination in human infancy. I. An autopsy study of myelination. J Neuropathol Exp Neurol 46(3):283–301

Cai J, Qi Y, Hu X, Tan M, Liu Z, Zhang J, Li Q, Sander M, Qiu M (2005) Generation of oligodendrocyte precursor cells from mouse dorsal spinal cord independent of Nkx6 regulation and Shh signaling. Neuron 45(1):41–53

Campagnoni AT, Macklin WB (1988) Cellular and molecular aspects of myelin protein gene expression. Mol Neurobiol 2(1):41–89

Carson MJ, Behringer RR, Brinster RL, McMorris FA (1993) Insulin-like growth factor I increases brain growth and central nervous system myelination in transgenic mice. Neuron 10(4):729–740

Cerdan S, Kunnecke B, Seelig J (1990) Cerebral metabolism of [1,2-$^{13}C_2$]acetate as detected by in vivo and in vitro ^{13}C NMR. J Biol Chem 265(22):12916–12926

Cesar M, Hamprecht B (1995) Immunocytochemical examination of neural rat and mouse primary cultures using monoclonal antibodies raised against pyruvate carboxylase. J Neurochem 64(5):2312–2318

Chakraborty G, Mekala P, Yahya D, Wu G, Ledeen RW (2001) Intraneuronal N-acetylaspartate supplies acetyl groups for myelin lipid synthesis: evidence for myelin-associated aspartoacylase. J Neurochem 78(4):736–745

Crang AJ, Gilson J, Blakemore WF (1998) The demonstration by transplantation of the very restricted remyelinating potential of post-mitotic oligodendrocytes. J Neurocytol 27(7):541–553

Crawford AH, Stockley JH, Tripathi RB, Richardson WD, Franklin RJ (2014) Oligodendrocyte progenitors: adult stem cells of the central nervous system? Exp Neurol 260:50–55

Cruz F, Scott SR, Barroso I, Santisteban P, Cerdan S (1998) Ontogeny and cellular localization of the pyruvate recycling system in rat brain. J Neurochem 70(6):2613–2619

Douvaras P, Fossati V (2015) Generation and isolation of oligodendrocyte progenitor cells from human pluripotent stem cells. Nat Protoc 10(8):1143–1154

Duchala CS, Asotra K, Macklin WB (1995) Expression of cell surface markers and myelin proteins in cultured oligodendrocytes from neonatal brain of rat and mouse: a comparative study. Dev Neurosci 17(2):70–80

Dusick JR, Glenn TC, Lee WN, Vespa PM, Kelly DF, Lee SM, Hovda DA, Martin NA (2007) Increased pentose phosphate pathway flux after clinical traumatic brain injury: a [1,2-13C2] glucose labeling study in humans. J Cereb Blood Flow Metab 27(9):1593–1602

Edmond J, Robbins RA, Bergstrom JD, Cole RA, de Vellis J (1987) Capacity for substrate utilization in oxidative metabolism by neurons, astrocytes, and oligodendrocytes from developing brain in primary culture. J Neurosci Res 18(4):551–561

Fancy SP, Kotter MR, Harrington EP, Huang JK, Zhao C, Rowitch DH, Franklin RJ (2010) Overcoming remyelination failure in multiple sclerosis and other myelin disorders. Exp Neurol 225(1):18–23

Fancy SP, Chan JR, Baranzini SE, Franklin RJ, Rowitch DH (2011) Myelin regeneration: a recapitulation of development? Annu Rev Neurosci 34:21–43

Ffrench-Constant C, Raff MC (1986) The oligodendrocyte-type-2 astrocyte cell lineage is specialized for myelination. Nature 323(6086):335–338

Francis JS, Strande L, Markov V, Leone P (2012) Aspartoacylase supports oxidative energy metabolism during myelination. J Cereb Blood Flow Metab 32(9):1725–1736

Franklin RJ, Ffrench-Constant C (2008) Remyelination in the CNS: from biology to therapy. Nat Rev Neurosci 9(11):839–855

Funfschilling U, Supplie LM, Mahad D, Boretius S, Saab AS, Edgar J, Brinkmann BG, Kassmann CM, Tzvetanova ID, Mobius W, Diaz F, Meijer D, Suter U, Hamprecht B, Sereda MW, Moraes CT, Frahm J, Goebbels S, Nave KA (2012) Glycolytic oligodendrocytes maintain myelin and long-term axonal integrity. Nature 485(7399):517–521

Galvao RP, Kasina A, McNeill RS, Harbin JE, Foreman O, Verhaak RG, Nishiyama A, Miller CR, Zong H (2014) Transformation of quiescent adult oligodendrocyte precursor cells into malignant glioma through a multistep reactivation process. Proc Natl Acad Sci U S A 111(40):E4214–E4223

Gamberino WC, Berkich DA, Lynch CJ, Xu B, LaNoue KF (1997) Role of pyruvate carboxylase in facilitation of synthesis of glutamate and glutamine in cultured astrocytes. J Neurochem 69(6):2312–2325

Garcia-Nogales P, Almeida A, Bolanos JP (2003) Peroxynitrite protects neurons against nitric oxide-mediated apoptosis. A key role for glucose-6-phosphate dehydrogenase activity in neuroprotection. J Biol Chem 278(2):864–874

Gautier HO, Evans KA, Volbracht K, James R, Sitnikov S, Lundgaard I, James F, Lao-Peregrin C, Reynolds R, Franklin RJ, Karadottir RT (2015) Neuronal activity regulates remyelination via glutamate signalling to oligodendrocyte progenitors. Nat Commun 6:8518

Gibson EM, Purger D, Mount CW, Goldstein AK, Lin GL, Wood LS, Inema I, Miller SE, Bieri G, Zuchero JB, Barres BA, Woo PJ, Vogel H, Monje M (2014) Neuronal activity promotes oligodendrogenesis and adaptive myelination in the mammalian brain. Science 344(6183):1252304

Goddard DR, Berry M, Butt AM (1999) In vivo actions of fibroblast growth factor-2 and insulin-like growth factor-I on oligodendrocyte development and myelination in the central nervous system. J Neurosci Res 57(1):74–85

Gorris R, Fischer J, Erwes KL, Kesavan J, Peterson DA, Alexander M, Nothen MM, Peitz M, Quandel T, Karus M, Brustle O (2015) Pluripotent stem cell-derived radial glia-like cells as stable intermediate for efficient generation of human oligodendrocytes. Glia 63(12):2152–2167

Griffiths I, Klugmann M, Anderson T, Thomson C, Vouyiouklis D, Nave KA (1998) Current concepts of PLP and its role in the nervous system. Microsc Res Tech 41(5):344–358

Guo F, Maeda Y, Ma J, Xu J, Horiuchi M, Miers L, Vaccarino F, Pleasure D (2010) Pyramidal neurons are generated from oligodendroglial progenitor cells in adult piriform cortex. J Neurosci 30(36):12036–12049

Haberg A, Qu H, Bakken IJ, Sande LM, White LR, Haraldseth O, Unsgard G, Aasly J, Sonnewald U (1998) In vitro and ex vivo ^{13}C-NMR spectroscopy studies of pyruvate recycling in brain. Dev Neurosci 20(4–5):389–398

Hines JH, Ravanelli AM, Schwindt R, Scott EK, Appel B (2015) Neuronal activity biases axon selection for myelination in vivo. Nat Neurosci 18(5):683–689

Hirrlinger J, Nave KA (2014) Adapting brain metabolism to myelination and long-range signal transduction. Glia 62(11):1749–1761

Huang JK, Phillips GR, Roth AD, Pedraza L, Shan W, Belkaid W, Mi S, Fex-Svenningsen A, Florens L, Yates JR III, Colman DR (2005) Glial membranes at the node of Ranvier prevent neurite outgrowth. Science 310(5755):1813–1817

Kang SH, Fukaya M, Yang JK, Rothstein JD, Bergles DE (2010) NG2+ CNS glial progenitors remain committed to the oligodendrocyte lineage in postnatal life and following neurodegeneration. Neuron 68(4):668–681

Karadottir R, Attwell D (2007) Neurotransmitter receptors in the life and death of oligodendrocytes. Neuroscience 145(4):1426–1438

Karadottir R, Hamilton NB, Bakiri Y, Attwell D (2008) Spiking and nonspiking classes of oligodendrocyte precursor glia in CNS white matter. Nat Neurosci 11(4):450–456

Kondo T, Raff M (2000) Oligodendrocyte precursor cells reprogrammed to become multipotential CNS stem cells. Science 289(5485):1754–1757

Kunnecke B, Cerdan S, Seelig J (1993) Cerebral metabolism of [1,2-$^{13}C_2$]glucose and [U-$^{13}C_4$]3-hydroxybutyrate in rat brain as detected by ^{13}C NMR spectroscopy. NMR Biomed 6(4):264–277

Lappe-Siefke C, Goebbels S, Gravel M, Nicksch E, Lee J, Braun PE, Griffiths IR, Nave KA (2003) Disruption of Cnp1 uncouples oligodendroglial functions in axonal support and myelination. Nat Genet 33(3):366–374

Lassmann H (2010) Axonal and neuronal pathology in multiple sclerosis: what have we learnt from animal models. Exp Neurol 225(1):2–8

Lee S, Leach MK, Redmond SA, Chong SY, Mellon SH, Tuck SJ, Feng ZQ, Corey JM, Chan JR (2012a) A culture system to study oligodendrocyte myelination processes using engineered nanofibers. Nat Methods 9(9):917–922

Lee Y, Morrison BM, Li Y, Lengacher S, Farah MH, Hoffman PN, Liu Y, Tsingalia A, Jin L, Zhang PW, Pellerin L, Magistretti PJ, Rothstein JD (2012b) Oligodendroglia metabolically support axons and contribute to neurodegeneration. Nature 487(7408):443–448

Levine JM, Reynolds R, Fawcett JW (2001) The oligodendrocyte precursor cell in health and disease. Trends Neurosci 24(1):39–47

Lewis KM, Petritsch C (2013) Asymmetric cell division: implications for glioma development and treatment. Transl Neurosci 4(4):484–503

Lundgaard I, Luzhynskaya A, Stockley JH, Wang Z, Evans KA, Swire M, Volbracht K, Gautier HO, Franklin RJ, Charles F-C, Attwell D, Karadottir RT (2013) Neuregulin and BDNF induce a switch to NMDA receptor-dependent myelination by oligodendrocytes. PLoS Biol 11(12):e1001743

Lyons SA, Kettenmann H (1998) Oligodendrocytes and microglia are selectively vulnerable to combined hypoxia and hypoglycemia injury in vitro. J Cereb Blood Flow Metab 18(5):521–530

McCarthy KD, de Vellis J (1980) Preparation of separate astroglial and oligodendroglial cell cultures from rat cerebral tissue. J Cell Biol 85(3):890–902

McKenna MC, Tildon JT, Stevenson JH, Huang X, Kingwell KG (1995) Regulation of mitochondrial and cytosolic malic enzymes from cultured rat brain astrocytes. Neurochem Res 20(12):1491–1501

McKenna MC, Stevenson JH, Huang X, Tildon JT, Zielke CL, Hopkins IB (2000) Mitochondrial malic enzyme activity is much higher in mitochondria from cortical synaptic terminals compared with mitochondria from primary cultures of cortical neurons or cerebellar granule cells. Neurochem Int 36(4–5):451–459

McKenna M, Gruetter R, Sonnewald U, Waagepetersen HS, Schousboe A (2012) Energy metabolism of the brain. In: Brady S, Siegel G, Albers RW, Price D (eds) Basic neurochemistry: principles of molecular, cellular, and medical neurobiology. Elsevier Academic, Oxford, pp 200–231

Melo TM, Nehlig A, Sonnewald U (2005) Metabolism is normal in astrocytes in chronically epileptic rats: a (13)C NMR study of neuronal-glial interactions in a model of temporal lobe epilepsy. J Cereb Blood Flow Metab 25(10):1254–1264

Mensch S, Baraban M, Almeida R, Czopka T, Ausborn J, El Manira A, Lyons DA (2015) Synaptic vesicle release regulates myelin sheath number of individual oligodendrocytes in vivo. Nat Neurosci 18(5):628–630

Miron VE, Kuhlmann T, Antel JP (2011) Cells of the oligodendroglial lineage, myelination, and remyelination. Biochim Biophys Acta 1812(2):184–193

Mitew S, Hay CM, Peckham H, Xiao J, Koenning M, Emery B (2014) Mechanisms regulating the development of oligodendrocytes and central nervous system myelin. Neuroscience 276:29–47

Moffett JR, Ross B, Arun P, Madhavarao CN, Namboodiri AM (2007) N-Acetylaspartate in the CNS: from neurodiagnostics to neurobiology. Prog Neurobiol 81(2):89–131

Moffett JR, Arun P, Ariyannur PS, Garbern JY, Jacobowitz DM, Namboodiri AM (2011) Extensive aspartoacylase expression in the rat central nervous system. Glia 59(10):1414–1434

Morken TS, Brekke E, Haberg A, Wideroe M, Brubakk AM, Sonnewald U (2014) Altered astrocyte-neuronal interactions after hypoxia-ischemia in the neonatal brain in female and male rats. Stroke 45(9):2777–2785

Muir D, Berl S, Clarke DD (1986) Acetate and fluoroacetate as possible markers for glial metabolism in vivo. Brain Res 380(2):336–340

Murin R, Cesar M, Kowtharapu BS, Verleysdonk S, Hamprecht B (2009) Expression of pyruvate carboxylase in cultured oligodendroglial, microglial and ependymal cells. Neurochem Res 34(3):480–489

Nave KA (2010a) Myelination and support of axonal integrity by glia. Nature 468(7321):244–252

Nave KA (2010b) Myelination and the trophic support of long axons. Nat Rev Neurosci 11(4):275–283

Nave KA, Ehrenreich H (2014) Myelination and oligodendrocyte functions in psychiatric diseases. JAMA Psychiatry 71(5):582–584

Nave KA, Trapp BD (2008) Axon-glial signaling and the glial support of axon function. Annu Rev Neurosci 31:535–561

Nilsen LH, Witter MP, Sonnewald U (2014) Neuronal and astrocytic metabolism in a transgenic rat model of Alzheimer's disease. J Cereb Blood Flow Metab 34(5):906–914

Norenberg MD, Martinez-Hernandez A (1979) Fine structural localization of glutamine synthetase in astrocytes of rat brain. Brain Res 161:303–310

Olstad E, Olsen GM, Qu H, Sonnewald U (2007) Pyruvate recycling in cultured neurons from cerebellum. J Neurosci Res 85(15):3318–3325

Patel MS (1974) The relative significance of CO_2-fixing enzymes in the metabolism of rat brain. J Neurochem 22(5):717–724

Psachoulia K, Jamen F, Young KM, Richardson WD (2009) Cell cycle dynamics of NG2 cells in the postnatal and ageing brain. Neuron Glia Biol 5(3–4):57–67

Raff MC, Miller RH, Noble M (1983) A glial progenitor cell that develops in vitro into an astrocyte or an oligodendrocyte depending on culture medium. Nature 303(5916):390–396

Richardson WD, Pringle NP, Yu WP, Hall AC (1997) Origins of spinal cord oligodendrocytes: possible developmental and evolutionary relationships with motor neurons. Dev Neurosci 19(1):58–68

Richardson WD, Young KM, Tripathi RB, McKenzie I (2011) NG2-glia as multipotent neural stem cells: fact or fantasy? Neuron 70(4):661–673

Rinholm JE, Hamilton NB, Kessaris N, Richardson WD, Bergersen LH, Attwell D (2011) Regulation of oligodendrocyte development and myelination by glucose and lactate. J Neurosci 31(2):538–548

Rivers LE, Young KM, Rizzi M, Jamen F, Psachoulia K, Wade A, Kessaris N, Richardson WD (2008) PDGFRA/NG2 glia generate myelinating oligodendrocytes and piriform projection neurons in adult mice. Nat Neurosci 11(12):1392–1401

Rowitch DH (2004) Glial specification in the vertebrate neural tube. Nat Rev Neurosci 5(5):409–419

Sa Santos S, Sonnewald U, Carrondo MJ, Alves PM (2011) The role of glia in neuronal recovery following anoxia: in vitro evidence of neuronal adaptation. Neurochem Int 58(6):665–675

Sanchez-Abarca LI, Tabernero A, Medina JM (2001) Oligodendrocytes use lactate as a source of energy and as a precursor of lipids. Glia 36(3):321–329

Schoenfeld R, Wong A, Silva J, Li M, Itoh A, Horiuchi M, Itoh T, Pleasure D, Cortopassi G (2010) Oligodendroglial differentiation induces mitochondrial genes and inhibition of mitochondrial function represses oligodendroglial differentiation. Mitochondrion 10(2):143–150

Shank RP, Bennett GS, Freytag SO, Campbell GL (1985) Pyruvate carboxylase: an astrocyte-specific enzyme implicated in the replenishment of amino acid neurotransmitter pools. Brain Res 329(1–2):364–367

Siegel GJ, Albers RW (2006) Basic neurochemistry: molecular, cellular, and medical aspects. Elsevier Academic, San Diego

Snaidero N, Mobius W, Czopka T, Hekking LH, Mathisen C, Verkleij D, Goebbels S, Edgar J, Merkler D, Lyons DA, Nave KA, Simons M (2014) Myelin membrane wrapping of CNS axons by PI(3,4,5)P3-dependent polarized growth at the inner tongue. Cell 156(1–2):277–290

Sonnewald U (2014) Glutamate synthesis has to be matched by its degradation—where do all the carbons go? J Neurochem 131(4):399–406

Sonnewald U, Rae C (2010) Pyruvate carboxylation in different model systems studied by ^{13}C MRS. Neurochem Res 35(12):1916–1921

Sonnewald U, Westergaard N, Hassel B, Muller TB, Unsgard G, Fonnum F, Hertz L, Schousboe A, Petersen SB (1993) NMR spectroscopic studies of 13C acetate and 13C glucose metabolism in neocortical astrocytes: evidence for mitochondrial heterogeneity. Dev Neurosci 15(3–5):351–358

Stacpoole SR, Spitzer S, Bilican B, Compston A, Karadottir R, Chandran S, Franklin RJ (2013) High yields of oligodendrocyte lineage cells from human embryonic stem cells at physiological oxygen tensions for evaluation of translational biology. Stem Cell Rep 1(5):437–450

Sugiarto S, Persson AI, Munoz EG, Waldhuber M, Lamagna C, Andor N, Hanecker P, Ayers-Ringler J, Phillips J, Siu J, Lim DA, Vandenberg S, Stallcup W, Berger MS, Bergers G, Weiss WA, Petritsch C (2011) Asymmetry-defective oligodendrocyte progenitors are glioma precursors. Cancer Cell 20(3):328–340

Sykes JE, Lopes-Cardozo M, Van Den Bergh SG (1986) Relationship between the pentose-phosphate pathway and the de novo synthesis of fatty acids and cholesterol in oligodendrocyte-enriched glial cultures. Neurochem Int 8(1):77–82

Tatsumi K, Takebayashi H, Manabe T, Tanaka KF, Makinodan M, Yamauchi T, Makinodan E, Matsuyoshi H, Okuda H, Ikenaka K, Wanaka A (2008) Genetic fate mapping of Olig2 progenitors in the injured adult cerebral cortex reveals preferential differentiation into astrocytes. J Neurosci Res 86(16):3494–3502

Trajkovic K, Dhaunchak AS, Goncalves JT, Wenzel D, Schneider A, Bunt G, Nave KA, Simons M (2006) Neuron to glia signaling triggers myelin membrane exocytosis from endosomal storage sites. J Cell Biol 172(6):937–948

Tripathi RB, Clarke LE, Burzomato V, Kessaris N, Anderson PN, Attwell D, Richardson WD (2011) Dorsally and ventrally derived oligodendrocytes have similar electrical properties but myelinate preferred tracts. J Neurosci 31(18):6809–6819

Vallstedt A, Klos JM, Ericson J (2005) Multiple dorsoventral origins of oligodendrocyte generation in the spinal cord and hindbrain. Neuron 45(1):55–67

van den Berg CJ, Garfinkel D (1971) A stimulation study of brain compartments. Metabolism of glutamate and related substances in mouse brain. Biochem J 123(2):211–218

Waagepetersen HS, Sonnewald U, Larsson OM, Schousboe A (2001) Multiple compartments with different metabolic characteristics are involved in biosynthesis of intracellular and released glutamine and citrate in astrocytes. Glia 35(3):246–252

Wake H, Lee PR, Fields RD (2011) Control of local protein synthesis and initial events in myelination by action potentials. Science 333(6049):1647–1651

Waxman SG, Ritchie JM (1993) Molecular dissection of the myelinated axon. Ann Neurol 33(2):121–136

Yan H, Rivkees SA (2006) Hypoglycemia influences oligodendrocyte development and myelin formation. Neuroreport 17(1):55–59

Ye P, Carson J, D'Ercole AJ (1995) In vivo actions of insulin-like growth factor-I (IGF-I) on brain myelination: studies of IGF-I and IGF binding protein-1 (IGFBP-1) transgenic mice. J Neurosci 15(11):7344–7356

Yu AC, Drejer J, Hertz L, Schousboe A (1983) Pyruvate carboxylase activity in primary cultures of astrocytes and neurons. J Neurochem 41:1484–1487

Zawadzka M, Rivers LE, Fancy SP, Zhao C, Tripathi R, Jamen F, Young K, Goncharevich A, Pohl H, Rizzi M, Rowitch DH, Kessaris N, Suter U, Richardson WD, Franklin RJ (2010) CNS-resident glial progenitor/stem cells produce Schwann cells as well as oligodendrocytes during repair of CNS demyelination. Cell Stem Cell 6(6):578–590

Ziabreva I, Campbell G, Rist J, Zambonin J, Rorbach J, Wydro MM, Lassmann H, Franklin RJ, Mahad D (2010) Injury and differentiation following inhibition of mitochondrial respiratory chain complex IV in rat oligodendrocytes. Glia 58(15):1827–1837

Chapter 11
Dysregulation of Glutamate Cycling Mediates Methylmercury-Induced Neurotoxicity

Megan Culbreth and Michael Aschner

Abstract To examine the toxicological implications of glutamate, this chapter will focus specifically on its impact in the brain. More explicitly, it will illustrate the role glutamate plays in mediating methylmercury (MeHg)-induced neurotoxicity. In this chapter, one intends to highlight the processes that occur prior to glutamate-stimulated excitotoxicity and subsequent neurodegeneration. As such, it will emphasize three main routes by which MeHg alters glutamate homeostasis. It is essential to recognize that these effects are not mutually exclusive, and that they synergistically influence glutamate dysregulation. Furthermore, the consequences of MeHg exposure will be presented here as a direct pathway; however, it must be noted these effects occur simultaneously. First, glutamate uptake will be reviewed emphasizing the function of astrocytes. Next, the induction of oxidative stress by MeHg exposure will be discussed. This process has a two-fold effect on glutamate homeostasis by (1) inhibiting extracellular glutamate uptake and (2) altering transcription of genes vital to glutamate cycling. Finally, the impact glutamate dysregulation has on glutathione synthesis will be examined. Although this chapter centers on the link between glutamate and MeHg toxicity, it is imperative that the reader acknowledges the processes discussed here can be extended to any pro-oxidant.

Keywords Glutamate • Homeostasis • Toxicity • Cycling • Methylmercury

11.1 Introduction

Glutamate is a nonessential amino acid, deemed the primary excitatory neurotransmitter in the mammalian nervous system. Simply, glutamate is synthesized in neurons and released via exocytosis into the extracellular space, where it activates specific receptors that will propagate electrical signals. This process is essential for

M. Culbreth (✉) • M. Aschner
Department of Molecular Pharmacology, Albert Einstein College of Medicine, Forchheimer Building, 1300 Morris Park Avenue, Bronx, NY 10461, USA
e-mail: megan.culbreth@phd.einstein.yu.edu

A. Schousboe, U. Sonnewald (eds.), *The Glutamate/GABA-Glutamine Cycle*, Advances in Neurobiology 13, DOI 10.1007/978-3-319-45096-4_11

normal brain function. In order to prevent continual stimulation, glutamate must be removed from the extracellular space. Several mechanisms have evolved to mediate glutamate uptake. This chapter will focus primarily on the glutamate-aspartate transporter (GLAST or EAAT1) and the glutamate transporter-1 (GLT-1 or EAAT2). Both are predominately expressed in astrocytes (Rothstein et al. 1994; Lehre et al. 1995), the main cell responsible for glutamate uptake (Rothstein et al. 1996). Under normal conditions, glutamate levels are much higher intracellularly than extracellularly (Hamberger and Nystrom 1984). Neurotransmission results in glutamate release and subsequent activation of post-synaptic glutamate receptors. Once the signal has been disseminated, synapse-adjacent astrocytes must remove excess glutamate via GLAST and GLT-1. MeHg elicits toxicity by indirectly inhibiting GLAST and GLT-1 (Brookes and Kristt 1989; Aschner et al. 1990, 1993; Qu et al. 2003).

MeHg is a potent environmental neurotoxicant. Humans are primarily exposed to MeHg via fish consumption. Although extensive regulation has virtually eliminated high-dose MeHg exposure, there continues to be significant risk associated with low-dose exposure because the mechanism of MeHg toxicity is not fully understood. Accidental high-dose exposures in Japan (Harada 1995; Eto 2000) demonstrated the severe developmental consequences of MeHg, motivating the scientific community to further investigate its mechanism of action. Moreover, several more recent epidemiological studies of populations in which fish is a dietary staple have correlated low-dose in utero MeHg exposure to long-term developmental consequences (Davidson et al. 2008, 2010, 2011; Grandjean et al. 1997; Debes et al. 2006), further prompting the need to research MeHg toxicity. In light of these events, as well as the epidemiological studies, researchers have continued to use in vivo and in vitro models to explore the mechanism of MeHg toxicity.

It is widely accepted that MeHg exposure generates reactive oxygen species (ROS), initiating an oxidative stress response in the brain (Shanker et al. 2004). The primary ROS produced subsequent to MeHg exposure is hydrogen peroxide (H_2O_2). Increase in H_2O_2 is the direct result of MeHg inhibiting glutathione peroxidase, the enzyme responsible for detoxifying H_2O_2 to water (Franco et al. 2009). Additionally, MeHg indirectly enhances H_2O_2 production by interacting with the mitochondrial electron transfer chain, causing sporadic ROS generation (Mori et al. 2011). H_2O_2 has been shown to inhibit GLAST and GLT-1 (Sorg et al. 1997). Moreover, MeHg is lipophilic and promotes lipid peroxidation. It activates phospholipase A_2, stimulating release of arachidonic acid (AA) (Shanker et al. 2002). In turn, AA inhibits GLAST and GLT-1 (Barbour et al. 1989; Lundy and McBean 1996).

The pro-oxidant properties of MeHg are not limited to the effects on glutamate uptake. In generating reactive intermediates, MeHg activates nuclear factor erythroid 2-related factor 2 (Nrf2) (Wang et al. 2009). Nrf2 is a transcription factor that translocates to the nucleus in response to reactive oxygen species generation and upregulates phase II detoxifying enzymes and antioxidant proteins. In unstressed conditions, Nrf2 is retained in the cytosol, bound to kelch-like ECH-associated protein 1 (Keap1) (Itoh et al. 1999). Keap1 serves as an adaptor for E3-ubiquitin ligases (Kobayashi et al. 2004), which will ubiquitinate Nrf2, promoting its proteasomal degradation. Keap1 further functions as an oxidative stress sensor (Kobayashi et al. 2006). Upon ROS generation, cysteine residues on Keap1 are modified, allowing the dissociation and subsequent nuclear translocation of Nrf2. Altering Nrf2 activity has significant implications on glutamate homeostasis. Nrf2 promotes glutathione peroxidase transcription, which as mentioned above is the enzyme required to detoxify H_2O_2.

Nrf2 further functions in glutamate homeostasis by stimulating transcription of the cystine/glutamate exchanger (X_c^-) (Sasaki et al. 2002). This system exchanges intracellular glutamate for extracellular cystine, which is necessary for glutathione (GSH) synthesis (Bannai and Kitamura 1980). GSH is essential for detoxification (Lu 2009). MeHg induces ROS generation, which in turn prevents glutamate uptake (Allen et al. 2001a). The resultant increase in extracellular glutamate inhibits X_c^-, thereby decreasing GSH synthesis and subsequent detoxification.

Although the processes discussed in this chapter are by no means exhaustive of the toxicological implications of glutamate dysregulation, they do provide a flavor for the potential effects of pro-oxidants in the brain.

11.2 Methylmercury Inhibits Glutamate Uptake

As mentioned previously, glutamate is an excitatory amino acid. Extracellular glutamate levels must be tightly regulated in order to sustain normal brain function. Astrocytes are the primary cells responsible for glutamate uptake from the extracellular space (Rothstein et al. 1996). The glutamate/aspartate transporter (GLAST) and glutamate transporter-1 (GLT-1) are the main channels through which glutamate uptake occurs in astrocytes. Both GLAST and GLT-1 are expressed ubiquitously throughout the brain and are located almost exclusively in astrocytes (Rothstein et al. 1994; Lehre et al. 1995). In vitro, glutamate stimulates concentration- and time-dependent uptake via GLAST and GLT-1 in astrocytes. Furthermore, it increases GLAST expression at the astrocyte cell surface (Duan et al. 1999).

Inorganic mercury has been found to preferentially accumulate in astrocytes subsequent to MeHg exposure (Charleston et al. 1995). As such, researchers wanted to investigate whether MeHg has any effects on glutamate homeostasis. MeHg causes a concentration- and time-dependent decrease in glutamate uptake in astrocytes (Brookes and Kristt 1989; Aschner et al. 1990; Allen et al. 2001a; Qu et al. 2003). Interestingly, intracellular glutamate concentration remained unchanged (Yin et al. 2009). MeHg also increases glutamate efflux from astrocytes (Aschner et al. 1993).

In order to understand the mechanism by which MeHg could alter glutamate uptake and efflux, GLAST and GLT-1 expression subsequent to exposure was explored. It was found that MeHg significantly increases GLT-1 mRNA expression with no effect on GLAST. Interestingly, GLT-1 protein levels were decreased, while GLAST protein levels increased (Mutkus et al. 2005). These data do not definitively reveal a link to glutamate homeostasis and MeHg toxicity, thus further investigation is warranted. It has been postulated that MeHg could alter GLT-1 and GLAST function due to its high affinity for sulfhydryl groups on cysteine residues.

11.3 Methylmercury and Glutamate Generate Reactive Oxygen Species

ROS are continually being generated intracellularly. Oxidative stress results from an imbalance between ROS production and the ability to detoxify them. It has become increasingly evident that oxidative stress is complicit in several neurodegenerative diseases including Alzheimer's disease (Liu et al. 2015), Parkinson's disease (Blesa et al. 2015), and amyotrophic lateral sclerosis (ALS) (Palomo and Manfredi 2015). ROS have potential to damage DNA, proteins, and lipids, which could be detrimental to cell function.

MeHg exposure increases ROS generation. It has been shown that intracellular H_2O_2 rises subsequent to exposure in astrocytes (Shanker et al. 2004). In addition, mitochondrial ROS formation increases. The mechanism(s) by which MeHg effects ROS production is not entirely known. MeHg has been found to decrease mitochondrial membrane potential (Yin et al. 2007). Loss of membrane potential would result in spurious electron escape from the electron transfer chain and resultant increase in ROS. MeHg has also been observed to alter the protein complexes of the electron transfer chain (Mori et al. 2011), impacting the cells' ability to produce ATP. In addition, MeHg inhibits glutathione peroxidase (GPx) activity (Franco et al. 2009). GPx reduces H_2O_2 to non-toxic H_2O.

Glutamate also enhances ROS formation. It has been demonstrated that exposure to *N*-methyl-D-aspartate (NMDA), the agonist for a neuronal glutamate receptor, increases ROS production (Dugan et al. 1995). Furthermore, glutamate was found to enrich ROS depending on NMDA receptor activation and Ca^{2+} influx (Reynolds and Hastings 1995). It is proposed that mitochondrial dysfunction is critical to glutamate-induced ROS generation.

11.4 Reactive Oxygen Species Inhibit Glutamate Uptake

As mentioned above, MeHg is a pro-oxidant, which induces a concentration- and time-dependent oxidative stress response. MeHg increases ROS generation by compromising mitochondrial membrane integrity. H_2O_2 is the most abundant ROS produced subsequent to MeHg exposure. It has been demonstrated that antioxidants including *n*-propyl gallate (Shanker et al. 2005) and superoxide dismutase (Shanker and Aschner 2003) attenuate MeHg toxicity. Furthermore, co-application of MeHg and glutamate has been shown to augment ROS formation (Amonpatumrat et al. 2008).

MeHg is also highly lipophilic. It has been found to increase cytosolic phospholipase A_2 ($cPLA_2$) mRNA and protein expression (Shanker et al. 2002). $cPLA_2$ assists in cell membrane phospholipid maintenance. Moreover, MeHg activates $cPLA_2$ resulting in AA release (Shanker et al. 2002). AA promotes ROS generation (Trotti et al. 1998; Cocco et al. 1999). The effects of MeHg on $cPLA_2$ activation and AA release are tempered by arachidonyl trifluoromethyl ketone, a specific $cPLA_2$ inhibitor (Shanker and Aschner 2003). Glutamate has also been shown to stimulate AA release (Dumuis et al. 1988; Stella et al. 1994).

As mentioned above, H_2O_2 is readily generated subsequent to MeHg exposure. H_2O_2 has been found to inhibit glutamate uptake (Sorg et al. 1997). Furthermore, catalase, an antioxidant, significantly reduced MeHg-stimulated inhibition of the uptake of ^{3}H-D-aspartate, a glutamate analog (Allen et al. 2001a). Catalase reduces H_2O_2 to non-toxic H_2O and O_2. AA has also been found to directly inhibit glutamate uptake (Barbour et al. 1989; Lundy and McBean 1996).

11.5 Reactive Oxygen Species Activate Nrf2

Nuclear factor erythroid 2-related factor 2 (Nrf2) is a transcription factor that translocates to the nucleus in response to oxidative stress. Nrf2 promotes transcription of phase II detoxifying enzymes (Itoh et al. 1997). Under normal conditions, Nrf2 is bound to kelch-like ECH-associated protein 1 (Keap1) in the cytosol. Keap1 blocks Nrf2 nuclear activation (Itoh et al. 1999). Furthermore, Keap1 serves as an adaptor for E3 ligases (Kobayashi et al. 2004), which will ubiquitinate Nrf2, marking it for proteasomal degradation. Keap1 also acts as an oxidative stress sensor (Kobayashi et al. 2006). Modification of cysteine residues on Keap1 by ROS allows for partial dissociation of Nrf2 from Keap1. Nrf2 must be further phosphorylated by protein kinase C (PKC) (Huang et al. 2002) in order to completely detach from Keap1 and move into the nucleus. For a review of Nrf2 regulation, refer to Niture et al. (2014), and see Fig. 11.1.

Nrf2 activation is crucial to the cells' ability to respond to oxidative stress. It has been demonstrated that Nrf2 knockout mice are more susceptible to oxidant-induced injury (Jiang et al. 2009; Kurzatkowski and Trombetta 2013). Furthermore, Nrf2 activation has been found to be protective in neurodegenerative models of Parkinson's

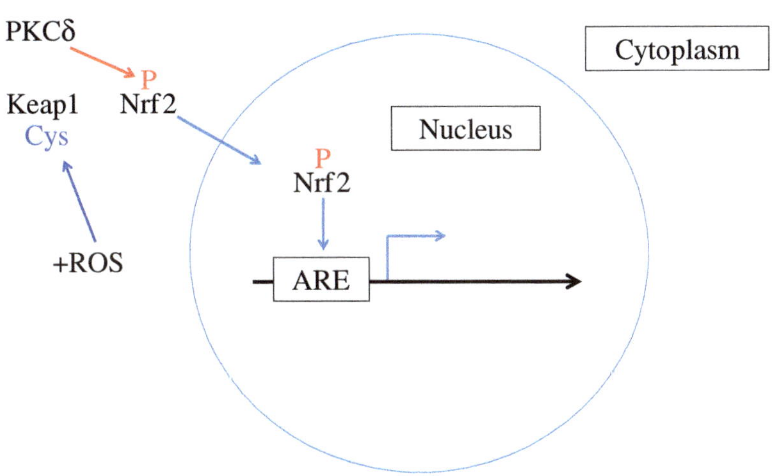

Fig. 11.1 Reactive oxygen species induce Nrf2 activation. Adapted from Niture et al. (2014). *ROS* reactive oxygen species, *Cys* cysteine, *Keap1* kelch-like ECH-associated protein 1, *Nrf2* nuclear factor erythroid 2-related factor 2, *PKCδ* protein kinase C delta, *ARE* antioxidant response element, *P* phosphorylation

(Jakel et al. 2007) and ALS (Vargas et al. 2008). It has also been shown that mutant Huntingtin gene expression significantly reduces Nrf2 activity (Jin et al. 2013). Interestingly, astrocytes were found to be the preferential site of Nrf2 activation (Kraft et al. 2004).

As mentioned previously, MeHg induces an oxidative stress response in the brain. Nrf2 has been found to partly mediate this reaction. It has been demonstrated that MeHg promotes Nrf2 nuclear translocation and subsequent upregulation of phase II detoxifying enzymes in both microglia (Ni et al. 2010) and astrocytes (Wang et al. 2009). Furthermore, Nrf2 knockout significantly increased vulnerability to MeHg with a concurrent decrease in phase II detoxifying enzyme expression (Ni et al. 2011).

11.6 Nrf2 Induces Cystine/Glutamate Antiporter Gene Expression

Nrf2 promotes the transcription of various genes required for cellular defense against oxidative stress. One such gene is xCT (Sasaki et al. 2002), the transporter component of the cystine/glutamate antiporter (X_c^-). X_c^- exchanges intracellular glutamate for extracellular cystine at a 1:1 ratio (Bannai 1986). Cystine is readily converted to cysteine, the rate-limiting precursor for glutathione (GSH) synthesis, intracellularly (Bannai and Kitamura 1980). GSH is essential for cellular detoxification (Lu 2009). Astrocytes readily transport cystine (Sagara et al. 1993) and maintain high intracellular GSH levels (Yudkoff et al. 1990).

As previously mentioned, MeHg induces Nrf2 nuclear translocation and subsequent upregulation of phase II detoxifying enzymes. It has been demonstrated that MeHg exposure stimulates xCT mRNA expression. This effect is tempered by Nrf2 knockdown in astrocytes (Wang et al. 2009; Ni et al. 2011).

Although MeHg stimulates xCT transcription, it also concurrently inhibits cystine uptake. It has been found that MeHg acts on the glutamate/aspartate transporter (X_{AG}^-) to inhibit cystine uptake in astrocytes (Allen et al. 2001b). X_{AG}^- is the primary glutamate transporter in astrocytes (Cho and Bannai 1990). It has also been demonstrated that co-application of an X_{AG}^- inhibitor and MeHg hinders cysteine uptake (Shanker et al. 2001). It must be noted that high extracellular glutamate concentration inhibits both X_c^- (Murphy et al. 1989) and X_{AG}^- (Murphy et al. 1990) transport.

11.7 Conclusion

Glutamate is the predominant excitatory neurotransmitter in the brain. As such, it must be tightly regulated in order to avoid excitotoxic injury and subsequent neurodegeneration. Several transport mechanisms have evolved to maintain glutamate homeostasis. GLAST and GLT-1 are the main channels through which glutamate uptake occurs (Zhou and Danbolt 2014). These transporters are expressed almost exclusively in astrocytes (Rothstein et al. 1994; Lehre et al. 1995). Astrocytes are also the preferential site of MeHg accumulation in the brain (Charleston et al. 1995), rendering glutamate homeostasis a likely target for MeHg-induced toxicity.

MeHg has been observed to inhibit glutamate uptake (Brookes and Kristt 1989; Aschner et al. 1990; Allen et al. 2001a; Qu et al. 2003). This might occur by MeHg reacting with sulfhydryl groups on cysteine residues in GLAST and GLT-1, thus altering transporter function. Further research is required to confirm this hypothesis. MeHg also enhances ROS generation (Shanker et al. 2004) and AA release (Shanker et al. 2002). Both ROS (Sorg et al. 1997; Allen et al. 2001a) and AA (Barbour et al. 1989; Lundy and McBean 1996) have been demonstrated to inhibit glutamate uptake.

ROS activate Nrf2-dependent phase II detoxifying enzyme transcription. Nrf2 promotes xCT, the transporter component of the cystine/glutamate antiporter (Sasaki et al. 2002). MeHg has been shown to activate Nrf2, consequently increasing xCT transcription (Wang et al. 2009; Ni et al. 2011).

MeHg has also been found to inhibit cystine uptake via the glutamate/aspartate transporter (Allen et al. 2001b). This is mostly due to high extracellular glutamate levels as a result of MeHg-induced glutamate uptake inhibition. Cystine is a necessary precursor for GSH synthesis, the principal cellular antioxidant (Bannai and Kitamura 1980). The dynamic relationship between glutamate and MeHg is summarized in Fig. 11.2.

This chapter only highlights a portion of the potential toxicological effects of glutamate homeostasis dysfunction. However, it is important to recognize that much of what is discussed here can be applied to any pro-oxidant. Furthermore, it underscores glutamate itself as the key mediator of toxicity.

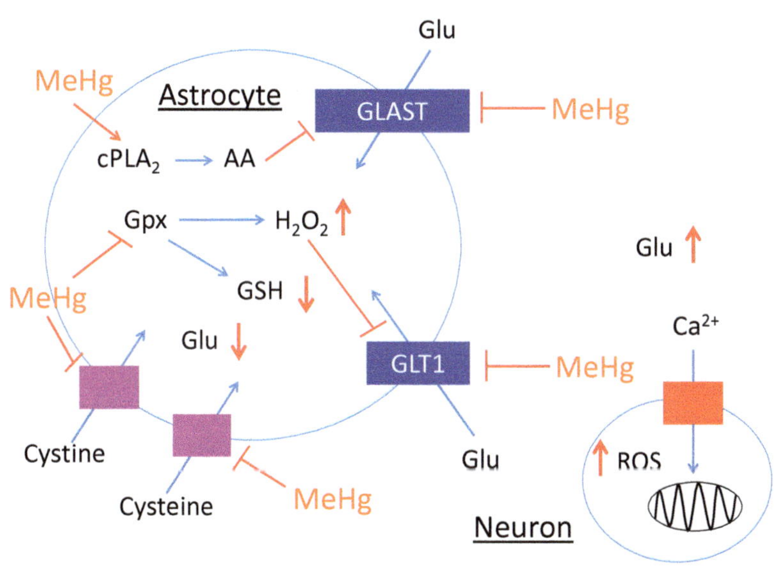

Fig. 11.2 Methylmercury alters glutamate homeostasis. Adapted from Aschner et al. (2007). *MeHg* methylmercury, *cPLA₂* cytosolic phospholipase A_2, *AA* arachidonic acid, *Gpx* glutathione peroxidase, H_2O_2 hydrogen peroxide, *GSH* glutathione, *Glu* glutamate, *GLAST* glutamate/aspartate transporter, *GLT-1* glutamate transporter-1, *ROS* reactive oxygen species, Ca^{2+} calcium

Acknowledgments The review was supported in part by NIH grants from the National Institute of Environmental Health Sciences ES R01 07331 and R01 ES 020852.

References

Allen JW, Mutkus LA, Aschner M (2001a) Methylmercury-mediated inhibition of 3H-D-aspartate transport in cultured astrocytes is reversed by the antioxidant catalase. Brain Res 902(1):92–100

Allen JW, Shanker G, Aschner M (2001b) Methylmercury inhibits the in vitro uptake of the glutathione precursor, cystine, in astrocytes, but not in neurons. Brain Res 894(1):131–140

Amonpatumrat S, Sakurai H, Wiriyasermkul P, Khunweeraphong N, Nagamori S, Tanaka H, Piyachaturawat P, Kanai Y (2008) L-glutamate enhances methylmercury toxicity by synergistically increasing oxidative stress. J Pharmacol Sci 108(3):280–289

Aschner M, Eberle NB, Miller K, Kimelberg HK (1990) Interactions of methylmercury with rat primary astrocyte cultures: inhibition of rubidium and glutamate uptake and induction of swelling. Brain Res 530(2):245–250

Aschner M, Du YL, Gannon M, Kimelberg HK (1993) Methylmercury-induced alterations in excitatory amino acid transport in rat primary astrocyte cultures. Brain Res 602(2):181–186

Aschner M, Syversen T, Souza DO, Rocha JB, Farina M (2007) Involvement of glutamate and reactive oxygen species in methylmercury neurotoxicity. Braz J Med Biol Res 40(3):285–291

Bannai S (1986) Exchange of cystine and glutamate across plasma membrane of human fibroblasts. J Biol Chem 261(5):2256–2263

Bannai S, Kitamura E (1980) Transport interaction of L-cystine and L-glutamate in human diploid fibroblasts in culture. J Biol Chem 255(6):2372–2376

Barbour B, Szatkowski M, Ingledew N, Attwell D (1989) Arachidonic acid induces a prolonged inhibition of glutamate uptake into glial cells. Nature 342(6252):918–920. doi:10.1038/342918a0

Blesa J, Trigo-Damas I, Quiroga-Varela A, Jackson-Lewis VR (2015) Oxidative stress and Parkinson's disease. Front Neuroanat 9:91. doi:10.3389/fnana.2015.00091

Brookes N, Kristt DA (1989) Inhibition of amino acid transport and protein synthesis by $HgCl_2$ and methylmercury in astrocytes: selectivity and reversibility. J Neurochem 53(4):1228–1237

Charleston JS, Body RL, Mottet NK, Vahter ME, Burbacher TM (1995) Autometallographic determination of inorganic mercury distribution in the cortex of the calcarine sulcus of the monkey Macaca fascicularis following long-term subclinical exposure to methylmercury and mercuric chloride. Toxicol Appl Pharmacol 132(2):325–333. doi:10.1006/taap.1995.1114

Cho Y, Bannai S (1990) Uptake of glutamate and cysteine in C-6 glioma cells and in cultured astrocytes. J Neurochem 55(6):2091–2097

Cocco T, Di Paola M, Papa S, Lorusso M (1999) Arachidonic acid interaction with the mitochondrial electron transport chain promotes reactive oxygen species generation. Free Radic Biol Med 27(1–2):51–59

Davidson PW, Jean Sloane R, Myers GJ, Hansen ON, Huang LS, Georger LA, Cox C, Thurston SW, Shamlaye CF, Clarkson TW (2008) Association between prenatal exposure to methylmercury and visuospatial ability at 10.7 years in the Seychelles child development study. Neurotoxicology 29(3):453–459. doi:10.1016/j.neuro.2008.02.009

Davidson PW, Leste A, Benstrong E, Burns CM, Valentin J, Sloane-Reeves J, Huang LS, Miller WA, Gunzler D, van Wijngaarden E, Watson GE, Zareba G, Shamlaye CF, Myers GJ (2010) Fish consumption, mercury exposure, and their associations with scholastic achievement in the Seychelles Child Development Study. Neurotoxicology 31(5):439–447. doi:10.1016/j.neuro.2010.05.010

Davidson PW, Cory-Slechta DA, Thurston SW, Huang LS, Shamlaye CF, Gunzler D, Watson G, van Wijngaarden E, Zareba G, Klein JD, Clarkson TW, Strain JJ, Myers GJ (2011) Fish consumption and prenatal methylmercury exposure: cognitive and behavioral outcomes in the

main cohort at 17 years from the Seychelles child development study. Neurotoxicology 32(6):711–717. doi:10.1016/j.neuro.2011.08.003
Debes F, Budtz-Jorgensen E, Weihe P, White RF, Grandjean P (2006) Impact of prenatal methylmercury exposure on neurobehavioral function at age 14 years. Neurotoxicol Teratol 28(5):536–547. doi:10.1016/j.ntt.2006.02.005
Duan S, Anderson CM, Stein BA, Swanson RA (1999) Glutamate induces rapid upregulation of astrocyte glutamate transport and cell-surface expression of GLAST. J Neurosci 19(23):10193–10200
Dugan LL, Sensi SL, Canzoniero LM, Handran SD, Rothman SM, Lin TS, Goldberg MP, Choi DW (1995) Mitochondrial production of reactive oxygen species in cortical neurons following exposure to N-methyl-D-aspartate. J Neurosci 15(10):6377–6388
Dumuis A, Sebben M, Haynes L, Pin JP, Bockaert J (1988) NMDA receptors activate the arachidonic acid cascade system in striatal neurons. Nature 336(6194):68–70. doi:10.1038/336068a0
Eto K (2000) Minamata disease. Neuropathology 20(Suppl):S14–S19
Franco JL, Posser T, Dunkley PR, Dickson PW, Mattos JJ, Martins R, Bainy AC, Marques MR, Dafre AL, Farina M (2009) Methylmercury neurotoxicity is associated with inhibition of the antioxidant enzyme glutathione peroxidase. Free Radic Biol Med 47(4):449–457. doi:10.1016/j.freeradbiomed.2009.05.013
Grandjean P, Weihe P, White RF, Debes F, Araki S, Yokoyama K, Murata K, Sorensen N, Dahl R, Jorgensen PJ (1997) Cognitive deficit in 7-year-old children with prenatal exposure to methylmercury. Neurotoxicol Teratol 19(6):417–428
Hamberger A, Nystrom B (1984) Extra- and intracellular amino acids in the hippocampus during development of hepatic encephalopathy. Neurochem Res 9(9):1181–1192
Harada M (1995) Minamata disease: methylmercury poisoning in Japan caused by environmental pollution. Crit Rev Toxicol 25(1):1–24. doi:10.3109/10408449509089885
Huang HC, Nguyen T, Pickett CB (2002) Phosphorylation of Nrf2 at Ser-40 by protein kinase C regulates antioxidant response element-mediated transcription. J Biol Chem 277(45):42769–42774. doi:10.1074/jbc.M206911200
Itoh K, Chiba T, Takahashi S, Ishii T, Igarashi K, Katoh Y, Oyake T, Hayashi N, Satoh K, Hatayama I, Yamamoto M, Nabeshima Y (1997) An Nrf2/small Maf heterodimer mediates the induction of phase II detoxifying enzyme genes through antioxidant response elements. Biochem Biophys Res Commun 236(2):313–322
Itoh K, Wakabayashi N, Katoh Y, Ishii T, Igarashi K, Engel JD, Yamamoto M (1999) Keap1 represses nuclear activation of antioxidant responsive elements by Nrf2 through binding to the amino-terminal Neh2 domain. Genes Dev 13(1):76–86
Jakel RJ, Townsend JA, Kraft AD, Johnson JA (2007) Nrf2-mediated protection against 6-hydroxydopamine. Brain Res 1144:192–201. doi:10.1016/j.brainres.2007.01.131
Jiang T, Huang Z, Chan JY, Zhang DD (2009) Nrf2 protects against As(III)-induced damage in mouse liver and bladder. Toxicol Appl Pharmacol 240(1):8–14. doi:10.1016/j.taap.2009.06.010
Jin YN, Yu YV, Gundemir S, Jo C, Cui M, Tieu K, Johnson GV (2013) Impaired mitochondrial dynamics and Nrf2 signaling contribute to compromised responses to oxidative stress in striatal cells expressing full-length mutant huntingtin. PLoS One 8(3):e57932. doi:10.1371/journal.pone.0057932
Kobayashi A, Kang MI, Okawa H, Ohtsuji M, Zenke Y, Chiba T, Igarashi K, Yamamoto M (2004) Oxidative stress sensor Keap1 functions as an adaptor for Cul3-based E3 ligase to regulate proteasomal degradation of Nrf2. Mol Cell Biol 24(16):7130–7139. doi:10.1128/MCB.24.16.7130-7139.2004
Kobayashi A, Kang MI, Watai Y, Tong KI, Shibata T, Uchida K, Yamamoto M (2006) Oxidative and electrophilic stresses activate Nrf2 through inhibition of ubiquitination activity of Keap1. Mol Cell Biol 26(1):221–229. doi:10.1128/MCB.26.1.221-229.2006
Kraft AD, Johnson DA, Johnson JA (2004) Nuclear factor E2-related factor 2-dependent antioxidant response element activation by tert-butylhydroquinone and sulforaphane occurring preferentially in astrocytes conditions neurons against oxidative insult. J Neurosci 24(5):1101–1112. doi:10.1523/JNEUROSCI.3817-03.2004

Kurzatkowski DM, Trombetta LD (2013) Maneb causes pro-oxidant effects in the hippocampus of Nrf2 knockout mice. Environ Toxicol Pharmacol 36(2):427–436. doi:10.1016/j.etap.2013.04.016

Lehre KP, Levy LM, Ottersen OP, Storm-Mathisen J, Danbolt NC (1995) Differential expression of two glial glutamate transporters in the rat brain: quantitative and immunocytochemical observations. J Neurosci 15(3 Pt 1):1835–1853

Liu Z, Li T, Li P, Wei N, Zhao Z, Liang H, Ji X, Chen W, Xue M, Wei J (2015) The ambiguous relationship of oxidative stress, Tau hyperphosphorylation, and autophagy dysfunction in Alzheimer's disease. Oxid Med Cell Longev 2015:352723. doi:10.1155/2015/352723

Lu SC (2009) Regulation of glutathione synthesis. Mol Aspects Med 30(1–2):42–59. doi:10.1016/j.mam.2008.05.005

Lundy DF, McBean GJ (1996) Inhibition of the high-affinity uptake of D-[3H]aspartate in rate by L-alpha-aminoadipate and arachidonic acid. J Neurol Sci 139(Suppl):1–9

Mori N, Yasutake A, Marumoto M, Hirayama K (2011) Methylmercury inhibits electron transport chain activity and induces cytochrome c release in cerebellum mitochondria. J Toxicol Sci 36(3):253–259

Murphy TH, Miyamoto M, Sastre A, Schnaar RL, Coyle JT (1989) Glutamate toxicity in a neuronal cell line involves inhibition of cystine transport leading to oxidative stress. Neuron 2(6):1547–1558

Murphy TH, Schnaar RL, Coyle JT (1990) Immature cortical neurons are uniquely sensitive to glutamate toxicity by inhibition of cystine uptake. FASEB J 4(6):1624–1633

Mutkus L, Aschner JL, Syversen T, Aschner M (2005) Methylmercury alters the in vitro uptake of glutamate in GLAST- and GLT-1-transfected mutant CHO-K1 cells. Biol Trace Elem Res 107(3):231–245. doi:10.1385/BTER:107:3:231

Ni M, Li X, Yin Z, Jiang H, Sidoryk-Wegrzynowicz M, Milatovic D, Cai J, Aschner M (2010) Methylmercury induces acute oxidative stress, altering Nrf2 protein level in primary microglial cells. Toxicol Sci 116(2):590–603. doi:10.1093/toxsci/kfq126

Ni M, Li X, Yin Z, Sidoryk-Wegrzynowicz M, Jiang H, Farina M, Rocha JB, Syversen T, Aschner M (2011) Comparative study on the response of rat primary astrocytes and microglia to methylmercury toxicity. Glia 59(5):810–820. doi:10.1002/glia.21153

Niture SK, Khatri R, Jaiswal AK (2014) Regulation of Nrf2-an update. Free Radic Biol Med 66:36–44. doi:10.1016/j.freeradbiomed.2013.02.008

Palomo GM, Manfredi G (2015) Exploring new pathways of neurodegeneration in ALS: the role of mitochondria quality control. Brain Res 1607:36–46. doi:10.1016/j.brainres.2014.09.065

Qu H, Syversen T, Aschner M, Sonnewald U (2003) Effect of methylmercury on glutamate metabolism in cerebellar astrocytes in culture. Neurochem Int 43(4–5):411–416

Reynolds IJ, Hastings TG (1995) Glutamate induces the production of reactive oxygen species in cultured forebrain neurons following NMDA receptor activation. J Neurosci 15(5 Pt 1):3318–3327

Rothstein JD, Martin L, Levey AI, Dykes-Hoberg M, Jin L, Wu D, Nash N, Kuncl RW (1994) Localization of neuronal and glial glutamate transporters. Neuron 13(3):713–725

Rothstein JD, Dykes-Hoberg M, Pardo CA, Bristol LA, Jin L, Kuncl RW, Kanai Y, Hediger MA, Wang Y, Schielke JP, Welty DF (1996) Knockout of glutamate transporters reveals a major role for astroglial transport in excitotoxicity and clearance of glutamate. Neuron 16(3):675–686

Sagara JI, Miura K, Bannai S (1993) Maintenance of neuronal glutathione by glial cells. J Neurochem 61(5):1672–1676

Sasaki H, Sato H, Kuriyama-Matsumura K, Sato K, Maebara K, Wang H, Tamba M, Itoh K, Yamamoto M, Bannai S (2002) Electrophile response element-mediated induction of the cystine/glutamate exchange transporter gene expression. J Biol Chem 277(47):44765–44771. doi:10.1074/jbc.M208704200

Shanker G, Aschner M (2003) Methylmercury-induced reactive oxygen species formation in neonatal cerebral astrocytic cultures is attenuated by antioxidants. Brain Res Mol Brain Res 110(1):85–91

Shanker G, Allen JW, Mutkus LA, Aschner M (2001) Methylmercury inhibits cysteine uptake in cultured primary astrocytes, but not in neurons. Brain Res 914(1–2):159–165

Shanker G, Mutkus LA, Walker SJ, Aschner M (2002) Methylmercury enhances arachidonic acid release and cytosolic phospholipase A2 expression in primary cultures of neonatal astrocytes. Brain Res Mol Brain Res 106(1–2):1–11

Shanker G, Aschner JL, Syversen T, Aschner M (2004) Free radical formation in cerebral cortical astrocytes in culture induced by methylmercury. Brain Res Mol Brain Res 128(1):48–57. doi:10.1016/j.molbrainres.2004.05.022

Shanker G, Syversen T, Aschner JL, Aschner M (2005) Modulatory effect of glutathione status and antioxidants on methylmercury-induced free radical formation in primary cultures of cerebral astrocytes. Brain Res Mol Brain Res 137(1–2):11–22. doi:10.1016/j.molbrainres.2005.02.006

Sorg O, Horn TF, Yu N, Gruol DL, Bloom FE (1997) Inhibition of astrocyte glutamate uptake by reactive oxygen species: role of antioxidant enzymes. Mol Med 3(7):431–440

Stella N, Tence M, Glowinski J, Premont J (1994) Glutamate-evoked release of arachidonic acid from mouse brain astrocytes. J Neurosci 14(2):568–575

Trotti D, Danbolt NC, Volterra A (1998) Glutamate transporters are oxidant-vulnerable: a molecular link between oxidative and excitotoxic neurodegeneration? Trends Pharmacol Sci 19(8): 328–334

Vargas MR, Johnson DA, Sirkis DW, Messing A, Johnson JA (2008) Nrf2 activation in astrocytes protects against neurodegeneration in mouse models of familial amyotrophic lateral sclerosis. J Neurosci 28(50):13574–13581. doi:10.1523/JNEUROSCI.4099-08.2008

Wang L, Jiang H, Yin Z, Aschner M, Cai J (2009) Methylmercury toxicity and Nrf2-dependent detoxification in astrocytes. Toxicol Sci 107(1):135–143. doi:10.1093/toxsci/kfn201

Yin Z, Milatovic D, Aschner JL, Syversen T, Rocha JB, Souza DO, Sidoryk M, Albrecht J, Aschner M (2007) Methylmercury induces oxidative injury, alterations in permeability and glutamine transport in cultured astrocytes. Brain Res 1131(1):1–10. doi:10.1016/j.brainres.2006.10.070

Yin Z, Albrecht J, Syversen T, Jiang H, Summar M, Rocha JB, Farina M, Aschner M (2009) Comparison of alterations in amino acids content in cultured astrocytes or neurons exposed to methylmercury separately or in co-culture. Neurochem Int 55(1–3):136–142. doi:10.1016/j.neuint.2009.01.015

Yudkoff M, Pleasure D, Cregar L, Lin ZP, Nissim I, Stern J, Nissim I (1990) Glutathione turnover in cultured astrocytes: studies with [15N]glutamate. J Neurochem 55(1):137–145

Zhou Y, Danbolt NC (2014) Glutamate as a neurotransmitter in the healthy brain. J Neural Transm (Vienna) 121(8):799–817. doi:10.1007/s00702-014-1180-8

Chapter 12
Astroglia, Glutamatergic Transmission and Psychiatric Diseases

Alexei Verkhratsky, Luca Steardo, Liang Peng, and Vladimir Parpura

Abstract Astrocytes are primary homeostatic cells of the central nervous system. They regulate glutamatergic transmission through the removal of glutamate from the extracellular space and by supplying neurons with glutamine. Glutamatergic transmission is generally believed to be significantly impaired in the contexts of all major neuropsychiatric diseases. In most of these neuropsychiatric diseases, astrocytes show signs of degeneration and atrophy, which is likely to be translated into reduced homeostatic capabilities. Astroglial glutamate uptake/release and glutamate homeostasis are affected in all forms of major psychiatric disorders and represent a common mechanism underlying neurotransmission disbalance, aberrant connectome and overall failure on information processing by neuronal networks, which underlie pathogenesis of neuropsychiatric diseases.

Keywords Astrocyte • Glutamatergic transmission • Neurotransmitter receptors • Glutamate synthesis • Neuropathology • Psychiatric diseases • Major depression • Schizophrenia

---

A. Verkhratsky (✉)
Faculty of Life Sciences, The University of Manchester, Manchester M13 9PT, UK

Achucarro Center for Neuroscience, IKERBASQUE, Basque Foundation for Science, Bilbao 48011, Spain

Department of Neurosciences, University of the Basque Country UPV/EHU, Leioa 48940, Spain
e-mail: Alexej.Verkhratsky@manchester.ac.uk

L. Steardo
Department of Psychiatry, University of Naples SUN, Largo Madonna delle Grazie, Naples, Italy

L. Peng
Laboratory of Metabolic Brain Diseases, Institute of Metabolic Disease Research and Drug Development, China Medical University, Shenyang, P. R. China

V. Parpura
Department of Neurobiology, Center for Glial Biology in Medicine, Atomic Force Microscopy & Nanotechnology Laboratories, Civitan International Research Center, Evelyn F. McKnight Brain Institute, University of Alabama, Birmingham, AL 35294, USA

A. Schousboe, U. Sonnewald (eds.), *The Glutamate/GABA-Glutamine Cycle*, Advances in Neurobiology 13, DOI 10.1007/978-3-319-45096-4_12

12.1 The Multicellular Nature of Neural Tissue: The Neuron Doctrine Challenged

The human brain crowns hundreds of millions of years of the evolution of the nervous system, which occurred in a form of a relatively simple synaptically connected network of neurons, evenly distributed through the body of antediluvian Cnidarians and Ctenophores (Roth and Dicke 2013). This primeval nervous system bestowed a fundamental evolutionary advantage of the coordination of body parts that greatly increased the adaptability of these creatures. The evolution proceeded through the occurrence of concentrated neuronal masses or ganglia to the centralised nervous system, which coincided with the appearance of bilateral symmetry and emergence of Bilateralia. At early evolutionary stages (in, for example, nematodes), the centralised nervous system appears in a form of several ganglia localised around the oral orifice; in more advanced insects or in crustacean, the central nervous system is present in the form of a polyganglionic brain. In the more advanced Echinoderms, Hemichordate and Chordate the layered nervous system becomes the main trait, and this layered organisation characterises the brain of humans.

The connectivity between neurons is accomplished through evolutionary conserved specialised structures known as synapses. Whilst synapses come in two flavours, chemical and electrical, we shall focus our attention to chemical synapses and transmission. The presynaptic elements, mainly represented by axonal terminals, contain neurotransmitter vesicles from which neurotransmitter is released upon the action potential arrival, whereas the postsynaptic elements (often represented by dendritic spines) contain specific receptors and highly complex molecular machinery responsible for short- and long-term regulation of synaptic efficacy. Since the days of Ramón y Cajal, the synaptically connected neuronal networks are considered to represent the substrate for the brain function; this concept is known as the neuron doctrine (Glickstein 2006). The nervous system comprised from the neuronal networks and interneuronal synapses as the system for information processing and decision making epitomises the neuron doctrine that governs neuroscientific thoughts for the last hundred or so years.

The evolutionary history of the brain is, however, more complex and multifaceted. The appearance of the centralised nervous system coincided with the emergence of a second class of neural cells, the neuroglia. This event signalled the cell specialisation and division of labour between distinct elements of the nervous system. Neurons went on by perfecting the fast signalling apparatus, whereas the homeostatic functions were shifted to neuroglia. This, in the course of time, created a highly complex network of specialised neural cells which work in concert towards the brain function. The neuroglial cells by themselves specialised: some of them became brain homeostatic cells (astrocytes) and some evolved into producing the myelin sheath (oligodendrocytes), which is indispensable for making the compact interconnected brain of higher vertebrates (Parpura et al. 2012; Verkhratsky and Butt 2013). The evolving neuroglia also created the barrier between the brain and the blood, thus imposing the immune privilege on the nervous tissue; this in turn called for resident immune cells which appeared in the form of microglia. Once the immune cells invade the brain parenchyma, they undergo the most remarkable metamorphosis which transforms the round foetal macrophages into surveillance microglia which has almost neuron-like appearance (small body with long ramified processes) and expresses the wealth of neurotransmitter receptors (Kettenmann et al. 2011). The almost revolutionary explosion of the neuroglial field in the last two decades challenged the neuron doctrine and moved attention towards a more inclusive multicellular theory of the organisation of the brain. Moreover, it became obvious that neuroglial cells contribute to all neuropathological processes and often define the progression and outcome of neurological diseases. In this short essay, we overview the role of astrocytes in neuropsychiatric disorders, specifically from the point of view of astroglial regulation of glutamatergic neurotransmission.

12.2 The Glutamatergic Synapse: The Role for Astroglial Cradle

The majority (~60–70 %) of synapses in the CNS are organised from several compartments, that include (1) the presynaptic terminal, (2) the postsynaptic neuronal part, (3) the perisynaptic astroglial process, (4) the microglial process that periodically contacts the synapse and (5) the extracellular matrix (Derouiche et al. 2002; Reichenbach et al. 2010; Dityatev and Rusakov 2011); this complex structure is often being referred to as a "multipartite" synapse (a term which represents an evolution of the tripartite and tetrapartite variants—see (Araque et al. 1999; De Leo et al. 2006; Halassa et al. 2007; Dityatev and Rusakov 2011; Nedergaard and Verkhratsky 2012)). In the complex synaptic arrangement, each component has a clearly defined role. The presynaptic terminal is packed with synaptic vesicles and proteins that underlie fast exocytosis of the neurotransmitters. The postsynaptic compartment provides a dwelling for neurotransmitter receptors and highly diverse biochemical machinery responsible for synaptic responses. The extracellular matrix exerts regulatory effects on both pre- and post-synaptic elements, whereas microglial process monitors the synaptic well-being. The role for astroglial perisynaptic process has emerged only recently, as it appeared that the astroglial compartment (dubbed the astroglial cradle (Nedergaard and Verkhratsky 2012)) controls all aspects of synaptic life, from synaptogenesis, to synaptic maturation, maintenance and extinction (Eroglu and Barres 2010; Verkhratsky and Nedergaard 2014). In particular, the perisynaptic astroglial processes contain exceptionally diverse complement of transporters that control numerous aspects of synaptic cleft homeostasis (Kirischuk et al. 2012; Parpura and Verkhratsky 2012). These transporters buffer extracellular potassium, control extracellular pH, provide reactive oxygen species scavengers and, most importantly in the context of present essay, they control the extracellular presence of neurotransmitters. Astrocytes are fundamental for the homeostasis and metabolism of three key neurotransmitters: adenosine, glutamate and GABA (Danbolt 2001; Boison et al. 2010; Parpura and Verkhratsky 2012).

In particular, astroglial cells are indispensable elements of the glutamatergic synapse. About 80 % of glutamate, released in the course of synaptic transmission is taken up by astrocytes through the operation of plasmalemmal glutamate transporters that belong to the extended family of solute carriers (SLC). Astroglial plasmalemmal glutamate transporters are represented by two subtypes: the excitatory amino acid transporters 1 and 2 (EAAT1/SLCA3 and EAAT2/SLC1A2). These transporters utilise transmembrane gradient of Na^+ ions for glutamate translocation; transport of a single glutamate molecule (monovalent anion at physiological pH) is associated with an influx of three Na^+ ions, one H^+ ion and efflux of one K^+ ion. This stoichiometry stipulates the electrogenic effects of glutamate uptake (which results in a net inward current); the operation of plasmalemmal glutamate transporters leads to a substantial Na^+ influx that may significantly increase cytosolic Na^+ concentration hence contributing to astroglial Na^+ signalling (Kirischuk et al. 2012). Of note, the reversal potential of plasmalemmal glutamate transporters is more positive than +50 mV or even +60 mV (reflecting transmembrane distribution of relevant ions) and hence they cannot revert in physiological conditions (Kirischuk et al. 2007).

Cytosolic glutamate in astrocytes can be converted into glutamine, this reaction being catalysed (at the expanse of a single ATP) by glutamine synthetase, the astroglia-specific enzyme (Martinez-Hernandez et al. 1977), which is also fundamental for nitrogen metabolism (Rose et al. 2013). Glutamate taken up into astrocytes is not only metabolised to glutamine but, to a significant extent, is oxidatively metabolised, a process that requires the activity of glutamate dehydrogenase and the tricarboxylic acid cycle enzymes (reviewed in Kreft et al. 2012; Obel et al. 2012; McKenna 2013). Pyruvate carboxylase, an enzyme within the cycle and otherwise expressed in astrocytes but not neurons, endows these glial cells with the de novo production of glutamate from pyruvate; the latter being sourced from energy substrates, glucose and lactate (reviewed in Hertz and Zielke 2004). Nonetheless, glutamine synthetase represents a central enzyme of glutamine–glutamate shuttle; that is a coordinated system of glutamate uptake (see earlier) glutamate conversion into glutamine and glutamine transfer from astrocytes to neurons. The later transfer occurs via the glutamine transport which involves two distinct systems of amino acid transporters. The system N, comprised of Na^+-dependent neutral amino acid transporters SN1/SNAT3/SLC38A3 and SN2/SNAT5/SLC38A5, is expressed in astrocytes, where it mediates the efflux of glutamine (Nakanishi et al. 2001; Chaudhry et al. 2002). Neurons accumulate glutamine through the operation of the system A, which includes Na^+-dependent neutral amino acid transporters ATA1/SNAT1/SLC38A1 and ATA2/SNAT2/SLC38A2 (Chaudhry et al. 2002). Importantly, astroglial release of glutamine is controlled by neuronal activity, the increase of which stimulates the astrocytic secretion of glutamine (Martinez-Lozada et al. 2013). After entering the presynaptic terminals glutamine is principally hydrolysed to glutamate by phosphate-activated glutaminase, an energetically neutral conversion. Subsequently, glutamate is accumulated within synaptic vesicles by the coordinated activity of the proton pump V-ATPase and vesicular glutamate transporters VGLUT1-3 (SLC17A7, SLC17A6 and SLC17A8, respectively), the latter linked to the chloride flux. Like neurons, astrocytes express, albeit to a lesser degree, all the three known isoforms of VGLUTs, which mediate packaging of glutamate into astroglial secretory vesicles (Montana et al. 2004); VGLUT3 emerged as a regulator of vesicular packaging of glutamate in astrocytes (Ni and Parpura 2009). The intravesicular concentration of glutamate likely reaches ~60 mM in neurons and ~20 mM in astrocytes (Montana et al. 2006). Ca^{2+}-dependent exocytosis at presynaptic terminal raises the concentration of glutamate in the synaptic cleft to ~1.1 mM (Clements et al. 1992), while in astrocytes it results in release with localised extracellular glutamate accumulation of 1–100 μM (Innocenti et al. 2000); the latter gliotransmission can lead to modulation of synaptic transmission and plasticity (Araque et al. 1999; Perez-Alvarez and Araque 2013).

12.3 The Psychiatric Diseases as Synaptic Malfunction: The Role for Glutamatergic Transmission

The human brain contains ~15 trillion of synapses, with glutamatergic and GABA-ergic synapses being the majority accounting for over 85 % of all synaptic contacts in the brain (Douglas and Martin 2007). Importantly, however, many synapses utilise more than one neurotransmitter (the concept of co-transmission (Burnstock 2004)), and all regions of the brain receive noradrenergic innervation (from the locus coeruleus) that controls functional activity of neurons and neuroglia alike. This intricate neurochemical concoction with its ever changing formulation is the substrate for the brain function. The psychiatric diseases reflect the general disbalance of the neurochemistry in the brain, in which the impairment of glutamatergic transmission has been receiving an increased attention, which lead to the emergence of the "glutamate" hypothesis of major depression and schizophrenia (Pittenger and Duman 2008; Racagni and Popoli 2008; Sanacora et al. 2012). At the core of this hypothesis lies an idea that glia affects glutamatergic synapses, which otherwise mediate learning, memory, cognition and emotional reactions.

Levels of glutamate have been found elevated in the cerebrospinal fluid and in the brain of patients with various forms of mood disorders, including those suffering from major depression (Altamura et al. 1993, 1995; Mitani et al. 2006; Kucukibrahimoglu et al. 2009). Levels of glutamate were also found to be increased in the post-mortem specimens of frontal and dorsolateral prefrontal cortices of depressed and bipolar patients (Hashimoto et al. 2007; Lan et al. 2009). Importantly, the treatment with antidepressants correlated with decreased levels of glutamate in the plasma (Maes et al. 1998). An impaired level of glutamate-related metabolites (the so-called *Glx index*, which included glutamate and glutamine) was found in the brains of patients with mood disorders by magnetic resonance spectroscopy. It seems that the Glx index decreases in frontal and cingulate cortices and increases in occipital and parieto-occipital regions (Rosenberg et al. 2004, 2005; Sanacora et al. 2004; Yuksel and Ongur 2010; Taylor et al. 2011), see also in depth meta-analysis supporting impaired astroglial glutamate–glutamine shuttling in depression (Arnone et al. 2015).

The glutamatergic model of pathophysiology of schizophrenia was prompted by the discovery of the N-methyl-D-aspartate (NMDA) type of glutamate receptors as a molecular target for psychotomimetic compounds such as ketamine or phencyclidine (Javitt 1987; Javitt and Zukin 1991). It appeared that the inhibition of NMDA receptors can transiently reproduce some symptoms of schizophrenia, and this pharmacological model has become widely used (Olney and Farber 1995; Coyle and Tsai 2004). Considering astrocytes as the major component regulating glutamatergic transmission, their involvement in pathogenesis of psychiatric diseases seems to be highly probable.

12.4 Astroglia in Major Psychiatric Disorders

12.4.1 Psychosis Associated with Toxic Encephalopathies

Numerous toxic lesions to the brain trigger neuropsychiatric symptomatology, which can range from an acute state of confusion (the delirium) to delusions, behavioural abnormalities and impaired consciousness. Accumulating evidence links these neurotoxic disorders to impaired glutamate homeostasis associated with an astroglial failure. In encephalopathies, resulting from the exposure to heavy metals, such as lead, manganese, aluminium and/or mercury, astrocytes appear to be the primary target. These heavy metals, being accumulated into astroglial cells, lead to a substantial down-regulation of glial plasmalemmal glutamate transporters and hence to the reduction of astroglial glutamate uptake, which affects neurotransmission and causes excitotoxic neuronal damage (Suarez-Fernandez et al. 1999; Struys-Ponsar et al. 2000; Yin et al. 2007; Verkhratsky et al. 2013). Similarly, impaired glutamate homeostasis and glutamine synthesis emerge as a leading mechanism of hepatic encephalopathy, in which increased levels of ammonium in the brain affect astroglia and cause generalised homeostatic failure (Norenberg 1987; Albrecht et al. 2010; Rose et al. 2013; Haack et al. 2014; Obara-Michlewska et al. 2014). Impaired glutamate homeostasis also represents the main pathogenic factor in the development of Korsakoff syndrome (ante- and retrograde amnesia and confabulations), which is usually linked to Wernicke encephalopathy. In fact, these two disorders are so closely related that the patients are commonly diagnosed with Wernicke–Korsakoff syndrome, also called wet brain, which is a thiamine (vitamin B_1) deficiency associated with a severe excitotoxic damage secondary to alcoholism. The Wernicke encephalopathy is a primary astrogliopathy; the key pathogenetic step is a prominent down-regulation (by ~70 %) of the expression of astroglial plasmalemmal glutamate transporters, which leads to impaired glutamate transmission and glutamate excitotoxicity (Hazell 2009; Hazell et al. 2009).

12.4.2 Schizophrenia

In contrast to many neurological conditions, the pathophysiology of major psychiatric disorders (schizophrenia, major depressive disorder and bipolar disorder) does not include reactive astrogliosis. Rather, it seems that glial atrophy and astrodegeneration prevail in major psychiatric disorders (Verkhratsky et al. 2014). The emergence of pathologically remodelled astroglia is the common feature of the post-mortem brain of schizophrenic patients. In various brain regions a substantial decrease in astroglial numbers and glial fibrillary acidic protein (GFAP) expression along with the appearance of atrophic astroglial profiles is widely documented (Falkai and Bogerts 1986; Webster et al. 2001; Rajkowska et al. 2002; Schmitt et al. 2009). Similarly, many proteins fundamental for homeostatic function of astroglia, such as deiodinase type II, aquaporin 4, S100 protein, glutamine synthetase, plasmalemmal glutamate transporters and thrombospondin, are down-regulated in the schizophrenic brain (Xia et al. 2014). At the same time, levels of S100 protein in the cerebrospinal fluid of schizophrenic patients were often increased (Rothermundt et al. 2004, 2007, 2009); this increase may indicate the astroglial damage, with an increased leakage of S100 protein (Takahashi and Sakurai 2013). A very substantial increase in astroglial synthesis of chondroitin sulphate proteoglycans that constitute a main component of the extracellular matrix has also been identified in the amygdala and entorhinal cortex; this may affect synaptic transmission (Pantazopoulos et al. 2010).

Astroglial plasmalemmal glutamate transporters EAAT1/2 were decreased in the prefrontal cortex and hippocampus (Ohnuma et al. 2000; Bauer et al. 2008, 2010; Shan et al. 2013). At the same time, animals in which EAAT1 was genetically deleted demonstrated aberrant social behaviour and locomotor hyperactivity, resembling certain symptoms of schizophrenia (Karlsson et al. 2008, 2009). Expression of the plasmalemmal cystine-glutamate exchanger (Sxc-), which controls extrasynaptic concentration of glutamate in the CNS (Bridges et al. 2012) has been found to be up-regulated in the rodent phencyclidine model of schizophrenia (Baker et al. 2008). Aberrations in astroglia-dependent control over extracellular glutamate may instigate a profound synaptic remodelling hence affecting the spatial organisation of synaptic transmission and contributing to an overall confused information processing, which is characteristic for patients suffering from schizophrenia (Shan et al. 2014).

Pathophysiology of schizophrenia is also linked to changes in the biogenesis and catabolism of D-serine, a positive modulator of NMDA receptors. D-serine is released by both neurons and astroglia and by binding to the glycine site of NMDA receptors facilitates the receptor activation by glutamate (Wolosker 2007; Oliet and Mothet 2009). The proportional cellular sourcing (astrocytes vs. neurons) of D-serine remains a matter of debate (Balu et al. 2014). In schizophrenia, D-serine levels are decreased (Bendikov et al. 2007). Pharmacological, biochemical and genetic investigations demonstrate that, in addition to mutations and decreased expression of the NMDA receptor itself, changes may occur in the metabolism and levels of D-serine in patients with schizophrenia (Balu and Coyle 2015). This may be linked to the reduced expression of serine racemase, an enzyme claimed to be associated with the risk of schizophrenia development (Labrie et al. 2009). A down-regulation of serine racemase has been observed in transgenic animals expressing mutant Disrupted-In-Schizophrenia-1 (DISC1) gene specifically in astrocytes (Ma et al. 2013). Alteration in the interplay of DISC1 with serine racemase leads to depletion of D-serine and consequently may instigate schizophrenia-like behaviour. Accordingly, prevailing efforts are presently directed to normalise NMDAR occupancy by enhancing D-serine levels though the administration of the co-agonist itself or by raising its synaptic availability (Balu and Coyle 2015). Along this line, preclinical and clinical investigations demonstrate that D-serine or benzoate, an inhibitor of D-serine degrading enzyme, D-amino acid oxidase (DAAO), are able to ameliorate negative and cognitive symptoms (Kantrowitz et al. 2010; Lane et al. 2013). Furthermore, peripheral levels of D-serine and the D/L-serine ratio significantly increase in patients with schizophrenia after treatment with clozapine (Yamamori et al. 2014).

Astrocytes also produce kynurenic acid, which has been implicated in pathogenesis of schizophrenia. Kynurenic acid is an endogenous competitive inhibitor of α7-containing nicotinic acetylcholine receptors and of NMDA glutamate receptors in the CNS. Kynurenic acid is synthesised from tryptophan by kynurenine aminotransferase II, which, in the CNS, is mainly expressed in astroglia (Schwarcz et al. 2012). Genetic deletion of kynurenine transferase II augmented synaptic plasticity and improved cognitive function (Potter et al. 2010). Being an NMDA receptor antagonist, kynurenic acid shares pharmacological properties with phencyclidine and ketamine, substances reported to provoke psychotic symptoms in healthy subjects, highly resembling those observed in patients (Krystal et al. 1994). The levels of kynurenic acid were increased in the post-mortem brains of psychotics (Holtze et al. 2012). The kynurenine pathway regulates the dopaminergic system, which for decades is considered to play a prominent role in the pathogenesis of schizophrenia. Increase in kynurenic acid level in the rodent brain potentiates midbrain dopaminergic neurons firing, as well as dopamine release and locomotor activity (Nilsson et al. 2006). Notably, the metabolism of kynurenic acid appears as a causal link between *Toxoplasma gondii* infection and an increased risk of schizophrenia. Seemingly, *T. gondii* infects predominantly astrocytes, and this infection results in an increased production of kynurenic acid (Schwarcz and Hunter 2007).

12.4.3 Mood Disorders

There are no major signs of glial reactivity in mood disorders; rather, and similarly to schizophrenia, astroglial atrophy and astrodegeneration dominate. In major depressive disorder as well as in bipolar disorder numbers of astrocytes are found to be reduced in the fronto-limbic brain, including orbito-frontal area, anterior cingulate, and prefrontal, entorhinal and subgenual cortices, as well as the amygdala

(Ongur et al. 1998; Rajkowska et al. 1999; Cotter et al. 2001, 2002; Bowley et al. 2002). Significant decreases in astroglial numbers were identified in prefrontal and anterior cingulate cortices in familial forms of both major depressive and bipolar disorders (Ongur et al. 1998; Cotter et al. 2001; Gittins and Harrison 2011). Expression of GFAP has been reduced in several brain regions in post-mortem tissues from depressed patients, including the prefrontal cortex (Johnston-Wilson et al. 2000; Webster et al. 2001; Si et al. 2004), amygdala (Altshuler et al. 2010), hippocampus (Muller et al. 2001) and cerebellum (Fatemi et al. 2004). Expression of GFAP was markedly depressed in the thalamus and nucleus caudatus of suicidal patients (Torres-Platas et al. 2015). Decrease in astroglial numbers and GFAP expression in the brains of patients with major depressive disorders paralleled with down-regulation of astroglia-specific proteins, in particular, those critically important for astroglial homeostatic function. Notably, expression of *SLC1A2* and *SLC1A3* genes encoding EAAT1 and EAAT2 plasmalemmal glutamate transporters as well as *GLUL* (Glutamate-Ammonia Ligase) gene encoding glutamine synthetase was found to be markedly decreased (Choudary et al. 2005), indicating impairment of astroglia-dependent glutamate homeostasis. Other astroglial genes, including those encoding S100 protein, connexins Cx30 and Cx43, and aquaporin 4 were also down-regulated in the brain tissue of patients with major depression (Bernard et al. 2011). The expression of astrocyte-specific tropomyosin-related kinase-B receptor 1, as well as connexins was reduced in orbitofrontal and prefrontal cortices of suicide victims (Ernst et al. 2009, 2011).

Similarly, decrease in astroglial numbers as well as decrease in GFAP immunoreactivity, number of GFAP-positive astroglial profiles and expression of glial-specific proteins (aquaporin 4, connexins, astroglial plasmalemmal glutamate transporters and glutamine synthetase) was observed in animal models of mood disorders (Czeh et al. 2006; Barley et al. 2009; Braun et al. 2009; Sequeira et al. 2009; Bernard et al. 2011; Rajkowska and Stockmeier 2013). In particular, chronic stress down-regulated expression of astroglia-specific Cx43 that was reflected by a decreased dye coupling between astrocytes in the prefrontal cortex. In parallel, the pharmacological blockade of gap junctions (treatment with carbenoxolone or with Cx43 mimetic peptides Gap27 and Gap26) induced anhedonia that is generally considered as one of the major symptoms of depression (Sun et al. 2012). Anhedonia was also induced by treating rats with dihydrokainic acid, an inhibitor of astroglial glutamate uptake via EAATs; this treatment also affected spatial memory (Bechtholt-Gompf et al. 2010). Finally, a treatment with antidepressants was found to affect not only neurons, but also astrocytes, in particular affecting astroglial expression of neurotransmitter receptors, ion channels and homeostatic transporters (Czeh and Di Benedetto 2013; Dong et al. 2015; Peng et al. 2015; Ren et al. 2015).

The causal role for astroglia-associated pathological changes in mood disorders was highlighted in experiments with the direct ablation of astroglial cells from the medial-prefrontal cortex of mice with L alpha-aminoadipic acid (L-AAA), an astrocyte-selective toxin, which resulted in the emergence of depressive phenotype similar to that induced by chronic stress (Banasr and Duman 2008).

Notably, the unique pattern of cell pathology in major depressive disorder characterised by prominent reductions in the density of astrocytes and expression of astroglial markers is further confirmed by in vivo brain imaging using proton magnetic resonance technique, a non-invasive procedure that allows direct detection of brain metabolites. In these studies, lower levels of myo-inositol, commonly regarded as an astroglial marker, were in the anterior cingulate cortex of depressed patients; the levels of myo-inositol increased following the treatment only in subjects responsive to therapy (Sterado, unpublished observations). These results further corroborate the role of astroglial dysfunction in the pathogenesis of depressive disorders.

12.4.4 Addictive Disorders

Astroglia contribute to addictive disorders. Morphometric studies have identified a complex astroglial response to various forms of addiction with evidence for both astrodegenerative, i.e. atrophic changes, and astroglial reactivity (Oehmichen et al. 1996; Suarez et al. 2000; Fattore et al. 2002; Armstrong et al. 2004; Büttner and Weis 2006; Weber et al. 2013). This dichotomy in astroglial reactions may reflect the length and severity of the disease. For example, astroglial numbers were reduced and GFAP-positive astroglial profiles were diminished in patients with relatively short-lived alcoholism, whereas the long-lasting alcohol dependence was associated with astroglial reactivity, manifested by increased numbers of astroglial cells, i.e. proliferation, and increased GFAP expression (Miguel-Hidalgo et al. 2002, 2006; Skuja et al. 2012). In post-mortem tissues obtained from alcoholics, reactive and hypertrophic astrocytes were observed together with areas in which numbers of GFAP-positive astrocytes were substantially decreased (Cullen and Halliday 1994; Miguel-Hidalgo 2005). Astroglial reactions were also region specific. For example, in alcoholism, the number of astrocytes was decreased in the prefrontal cortex from post-mortem tissues (Miguel-Hidalgo et al. 2006).

Similarly, complex/dichotomous astroglial responses were found in animals treated with drugs of addiction. Reactive astrogliosis and GFAP increase have been identified in animals treated with cocaine, methamphetamine and morphine (Song and Zhao 2001; Fattore et al. 2002; Bowers and Kalivas 2003; Guilarte et al. 2003; Friend and Keefe 2013). Decreased expression of GFAP and astroglial atrophy was, however, observed in animals chronically exposed to alcohol (Franke 1995; Rintala et al. 2001). Astroglial density and GFAP expression were significantly decreased in the prelimbic cortex of rats addicted to ethanol (Miguel-Hidalgo 2005). Incidentally, binge-like ethanol administration reduced astrocytic density in the adult rat medial prefrontal cortex of males but not females (Koss et al. 2012). The role of astroglia in addiction was also inferred by experiments with the astrocyte ablation using L-AAA; removing astrocytes from the prelimbic area of the prefrontal cortex increased alcohol-seeking behaviour. Same effect was produced by the inhibition of astroglial connectivity following injection of 8-alpha-glycyrrhetinic acid, a connexon/gap-junction inhibitor (Miguel-Hidalgo 2009).

Morphological impairment of astroglia in the addictive disorders was also complemented by their functional failure, i.e. aberrant glutamate clearance and metabolism. In the context of addiction, astroglial cells down-regulated the expression of EAAT2 and Sxc^-, although overall extracellular glutamate concentration was found to be elevated probably because of disbalance between glutamate uptake (EAAT2) and glutamate release (Sxc^-) (Reissner and Kalivas 2010, 2014; Moussawi et al. 2011). Down-regulation of astroglial EAAT2 was found in various animal models of addiction, including that to cocaine, nicotine or alcohol (Knackstedt et al. 2010; Gipson et al. 2013). Increase in astroglial expression of plasmalemmal glutamate transporters following treatment with ceftriaxone, a beta-lactam antibiotic, reduced alcohol seeking in ethanol-preferring animals (Qrunfleh et al. 2013; Sari et al. 2013).

12.4.5 Huntington's Disease

Huntington's disease (HD) is an autosomal dominant inherited disease caused by the repeat of a nucleotide triplet in the exon 1 of huntingtin gene, which results in the synthesis of mutant huntingtin protein (mhtt) containing an expanded polyglutamine section in its N-terminal portion. The neurodegeneration in HD mainly occurs in striatal medium spiny neurons and in cortical pyramidal neurons (Vonsattel et al. 1985; Hedreen et al. 1991). HD shows both neurologic and psychiatric components and can represent a challenge to psychiatric services (Leroi and Michalon 1998).

Mhtt is present in neurons and astrocytes in brain tissue from human patients (Singhrao et al. 1998) and in mouse HD models (Shin et al. 2005; Faideau et al. 2010). Be that in human tissue or animal models, astrocytes contribute to HD through a prominent astrogliosis, which correlates with HD severity (Vonsattel et al. 1985; Gu et al. 2005; Faideau et al. 2010). Astrocytes also contribute to HD through the compromised glutamate and potassium uptake as well as an enhanced calcium-dependent glutamate release. Some or all of these contributions likely result in excitotoxicity, an important factor for neuronal death in HD (Gray 2014). Thus, EAAT2 mRNA and protein levels are decreased in human HD tissues (Cross et al. 1986; Arzberger et al. 1997), which correlate with the severity of the disease (Faideau et al. 2010). Astrocytes isolated from the cortex of the BACHD mouse model (Lee et al. 2013) showed an enhanced exocytotic glutamate release that was allied with an increased expression of pyruvate carboxylate and hence an increased *de novo* synthesis of glutamate. HD astrocytes of R6/2 and Q175 HD mouse models also have compromised potassium buffering due to a decreased expression of $K_{ir}4.1$ K^+ channels (Tong et al. 2014); this resulted in an increase of medium spiny neurons excitability in vitro. Taken together the dysregulation of glutamate and potassium homeostasis in astrocytes contributes to pathology seen in HD.

12.5 Concluding Remarks

Down-regulation of molecular pathways responsible for astroglial control over glutamate homeostasis in the brain seems to be a common feature of all major neuropsychiatric disorders. Indeed, decrease in astroglial glutamate uptake is identified in human tissue in toxic neuroencephalopathies, schizophrenia, mood and addictive disorders. Similar decrease is observed in various animal models of neuropsychiatric pathology. Failed astroglial glutamate homeostasis may therefore represent an important pathogenic step responsible for overall disbalance of neurotransmission as well as for synaptic and neuronal damage which all underlie psychiatric symptomatology; astrocytes and astroglia-specific molecules may therefore be regarded as a promising therapeutic target.

Acknowledgements Authors research was supported by Alzheimer's Research Trust (UK) Programme Grant (ART/PG2004A/1) to AV and by the National Institutes of Health grant (HD078678) to VP.

References

Albrecht J, Zielinska M, Norenberg MD (2010) Glutamine as a mediator of ammonia neurotoxicity: a critical appraisal. Biochem Pharmacol 80:1303–1308

Altamura CA, Mauri MC, Ferrara A, Moro AR, D'Andrea G, Zamberlan F (1993) Plasma and platelet excitatory amino acids in psychiatric disorders. Am J Psychiatry 150:1731–1733

Altamura C, Maes M, Dai J, Meltzer HY (1995) Plasma concentrations of excitatory amino acids, serine, glycine, taurine and histidine in major depression. Eur Neuropsychopharmacol 5(Suppl):71–75

Altshuler LL, Abulseoud OA, Foland-Ross L, Bartzokis G, Chang S, Mintz J, Hellemann G, Vinters HV (2010) Amygdala astrocyte reduction in subjects with major depressive disorder but not bipolar disorder. Bipolar Disord 12:541–549

Araque A, Parpura V, Sanzgiri RP, Haydon PG (1999) Tripartite synapses: glia, the unacknowledged partner. Trends Neurosci 22:208–215

Armstrong V, Reichel CM, Doti JF, Crawford CA, McDougall SA (2004) Repeated amphetamine treatment causes a persistent elevation of glial fibrillary acidic protein in the caudate-putamen. Eur J Pharmacol 488:111–115

Arnone D, Mumuni AN, Jauhar S, Condon B, Cavanagh J (2015) Indirect evidence of selective glial involvement in glutamate-based mechanisms of mood regulation in depression: meta-analysis of absolute prefrontal neuro-metabolic concentrations. Eur Neuropsychopharmacol 25:1109–1117

Arzberger T, Krampfl K, Leimgruber S, Weindl A (1997) Changes of NMDA receptor subunit (NR1, NR2B) and glutamate transporter (GLT1) mRNA expression in Huntington's disease—an in situ hybridization study. J Neuropathol Exp Neurol 56:440–454

Baker DA, Madayag A, Kristiansen LV, Meador-Woodruff JH, Haroutunian V, Raju I (2008) Contribution of cystine-glutamate antiporters to the psychotomimetic effects of phencyclidine. Neuropsychopharmacology 33:1760–1772

Balu DT, Coyle JT (2015) The NMDA receptor 'glycine modulatory site' in schizophrenia: D-serine, glycine, and beyond. Curr Opin Pharmacol 20:109–115

Balu DT, Takagi S, Puhl MD, Benneyworth MA, Coyle JT (2014) D-serine and serine racemase are localized to neurons in the adult mouse and human forebrain. Cell Mol Neurobiol 34:419–435

Banasr M, Duman RS (2008) Glial loss in the prefrontal cortex is sufficient to induce depressive-like behaviors. Biol Psychiatry 64:863–870

Barley K, Dracheva S, Byne W (2009) Subcortical oligodendrocyte- and astrocyte-associated gene expression in subjects with schizophrenia, major depression and bipolar disorder. Schizophr Res 112:54–64

Bauer D, Gupta D, Harotunian V, Meador-Woodruff JH, McCullumsmith RE (2008) Abnormal expression of glutamate transporter and transporter interacting molecules in prefrontal cortex in elderly patients with schizophrenia. Schizophr Res 104:108–120

Bauer D, Haroutunian V, Meador-Woodruff JH, McCullumsmith RE (2010) Abnormal glycosylation of EAAT1 and EAAT2 in prefrontal cortex of elderly patients with schizophrenia. Schizophr Res 117:92–98

Bechtholt-Gompf AJ, Walther HV, Adams MA, Carlezon WA Jr, Ongur D, Cohen BM (2010) Blockade of astrocytic glutamate uptake in rats induces signs of anhedonia and impaired spatial memory. Neuropsychopharmacology 35:2049–2059

Bendikov I, Nadri C, Amar S, Panizzutti R, De Miranda J, Wolosker H, Agam G (2007) A CSF and postmortem brain study of D-serine metabolic parameters in schizophrenia. Schizophr Res 90:41–51

Bernard R, Kerman IA, Thompson RC, Jones EG, Bunney WE, Barchas JD, Schatzberg AF, Myers RM, Akil H, Watson SJ (2011) Altered expression of glutamate signaling, growth factor, and glia genes in the locus coeruleus of patients with major depression. Mol Psychiatry 16:634–646

Boison D, Chen JF, Fredholm BB (2010) Adenosine signaling and function in glial cells. Cell Death Differ 17:1071–1082

Bowers MS, Kalivas PW (2003) Forebrain astroglial plasticity is induced following withdrawal from repeated cocaine administration. Eur J Neurosci 17:1273–1278

Bowley MP, Drevets WC, Ongur D, Price JL (2002) Low glial numbers in the amygdala in major depressive disorder. Biol Psychiatry 52:404–412

Braun K, Antemano R, Helmeke C, Buchner M, Poeggel G (2009) Juvenile separation stress induces rapid region- and layer-specific changes in S100ss- and glial fibrillary acidic protein-immunoreactivity in astrocytes of the rodent medial prefrontal cortex. Neuroscience 160:629–638

Bridges RJ, Natale NR, Patel SA (2012) System xc$^-$ cystine/glutamate antiporter: an update on molecular pharmacology and roles within the CNS. Br J Pharmacol 165:20–34

Burnstock G (2004) Cotransmission. Curr Opin Pharmacol 4:47–52

Büttner A, Weis S (2006) Neuropathological alterations in drug abusers. The involvement of neurons, glial and vascular systems. Forensic Sci Med Pathol 2:115–126

Chaudhry FA, Reimer RJ, Edwards RH (2002) The glutamine commute: take the N line and transfer to the A. J Cell Biol 157:349–355

Choudary PV, Molnar M, Evans SJ, Tomita H, Li JZ, Vawter MP, Myers RM, Bunney WE Jr, Akil H, Watson SJ, Jones EG (2005) Altered cortical glutamatergic and GABAergic signal transmission with glial involvement in depression. Proc Natl Acad Sci U S A 102:15653–15658

Clements JD, Lester RA, Tong G, Jahr CE, Westbrook GL (1992) The time course of glutamate in the synaptic cleft. Science 258:1498–1501

Cotter D, Mackay D, Landau S, Kerwin R, Everall I (2001) Reduced glial cell density and neuronal size in the anterior cingulate cortex in major depressive disorder. Arch Gen Psychiatry 58:545–553

Cotter D, Mackay D, Chana G, Beasley C, Landau S, Everall IP (2002) Reduced neuronal size and glial cell density in area 9 of the dorsolateral prefrontal cortex in subjects with major depressive disorder. Cereb Cortex 12:386–394

Coyle JT, Tsai G (2004) The NMDA receptor glycine modulatory site: a therapeutic target for improving cognition and reducing negative symptoms in schizophrenia. Psychopharmacology (Berl) 174:32–38

Cross AJ, Slater P, Reynolds GP (1986) Reduced high-affinity glutamate uptake sites in the brains of patients with Huntington's disease. Neurosci Lett 67:198–202

Cullen KM, Halliday GM (1994) Chronic alcoholics have substantial glial pathology in the forebrain and diencephalon. Alcohol Alcohol Suppl 2:253–257

Czeh B, Di Benedetto B (2013) Antidepressants act directly on astrocytes: evidences and functional consequences. Eur Neuropsychopharmacol 23:171–185

Czeh B, Simon M, Schmelting B, Hiemke C, Fuchs E (2006) Astroglial plasticity in the hippocampus is affected by chronic psychosocial stress and concomitant fluoxetine treatment. Neuropsychopharmacology 31:1616–1626

Danbolt NC (2001) Glutamate uptake. Progr Neurobiol 65:1–105

De Leo JA, Tawfik VL, LaCroix-Fralish ML (2006) The tetrapartite synapse: path to CNS sensitization and chronic pain. Pain 122:17–21

Derouiche A, Anlauf E, Aumann G, Muhlstadt B, Lavialle M (2002) Anatomical aspects of glia-synapse interaction: the perisynaptic glial sheath consists of a specialized astrocyte compartment. J Physiol Paris 96:177–182

Dityatev A, Rusakov DA (2011) Molecular signals of plasticity at the tetrapartite synapse. Curr Opin Neurobiol 21:353–359

Dong L, Li B, Verkhratsky A, Peng L (2015) Cell type-specific *in vivo* expression of genes encoding signalling molecules in the brain in response to chronic mild stress and chronic treatment with fluoxetine. Psychopharmacology (Berl)

Douglas RJ, Martin KA (2007) Mapping the matrix: the ways of neocortex. Neuron 56:226–238

Ernst C, Chen ES, Turecki G (2009) Histone methylation and decreased expression of TrkB.T1 in orbital frontal cortex of suicide completers. Mol Psychiatry 14:830–832

Ernst C, Nagy C, Kim S, Yang JP, Deng X, Hellstrom IC, Choi KH, Gershenfeld H, Meaney MJ, Turecki G (2011) Dysfunction of astrocyte connexins 30 and 43 in dorsal lateral prefrontal cortex of suicide completers. Biol Psychiatry 70:312–319

Eroglu C, Barres BA (2010) Regulation of synaptic connectivity by glia. Nature 468:223–231

Faideau M, Kim J, Cormier K, Gilmore R, Welch M, Auregan G, Dufour N, Guillermier M, Brouillet E, Hantraye P, Deglon N, Ferrante RJ, Bonvento G (2010) *In vivo* expression of polyglutamine-expanded huntingtin by mouse striatal astrocytes impairs glutamate transport: a correlation with Huntington's disease subjects. Hum Mol Genet 19:3053–3067

Falkai P, Bogerts B (1986) Cell loss in the hippocampus of schizophrenics. Eur Arch Psychiatry Neurol Sci 236:154–161

Fatemi SH, Laurence JA, Araghi-Niknam M, Stary JM, Schulz SC, Lee S, Gottesman II (2004) Glial fibrillary acidic protein is reduced in cerebellum of subjects with major depression, but not schizophrenia. Schizophr Res 69:317–323

Fattore L, Puddu MC, Picciau S, Cappai A, Fratta W, Serra GP, Spiga S (2002) Astroglial in vivo response to cocaine in mouse dentate gyrus: a quantitative and qualitative analysis by confocal microscopy. Neuroscience 110:1–6

Franke H (1995) Influence of chronic alcohol treatment on the GFAP-immunoreactivity in astrocytes of the hippocampus in rats. Acta Histochem 97:263–271

Friend DM, Keefe KA (2013) Glial reactivity in resistance to methamphetamine-induced neurotoxicity. J Neurochem 125:566–574

Gipson CD, Reissner KJ, Kupchik YM, Smith AC, Stankeviciute N, Hensley-Simon ME, Kalivas PW (2013) Reinstatement of nicotine seeking is mediated by glutamatergic plasticity. Proc Natl Acad Sci U S A 110:9124–9129

Gittins RA, Harrison PJ (2011) A morphometric study of glia and neurons in the anterior cingulate cortex in mood disorder. J Affect Disord 133:328–332

Glickstein M (2006) Golgi and Cajal: The neuron doctrine and the 100th anniversary of the 1906 Nobel Prize. Curr Biol 16:R147–151

Gray M (2014) The role of astrocytes in Huntington's disease. In: Parpura V, Verkhratsky A (eds) Pathological potentual of neuroglia. Possible new targets for medical intervention. Springer, New York, pp 213–229

Gu X, Li C, Wei W, Lo V, Gong S, Li SH, Iwasato T, Itohara S, Li XJ, Mody I, Heintz N, Yang XW (2005) Pathological cell-cell interactions elicited by a neuropathogenic form of mutant Huntingtin contribute to cortical pathogenesis in HD mice. Neuron 46:433–444

Guilarte TR, Nihei MK, McGlothan JL, Howard AS (2003) Methamphetamine-induced deficits of brain monoaminergic neuronal markers: distal axotomy or neuronal plasticity. Neuroscience 122:499–513

Haack N, Dublin P, Rose CR (2014) Dysbalance of astrocyte calcium under hyperammonemic conditions. PLoS One 9, e105832
Halassa MM, Fellin T, Haydon PG (2007) The tripartite synapse: roles for gliotransmission in health and disease. Trends Mol Med 13:54–63
Hashimoto K, Sawa A, Iyo M (2007) Increased levels of glutamate in brains from patients with mood disorders. Biol Psychiatry 62:1310–1316
Hazell AS (2009) Astrocytes are a major target in thiamine deficiency and Wernicke's encephalopathy. Neurochem Int 55:129–135
Hazell AS, Sheedy D, Oanea R, Aghourian M, Sun S, Jung JY, Wang D, Wang C (2009) Loss of astrocytic glutamate transporters in Wernicke encephalopathy. Glia 58:148–156
Hedreen JC, Peyser CE, Folstein SE, Ross CA (1991) Neuronal loss in layers V and VI of cerebral cortex in Huntington's disease. Neurosci Lett 133:257–261
Hertz L, Zielke HR (2004) Astrocytic control of glutamatergic activity: astrocytes as stars of the show. Trends Neurosci 27:735–743
Holtze M, Saetre P, Engberg G, Schwieler L, Werge T, Andreassen OA, Hall H, Terenius L, Agartz I, Jonsson EG, Schalling M, Erhardt S (2012) Kynurenine 3-monooxygenase polymorphisms: relevance for kynurenic acid synthesis in patients with schizophrenia and healthy controls. J Psychiatry Neurosci 37:53–57
Innocenti B, Parpura V, Haydon PG (2000) Imaging extracellular waves of glutamate during calcium signaling in cultured astrocytes. J Neurosci 20:1800–1808
Javitt DC (1987) Negative schizophrenic symptomatology and the PCP (phencyclidine) model of schizophrenia. Hillside J Clin Psychiatry 9:12–35
Javitt DC, Zukin SR (1991) Recent advances in the phencyclidine model of schizophrenia. Am J Psychiatry 148:1301–1308
Johnston-Wilson NL, Sims CD, Hofmann JP, Anderson L, Shore AD, Torrey EF, Yolken RH (2000) Disease-specific alterations in frontal cortex brain proteins in schizophrenia, bipolar disorder, and major depressive disorder. The Stanley Neuropathology Consortium. Mol Psychiatry 5:142–149
Kantrowitz JT, Malhotra AK, Cornblatt B, Silipo G, Balla A, Suckow RF, D'Souza C, Saksa J, Woods SW, Javitt DC (2010) High dose D-serine in the treatment of schizophrenia. Schizophr Res 121:125–130
Karlsson RM, Tanaka K, Heilig M, Holmes A (2008) Loss of glial glutamate and aspartate transporter (excitatory amino acid transporter 1) causes locomotor hyperactivity and exaggerated responses to psychotomimetics: rescue by haloperidol and metabotropic glutamate 2/3 agonist. Biol Psychiatry 64:810–814
Karlsson RM, Tanaka K, Saksida LM, Bussey TJ, Heilig M, Holmes A (2009) Assessment of glutamate transporter GLAST (EAAT1)-deficient mice for phenotypes relevant to the negative and executive/cognitive symptoms of schizophrenia. Neuropsychopharmacology 34:1578–1589
Kettenmann H, Hanisch UK, Noda M, Verkhratsky A (2011) Physiology of microglia. Physiol Rev 91:461–553
Kirischuk S, Kettenmann H, Verkhratsky A (2007) Membrane currents and cytoplasmic sodium transients generated by glutamate transport in Bergmann glial cells. Pflugers Arch 454:245–252
Kirischuk S, Parpura V, Verkhratsky A (2012) Sodium dynamics: another key to astroglial excitability? Trends Neurosci 35:497–506
Knackstedt LA, Melendez RI, Kalivas PW (2010) Ceftriaxone restores glutamate homeostasis and prevents relapse to cocaine seeking. Biol Psychiatry 67:81–84
Koss WA, Sadowski RN, Sherrill LK, Gulley JM, Juraska JM (2012) Effects of ethanol during adolescence on the number of neurons and glia in the medial prefrontal cortex and basolateral amygdala of adult male and female rats. Brain Res 1466:24–32
Kreft M, Bak LK, Waagepetersen HS, Schousboe A (2012) Aspects of astrocyte energy metabolism, amino acid neurotransmitter homoeostasis and metabolic compartmentation. ASN Neuro 4
Krystal JH, Karper LP, Seibyl JP, Freeman GK, Delaney R, Bremner JD, Heninger GR, Bowers MB Jr, Charney DS (1994) Subanesthetic effects of the noncompetitive NMDA antagonist, ketamine, in humans. Psychotomimetic, perceptual, cognitive, and neuroendocrine responses. Arch Gen Psychiatry 51:199–214

Kucukibrahimoglu E, Saygin MZ, Caliskan M, Kaplan OK, Unsal C, Goren MZ (2009) The change in plasma GABA, glutamine and glutamate levels in fluoxetine- or S-citalopram-treated female patients with major depression. Eur J Clin Pharmacol 65:571–577

Labrie V, Fukumura R, Rastogi A, Fick LJ, Wang W, Boutros PC, Kennedy JL, Semeralul MO, Lee FH, Baker GB, Belsham DD, Barger SW, Gondo Y, Wong AH, Roder JC (2009) Serine racemase is associated with schizophrenia susceptibility in humans and in a mouse model. Hum Mol Genet 18:3227–3243

Lan MJ, McLoughlin GA, Griffin JL, Tsang TM, Huang JT, Yuan P, Manji H, Holmes E, Bahn S (2009) Metabonomic analysis identifies molecular changes associated with the pathophysiology and drug treatment of bipolar disorder. Mol Psychiatry 14:269–279

Lane HY, Lin CH, Green MF, Hellemann G, Huang CC, Chen PW, Tun R, Chang YC, Tsai GE (2013) Add-on treatment of benzoate for schizophrenia: a randomized, double-blind, placebo-controlled trial of D-amino acid oxidase inhibitor. JAMA Psychiatry 70:1267–1275

Lee W, Reyes RC, Gottipati MK, Lewis K, Lesort M, Parpura V, Gray M (2013) Enhanced Ca^{2+}-dependent glutamate release from astrocytes of the BACHD Huntington's disease mouse model. Neurobiol Dis 58:192–199

Leroi I, Michalon M (1998) Treatment of the psychiatric manifestations of Huntington's disease: a review of the literature. Can J Psychiatry 43:933–940

Ma TM, Abazyan S, Abazyan B, Nomura J, Yang C, Seshadri S, Sawa A, Snyder SH, Pletnikov MV (2013) Pathogenic disruption of DISC1-serine racemase binding elicits schizophrenia-like behavior via D-serine depletion. Mol Psychiatry 18:557–567

Maes M, Verkerk R, Vandoolaeghe E, Lin A, Scharpe S (1998) Serum levels of excitatory amino acids, serine, glycine, histidine, threonine, taurine, alanine and arginine in treatment-resistant depression: modulation by treatment with antidepressants and prediction of clinical responsivity. Acta Psychiatr Scand 97:302–308

Martinez-Hernandez A, Bell KP, Norenberg MD (1977) Glutamine synthetase: glial localization in brain. Science 195:1356–1358

Martinez-Lozada Z, Guillem AM, Flores-Mendez M, Hernandez-Kelly LC, Vela C, Meza E, Zepeda RC, Caba M, Rodriguez A, Ortega A (2013) GLAST/EAAT1-induced glutamine release via SNAT3 in Bergmann glial cells: evidence of a functional and physical coupling. J Neurochem 125:545–554

McKenna MC (2013) Glutamate pays its own way in astrocytes. Front Endocrinol (Lausanne) 4:191

Miguel-Hidalgo JJ (2005) Lower packing density of glial fibrillary acidic protein-immunoreactive astrocytes in the prelimbic cortex of alcohol-naive and alcohol-drinking alcohol-preferring rats as compared with alcohol-nonpreferring and Wistar rats. Alcohol Clin Exp Res 29:766–772

Miguel-Hidalgo JJ (2009) The role of glial cells in drug abuse. Curr Drug Abuse Rev 2:76–82

Miguel-Hidalgo JJ, Wei J, Andrew M, Overholser JC, Jurjus G, Stockmeier CA, Rajkowska G (2002) Glia pathology in the prefrontal cortex in alcohol dependence with and without depressive symptoms. Biol Psychiatry 52:1121–1133

Miguel-Hidalgo JJ, Overholser JC, Meltzer HY, Stockmeier CA, Rajkowska G (2006) Reduced glial and neuronal packing density in the orbitofrontal cortex in alcohol dependence and its relationship with suicide and duration of alcohol dependence. Alcohol Clin Exp Res 30:1845–1855

Mitani H, Shirayama Y, Yamada T, Maeda K, Ashby CR Jr, Kawahara R (2006) Correlation between plasma levels of glutamate, alanine and serine with severity of depression. Prog Neuropsychopharmacol Biol Psychiatry 30:1155–1158

Montana V, Ni Y, Sunjara V, Hua X, Parpura V (2004) Vesicular glutamate transporter-dependent glutamate release from astrocytes. J Neurosci 24:2633–2642

Montana V, Malarkey EB, Verderio C, Matteoli M, Parpura V (2006) Vesicular transmitter release from astrocytes. Glia 54:700–715

Moussawi K, Riegel A, Nair S, Kalivas PW (2011) Extracellular glutamate: functional compartments operate in different concentration ranges. Front Syst Neurosci 5:94

Muller MB, Lucassen PJ, Yassouridis A, Hoogendijk WJ, Holsboer F, Swaab DF (2001) Neither major depression nor glucocorticoid treatment affects the cellular integrity of the human hippocampus. Eur J Neurosci 14:1603–1612

Nakanishi T, Kekuda R, Fei YJ, Hatanaka T, Sugawara M, Martindale RG, Leibach FH, Prasad PD, Ganapathy V (2001) Cloning and functional characterization of a new subtype of the amino acid transport system N. Am J Physiol Cell Physiol 281:C1757–1768

Nedergaard M, Verkhratsky A (2012) Artifact versus reality—how astrocytes contribute to synaptic events. Glia 60:1013–1023

Ni Y, Parpura V (2009) Dual regulation of Ca2+-dependent glutamate release from astrocytes: vesicular glutamate transporters and cytosolic glutamate levels. Glia 57:1296–1305

Nilsson LK, Linderholm KR, Erhardt S (2006) Subchronic treatment with kynurenine and probenecid: effects on prepulse inhibition and firing of midbrain dopamine neurons. J Neural Transm (Vienna) 113:557–571

Norenberg MD (1987) The role of astrocytes in hepatic encephalopathy. Neurochem Pathol 6:13–33

Obara-Michlewska M, Ruszkiewicz J, Zielinska M, Verkhratsky A, Albrecht J (2014) Astroglial NMDA receptors inhibit expression of K4.1 channels in glutamate-overexposed astrocytes in vitro and in the brain of rats with acute liver failure. Neurochem Int 88:20–5

Obel LF, Andersen KM, Bak LK, Schousboe A, Waagepetersen HS (2012) Effects of adrenergic agents on intracellular Ca2+ homeostasis and metabolism of glucose in astrocytes with an emphasis on pyruvate carboxylation, oxidative decarboxylation and recycling: implications for glutamate neurotransmission and excitotoxicity. Neurotox Res 21:405–417

Oehmichen M, Meissner C, Reiter A, Birkholz M (1996) Neuropathology in non-human immunodeficiency virus-infected drug addicts: hypoxic brain damage after chronic intravenous drug abuse. Acta Neuropathol 91:642–646

Ohnuma T, Tessler S, Arai H, Faull RL, McKenna PJ, Emson PC (2000) Gene expression of metabotropic glutamate receptor 5 and excitatory amino acid transporter 2 in the schizophrenic hippocampus. Brain Res Mol Brain Res 85:24–31

Oliet SH, Mothet JP (2009) Regulation of N-methyl-D-aspartate receptors by astrocytic D-serine. Neuroscience 158:275–283

Olney JW, Farber NB (1995) Glutamate receptor dysfunction and schizophrenia. Arch Gen Psychiatry 52:998–1007

Ongur D, Drevets WC, Price JL (1998) Glial reduction in the subgenual prefrontal cortex in mood disorders. Proc Natl Acad Sci U S A 95:13290–13295

Pantazopoulos H, Woo TU, Lim MP, Lange N, Berretta S (2010) Extracellular matrix-glial abnormalities in the amygdala and entorhinal cortex of subjects diagnosed with schizophrenia. Arch Gen Psychiatry 67:155–166

Parpura V, Verkhratsky A (2012) Homeostatic function of astrocytes: Ca(2+) and Na(+) signalling. Transl Neurosci 3:334–344

Parpura V, Heneka MT, Montana V, Oliet SH, Schousboe A, Haydon PG, Stout RF Jr, Spray DC, Reichenbach A, Pannicke T, Pekny M, Pekna M, Zorec R, Verkhratsky A (2012) Glial cells in (patho)physiology. J Neurochem 121:4–27

Peng L, Verkhratsky A, Gu L, Li B (2015) Targeting astrocytes in major depression. Expert Rev Neurother 15(11):1299–1306

Perez-Alvarez A, Araque A (2013) Astrocyte-neuron interaction at tripartite synapses. Curr Drug Targets 14:1220–1224

Pittenger C, Duman RS (2008) Stress, depression, and neuroplasticity: a convergence of mecha nisms. Neuropsychopharmacology 33:88–109

Potter MC, Elmer GI, Bergeron R, Albuquerque EX, Guidetti P, Wu HQ, Schwarcz R (2010) Reduction of endogenous kynurenic acid formation enhances extracellular glutamate, hippocampal plasticity, and cognitive behavior. Neuropsychopharmacology 35:1734–1742

Qrunfleh AM, Alazizi A, Sari Y (2013) Ceftriaxone, a beta-lactam antibiotic, attenuates relapse-like ethanol-drinking behavior in alcohol-preferring rats. J Psychopharmacol 27:541–549

Racagni G, Popoli M (2008) Cellular and molecular mechanisms in the long-term action of antidepressants. Dialog Clin Neurosci 10:385–400
Rajkowska G, Stockmeier CA (2013) Astrocyte pathology in major depressive disorder: insights from human postmortem brain tissue. Curr Drug Targets 14:1225–1236
Rajkowska G, Miguel-Hidalgo JJ, Wei J, Dilley G, Pittman SD, Meltzer HY, Overholser JC, Roth BL, Stockmeier CA (1999) Morphometric evidence for neuronal and glial prefrontal cell pathology in major depression. Biol Psychiatry 45:1085–1098
Rajkowska G, Miguel-Hidalgo JJ, Makkos Z, Meltzer H, Overholser J, Stockmeier C (2002) Layer-specific reductions in GFAP-reactive astroglia in the dorsolateral prefrontal cortex in schizophrenia. Schizophr Res 57:127–138
Reichenbach A, Derouiche A, Kirchhoff F (2010) Morphology and dynamics of perisynaptic glia. Brain Res Rev 63:11–25
Reissner KJ, Kalivas PW (2010) Using glutamate homeostasis as a target for treating addictive disorders. Behav Pharmacol 21:514–522
Reissner KJ, Kalivas PW (2014) Emerging roles for glial pathology in addiction. In: Parpura V, Verkhratsky A (eds) Pathological potential of neuroglia: possible new targets for medical intervention. Springer, New York, pp 397–418
Ren J, Song D, Bai Q, Verkhratsky A, Peng L (2015) Fluoxetine induces alkalinization of astroglial cytosol through stimulation of sodium-hydrogen exchanger 1: dissection of intracellular signaling pathways. Front Cell Neurosci 9:61
Rintala J, Jaatinen P, Kiianmaa K, Riikonen J, Kemppainen O, Sarviharju M, Hervonen A (2001) Dose-dependent decrease in glial fibrillary acidic protein-immunoreactivity in rat cerebellum after lifelong ethanol consumption. Alcohol 23:1–8
Rose CF, Verkhratsky A, Parpura V (2013) Astrocyte glutamine synthetase: pivotal in health and disease. Biochem Soc Trans 41:1518–1524
Rosenberg DR, Mirza Y, Russell A, Tang J, Smith JM, Banerjee SP, Bhandari R, Rose M, Ivey J, Boyd C, Moore GJ (2004) Reduced anterior cingulate glutamatergic concentrations in childhood OCD and major depression versus healthy controls. J Am Acad Child Adolesc Psychiatry 43:1146–1153
Rosenberg DR, Macmaster FP, Mirza Y, Smith JM, Easter PC, Banerjee SP, Bhandari R, Boyd C, Lynch M, Rose M, Ivey J, Villafuerte RA, Moore GJ, Renshaw P (2005) Reduced anterior cingulate glutamate in pediatric major depression: a magnetic resonance spectroscopy study. Biol Psychiatry 58:700–704
Roth G, Dicke U (2013) Evolution of nervous systems and brains. In: Galizia CG, Lledo P-M (eds) Neurosciences—from molecule to behavior: a university textbook. Springer, Berlin, pp 19–45
Rothermundt M, Falkai P, Ponath G, Abel S, Burkle H, Diedrich M, Hetzel G, Peters M, Siegmund A, Pedersen A, Maier W, Schramm J, Suslow T, Ohrmann P, Arolt V (2004) Glial cell dysfunction in schizophrenia indicated by increased S100B in the CSF. Mol Psychiatry 9:897–899
Rothermundt M, Ohrmann P, Abel S, Siegmund A, Pedersen A, Ponath G, Suslow T, Peters M, Kaestner F, Heindel W, Arolt V, Pfleiderer B (2007) Glial cell activation in a subgroup of patients with schizophrenia indicated by increased S100B serum concentrations and elevated myo-inositol. Prog Neuropsychopharmacol Biol Psychiatry 31:361–364
Rothermundt M, Ahn JN, Jorgens S (2009) S100B in schizophrenia: an update. Gen Physiol Biophys 28:76–81
Sanacora G, Gueorguieva R, Epperson CN, Wu YT, Appel M, Rothman DL, Krystal JH, Mason GF (2004) Subtype-specific alterations of gamma-aminobutyric acid and glutamate in patients with major depression. Arch Gen Psychiatry 61:705–713
Sanacora G, Treccani G, Popoli M (2012) Towards a glutamate hypothesis of depression: an emerging frontier of neuropsychopharmacology for mood disorders. Neuropharmacology 62:63–77
Sari Y, Sreemantula SN, Lee MR, Choi DS (2013) Ceftriaxone treatment affects the levels of GLT1 and ENT1 as well as ethanol intake in alcohol-preferring rats. J Mol Neurosci 51:779–787
Schmitt A, Steyskal C, Bernstein HG, Schneider-Axmann T, Parlapani E, Schaeffer EL, Gattaz WF, Bogerts B, Schmitz C, Falkai P (2009) Stereologic investigation of the posterior part of the hippocampus in schizophrenia. Acta Neuropathol 117:395–407

Schwarcz R, Hunter CA (2007) Toxoplasma gondii and schizophrenia: linkage through astrocyte-derived kynurenic acid? Schizophr Bull 33:652–653

Schwarcz R, Bruno JP, Muchowski PJ, Wu HQ (2012) Kynurenines in the mammalian brain: when physiology meets pathology. Nat Rev Neurosci 13:465–477

Sequeira A, Mamdani F, Ernst C, Vawter MP, Bunney WE, Lebel V, Rehal S, Klempan T, Gratton A, Benkelfat C, Rouleau GA, Mechawar N, Turecki G (2009) Global brain gene expression analysis links glutamatergic and GABAergic alterations to suicide and major depression. PLoS One 4, e6585

Shan D, Lucas EK, Drummond JB, Haroutunian V, Meador-Woodruff JH, McCullumsmith RE (2013) Abnormal expression of glutamate transporters in temporal lobe areas in elderly patients with schizophrenia. Schizophr Res 144:1–8

Shan D, Yates S, Roberts RC, McCullumsmith RE (2014) Astroglia and severe mental illness: a role for glutamate microdomains. In: Parpura V, Verkhratsky A (eds) Pathological potential of neuroglia: possible new targets for medical intervention. Springer, New York, pp 373–395

Shin JY, Fang ZH, Yu ZX, Wang CE, Li SH, Li XJ (2005) Expression of mutant huntingtin in glial cells contributes to neuronal excitotoxicity. J Cell Biol 171:1001–1012

Si X, Miguel-Hidalgo JJ, O'Dwyer G, Stockmeier CA, Rajkowska G (2004) Age-dependent reductions in the level of glial fibrillary acidic protein in the prefrontal cortex in major depression. Neuropsychopharmacology 29:2088–2096

Singhrao SK, Thomas P, Wood JD, MacMillan JC, Neal JW, Harper PS, Jones AL (1998) Huntingtin protein colocalizes with lesions of neurodegenerative diseases: an investigation in Huntington's, Alzheimer's, and Pick's diseases. Exp Neurol 150:213–222

Skuja S, Groma V, Smane L (2012) Alcoholism and cellular vulnerability in different brain regions. Ultrastruct Pathol 36:40–47

Song P, Zhao ZQ (2001) The involvement of glial cells in the development of morphine tolerance. Neurosci Res 39:281–286

Struys-Ponsar C, Guillard O, van den Bosch de Aguilar P (2000) Effects of aluminum exposure on glutamate metabolism: a possible explanation for its toxicity. Exp Neurol 163:157–164

Suarez I, Bodega G, Ramos JA, Fernandez-Ruiz JJ, Fernandez B (2000) Neuronal and astroglial response to pre- and perinatal exposure to delta-9-tetra- hydrocannabinol in the rat substantia nigra. Dev Neurosci 22:253–263

Suarez-Fernandez MB, Soldado AB, Sanz-Medel A, Vega JA, Novelli A, Fernandez-Sanchez MT (1999) Aluminum-induced degeneration of astrocytes occurs via apoptosis and results in neuronal death. Brain Res 835:125–136

Sun JD, Liu Y, Yuan YH, Li J, Chen NH (2012) Gap junction dysfunction in the prefrontal cortex induces depressive-like behaviors in rats. Neuropsychopharmacology 37:1305–1320

Takahashi N, Sakurai T (2013) Roles of glial cells in schizophrenia: possible targets for therapeutic approaches. Neurobiol Dis 53:49–60

Taylor MJ, Mannie ZN, Norbury R, Near J, Cowen PJ (2011) Elevated cortical glutamate in young people at increased familial risk of depression. Int J Neuropsychopharmacol 14:255–259

Tong X, Ao Y, Faas GC, Nwaobi SE, Xu J, Haustein MD, Anderson MA, Mody I, Olsen ML, Sofroniew MV, Khakh BS (2014) Astrocyte Kir4.1 ion channel deficits contribute to neuronal dysfunction in Huntington's disease model mice. Nat Neurosci 17:694–703

Torres-Platas SG, Nagy C, Wakid M, Turecki G, Mechawar N (2015) Glial fibrillary acidic protein is differentially expressed across cortical and subcortical regions in healthy brains and downregulated in the thalamus and caudate nucleus of depressed suicides., Mol Psychiatry

Verkhratsky A, Butt AM (2013) Glial physiology and pathophysiology. Wiley-Blackwell, Chichester

Verkhratsky A, Nedergaard M (2014) Astroglial cradle in the life of the synapse. Philos Trans R Soc Lond B Biol Sci 369:20130595

Verkhratsky A, Rodriguez JJ, Parpura V (2013) Astroglia in neurological diseases. Future Neurol 8:149–158

Verkhratsky A, Rodriguez JJ, Steardo L (2014) Astrogliopathology: a central element of neuropsychiatric diseases? Neuroscientist 20:576–588

Vonsattel JP, Myers RH, Stevens TJ, Ferrante RJ, Bird ED, Richardson EP Jr (1985) Neuropathological classification of Huntington's disease. J Neuropathol Exp Neurol 44:559–577

Weber M, Scherf N, Kahl T, Braumann UD, Scheibe P, Kuska JP, Bayer R, Buttner A, Franke H (2013) Quantitative analysis of astrogliosis in drug-dependent humans. Brain Res 1500:72–87

Webster MJ, Knable MB, Johnston-Wilson N, Nagata K, Inagaki M, Yolken RH (2001) Immunohistochemical localization of phosphorylated glial fibrillary acidic protein in the prefrontal cortex and hippocampus from patients with schizophrenia, bipolar disorder, and depression. Brain Behav Immun 15:388–400

Wolosker H (2007) NMDA receptor regulation by D-serine: new findings and perspectives. Mol Neurobiol 36:152–164

Xia M, Abazyan S, Jouroukhin Y, Pletnikov M (2014) Behavioral sequelae of astrocyte dysfunction: focus on animal models of schizophrenia., Schizophr Res

Yamamori H, Hashimoto R, Fujita Y, Numata S, Yasuda Y, Fujimoto M, Ohi K, Umeda-Yano S, Ito A, Ohmori T, Hashimoto K, Takeda M (2014) Changes in plasma D-serine, L-serine, and glycine levels in treatment-resistant schizophrenia before and after clozapine treatment. Neurosci Lett 582:93–98

Yin Z, Milatovic D, Aschner JL, Syversen T, Rocha JB, Souza DO, Sidoryk M, Albrecht J, Aschner M (2007) Methylmercury induces oxidative injury, alterations in permeability and glutamine transport in cultured astrocytes. Brain Res 1131:1–10

Yuksel C, Ongur D (2010) Magnetic resonance spectroscopy studies of glutamate-related abnormalities in mood disorders. Biol Psychiatry 68:785–794

Chapter 13
Glutamine Synthetase: Role in Neurological Disorders

Arumugam R. Jayakumar and Michael D. Norenberg

Abstract Glutamine synthetase (GS) is an ATP-dependent enzyme found in most species that synthesizes glutamine from glutamate and ammonia. In brain, GS is exclusively located in astrocytes where it serves to maintain the glutamate–glutamine cycle, as well as nitrogen metabolism. Changes in the activity of GS, as well as its gene expression, along with excitotoxicity, have been identified in a number of neurological conditions. The literature describing alterations in the activation and gene expression of GS, as well as its involvement in different neurological disorders, however, is incomplete. This review summarizes changes in GS gene expression/activity and its potential contribution to the pathogenesis of several neurological disorders, including hepatic encephalopathy, ischemia, epilepsy, Alzheimer's disease, amyotrophic lateral sclerosis, traumatic brain injury, Parkinson's disease, and astroglial neoplasms. This review also explores the possibility of targeting GS in the therapy of these conditions.

Keywords Ammonia • Astrocytes • Glutamine synthetase • Glutamine • Excitotoxicity • Neurological disorders • Neurotoxicity • Therapy

A.R. Jayakumar
Laboratory of Neuropathology, Veterans Affairs Medical Center, Miami, FL, USA

M.D. Norenberg (✉)
Laboratory of Neuropathology, Veterans Affairs Medical Center, Miami, FL, USA

Departments of Pathology, University of Miami School of Medicine,
P.O. Box 016960, Miami, FL 33101, USA

Departments of Biochemistry & Molecular Biology, University of Miami School of Medicine, Miami, FL, USA
e-mail: mnorenbe@med.miami.edu

A. Schousboe, U. Sonnewald (eds.), *The Glutamate/GABA-Glutamine Cycle*,
Advances in Neurobiology 13, DOI 10.1007/978-3-319-45096-4_13

Abbreviations

AD	Alzheimer's disease
ALS	Amyotrophic lateral sclerosis
Aβ	Beta amyloid
COX-2	Cyclooxygenase-2
EAAT	Excitatory amino acid transporter
GABA	γ-aminobutyric acid
GBM	Glioblastoma multiforme
GFAP	Glial fibrillary acidic protein
GLAST	Glutamate-aspartate transporter
GLT-1	Glutamate transporter-1
GS	Glutamine synthetase
HD	Huntington's disease
HE	Hepatic encephalopathy
iNOS	Inducible nitric oxide synthase
MSO	Methionine sulfoximine
NMDA	*N*-methyl-D-aspartate
PCA	Portacaval anastomosis
TBI	Traumatic brain injury

13.1 Introduction

Glutamine synthetase (GS) is an enzyme that is predominantly located in brain, kidney, and liver. In brain it participates in the recycling of the neurotransmitter glutamate and in the detoxification of ammonia. Glutamate, which is released by glutamatergic neurons, is taken up by astrocytes where it is converted by GS to the nontoxic amino acid glutamine. Glutamine is then released by astrocytes and taken up by neurons, where it is hydrolyzed by glutaminase to form glutamate (the "glutamate–glutamine cycle") (Hertz 2013; Van den Berg 1970; Walls et al. 2015). In this manner, GS prevents the excessive accumulation of glutamate and ammonia in synaptic regions, thereby inhibiting the development of glutamate/ammonia neurotoxicity.

Glutamate neurotoxicity has been implicated in the process of neuronal degeneration in a large number of neurological disorders. Many of these conditions are accompanied by a decline in GS expression and/or activity (see later for a detailed discussion), although contradictory observations have also been reported (see later). Considering the role of GS in glutamate and ammonia detoxification, a decline in GS expression in different neurological conditions may markedly aggravate the process of neuronal degeneration. Recent studies have indeed demonstrated that hormonal induction of GS gene expression, or the exogenous administration of purified GS to traumatized retinal glial cells, reduced the neuronal degeneration, whereas inhibition of GS activity by methionine sulfoximine (MSO) led to neural cell death (Gorovits et al. 1997). These findings suggest that astrocytic GS is an efficacious neuroprotectant by activating a number of signaling systems that may protect neurons from the deleterious effects of both glutamate and ammonia.

13.1.1 Biochemical Properties of GS

GS catalyzes the ATP-dependent condensation of glutamate with ammonia to yield glutamine. ATP initially phosphorylates glutamate by forming an acyl-phosphate intermediate, γ-glutamyl phosphate, which then reacts with ammonia to form glutamine and inorganic phosphate. The ammonium ion binds strongly to GS only if the acyl-phosphate moiety intermediate is present. The ammonium ion, rather than ammonia, binds to GS as the binding site is polar (Liaw and Eisenberg 1994).

13.1.2 Cellular Localization of GS in the CNS

GS in brain is normally found in astrocytes (Norenberg 1979, 1983; Norenberg and Martinez-Hernandez 1979). However, in some settings (e.g., Alzheimer's disease, suicide victims, individuals with schizophrenia) reports indicate that GS may also be present in other CNS cells, including neurons, microglia, and oligodentrocytes (for review, see Anlauf and Derouiche 2013 and references therein; Bernstein et al. 2013; Nakajima et al. 2015; Robinson 2000; Yamamoto et al. 1989). The existence, as well as the significance of GS in CNS cells other than astrocytes, remains to be determined.

13.1.3 GS Activity

GS activity is regulated by adenylation (a specific tyrosine residue in each subunit of GS is modified) (Stryer et al. 2007). The rate of adenylation depends on the ratio of glutamine to α-ketoglutarate. Adenylation of GS, which is catalyzed by adenylyltransferase, involves the phosphodiester bond between the hydroxyl group of the tyrosine residue in GS and the phosphate group of an AMP nucleotide. A complex of adenylyl transferase and a regulatory protein known as PII, which may exist unmodified as PII (also known as P_A), and uridylylated PII-UMP (also known as P_D), causes an AMP molecule to either attach to, or to be removed from GS, respectively (Itzen et al. 2011). A complex of PII with adenylyl transferase that catalyzes the attachment of an AMP molecule to GS forms adenylated GS that is inactive. Alternatively, a complex of PII-UMP and adenylyl transferase may activate deadenylation which removes AMP from GS, thereby creating deadenylated GS that is active (Garrett and Grisham 2007; Itzen et al. 2011).

13.1.4 Regulation of GS Gene Expression

Recent findings indicate that the GS gene is a novel forkhead box O (FoxO) transcriptional target (van der Vos et al. 2012; van der Vos and Coffer 2012). Removal of various growth factors (e.g., basic fibroblast growth factor, transforming growth factor beta 1 and 2) in astrocytes was shown to increase GS expression and activity through the activation of the FoxO transcription factor (Chao et al. 1992; Kruchkova et al. 2001; van der Vos et al. 2012; van der Vos and Coffer 2012). Conversely, GS expression is inhibited by stimulation of the earlier growth factors, or by the selective activation of either phosphatidylinositol 3-kinase (PI3K), or by the serine/threonine-protein kinase-1 (AKT1) signaling pathway (van der Vos and Coffer 2012). Alternately, inhibition of PI3K activity induces the binding of activated FoxO to the GS promoter which increases GS transcription (van der Vos and Coffer 2012). Additionally, increased GS levels inhibit the mammalian target of rapamycin signaling by blocking its lysosomal translocation. Glucocorticoid receptor-mediated cell signaling also appears to be involved in the regulation of GS gene expression (Gorovits et al. 1996; Vardimon et al. 1988; Zhang and Young 1991).

13.2 Role of GS in Neurological Conditions

13.2.1 Hepatic Encephalopathy

Hepatic encephalopathy (HE) is a major neurological complication in patients with severe liver disease, which presents in acute and chronic forms. Acute HE usually occurs following massive liver necrosis due to viral hepatitis, acetaminophen toxicity, or exposure to other hepatotoxins. It presents with severe brain edema (Ede and Williams 1986; Hoofnagle et al. 1995), increase in intracranial pressure, and brain herniation resulting in a high mortality rate (Capocaccia and Angelico 1991). Chronic HE is generally a consequence of cirrhosis of the liver and presents with a wide range of neurological symptoms, including altered mood, increased irritability, alterations in the sleep/wake cycle, changes in muscle tone, and cognitive impairments (Jones and Weissenborn 1997; Lockwood et al. 1991). While the molecular basis underlying HE remains incompletely understood, astrocytes are generally recognized to play a critical role in its pathogenesis (Norenberg 1977; Norenberg et al. 2007). It is generally accepted that elevated blood and brain ammonia levels, due to the inability of the injured liver to adequately detoxify ammonia by the synthesis of urea, is a crucial factor in the pathogenesis of HE. Once in brain, ammonia is first metabolized by GS to form glutamine in astrocytes. A number of events are then triggered that ultimately culminates into HE (for review, see Norenberg et al. 2007).

Evidence for the involvement of GS in ammonia toxicity was first demonstrated by Warren and Schenker (1964), who observed that methionine sulfoximine (MSO), an inhibitor of GS, protected mice from the development of ammonia-induced coma. MSO was also shown to reverse the biochemical manifestations of ammonia toxicity (e.g., depressed glucose consumption and increased blood–brain barrier transport of aromatic amino acids in hyperammonemic rats) (Hawkins et al. 1993). Pathophysiological manifestations of HE, including increased cerebral blood flow (Master et al. 1999), increased brain water content and intracranial pressure (Blei et al. 1994; Tanigami et al. 2005), astrocytic control of extracellular K^+ activity (Sugimoto et al. 1997), and impaired vascular reactivity to CO_2 (Hirata et al. 1996, 1999; Takahashi et al. 1992) were also abated by MSO, presumably by preventing the accumulation of glutamine.

Additionally, studies have shown that GS levels are increased in ammonia-treated cultured astrocytes (Suarez et al. 2002; Waniewski 1992), and in animal models of acute HE (Suarez et al. 2002; Takahashi et al. 1991). Increased GS mRNA was also reported in the thioacetamide-induced rat model of acute HE (Thomas et al. 1988). Moreover, inhibition of GS was found to be an effective means of reducing the ammonia-induced astrocyte swelling, as well as the brain edema in rats with acute HE (Bender and Norenberg 1996; Tanigami et al. 2005; Willard-Mack et al. 1996).

These findings suggest that increased GS expression and/or activity are involved in the development of brain edema in acute HE, and that inhibition of GS with MSO may serve as an important treatment modality for hyperammonemia. However, the use of MSO is problematic as MSO is well known to induce seizures likely by inducing hyperammonemia (see also Section c, epilepsy). The potential seizure-inducing effect of MSO is well established, as this agent, especially the L–S isomer, binds to the active sites of GS leading to an irreversible inhibition of the enzyme (Rowe and Meister 1970). The effect of MSO on seizure induction may also be mediated by means other than GS inhibition, as MSO is also known to inhibit glutamate cysteine ligase (Richman et al. 1973; Schatz and Sellinger 1975), and to increase ornithine decarboxylase activity (Di Giacomo et al. 1997), both of which have been shown to be associated with epileptogenesis (Baudry et al. 1986; Cárdenas-Rodríguez et al. 2014; Herberg et al. 1992; Jarrett et al. 2008). While MSO is beneficial in preventing glutamine accumulation and the subsequent increase in brain edema/intracranial pressure/coma in HE, MSO alone cannot be used to treat HE. It is possible that a combinatorial approach (MSO plus an anticonvulsant) may be more efficacious in diminishing the symptoms of HE.

While the earlier findings strongly suggest a critical role of increased GS in the pathogenesis of acute HE, a few studies, however, indicate that GS protein levels or activity are decreased in ammonia-treated astrocyte cultures (Leite et al. 2006) as well as in chronic and acute HE (Desjardins et al. 1999; Girard et al. 1993; Hilgier 1981; Kanamori et al. 1993; Lavoie et al. 1987), and that such decreased activity was shown to be promoted, in part, through increased protein tyrosine nitration of GS by ammonia, the peripheral benzodiazepine receptor and cytokines (Häussinger et al. 2005).

The reason for the differences in GS protein expression and activity (increased or decreased) in HE is unclear. It is possible that the duration of hyperammonemia may have varied in these studies, resulting in differences in GS protein expression and activity. In this regard, Desjardins et al. (1999) found decreased GS activity in cerebral cortex at 4 weeks after portacaval anastomosis (PCA, a model of hyperammonemia), whereas increased GS-immunoreactivity was demonstrated after long-term PCA (6 months) (Suárez et al. 1996, 1997). It should be emphasized that brain ammonia concentrations were different in these two studies (0.5–0.1 mM, respectively) that may, in part, explain the differential expression and/or activity of GS in HE. Additionally, while glutamate and ammonia bind to active sites of several GS subunits for the production of glutamine, these subunits also contain binding sites for other amino acids (glycine, alanine, histidine, tryptophan), as well as for nucleotides (including adenosine monophosphate and cytidine triphosphate). Binding of these molecules inhibits GS activity. It is thus possible that a rise in these molecules in various neurological conditions (see later) may contribute to the inhibition of GS.

GS present in organs other than brain (e.g., skeletal muscle) has also been considered as potential ammonia-lowering strategies for HE. For instance, GS in skeletal muscle has the capacity to remove ammonia (Chatauret et al. 2006; Desjardins et al. 1999; Dam et al. 2011; Jalan et al. 2007; Rose et al. 1998, 1999; Zieve et al. 1986). Additionally, a gene therapy approach (delivery of the GS gene by a viral vector) in muscle tissue in acute hyperammonemic rats was shown to prevent the development of hyperammonemic symptoms (Torres-Vega et al. 2015).

The earlier findings strongly suggest that GS protein expression or activity is altered in HE. It is possible that increased GS may detoxify ammonia and glutamate resulting in high levels of glutamine which may contribute to the astrocyte swelling/brain edema. However, inhibiting GS with MSO to prevent edema may increase the concentrations of ammonia in brain, resulting in seizure activity (Fig. 13.1). Accordingly, when treating the brain edema associated with acute HE with MSO, the possibility of concurrently using an anticonvulsant should be considered.

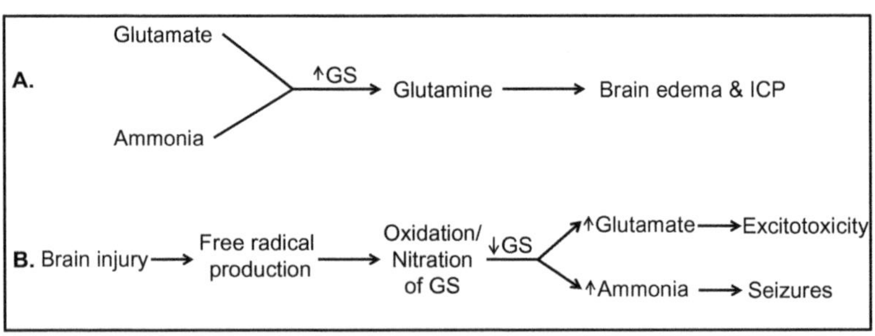

Fig. 13.1 Altered glutamine synthetase (GS) expression/activity in neurological disorders. (**a**) Increased GS detoxifies ammonia and glutamate resulting in high levels of glutamine, which may contribute to the astrocyte swelling/brain edema in various neurological conditions, including hepatic encephalopathy, ischemic stroke, and traumatic brain injury. (**b**) In conditions with a decrease in GS, or inhibiting GS with MSO to prevent the ammonia-mediated edema, may increase the concentrations of ammonia and glutamate in brain, resulting in seizure activity

13.2.2 Ischemia

Brain ischemia leads to decreased brain energy metabolism, a reduction in metabolic rates, and energy crisis. During ischemia, a massive release of glutamate from synaptic vesicles is known to occur (Benveniste et al. 1984; Drejer et al. 1985), and such increase was shown to be a major cause of neuronal degeneration (by excitotoxicity) (Dugan and Choi 1999). The accumulation of synaptically released glutamate was prevented, in part, by the astrocytic glutamate uptake system, wherein GS plays a critical role.

Immunohistochemical analysis of the gerbil hippocampus following transient ischemia showed that on day 2 after ischemia, and more prominently on day 3, astrocytes intensely stained for GS, especially in the CA1 region (Tanaka et al. 1992), and such increase was associated with increased GS protein levels (Hertz and Zielke 2004; Fujioka et al. 2003; Lee et al. 2003). Spectrophotometric analysis of frozen sections of cerebellum from children who died with acute or chronic hypoxic–ischemic insults also showed that GS activities were elevated. GS was also upregulated in retinal Müller cells after middle cerebral artery occlusion (Allen et al. 2014). Additionally, following ischemia, treatment of rats with total oligomeric flavonoids (e.g., quercetin, polyphenols), or telmisartan (an angiotensin II receptor antagonist), resulted in a decrease in GS activity, reduction in neurological deficits, and a reversal of anxiogenic behavior (Sunil et al. 2011; Justin et al. 2014). The earlier findings suggest that GS protein expression and activity are increased in brains of experimental animals and in patients with brain ischemia, and that such increase in GS activity likely contributes to the development of brain injury, presumably by increasing glutamine levels. It should be noted that increased glutamine is known to initiate a number of signaling pathways (e.g., free radical generation, activation of mitogen-activated protein kinases, as well as various transcription factors), all of which have been shown to result in astrocyte dysfunction (Norenberg et al. 2007; Rama Rao et al. 2012), that may also have contributed the pathogenesis of ischemic brain injury.

However, there is also evidence indicating that a reduction in GS activity may contribute to neuronal injury postischemia. Exposure of spinal cords to high amount of GS (15 Units) was shown to prevent GABA-A receptor-mediated axonal depression after ischemia, possibly by an ischemia-induced accumulation of ammonia (Matsumoto et al. 2008). Excess ammonia is known to potentiate the $GABA_A$ receptor response due to GABA released from the hypoxia-injured spinal cord (Takahashi et al. 1993). Oxygen-glucose deprivation (OGD) was also shown to decrease GS expression, as well as to increase extracellular glutamate level, both of which can lead to excitotoxicity (Lee et al. 2010; Wang et al. 2013). Additionally, GS expression and activity were decreased after middle cerebral artery occlusion in rats, and memantine, a highly efficacious NMDA receptor channel blocker, as well as ceftriaxone, an activator of the GLT-1 subtype of glutamate transporter expression, were both shown to prevent such reduction in GS expression/activity. The memantine and ceftriaxone effect in reversing GS expression and activity after middle cerebral artery occlusion in rats was also associated with a reversal of altered neurological and behavioral deficits following ischemia (Babu and Ramanathan 2009; Verma et al. 2010). These findings suggest that a reduction in GS expression and activity contributes to ischemic brain injury, likely by increasing both brain ammonia concentration and excitotoxicity.

A significant decrease in GS activity was also observed after retinal ischemic injury in rats (Fernandez et al. 2009). Similarly, Lee et al.. (2010) also found that GS was rapidly lost from astrocytes in regions susceptible to ischemic damage following hypoxia in neonatal pigs, while an examination of brain cortical and hippocampal sections showed that the loss of GS preceded any overt neuronal injury. Moreover, the decreased survival rate following ischemic brain injury was shown to be potentiated by pretreatment of ischemic rats with MSO (Di Giacomo et al. 1997), suggesting that inhibition of GS expression/activity contributes to the decreased survival of rats following ischemic brain injury. Altogether, these findings strongly suggest that decreased GS protein expression and activity, resulting in increased extracellular glutamate and ammonia levels, contribute to the development of ischemic brain injury.

While both increased and decreased GS protein expression and activity have been reported in ischemic brain injury, the reason for these differences, as well as their contribution to ischemic brain injury, remains unclear. The duration of ischemia (i.e., transient vs. prolonged) may be responsible for changes in GS expression/activity. For example, short- or long-term ischemic injury to brain is well known to cause free radical production, and such increase may have contributed to the altered GS expression/activity. In support of the view that the duration of ischemia may be responsible for the differences in GS activity (increase and decrease after short- or long-term ischemia, respectively), GS activity and free radical production were shown to be increased following transient ischemia (Petito et al. 1992; Phillis 1994). Since oxidative stress is known to increase GS activity (Bondy et al. 1998), it is possible that free radicals generated by transient ischemia may have enhanced GS expression.

However, ischemia-induced oxidative stress at later stages may have initiated alterations in key proteins, in particular GS, that may have contributed to the reduced GS expression and activity (Butterfield et al. 1997; Castegna et al. 2011; Oliver et al. 1990). Studies have indeed shown that GS is sensitive to severe oxidative damage (Floyd 1990; Nakamura and Stadtman 1984; Oliver et al. 1990), and that GS protein oxidation is known to be increased in brain by 75 % after long-term ischemic injury (Akinmoladun et al. 2015; Dorfman et al. 2013; Floyd 1990; Nakamura and Stadtman 1984; Oliver et al. 1990; Safwen et al. 2015) but not after transient ischemia (15 min) (Folbergrová et al. 1993; Pahlmark et al. 1993). In aggregate, these findings suggest that the duration of ischemic injury may partly determine the status of GS expression/activity following ischemic brain injury.

It should also be highlighted that a significant increase in brain ammonia levels, as well as glutamate concentrations, was observed after brain ischemic insult (Hołownia et al. 1994; Thorn and Heimann 1958), which may have played a role in the alterations in GS expression observed following CNS ischemia. Together, these findings suggest that regulation of GS expression/activity following ischemia may represent a useful therapeutic approach for the treatment of ischemic CNS injury.

13.2.3 Epilepsy

Epilepsy is a common neurological disorder globally affecting approximately 65 million people (Ngugi et al. 2010; Thurman et al. 2011). While the cause of most cases of epilepsy is unknown, brain trauma, stroke, tumors, and drug and alcohol abuse have been shown to significantly contribute to its development.

Dysfunctional astrocytes have been strongly implicated in seizure pathogenesis (Crunelli et al. 2015). The seizure producing effect of MSO, likely by the inhibition of GS, was first demonstrated by Reiner et al. 1950, and subsequently by Proler and Kellaway 1962, 1965; Warren and Schenker 1964; and Johnson et al. 1965. Further, studies have shown reduced GS activity in the hippocampus in association with temporal-lobe epilepsy (Eid et al. 2008), that MSO exacerbated epileptic seizures (Chung and Johnson 1984), and that continuous intra-hippocampal infusion of MSO in rats resulting in a loss of GS activity, led to increased extracellular glutamate concentrations which were shown to contribute to recurrent seizure activity (for review, see Eid et al. 2013; Wang et al. 2009).

Additional support implicating the involvement of GS in epileptogenesis is derived from findings showing reduced brain levels/activity of GS in rats with genetic absence epilepsy (Dutuit et al. 2000), in amygdala-kindled seizures (Tiffany-Castiglioni et al. 1990), as well as in hippocampal regions associated with neuron loss in patients with temporal lobe epilepsy (van der Hel et al. 2005). A reduction in GS activity was also shown to increase the susceptibility to experimental febrile seizures (van Gassen et al. 2009). Altogether, these findings indicate that reduced GS expression or activity is indeed involved in epileptogenesis (See Eid et al. 2013 for review).

However, there are reports that increased GS activity may also be associated with submaximal seizure activity (Tiffany-Castiglioni et al. 1990), and that a transient upregulation of GS in the dentate gyrus was associated with amygdala-kindling seizures in rats (Sun et al. 2013). While the reason for discrepancies in GS protein expression/activity associated with epileptogenesis is unclear, it is possible that the duration of seizures may have varied in these studies, which might have generated the production of free radicals and increased brain lactate concentration, both of which may have resulted in a neurotransmitter imbalance (e.g., NMDA; see ALS section) (Kosenko et al. 2003) that may have altered GS expression. Differences in the experimental models used in these studies may also be responsible for the variations in GS expression/activity relative to epileptogenesis.

As noted in the "Hepatic encephalopathy section," the seizure-inducing effect of MSO is well established as this amino acid binds to the active sites of GS and thereby irreversibly inhibits the enzyme, resulting in convulsions (Rowe and Meister 1970). As noted earlier in "HE section" (page 9) the effect of MSO on seizure induction may also be mediated by the inhibition of glutamate cysteine ligase (Griffith and Meister 1978, 1979) or by increase in ornithine decarboxylase activity (Di Giacomo et al. 1997, and references therein), both of which have been shown to be associated with epileptogenesis (Di Giacomo et al. 1997, and references therein; Griffith and Meister 1978, 1979). These studies collectively suggest that GS inhibition, along with reduced glutamate cysteine ligase and increased ornithine decarboxylase activities, is involved in epileptogenesis. Accordingly, the restoration of GS activity, possibly by treatment with memantine or ceftriaxone, agents known to reverse GS expression and activity in other conditions (e.g., ischemia) (Babu and Ramanathan 2009; Verma et al. 2010), may represent a novel approach for the prevention of epileptic seizure activity.

13.2.4 Traumatic Brain Injury

Traumatic brain injury (TBI) is a common neurological disorder that represents a leading cause of morbidity and mortality (Coronado et al. 2011; Harrison-Felix et al. 2006), which presents in acute and chronic forms. Brain edema and associated increased intracranial pressure in the early phase of TBI are major consequences of acute trauma, whereas more subtle neuronal injury, leading to neurobehavioral and cognitive impairments occurring years after repetitive episodes of head trauma, are major consequences of chronic TBI.

The molecular mechanisms responsible for TBI are incompletely understood. Following TBI, neuronal survival is strongly influenced by the functions of astrocytes, including glutamate uptake and release, and the release of substrates necessary for neuronal energy metabolism (e.g., lactate, glutamate, and GABA) (Bak et al. 2006; Pascual et al. 1998). Considering the role of GS in the conversion of glutamate to glutamine, increased or decreased GS expression in astrocytes may affect the concentration of glutamate (both intracellular and extracellular) potentially resulting in adverse effects on neurons caused by excitotoxicity.

A marked reduction in brain GS immunoreactivity was identified following fluid percussion injury in rats (Zhao et al. 2003). On the other hand, Gorovits and colleagues demonstrated that elevation of GS expression in astrocytes by cortisol treatment protected against neuronal degeneration in injured retinal tissue, whereas inhibition of GS activity (by MSO) led to increased neuronal cell death (Gorovits et al. 1997). These investigators also showed that the addition of purified GS to injured retinal tissue resulted in a decline in neuronal death. Of interest, a recent report by Zou et al. (2011) noted that suppression of astrocytic GS with siRNA increased astrocyte migration into the site of injury in a scratch model of TBI, and such migration may have contributed to the neuronal protection following TBI. The earlier findings suggest an important role of altered GS activity in TBI, and that increasing GS activity may represent a useful therapeutic approach for TBI.

13.2.5 Alzheimer's Disease

Alzheimer's disease (AD) is a neurodegenerative disorder characterized by cognitive dysfunction, memory loss, diminished orientation and judgment, inadequate language expression, and impaired calculation ability (Battaglia et al. 2007;

Cummings 2004; Selkoe 2001). Neurofibrillary tangles, intra- and extracellular accumulation of β-amyloid (Aβ) (neuritic plaques), as well as an increase in intraneuronal hyperphosphorylated Tau protein, are histopathological hallmarks of AD (Veerhuis 2011). Neuritic plaques are frequently associated with GFAP-positive activated astrocytes (Nagele et al. 2004). While the means by which Aβ causes neurotoxicity in AD is unclear, altered glutamatergic signaling, due to changes in the expression of glutamate transporters (e.g., GLAST, GLT-1), phosphate-activated glutaminase, changes in ionotropic and metabotropic glutamate receptors, as well as in GS, all appear to be critically involved (Burbaeva et al. 2005; Miguel-Hidalgo et al. 2002; Revett et al. 2013).

A recent finding suggests that astrocytes play an important role in the clearance of the Aβ peptides, potentially preventing plaque formation (Nielsen et al. 2009; Shibata et al. 2000; Wyss-Coray et al. 2003). Exposure of cultured astrocytes to Aβ peptides was also shown to decrease glutamate uptake, ultimately leading to oxidative stress and the subsequent activation of various intracellular signaling cascades (Agostinho et al. 2010; Matos et al. 2008). Considering the critical role of GS in glutamate metabolism, altered GS expression in astrocytes may affect the concentrations of glutamate and ammonia, both of which may adversely affect neurons, resulting in neuronal death.

Decreased GS activity and protein level were identified in postmortem brains from patients with AD (Butterfield et al. 1997; Le Prince et al. 1995; Olabarria et al. 2011; Robinson 2000). Additionally, Aβ peptides were shown to directly inhibit GS activity in vitro (Aksenov et al. 1996). Decreased GS activity in AD may also be due to the oxidation of GS by free radicals (Butterfield et al. 1997; Castegna et al. 2011; Oliver et al. 1990; Smith et al. 1991). Conversely, Aβ peptides were shown to enhance GS activity (Pike et al. 1996). While the reason for the differences in GS protein expression and activity (increased and decreased) in AD is unclear, it is likely that the concentration of Aβ peptides, the peptide sequences used (i.e., 1–42 vs. 35–42), or the duration of Aβ exposure to neuronal cells may have contributed to the differences in GS protein expression and activity reported in AD. The earlier findings suggest that alterations in GS expression/activity may provide valuable insights into the pathogenetic mechanisms of AD that may have therapeutic implications for this condition.

13.2.6 Amyotrophic Lateral Sclerosis

Amyotrophic lateral sclerosis (ALS) is a progressive disorder of motor neurons, in which both upper and lower motor neurons are affected leading to muscle atrophy. While the molecular mechanism responsible for ALS is unknown, astrocytes appear to play a role, as an SOD1 gene mutation has been identified in astrocytes in ALS which was shown to result in a disturbance in astrocyte–motor neuron interactions, resulting in a loss of motor neurons (Bame et al. 2012; Ghoddoussi et al. 2010).

Studies of brain tissue from patients with ALS, as well as from transgenic mouse models of ALS, have provided strong evidence that astroglial abnormalities precede the onset of clinical disease (for review, see Radford et al. 2015). These abnormalities include a loss of the astrocytic glutamate transporter GLT-1 protein expression and activity (up to 95 %) (Bristol and Rothstein 1996).

Increased expression of immune and inflammatory markers was identified in brains of patients with ALS (Chiu et al. 2008, 2009). Studies have shown that pro-inflammatory mediators (e.g., cytokines, iNOS, COX-2) released by microglia and astrocytes may result in NMDA-mediated neurotoxicity, possibly by phosphorylation of the NMDA receptor, resulting in the cell death of motor neurons (Hensley et al. 2006; Hewett et al. 1994). Dysfunctional astrocytes may also contribute to excitotoxic damage by decreasing glutamate transport or by actively releasing the excitotoxic amino acids, suggesting that astrocytic glutamate metabolism may be affected in ALS which may contribute to its pathogenesis.

Consistent with the earlier metabolic derangements in ALS, there is increasing evidence that elevated GS levels are involved in the mechanism of ALS, as MSO lowers brain glutamine levels, as well as improves the survival of transgenic mice that overexpresses the mutant human SOD1(G93A) gene (a mice model of ALS) (Bame et al. 2012; Ghoddoussi et al. 2010). Further, levels of GS were significantly increased in the CSF of patients with ALS (Tumani et al. 1999), suggesting that an astroglial dysfunction may have contributed to the development of ALS. While these studies examined the potential role of GS in ALS, a more comprehensive understanding of the interactions between motor neuron death and the precise role of astrocytic GS in this process is required.

13.2.7 Huntington's Disease

Huntington's disease (HD) is a fatal autosomal dominant neurodegenerative disease that results from an expanded DNA trinucleotide CAG repeats in the gene encoding for the huntingtin protein (for review, see Walker 2007). Neuropathological changes include atrophy of the caudate/putamen, along with a concomitant neuronal loss and astrogliosis.

The mutant huntingtin protein is known to aggregate in neurons of transgenic mouse models of HD (Davies et al. 1997). Recent evidence suggests that mutant huntingtin protein is also present in the nuclei of astrocytes, a phenomenon that becomes more prominent with age, and which is associated with a downregulation of glutamate transporters (especially GLT-1) in these cells (Shin et al. 2005).

A potential cause-and-effect relationship implicating astrocytes in the neurotoxicity associated with HD was reported following the observation that in a neuron–astrocyte coculture system, in which wild-type astrocytes were found to protect cortical neurons against the mutant huntingtin-mediated neurotoxicity, whereas astrocytes expressing the mutant huntingtin displayed an exacerbation of neuronal injury (Shin et al. 2005). These observations suggest that dysfunctional astrocytes containing the mutant huntingtin contribute to the neuronal abnormalities in HD, possibly due to a disturbance of glutamate–glutamine cycle, wherein GS plays a major role.

In support of the earlier mechanistic view, GS activity was found to be reduced in the frontal and temporal cortices, putamen and cerebellum of patients with HD (Carter 1981, 1982, 1983). Decreased GS immunoreactivity was also identified in transgenic animal models of HD (Behrens et al. 2002). These findings suggest that a reduction in astrocytic GS activity may contribute to the pathogenesis of HD. However, other studies have failed to identify significant changes in glutamate metabolic pathways, or in GS levels in brains from patients with HD (Behrens et al. 2002; Perry 1981). Additional information is needed regarding the precise role of astrocytic GS activity in the pathogenesis of HD.

13.2.8 CNS Neoplasms

GS has been strongly implicated in the growth of astrocytic neoplasms as immunohistochemical analysis of GS in autopsied brain tumor tissue showed that GS was overexpressed in astrocytomas, but not in other CNS neoplasms (Pilkington and Lantos 1982). Additionally, overexpression of GS correlated well with tumor progression (Loeppen et al. 2002; McCormick et al. 1990).

While the contribution of GS to tumor initiation or progression is unclear, the overexpression of GS has been shown to enhance mutations in β-catenin. As enhanced β-catenin activity has been strongly implicated in tumor progression (Thompson et al. 2011), it is possible that GS-mediated promotion of β-catenin mutations may contribute to the progression of astrocytomas.

In contrast to the increased GS activity in lower grade astrocytomas (Grades I–III), several reports indicate that in glioblastoma multiforme (GBM, astrocytoma Grade IV) GS levels are decreased (Rosati et al. 2009, 2013). This finding correlated with the development of seizures (Rosati et al. 2013). Even though epilepsy in patients with glioma is multifactorial, it is conceivable that a downregulation of GS may result in epileptogenesis in patients with GBM by the slowing of the glutamate–glutamine cycle. The regulation of GS activity might constitute a novel approach for the treatment of the seizure disorder associated with GBM.

13.2.9 Other Neurological Disorders

Inhibition of GS activity by MSO resulting in a dysfunction of the glutamate–glutamine cycle was shown to result in depressive-like behaviors in mice (Lee et al. 2013). Brain GS level, as well as its activity was also reduced in rats after ethanol exposure (Babu and Ramanathan 2009), as well as in the cerebral cortex from individuals with major depressive disorders (Choudary et al. 2005;

Miguel-Hidalgo et al. 2010; Rajkowska and Stockmeier 2013; Toro et al. 2006), and suicidal behavior (Kalkman 2011; Klempan et al. 2009; Sequeira et al. 2009). While the reason for the decreased GS protein expression and activity in these conditions is unclear, inflammation may have played a role as inflammation is a common factor involved in these conditions and inflammatory cytokines are well known to decrease GS expression/activity by increasing its oxidation/nitration (Häussinger et al. 2005).

While decreased GS expression/activity has been identified in many neurological conditions, increased expression/activity of GS was also found in some neurological disorders. Increased GS expression was identified in brains of patients with schizophrenia (Kim et al. 2007; Bruneau et al. 2005; Burbaeva et al. 2005), in rat brain following sleep deprivation (Levental et al. 1972; Sallanon et al. 1992), as well as in rat brain after exposure to valproic acid (VPA), an animal model of autism (Bristot Silvestrin et al. 2013). While there is substantial evidence that GS expression or activity is increased or decreased in these conditions, the precise significance of changes in GS in the mechanisms of these disorders remains to be determined.

13.3 Summary

GS plays a major role in the glutamate–glutamine cycle in the CNS, by preventing the excessive accumulation of glutamate and ammonia in synaptic regions, and thereby inhibiting the development of glutamate/ammonia neurotoxicity. Increased GS levels have been identified in number of neurological conditions, including hepatic encephalopathy, ischemic stroke, traumatic brain injury, epilepsy, neoplasms, Alzheimer's disease, amyotrophic lateral sclerosis, schizophrenia, sleep disorders, and autism spectrum disorders, while inhibiting GS activity with MSO was shown to improve some of these conditions. In contrast, decreased GS expression/activity was identified in hepatic encephalopathy, ischemic stroke, traumatic brain injury, epilepsy, brain neoplasms, and in Alzheimer's disease. Such reversal of GS activity by genetic manipulation was further shown to reduce the clinical progression of these conditions.

While the reason for the differences in GS expression/activity and their involvement in the development of various neurological disorders remain unclear, it is possible that a number of factors, including the duration or severity of insult or the concentration of agents used in these studies to induce a disease process (e.g., Aβ peptides, ammonia), or the animal model used, may have varied in these studies, possibly accounting for the differences in GS protein expression and activity identified. A better understanding of the precise role of GS in these neurological conditions and its involvement in disease progression should facilitate the identification of novel agents to treat these disorders.

With regard to the current treatment strategy to prevent GS-mediated neurological conditions, inhibition of GS by MSO was shown to diminish clinical features (e.g., brain edema, cognitive dysfunction) in various conditions (e.g., hepatic encephalopathy, traumatic brain injury, epilepsy, ischemia, and amyotrophic lateral sclerosis). Nevertheless, the use of MSO, however, is problematic as MSO is well known to induce seizures. It is possible, however, that a combinatorial therapy (MSO along with an anticonvulsant) may represent an efficacious approach for the treatment of these disorders. Additionally, when GS levels are found to be decreased in some conditions (e.g., traumatic brain injury, ischemia, Alzheimer's disease, epilepsy), induction of GS gene expression, or the exogenous administration of purified GS may provide a useful therapeutic approach for the management of these disorders (Fig. 13.1).

We suggest that alterations in GS expression/activity are critically involved in the pathogenesis of many neurological and neuropsychiatric conditions. A better understanding of the precise role of GS in these disorders may provide valuable insights into disease pathogenesis that hopefully will lead to improved therapeutic outcomes for these debilitating conditions.

Acknowledgements This work was supported by a Merit Review from the US Department of Veterans Affairs and by a National Institutes of Health grant DK063311.

References

Agostinho P, Cunha RA, Oliveira C (2010) Neuroinflammation, oxidative stress and the pathogenesis of Alzheimer's disease. Curr Pharm Des 16(25):2766–2778

Akinmoladun AC, Akinrinola BL, Olaleye MT, Farombi EO (2015) Kolaviron, a Garcinia kola biflavonoid complex, protects against ischemia/reperfusion injury: pertinent mechanistic insights from biochemical and physical evaluations in rat brain. Neurochem Res 40(4):777–787. doi:10.1007/s11064-015-1527-z

Aksenov MY, Aksenova MV, Butterfield DA, Hensley K, Vigo-Pelfrey C, Carney JM (1996) Glutamine synthetase-induced enhancement of beta-amyloid peptide A beta (1-40) neurotoxicity accompanied by abrogation of fibril formation and A beta fragmentation. J Neurochem 66(5):2050–2056

Allen RS, Sayeed I, Cale HA, Morrison KC, Boatright JH, Pardue MT, Stein DG (2014) Severity of middle cerebral artery occlusion determines retinal deficits in rats. Exp Neurol 254:206–215. doi:10.1016/j.expneurol.2014.02.005

Anlauf E, Derouiche A (2013) Glutamine synthetase as an astrocytic marker: its cell type and vesicle localization. Front Endocrinol (Lausanne) 4:144. doi:10.3389/fendo.2013.00144

Babu CS, Ramanathan M (2009) Pre-ischemic treatment with memantine reversed the neurochemical and behavioural parameters but not energy metabolites in middle cerebral artery occluded rats. Pharmacol Biochem Behav 92(3):424–432. doi:10.1016/j.pbb.2009.01.010

Bak LK, Schousboe A, Waagepetersen HS (2006) The glutamate/GABA-glutamine cycle: aspects of transport, neurotransmitter homeostasis and ammonia transfer. J Neurochem 98(3):641–653

Bame M, Pentiak PA, Needleman R, Brusilow WS (2012) Effect of sex on lifespan, disease progression, and the response to methionine sulfoximine in the SOD1 G93A mouse model for ALS. Gend Med 9(6):524–535. doi:10.1016/j.genm.2012.10.014

Battaglia F, Wang HY, Ghilardi MF, Gashi E, Quartarone A, Friedman E, Nixon RA (2007) Cortical plasticity in Alzheimer's disease in humans and rodents. Biol Psychiatry 62(12):1405–1412

Baudry M, Lynch G, Gall C (1986) Induction of ornithine decarboxylase as a possible mediator of seizure-elicited changes in genomic expression in rat hippocampus. J Neurosci 6(12):3430–3435

Behrens PF, Franz P, Woodman B, Lindenberg KS, Landwehrmeyer GB (2002) Impaired glutamate transport and glutamate-glutamine cycling: downstream effects of the Huntington mutation. Brain 125(Pt 8):1908–1922

Bender AS, Norenberg MD (1996) Effects of ammonia on L-glutamate uptake in cultured astrocytes. Neurochem Res 21:567–573

Benveniste H, Drejer J, Schousboe A, Diemer NH (1984) Elevation of the extracellular concentrations of glutamate and aspartate in rat hippocampus during transient cerebral ischemia monitored by intracerebral microdialysis. J Neurochem 43(5):1369–1374

Bernstein HG, Tausch A, Wagner R, Steiner J, Seeleke P, Walter M, Dobrowolny H, Bogerts B (2013) Disruption of glutamate-glutamine GABA cycle significantly impacts on suicidalbehaviour: survey of the literature and own findings on glutamine synthetase. CNS Neurol Disord Drug Targets 12(7):900–913

Blei AT, Olafsson S, Therrien G, Butterworth RF (1994) Ammonia-induced brain edema and intracranial hypertension in rats after portacaval anastomosis. Hepatology 19:1437–1444

Bondy SC, Ali SF, Guo-Ross S (1998) Aluminum but not iron treatment induces pro-oxidant events in the rat brain. Mol Chem Neuropathol 34(2-3):219–232

Bristol LA, Rothstein JD (1996) Glutamate transporter gene expression in amyotrophic lateral sclerosis motor cortex. Ann Neurol 39(5):676–679

Bristot Silvestrin R, Bambini-Junior V, Galland F, Daniele Bobermim L, Quincozes-Santos A, Torres Abib R, Zanotto C, Batassini C, Brolese G, Gonçalves CA, Riesgo R, Gottfried C (2013) Animal model of autism induced by prenatal exposure to valproate: altered glutamate metabolism in the hippocampus. Brain Res 1495:52–60. doi:10.1016/j.brainres.2012.11.048

Bruneau EG, McCullumsmith RE, Haroutunian V, Davis KL, Meador-Woodruff JH (2005) Increased expression of glutaminase and glutamine synthetase mRNA in the thalamus in schizophrenia. Schizophr Res 75(1):27–34

Burbaeva GS, Boksha IS, Tereshkina EB, Savushkina OK, Starodubtseva LI, Turishcheva MS (2005) Glutamate metabolizing enzymes in prefrontal cortex of Alzheimer's disease patients. Neurochem Res 30(11):1443–1451

Butterfield D, Hensley CP, Subramaniam R, Aksenov M, Aksenova M, Bummer PM, Haley BE, Carney JM (1997) Oxidatively induced structural alteration of glutamine synthetase assessed by analysis of spin label incorporation kinetics: relevance to Alzheimer's disease. J Neurochem 68(6):2451–2457

Capocaccia L, Angelico M (1991) Fulminant hepatic failure: clinical features, etiology, epidemiology, and current management. Dig Dis Sci 36:775–779

Cárdenas-Rodríguez N, Coballase-Urrutia E, Pérez-Cruz C, Montesinos-Correa H, Rivera-Espinosa L, Sampieri A 3rd, Carmona-Aparicio L (2014) Relevance of the glutathione system in temporal lobe epilepsy: evidence in human and experimental models. Oxid Med Cell Longev 2014:759293. doi:10.1155/2014/759293

Carter CJ (1981) Loss of glutamine synthetase activity in the brain in Huntington's disease. Lancet 1(8223):782–789

Carter CJ (1982) Glutamine synthetase activity in Huntington's disease. Life Sci 31(11):1151–1159

Carter CJ (1983) Glutamine synthetase and fructose-1, 6-diphosphatase activity in the putamen of control and Huntington's disease brain post mortem. Life Sci 32(17):1949–1955

Castegna A, Palmieri L, Spera I, Porcelli V, Palmieri F, Fabis-Pedrini MJ, Kean RB, Barkhouse DA, Curtis MT, Hooper DC (2011) Oxidative stress and reduced glutamine synthetase activity in the absence of inflammation in the cortex of mice with experimental allergic encephalomyelitis. Neuroscience 185:97–105. doi:10.1016/j.neuroscience.2011.04.041

Chao CC, Hu S, Tsang M, Weatherbee J, Molitor TW, Anderson WR, Peterson PK (1992) Effects of transforming growth factor-beta on murine astrocyte glutamine synthetase activity. Implications in neuronal injury. J Clin Invest 90(5):1786–1793

Chatauret N, Desjardins P, Zwingmann C, Rose C, Rao KV, Butterworth RF (2006) Direct molecular and spectroscopic evidence for increased ammonia removal capacity of skeletal muscle in acute liver failure. J Hepatol 44:1083–1088

Chiu IM, Chen A, Zheng Y, Kosaras B, Tsiftsoglou SA, Vartanian TK, Brown RH Jr, Carroll MC (2008) T lymphocytes potentiate endogenous neuroprotective inflammation in a mouse model of ALS. Proc Natl Acad Sci U S A 105(46):17913–17918. doi:10.1073/pnas.0804610105
Chiu IM, Phatnani H, Kuligowski M, Tapia JC, Carrasco MA, Zhang M, Maniatis T, Carroll MC (2009) Activation of innate and humoral immunity in the peripheral nervous system of ALS transgenic mice. Proc Natl Acad Sci U S A 106(49):20960–20965
Choudary PV, Molnar M, Evans SJ, Tomita H, Li JZ, Vawter MP, Myers RM, Bunney WE Jr, Akil H, Watson SJ, Jones EG (2005) Altered cortical glutamatergic and GABAergic signal transmission with glial involvement in depression. Proc Nat Acad Sci 102:15653–15658
Chung SH, Johnson MS (1984) Studies on sound-induced epilepsy in mice. Proc R Soc Lond B Biol Sci 221(1223):145–168
Coronado VG, Xu L, Basavaraju SV, McGuire LC, Wald MM, Faul MD, Guzman BR, Hemphill JD (2011) Surveillance for traumatic brain injury-related deaths—United States, 1997–2007. MMWR Surveill Summ 60(5):1–32
Crunelli V, Carmignoto G, Steinhäuser C (2015) Novel astrocyte targets: new avenues for the therapeutic treatment of epilepsy. Neuroscientist 21(1):62–83. doi:10.1177/1073858414523320
Cummings JL (2004) Alzheimer's disease. N Engl J Med 351(1):56–67
Dam G, Keiding S, Munk OL, Ott P, Buhl M, Vilstrup H, Bak LK, Waagepetersen HS, Schousboe A, Møller N, Sørensen M (2011) Branched-chain amino acids increase arterial blood ammonia in spite of enhanced intrinsic muscle ammonia metabolism in patients with cirrhosis and healthy subjects. Am J Physiol Gastrointest Liver Physiol 301:G269–G277. doi:10.1152/ajpgi.00062.2011
Davies SW, Turmaine M, Cozens BA, DiFiglia M, Sharp AH, Ross CA, Scherzinger E, Wanker EE, Mangiarini L, Bates GP (1997) Formation of neuronal intranuclear inclusions underlies the neurological dysfunction in mice transgenic for the HD mutation. Cell 90(3):537–548
Desjardins P, Rao KV, Michalak A, Rose C, Butterworth RF (1999) Effect of portacaval anastomosis on glutamine synthetase protein and gene expression in brain, liver and skeletal muscle. Metab Brain Dis 14(4):273–280
Di Giacomo C, Sorrenti V, Acquaviva R, Campisi A, Vanella G, Perez-Polo JR, Vanella A (1997) Ornithine decarboxylase activity in cerebral post-ischemic reperfusion damage: effect of methionine sulfoximine. Neurochem Res 22(9):1145–1150
Dorfman D, Fernandez DC, Chianelli M, Miranda M, Aranda ML, Rosenstein RE (2013) Post-ischemic environmental enrichment protects the retina from ischemic damage in adult rats. Exp Neurol 240:146–156. doi:10.1016/j.expneurol.2012.11.017
Drejer J, Benveniste H, Diemer NH, Schousboe A (1985) Cellular origin of ischemia-induced glutamate release from brain tissue in vivo and in vitro. J Neurochem 45(1):145–151
Dugan LL, Choi DW (1999) Excitotoxic injury in Hypoxia-Ischemia In Siegel GJ, Agranoff BW, Albers RW, Fisher SK, Uhler MD (eds) Molecular, cellular and medical aspects. Basic neurochemistry, 6th ed. Lippincott-Raven, Philadelphia
Dutuit M, Didier-Bazès M, Vergnes M, Mutin M, Conjard A, Akaoka H, Belin MF, Touret M (2000) Specific alteration in the expression of glial fibrillary acidic protein, glutamate dehydrogenase, and glutamine synthetase in rats with genetic absence epilepsy. Glia 32(1):15–24
Ede RJ, Williams RW (1986) Hepatic encephalopathy and cerebral edema. Semin Liver Dis 6(2):107–118
Eid T, Ghosh A, Wang Y, Beckström H, Zaveri HP, Lee TS, Lai JC, Malthankar-Phatak GH, de Lanerolle NC (2008) Recurrent seizures and brain pathology after inhibition of glutamine synthetase in the hippocampus in rats. Brain 131(Pt 8):2061–2070. doi:10.1093/brain/awn133
Eid T, Tu N, Lee TS, Lai JC (2013) Regulation of astrocyte glutamine synthetase in epilepsy. Neurochem Int 63(7):670–681. doi:10.1016/j.neuint.2013.06.008
Fernandez DC, Chianelli MS, Rosenstein RE (2009) Involvement of glutamate in retinal protection against ischemia/reperfusion damage induced by post-conditioning. J Neurochem 111(2):488–498. doi:10.1111/j.1471-4159.2009.06334.x
Floyd RA (1990) Role of oxygen free radicals in carcinogenesis and brain ischemia. FASEB J 4(9):2587–2597

Folbergrová J, Kiyota Y, Pahlmark K, Memezawa H, Smith ML, Siesjö BK (1993) Does ischemia with reperfusion lead to oxidative damage to proteins in the brain? J Cereb Blood Flow Metab 13(1):145–152
Fujioka M, Taoka T, Matsuo Y, Mishima K, Ogoshi K, Kondo Y, Tsuda M, Fujiwara M, Asano T, Sakaki T, Miyasaki A, Park D, Siesjö BK (2003) Magnetic resonance imaging shows delayed ischemic striatal neurodegeneration. Ann Neurol 54(6):732–747
Garrett RH, Grisham CM (2007) Biochemistry, 3rd edn. Thomas, Belmont
Ghoddoussi F, Galloway MP, Jambekar A, Bame M, Needleman R, Brusilow WS (2010) Methionine sulfoximine, an inhibitor of glutamine synthetase, lowers brain glutamine and glutamate in a mouse model of ALS. J Neurol Sci 290(1-2):41–47. doi:10.1016/j.jns.2009.11.013
Girard G, Giguère JF, Butterworth RF (1993) Region-selective reductions in activities of glutamine synthetase in rat brain following portacaval anastomosis. Metab Brain Dis 8(4):207–215
Gorovits R, Yakir A, Fox LE, Vardimon L (1996) Hormonal and non-hormonal regulation of glutamine synthetase in the developing neural retina. Brain Res Mol Brain Res 43(1-2):321–329
Gorovits R, Avidan N, Avisar N, Shaked I, Vardimon L (1997) Glutamine synthetase protects against neuronal degeneration in injured retinal tissue. Proc Natl Acad Sci U S A 94(13):7024–7029
Griffith OW, Meister A (1978) Differential inhibition of glutamine and γ-glutamylcysteine synthetases by α-alkyl analogs of methionine sulfoximine that induce convulsions. J Biol Chem 253:2333–2338
Griffith OW, Meister A (1979) Potent and specific inhibition of glutathione synthesis by buthionine sulfoximine (S-n-butyl homocysteine sulfoximine). J Biol Chem 254:7558–7560
Harrison-Felix C, Whiteneck G, Devivo MJ, Hammond FM, Jha A (2006) Causes of death following 1 year postinjury among individuals with traumatic brain injury. J Head Trauma Rehabil 21(1):22–33
Häussinger D, Görg B, Reinehr R, Schliess F (2005) Protein tyrosine nitration in hyperammonemia and hepatic encephalopathy. Metab Brain Dis 20(4):285–294
Hawkins RA, Jessy J, Mans AM, De Joseph MR (1993) Effect of reducing brain glutamine synthesis on metabolic symptoms of hepatic encephalopathy. J Neurochem 60:1000–1006
Hensley K, Mhatre M, Mou S, Pye QN, Stewart C, West M, Williamson KS (2006) On the relation of oxidative stress to neuroinflammation: lessons learned from the G93A-SOD1 mouse model of amyotrophic lateral sclerosis. Antioxid Redox Signal 8:2075–2087
Herberg LJ, Rose IC, de Belleroche JS, Mintz M (1992) Ornithine decarboxylase induction and polyamine synthesis in the kindling of seizures: the effect of alpha-difluoromethylornithine. Epilepsy Res 11(1):3–7
Hertz L (2013) The Glutamate Glutamine (GABA) Cycle: Importance of late postnatal development and potential reciprocal interactions between biosynthesis and degradation. Front Endocrinol (Lausanne) 4:59. doi:10.3389/fendo.2013.00059
Hertz L, Zielke HR (2004) Astrocytic control of glutamatergic activity: astrocytes as stars of the show. Trends Neurosci 27(12):735–743
Hewett SJ, Csernansky CA, Choi DW (1994) Selective potentiation of NMDA-induced neuronal injury following induction of astrocytic iNOS. Neuron 13:487–494
Hilgier W (1981) Ammonia content and glutamine synthetase and glutaminase activity in the brain in experimental hepatic coma. Neuropatol Pol 19(3):351–358
Hirata T, Koehler RC, Kawaguchi T, Brusilow SW, Traystman RJ (1996) Impaired pial arteriolar reactivity to hypercapnia during hyperammonemia depends on glutamine synthesis. Stroke 27(4):729–736
Hirata T, Kawaguchi T, Brusilow SW, Traystman RJ, Koehler RC (1999) Preserved hypocapnic pial arteriolar constriction during hyperammonemia by glutamine synthetase inhibition. Am J Physiol 276:H456–H463
Hołownia A, Chwiecko M, Farbiszewski R (1994) Accumulation of ammonia and changes in the activity of some ammonia metabolizing enzymes during brain ischemia/reperfusion injury in rats. Mater Med Pol 26(1):25–27
Hoofnagle JH, Carithers RL Jr, Shapiro C, Ascher N (1995) Fulminant hepatic failure: summary of a workshop. Hepatology 21:240–252

Itzen A, Blankenfeldt W, Goody RS (2011) Adenylylation: renaissance of a forgotten post-translational modification. Trends Biochem Sci 36(4):221–228. doi:10.1016/j.tibs.2010.12.004

Jalan R, Wright G, Davies NA, Hodges SJ (2007) L-Ornithine phenylacetate (OP): a novel treatment for hyperammonemia and hepatic encephalopathy. Med Hypotheses 69:1064–1069

Jarrett SG, Milder JB, Liang LP, Patel M (2008) The ketogenic diet increases mitochondrial glutathione levels. J Neurochem 106(3):1044–1051. doi:10.1111/j.1471-4159.2008.05460.x

Johnson WL, Goldring S, O'leary JL (1965) Behavioral, unit and slow potential changes in methionine sulfoximine seizures. Electroencephalogr Clin Neurophysiol 18:229–238

Jones EA, Weissenborn K (1997) Neurology and the liver. J Neurol Neurosurg Psychiatry 63:279–293

Justin A, Sathishkumar M, Sudheer A, Shanthakumari S, Ramanathan M (2014) Non-hypotensive dose of telmisartan and nimodipine produced synergistic neuroprotective effect in cerebral ischemic model by attenuating brain cytokine levels. Pharmacol Biochem Behav 122:61–73. doi:10.1016/j.pbb.2014.03.009

Kalkman HO (2011) Circumstantial evidence for a role of glutamine-synthetase in suicide. Med Hypotheses 76(6):905–907. doi:10.1016/j.mehy.2011.03.005

Kanamori K, Parivar F, Ross BD (1993) A 15N NMR study of in vivo cerebral glutamine synthesis in hyperammonemic rats. NMR Biomed 6(1):21–26

Kim S, Choi KH, Baykiz AF, Gershenfeld HK (2007) Suicide candidate genes associated with bipolar disorder and schizophrenia: an exploratory gene expression profiling analysis of post mortem prefrontal cortex. BMC Genomics 8:413

Klempan TA, Sequeira A, Canetti L, Lalovic A, Ernst C, French-Mullen J, Turecki G (2009) Altered expression of genes involved in ATP biosynthesis and GABAergic neurotransmission in the ventral prefrontal cortex of suicides with and without major depression Mol Psychiatry 14:175–189

Kosenko E, Llansola M, Montoliu C, Monfort P, Rodrigo R, Hernandez-Viadel M, Erceg S, Sanchez-Perez A, Felipo V (2003) Glutamine synthetase activity and glutamine content in brain: modulation by NMDA receptors and nitric oxide. Neurochem Int 43:493–499

Kruchkova Y, Ben-Dror I, Herschkovitz A, David M, Yayon A, Vardimon L (2001) Basic fibroblast growth factor: a potential inhibitor of glutamine synthetase expression in injured neural tissue. J Neurochem 77(6):1641–1649

Lavoie J, Giguère JF, Layrargues GP, Butterworth RF (1987) Activities of neuronal and astrocytic marker enzymes in autopsied brain tissue from patients with hepatic encephalopathy. Metab Brain Dis 2(4):283–290

Le Prince G, Delaere P, Fages C, Lefrançois T, Touret M, Salanon M, Tardy M (1995) Glutamine synthetase (GS) expression is reduced in senile dementia of the Alzheimer type. Neurochem Res 20(7):859–862

Lee DR, Helps SC, Gibbins IL, Nilsson M, Sims NR (2003) Losses of NG2 and NeuN immunoreactivity but not astrocytic markers during early reperfusion following severe focal cerebral ischemia. Brain Res 989(2):221–230

Lee A, Lingwood BE, Bjorkman ST, Miller SM, Poronnik P, Barnett NL, Colditz P, Pow DV (2010) Rapid loss of glutamine synthetase from astrocytes in response to hypoxia: implications for excitotoxicity. J Chem Neuroanat 39(3):211–220. doi:10.1016/j.jchemneu.2009.12.002

Lee Y, Son H, Kim G, Kim S, Lee DH, Roh GS, Kang SS, Cho GJ, Choi WS, Kim HJ (2013) Glutamine deficiency in the prefrontal cortex increases depressive-like behaviours in male mice. J Psychiatry Neurosci 38(3):183–191. doi:10.1503/jpn.120024

Leite MC, Brolese G, de Almeida LM, Piñero CC, Gottfried C, Gonçalves CA (2006) Ammonia-induced alteration in S100B secretion in astrocytes is not reverted by creatine addition. Brain Res Bull 70(2):179–185

Levental M, Rakić L, Rusić M (1972) Enzymes involved in the metabolism of glutamine in certain regions of the brain of paradoxical sleep-deprived rats. Arch Int Physiol Biochim 80(5):861–870

Liaw SH, Eisenberg D (1994) Structural model for the reaction mechanism of glutamine synthetase, based on five crystal structures of enzyme-substrate complexes. Biochemistry 33(3): 675–681

Lockwood AH, Yap EW, Wong WH (1991) Cerebral ammonia metabolism in patients with severe liver disease and minimal hepatic encephalopathy. J Cereb Blood Flow Metab 11:337–341

Loeppen S, Schneider D, Gaunitz F, Gebhardt R, Kurek R, Buchmann A, Schwarz M (2002) Overexpression of glutamine synthetase is associated with beta-catenin-mutations in mouse liver tumors during promotion of hepatocarcinogenesis by phenobarbital. Cancer Res 62(20):5685–5688

Master S, Gottstein J, Blei AT (1999) Cerebral blood flow and the development of ammonia-induced brain edema in rats after portacaval anastomosis. Hepatology 30:876–880

Matos M, Augusto E, Oliveira CR, Agostinho P (2008) Amyloid-beta peptide decreases glutamate uptake in cultured astrocytes: involvement of oxidative stress and mitogen-activated protein kinase cascades. Neuroscience 156(4):898–910. doi:10.1016/j.neuroscience.2008.08.022

Matsumoto M, Ichikawa T, Young W, Kodama N (2008) Glutamine synthetase protects the spinal cord against hypoxia-induced and GABA(A) receptor-activated axonal depressions. Surg Neurol 70(2):122–158. doi:10.1016/j.surneu.2007.07.011

McCormick D, McQuaid S, McCusker C, Allen IV (1990) A study of glutamine synthetase in normal human brain and intracranial tumours. Neuropathol Appl Neurobiol 16(3):205–211

Miguel-Hidalgo JJ, Alvarez XA, Cacabelos R, Quack G (2002) Neuroprotection by memantine against neurodegeneration induced by beta-amyloid(1-40). Brain Res 958(1):210–221

Miguel-Hidalgo JJ, Waltzer R, Whittom AA, Austin MC, Rajkowska G, Stockmeier CA (2010) Glial and glutamatergic markers in depression, alcoholism, and their comorbidity. J Affect Disord 127(1-3):230–240. doi:10.1016/j.jad.2010.06.003

Nagele RG, Wegiel J, Venkataraman V, Imaki H, Wang KC, Wegiel J (2004) Contribution of glial cells to the development of amyloid plaques in Alzheimer's disease. Neurobiol Aging 25(5):663–674

Nakajima K, Kanamatsu T, Takezawa Y, Kohsaka S (2015) Up-regulation of glutamine synthesis in microglia activated with endotoxin. Neurosci Lett 591:99–104. doi:10.1016/j.neulet.2015.02.021

Nakamura K, Stadtman ER (1984) Oxidative inactivation of glutamine synthetase subunits. Proc Natl Acad Sci USA 81:2011–2015

Ngugi AK, Bottomley C, Kleinschmidt I, Sander JW, Newton CR (2010) Estimation of the burden of active and life-time epilepsy: a meta-analytic approach. Epilepsia 51(5):883–890. doi:10.1111/j.1528-1167.2009.02481.x

Nielsen HM, Veerhuis R, Holmqvist B, Janciauskiene S (2009) Binding and uptake of A beta1-42 by primary human astrocytes in vitro. Glia 57:978–988. doi:10.1002/glia.20822

Norenberg MD (1977) A light and electron microscopic study of experimental portal-systemic (ammonia) encephalopathy. Progression and reversal of the disorder. Lab Invest 36:618–627

Norenberg MD (1979) The distribution of glutamine synthetase in the rat central nervous system. J Histochem Cytochem 27:756–762

Norenberg MD (1983) Immunohistochemistry of glutamine synthetase. In: Hertz L, Kvamme E, McGeer EG, Schousboe A, (eds) Glutamine, glutamate, and GABA in the central nervous system. Alan R. Liss, New York. pp 95–100

Norenberg MD, Martinez-Hernandez A (1979) Fine structural localization of glutamine synthetase in astrocytes of rat brain. Brain Res 161:303–310

Norenberg MD, Jayakumar AR, Rama Rao KV, Panickar KS (2007) New concepts in the mechanism of ammonia-induced astrocyte swelling. Metab Brain Dis 22(3-4):219–234

Olabarria M, Noristani HN, Verkhratsky A, Rodríguez JJ (2011) Age-dependent decrease in glutamine synthetase expression in the hippocampal astroglia of the triple transgenic Alzheimer's disease mouse model: mechanism for deficient glutamatergic transmission? Mol Neurodegener 30:6–55. doi:10.1186/1750-1326-6-55

Oliver CN, Starke-Reed PE, Stadtman ER, Liu GJ, Carney JM, Floyd RA (1990) Oxidative damage to brain proteins, loss of glutamine synthetase activity, and production of free radicals during ischemia/reperfusion-induced injury to gerbil brain. Proc Natl Acad Sci U S A 87(13):5144–5147

Pahlmark K, Folbergrová J, Smith ML, Siesjö BK (1993) Effects of dimethylthiourea on selective neuronal vulnerability in forebrain ischemia in rats. Stroke 24(5):731–736

Pascual JM, Carceller F, Roda JM, Cerdán S (1998) Glutamate, glutamine, and GABA as substrates for the neuronal and glial compartments after focal cerebral ischemia in rats. Stroke 29(5):1048–1056
Perry TL (1981) Post-mortem findings do not suggest a brain glutamine deficiency in Huntington's disease. Lancet 1(8232):1261–1262
Petito CK, Chung MC, Verkhovsky LM, Cooper AJ (1992) Brain glutamine synthetase increases following cerebral ischemia in the rat. Brain Res 569(2):275–280
Phillis JW (1994) A "radical" view of cerebral ischemic injury. Prog Neurobiol 42(4):441–448
Pike CJ, Ramezan-Arab N, Miller S, Cotman CW (1996) β-Amyloid increases enzyme activity and protein levels of glutamine synthetase in cultured astrocytes. Exp Neurol 139(1):167–171
Pilkington GJ, Lantos PL (1982) The role of glutamine synthetase in the diagnosis of cerebral tumours. Neuropathol Appl Neurobiol 8:227–236
Proler ML, Kellaway P (1962) The methionine sulfoximine syndrome in the cat. Epilepsia 3:117–130
Proler ML, Kellaway P (1965) Behavioral and convulsive phenomena induced in the cat by methionine sulfoximine. Neurology 15:931–940
Radford RA, Morsch M, Rayner SL, Cole NJ, Pountney DL, Chung RS (2015) The established and emerging roles of astrocytes and microglia in amyotrophic lateral sclerosis and frontotemporal dementia. Front Cell Neurosci 9:414
Rajkowska G, Stockmeier CA (2013) Astrocyte pathology in major depressive disorder: insights from human postmortem brain tissue. Curr Drug Targets 14(11):1225–1236
Rama Rao KV, Jayakumar AR, Norenberg MD (2012) Glutamine in the pathogenesis of acute hepatic encephalopathy. Neurochem Int 61(4):575–580. doi:10.1016/j.neuint.2012.01.012
Reiner L, Misant F, Weiss P (1950) Studies on nitrogen trichioride treated prolamines. VI Suppression of the development of convulsions with methionine Arch Biochem 25:447–454
Revett TJ, Baker GB, Jhamandas J, Kar S (2013) Glutamate system, amyloid ß peptides and tau protein: functional interrelationships and relevance to Alzheimer disease pathology. J Psychiatry Neurosci 38(1):6–23. doi:10.1503/jpn.110190
Richman PG, Orlowski M, Meister A (1973) Inhibition of gamma-glutamylcysteine synthetase by L-methionine-S-sulfoximine. J Biol Chem 248(19):6684–6690
Robinson SR (2000) Neuronal expression of glutamine synthetase in Alzheimer's disease indicates a profound impairment of metabolic interactions with astrocytes. Neurochem Int 36(4-5):471–482
Rosati A, Marconi S, Pollo B, Tomassini A, Lovato L, Maderna E, Maier K, Schwartz A, Rizzuto N, Padovani A, Bonetti B (2009) Epilepsy in glioblastoma multiforme: correlation with glutamine synthetase levels. J Neurooncol 93(3):319–324. doi:10.1007/s11060-008-9794-z
Rosati A, Poliani PL, Todeschini A, Cominelli M, Medicina D, Cenzato M, Simoncini EL, Magrini SM, Buglione M, Grisanti S, Padovani A (2013) Glutamine synthetase expression as a valuable marker of epilepsy and longer survival in newly diagnosed glioblastoma multiforme. Neuro Oncol 15(5):618–625
Rose C, Michalak A, Pannunzio P, Therrien G, Quack G, Kircheis G, Butterworth RF (1998) L-ornithine-L-aspartate in experimental portal-systemic encephalopathy: therapeutic efficacy and mechanism of action. Metab. Brain Dis 13:147–157
Rose C, Michalak A, Rao KV, Quack G, Kircheis G, Butterworth RF (1999) L-ornithine-L-aspartate lowers plasma and cerebrospinal fluid ammonia and prevents brain edema in rats with acute liver failure. Hepatology 30:636–640
Rowe WB, Meister A (1970) Identification of L-methionine-S-sulfoximine as the convulsant isomer of methionine sulfoximine. Proc Natl Acad Sci U S A 66(2):500–506
Safwen K, Selima S, Mohamed E, Ferid L, Pascal C, Mohamed A, Ezzedine A, Meherzia M (2015) Protective effect of grape seed and skin extract on cerebral ischemia in rat: implication of transition metals. Int J Stroke 10(3):415–424. doi:10.1111/ijs.12391
Sallanon M, Touret M, Didier-Bazes M, Belin MF, Tardy M, Jouvet M (1992) Paradoxical sleep deprivation increases glutamine synthetase in rat brain. C R Acad Sci III 315(3):109–114

Schatz RA, Sellinger OZ (1975) Effect of methionine and methionine sulphoximine on rat brain S-adenosyl methionine levels. J Neurochem 24(1):63–66
Selkoe DJ (2001) Alzheimer's disease: genes, proteins, and therapy. Physiol Rev 81(2):741–766
Sequeira A, Mamdani F, Ernst C, Vawter MP, Bunney WE, Lebel V, Rehal S, Klempan T, Gratton A, Benkelfat C, Rouleau GA, Mechawar N, Turecki G (2009) Global brain gene expression analysis links glutamatergic and GABAergic alterations to suicide and major depression. PLoS One 4, e6585. doi:10.1371/journal.pone.0006585
Shibata M, Yamada S, Kumar SR, Calero M, Bading J, Frangione B, Holtzman DM, Miller C, A, Strickland DK, Ghiso J, Zlokovic BV (2000) Clearance of Alzheimer's amyloid-ss(1-40) peptide from brain by LDL receptor-related protein-1 at the blood-brain barrier. J Clin Invest 106:1489–1499
Shin JY, Fang ZH, Yu ZX, Wang CE, Li SH, Li XJ (2005) Expression of mutant huntingtin in glial cells contributes to neuronal excitotoxicity. J Cell Biol 171(6):1001–1012. Erratum in: J Cell Biol 172(6):953
Smith CD, Carney JM, Starkereed PE, Oliver CN, Stadtman ER, Floyd RA, Markesbery WR (1991) Excess brain protein oxidation and enzyme dysfunction in normal aging and in Alzheimer-disease. Proc Natl Acad Sci USA 88:10540–10543
Stryer L, Berg JM, Tymoczko JL (2007). Biochemistry(6th ed.). W.H. Freeman, San Francisco. pp. 679–706
Suárez I, Bodega G, Arilla E, Fernández B (1996) Long-term changes in glial fibrillary acidic protein and glutamine synthetase immunoreactivities in the supraoptic nucleus of portacaval shunted rats. Metab Brain Dis 11:369–379
Suárez I, Bodega G, Arilla E, Fernández B (1997) Region-selective glutamine synthetase expression in the rat central nervous system following portocaval anastomosis. Neuropathol App Neurobiol 23:254–261
Suarez I, Bodega G, Fernandez B (2002) Glutamine synthetase in brain: effect of ammonia. Neurochem Int 41:123–142
Sugimoto H, Koehler RC, Wilson DA, Brusilow SW, Traystman RJ (1997) Methionine sulfoximine, a glutamine synthetase inhibitor, attenuates ncreased extracellularpotassium activity during acute hyperammonemia. J Cereb Blood Flow Metab 17(1):44–49
Sun HL, Zhang SH, Zhong K, Xu ZH, Feng B, Yu J, Fang Q, Wang S, Wu DC, Zhang JM, Chen Z (2013) A transient upregulation of glutamine synthetase in the dentate gyrus is involved in epileptogenesis induced by amygdala kindling in the rat. PLoS One 8(6), e66885
Sunil AG, Kesavanarayanan KS, Kalaivani P, Sathiya S, Ranju V, Priya RJ, Pramila B, Paul FD, Venkhatesh J, Babu CS (2011) Total oligomeric flavonoids of Cyperus rotundus ameliorates neurological deficits, excitotoxicity and behavioral alterations induced by cerebral ischemic-reperfusion injury in rats. Brain Res Bull 84(6):394–405. doi:10.1016/j.brainresbull.2011.01.008
Takahashi H, Koehler RC, Brusilow SW, Traystman RJ (1991) Inhibition of brain glutamine accumulation prevents cerebral edema in hyperammonemic rats. Am J Physiol 261(3 Pt 2):H825–H829
Takahashi H, Koehler RC, Brusilow SW, Traystman RJ (1992) Restoration of cerebrovascular CO2 responsivity by glutamine synthesis inhibition in hyperammonemic rats. Circ Res 71:1220–1230
Takahashi K, Kameda K, Kataoka M, Sanjou K, Harata N, Akaike N (1993) Ammonia potentiates GABAA response in dissociated rat cortical neurons. Neurosci Lett 151:51–54
Tanaka H, Araki M, Masuzawa T (1992) Reaction of astrocytes in the gerbil hippocampus following transient ischemia – immunohistochemical observations with antibodies against glial fibrillary acidic protein, glutamine synthetase, and S-100 protein. Exp Neurol 116:264–274
Tanigami H, Rebel A, Martin LJ, Chen TY, Brusilow SW, Traystman RJ, Koehler RC (2005) Effect of glutamine synthetase inhibition on astrocyte swelling and altered astroglial protein expression during hyperammonemia in rats. Neuroscience 131(2):437–449
Thomas JW, Banner C, Whitman J, Mullen KD, Freese E (1988) Changes in glutamate-cycle enzyme mRNA levels in a rat model of hepatic encephalopathy. Metab Brain Dis 3(2):81–90

Thompson MD, Dar MJ, Monga SP (2011) Pegylated interferon alpha targets Wnt signaling by inducing nuclear export of β-catenin. J Hepatol 54(3):506–512. doi:10.1016/j.jhep.2010.07.020
Thorn W, Heimann J (1958) The effects of anoxia, ischemia, asphyxia and hypothermia on the ammonia level in the brain, heart, liver, kidneys and muscles. J Neurochem 2(2-3):166–177
Thurman DJ, Beghi E, Begley CE, Berg AT, Buchhalter JR, Ding D, Hesdorffer DC, Hauser WA, Kazis L, Kobau R, Kroner B, Labiner D, Liow K, Logroscino G, Medina MT, Newton CR, Parko K, Paschal A, Preux PM, Sander JW, Selassie A, Theodore W, Tomson T, Wiebe S (2011) Standards for epidemiologic studies and surveillance of epilepsy. Epilepsia 52(Suppl. 7):2–26. doi:10.1111/j.1528-1167.2011.03121.x
Tiffany-Castiglioni EC, Peterson SL, Castiglioni AJ (1990) Alterations in glutamine synthetase activity by FeCl2-induced focal and kindled amygdaloid seizures. J Neurosci Res 25(2):223–228
Toro C, Hallak J, Dunham J, Deakin J (2006) Glial fibrillary acidic protein and glutamine synthetase in subregions of prefrontal cortex in schizophrenia and mood disorder. Neurosci Let 404:276–281
Torres-Vega MA, Vargas-Jerónimo RY, Montiel-Martínez AG, Muñoz-Fuentes RM, Zamorano-Carrillo A, Pastor AR, Palomares LA (2015) Delivery of glutamine synthetase gene by baculovirus vectors: a proof of concept for the treatment of acutehyperammonemia. Gene Ther 22(1):58–64. doi:10.1038/gt.2014.89
Tumani H, Shen G, Peter JB, Brück W (1999) Glutamine synthetase in cerebrospinal fluid, serum, and brain: a diagnostic marker for Alzheimer disease? Arch Neurol 56(10):1241–1246
Van den Berg CJ (1970) Glutamate and glutamine. In: Lajtha A, (ed) Handbook in neurochemistry. Plenum, New York, pp 355–379
van der Hel WS, Notenboom RG, Bos IW, van Rijen PC, van Veelen CW, de Graan PN (2005) Reduced glutamine synthetase in hippocampal areas with neuron loss in temporal lobe epilepsy.Neurology 64(2):326–333
van der Vos KE, Coffer PJ (2012) Glutamine metabolism links growth factor signaling to the regulation of autophagy. Autophagy 8(12):1862–1864. doi:10.4161/auto.22152
van der Vos KE, Eliasson P, Proikas-Cezanne T, Vervoort SJ, van Boxtel R, Putker M, van Zutphen IJ, Mauthe M, Zellmer S, Pals C, Verhagen LP, Groot Koerkamp MJ, Braat AK, Dansen TB, Holstege FC, Gebhardt R, Burgering BM, Coffer PJ (2012) Modulation of glutamine metabolism by the PI(3)K-PKB-FOXO network regulates autophagy. Nat Cell Biol 14(8):829–837. doi:10.1038/ncb2536
van Gassen KL, van der Hel WS, Hakvoort TB, Lamers WH, de Graan PN (2009) Haploin sufficiency of glutamine synthetase increases susceptibility to experimental febrile seizures. Genes Brain Behav 8(3):290–295. doi:10.1111/j.1601-183X.2008.00471.x
Vardimon L, Fox LL, Degenstein L, Moscona AA (1988) Cell contacts are required for induction by cortisol of glutamine synthetase gene transcription in the retina. Proc Natl Acad Sci U S A 85(16):59815985
Veerhuis R (2011) Histological and direct evidence for the role of complement in the neuroinflammation of AD. Curr Alzheimer Res 8(1):34–58
Verma R, Mishra V, Sasmal D, Raghubir R (2010) Pharmacological evaluation of glutamate transporter 1 (GLT-1) mediated neuroprotection following cerebral ischemia/reperfusion injury. Eur J Pharmacol 638(1-3):65–71. doi:10.1016/j.ejphar.2010.04.021
Walker FO (2007) Huntington's disease. Lancet 369(9557):218–228
Walls AB, Waagepetersen HS, Bak LK, Schousboe A, Sonnewald U (2015) The glutamine-glutamate/GABA cycle: function, regional differences in glutamate and GABA production and effects of interference with GABA metabolism. Neurochem Res 40(2):402–429. doi:10.1007/s11064-014-1473-1
Wang Y, Zaveri HP, Lee TS, Eid T (2009) The development of recurrent seizures after continuous intrahippocampal infusion of methionine sulfoximine in rats: a video-intracranial electroencephalographic study. Exp Neurol 220(2):293–302. doi:10.1016/j.expneurol.2009.08.034
Wang XF, Hu WW, Yan HJ, Tan L, Gao JQ, Tian YY, Shi XJ, Hou WW, Li J, Shen Y, Chen Z (2013) Modulation of astrocytic glutamine synthetase expression and cell viability by hista-

mine in cultured cortical astrocytes exposed to OGD insults. Neurosci Lett 549:69–73. doi:10.1016/j.neulet.2013.06.013

Waniewski RA (1992) Physiological levels of ammonia regulate glutamine synthesis from extracellular glutamate in astrocyte cultures. J Neurochem 58:167–174

Warren KS, Schenker S (1964) Effect of an inhibition of glutamine synthesis (methionine sulfoximine) on ammonia toxicity and metabolism. J Lab Clin Med 64:442–449

Willard-Mack CL, Koehler RC, Hirata T, Cork LC, Takahashi H, Traystman RJ, Brusilow SW (1996) Inhibition of glutamine synthetase reduces ammonia-induced astrocyte swelling in rat. Neuroscience 71:589–599

Wyss-Coray T, Loike JD, Brionne TC, Lu E, Anankov R, Yan F, Silverstein SC, Husemann J (2003) Adult mouse astrocytes degrade amyloid-beta in vitro and in situ. Nat Med 9:453–457

Yamamoto T, Iwasaki Y, Sato Y, Yamamoto H, Konno H (1989) Astrocytic pathology of methionine sulfoximine-induced encephalopathy. Acta Neuropathol 77(4):357–368

Zhang HY, Young AP (1991) A single upstream glucocorticoid response element juxtaposed to an AP1/ATF/CRE-like site renders the chicken glutamine synthetase gene hormonally inducible in transfected retina. J Biol Chem 266(36):24332–24338

Zhao X, Ahram A, Berman RF, Muizelaar JP, Lyeth BG (2003) Early loss of astrocytes after experimental traumatic brain injury. Glia 44(2):140–152

Zieve L, Lyftogt C, Raphael D (1986) Ammonia toxicity: comparative protective effect of various arginine and ornithine derivatives, aspartate, benzoate, and carbamyl glutamate. Metab Brain Dis 1:25–35

Zou J, Wang YX, Mu HJ, Xiang J, Wu W, Zhang B, Xie P (2011) Down-regulation of glutamine synthetase enhances migration of rat astrocytes after in vitro injury. Neurochem Int 58(3):404–413. doi:10.1016/j.neuint.2010.12.018

Chapter 14
The Glutamate–Glutamine Cycle in Epilepsy

Tore Eid, Shaun E. Gruenbaum, Roni Dhaher, Tih-Shih W. Lee, Yun Zhou, and Niels Christian Danbolt

Abstract Epilepsy is a complex, multifactorial disease characterized by spontaneous recurrent seizures and an increased incidence of comorbid conditions such as anxiety, depression, cognitive dysfunction, and sudden unexpected death. About 70 million people worldwide are estimated to suffer from epilepsy, and up to one-third of all people with epilepsy are expected to be refractory to current medications. Development of more effective and specific antiepileptic interventions is therefore requisite. Perturbations in the brain's glutamate–glutamine cycle, such as increased extracellular levels of glutamate, loss of astroglial glutamine synthetase, and changes in glutaminase and glutamate dehydrogenase, are frequently encountered in patients with epilepsy. Hence, manipulations of discrete glutamate–glutamine cycle components may represent novel approaches to treat the disease. The goal of his review is to discuss some of the glutamate–glutamine cycle components that are altered in epilepsy, particularly neurotransmitters and metabolites, enzymes, amino acid transporters, and glutamate receptors. We will also review approaches that potentially could be used in humans to target the glutamate–glutamine cycle. Examples of such approaches are treatment with glutamate receptor blockers, glutamate scavenging, dietary intervention, and hypothermia.

Keywords Epilepsy • Glutaminase • Glutamine synthetase • Glutamate receptors • Glutamine–glutamate cycle

T. Eid, M.D., Ph.D. (✉) • R. Dhaher, Ph.D.
Department of Laboratory Medicine, Yale School of Medicine,
330 Cedar Street, P.O. Box 208035, New Haven, CT 06520-8035, USA
e-mail: tore.eid@yale.edu

S.E. Gruenbaum, M.D.
Department of Anesthesiology, Yale School of Medicine, New Haven, CT, USA

T.-S.W. Lee, M.D., Ph.D.
Department of Psychiatry, Yale School of Medicine, New Haven, CT, USA

Y. Zhou, Ph.D. • N.C. Danbolt, M.D., Ph.D.
Department of Molecular Medicine, Institute for Basic Medical Sciences, University of Oslo, Oslo, Norway

A. Schousboe, U. Sonnewald (eds.), *The Glutamate/GABA-Glutamine Cycle*, Advances in Neurobiology 13, DOI 10.1007/978-3-319-45096-4_14

Abbreviations

AIDA	1-aminoindan-1,5-dicarboxylic acid
ALT	Alanine aminotransferase (alanine transaminase)
AMPA	α-amino-3-hydroxy-5-methyl-4-isoxazolepropionic acid
AST	Aspartate aminotransferase (aspartate transaminase)
BGT	Betaine/gamma-aminobutyric acid transporter
CA1-3	Cornu ammonis subfields 1-3 of the hippocampus
CGP 37849	(E)-(±)-2-amino-4-methyl-5-phosphono-3-pentenoic acid
CGP 39551	(E)-(±)-2-amino-4-methyl-5-phosphono-3-pentenoic acid ethyl ester
CGS 19755	cis-4-[phosphomethyl]-piperidine-2-carboxylic acid
CNQS	6-cyano-7-nitroquinoxaline-2,3-dione
CNS	Central nervous system
D-AP5 (D-APV APV)	D-2-amino-5-phosphonovaleric acid
DMCM	Methyl-6,7-dimethoxy-4-ethyl-beta--carboline-3-carboxylate
DNQX	6,7-dinitroquinoxaline-2,3-dione
EAAT	Excitatory amino acid transporters
GABA	Gamma-aminobutyric acid
GAD	Glutamic acid decarboxylase
GAT	Gamma-aminobutyric acid transporter
GLS	Glutaminase
GLUL	Glutamate ammonia ligase
GluR1-4	AMPA receptor subunits 1-4
GluR5-7	Kainate receptor subunits 5-7
GYKI 52466	4-(8-methyl-9H-1,3-dioxolo[4,5-h][2,3] benzodiazepin-5-yl)-benzenamine
GYKI 53773 (talampanel LY 300164)	(8R)-7-Acetyl-5-(4-aminophenyl)-8,9-dihydro-8-methyl-7H-1,3-dioxolo[4,5-h] [2,3]benzodiazepine
KA1-2	Kainate receptor subunits 1-2
LY 274614	((4-)3 SR 4a RS, 6 SR, 8a SR-6-(phosphonomethyl)-1,2,3,4,4a,5,6,7,8,8 a-decahydroisoquinoline-3-carboxylic acid

LY 341495	2-[(1S,2S)-2-carboxycyclopropyl]-3-(9H-xanthen-9-yl)-D-alanine
LY 382884	(3S,4aR,6S,8aR)-6-[(4-carboxyphenyl) methyl]-decahydroisoquinoline-3-carboxylic acid
mGluR	Metabotropic glutamate receptor
MK-801 (dizocilpine)	[5R,10S]-[+]-5-methyl-10,11- dihydro-5H-dibenzo[a,d]cyclohepten-5,10-imine
MPEP	2-methyl-6-(phenylethynyl)pyridine
MSG	Monosodium glutamate
MSOP	(RS)-α-methylserine-O-phosphate
MTLE	Mesial temporal lobe epilepsy
NBQX	2,3-dihydroxy-6-nitro-7-sulfamoyl-benzo[f]quinoxaline-2,3-dione
NMDA	n-methyl-D-aspartate
NPC 17742	2R,4R,5S-(2-amino-4,5-(1 2-cyclohexyl)-7-phosphonoheptanoic acid
NR1-3	*N*-methyl-D-aspartate receptor subunits
NVP-AAM077	[[[(1S)-1-(4-bromophenyl)ethyl]amino] (1,2,3,4-tetrahydro-2,3-dioxo-5-quinoxalinyl)methyl] phosphonic acid
PPDA	(2S*,3R*)-1-(Phenanthren-2-carbonyl) piperazine-2,3-dicarboxylic acid
Ro 63-1908 (Co 101244 PD 174494)	1-[2-(4-Hydroxyphenoxy)ethyl]-4-[(4-methylphenyl)methyl]-4-piperidinol
SDZ EAB-515	(S)-α-amino-2′-chloro-5-(phosphonomethyl)[1,1′-biphenyl]-3-propanoic acid
SIB-1893	(E)-2-methyl-6-(2-phenylethenyl) pyridine
SNAT	System-A-transporter
TCA	Tricarboxylic acid
UPB-302	2-{[3-[(2S)-2-amino-2-carboxyethyl]-2,6-dioxo-3,6-dihydropyrimidin-1(2H)-yl]methyl}benzoic acid
YM872	[2,3-dioxo-7-(1H-imidazol-1-yl)-6-nitro-1,2,3,4-tetrahydro-1-quinoxalinyl]-acetic acid

14.1 Introduction

14.1.1 Glutamate: From Umami Taste to Major Neurotransmitter

In his search for the chemical basis of the umami flavor, the Japanese chemist Kikunae Ikeda purified, in the late 1800s, a crystalline substance from brown seaweed (*Laminaria japonica*). The substance turned out to be monosodium glutamate (MSG)—the carboxylate sodium salt of the alpha-amino acid glutamic acid, which had been isolated by Ritthausen in 1866 (Ritthausen 1866). Ikeda was granted a patent for his discovery in 1908 (Patent 14805, Japan Patent Office, Tokyo, Japan) and industrial production of MSG began shortly after and eventually resulted in widespread use of the amino acids as an umami flavor enhancer in food production and cooking. The annual production of MSG was estimated to be two million metric tons per year in 2009 (Sano 2009).

Subsequent studies identified glutamate as one of the most abundant amino acids in biology, being present in high concentrations in nearly all prokaryotes and eukaryotes, including humans. Particularly high levels of the amino acid were found in skeletal muscle, kidney, liver, and brain, with concentrations in the latter ranging from 5 to 15 mmol/kg wet weight (Schousboe 1981). Approximately 50 years after Ikeda's discovery, Krebs and others demonstrated that glutamate was closely related to the tricarboxylic acid (TCA) cycle as glutamate can be interconverted to the TCA intermediate alpha-ketoglutarate via the reversible glutamate dehydrogenase reaction (Krebs 1935) (Fig. 14.2). Stern next discovered that brain tissue had an exceedingly high glutamate uptake activity, suggesting that the amino acid was actively shuttled between tissue compartments (Stern et al. 1949). Then, in 1954 Hayashi made the seminal discovery that glutamate excites neural tissue, a finding that initiated a new field of research by launching numerous studies aimed at understanding the role of glutamate in neurotransmission and brain pathology (Hayashi 1954).

The current state of knowledge is that glutamate is the primary small molecule neurotransmitter responsible for fast excitatory signaling in the CNS (Fonnum 1984). Glutamate is released preferentially from axon terminals into the synaptic cleft where the transmitter binds to several types of glutamate receptors. However, release from other cell types, particularly astrocytes may occur as well, although the mechanism of release is intensely debated (Li et al. 2013; Nedergaard and Verkhratsky 2012). To ensure high fidelity encoding of the synaptic glutamate response, glutamate is rapidly cleared from the synaptic cleft by diffusion into the surrounding extracellular space and by uptake via excitatory amino acid transporters (EAATs), which are present on the surface of several cell types, particularly astrocytes and neurons (Danbolt 2001; 2016a). Effective clearance of extracellular glutamate is also important because sustained high levels of the neurotransmitter can lead to seizures and neuronal death, a phenomenon known as excitotoxicity (Choi and Hartley 1993; Nadler et al. 1978; Olney et al. 1986).

14.1.2 *The Glutamate–Glutamine Cycle is Critical for Clearance and Synthesis of Neurotransmitter Glutamate*

The notion that glutamate needs to be effectively cleared from the extracellular space and replenished in axon terminals is underscored by a series of compartmentalized processes collectively known as the glutamate–glutamine cycle (Fig. 14.1). This cycle consists of the following principal components. (1) Exocytosis of glutamate from axon terminals into the synaptic cleft. (2) Binding of glutamate to postsynaptic glutamate receptors. (3) Diffusion of glutamate away from the synapse with possible binding to extrasynaptic receptors and uptake into astrocytes or neurons via EAATs. (4) Conversion of glutamate and ammonia to glutamine in the astroglial cytoplasm via the enzyme glutamine synthetase. (5) Shuttling of glutamine to neurons via glutamine transporters. (6) Conversion of glutamine to glutamate and ammonia via the enzyme glutaminase. And finally, (7) concentration of glutamate in synaptic vesicles via vesicular glutamate transporters. The biological advantages of the cycle are twofold. First, neurotransmitter glutamate is economically recycled from glutamine. Second, the

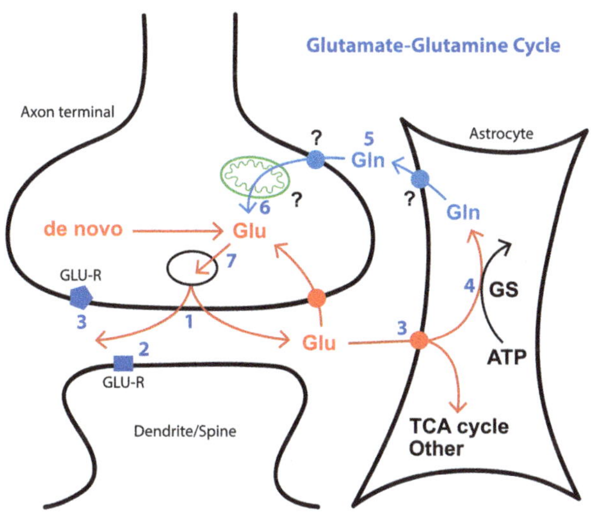

Fig 14.1 Simplified diagram of the glutamate–glutamine cycle, illustrating the flux of glutamate (Glu) and glutamine (Gln) between neurons and astrocytes. (1) Glutamate is released from vesicles in the axon terminals into the synaptic cleft (2) where the amino acid binds to glutamate receptors (GLU-R). (3) Glutamate diffuses rapidly away from the synapse followed by possible binding to extrasynaptic GLU-R and uptake into astrocytes and neurons via excitatory amino acid transporters. (4) Glutamate in astrocytes is converted to glutamine via the enzyme glutamine synthetase (GS). (5) Glutamine is next shuttled from astrocytes to neurons via glutamine transporters, which are not well defined. (6) Once in the neuron, glutamine is converted to glutamate by glutaminase, which is particularly enriched in mitochondria. (7) Finally, glutamate produced de novo or from glutamine is concentrated in synaptic vesicles via vesicular glutamate transporters, thus completing the cycle

relatively "inert" glutamine rather than the excitatory glutamate is shuttled back to the neuron, thus maintaining low extracellular levels of glutamate. However, it is important to note that the glutamate–glutamine cycle is not stoichiometric as a substantial portion of the absorbed glutamate is degraded via the TCA cycle, or modified by transaminases, rather than being converted to glutamine (McKenna 2007).

14.1.3 The Glutamate–Glutamine Cycle is Part of an Extensive Cellular, Molecular, and Metabolic Homeostatic Network

As detailed in several chapters throughout this volume, the glutamate–glutamine cycle does not operate isolated from other biological processes—it is affected by numerous metabolic and signaling pathways within and outside the CNS, as illustrated in Fig. 14.2. Brain metabolic pathways that are known to affect, depend on, or intercede with the glutamate–glutamine pathway include: the TCA cycle and glutamate dehydrogenase reaction (Hudson and Daniel 1993; Krebs 1935), the gamma-aminobutyric acid (GABA) synthetic pathway (Erlander et al. 1991), the metabolism of ammonia (Benjamin and Quastel 1975; Folbergrova et al. 1969; Krebs 1935), the synthesis of glutathione (Dringen 2000), the branched-chain amino acid pathway (Hutson et al. 1998; Leith et al. 2001), the alanine aminotransferase (a.k.a. alanine transaminase, ALT) and aspartate aminotransferase (a.k.a. aspartate transaminase, AST) pathways (Karmen et al. 1955), the pyruvate carboxylase pathway (Shank et al. 1985), the malic enzyme pathway (Hassel 2001), the glutaminase II (i.e., glutamine transaminase–ώ-amidase) pathway (Cooper and Kuhara 2014), and the isocitrate dehydrogenase pathway (Ceccarelli et al. 2002; Guo et al. 2011). Brain signaling events in the brain that are known to affect the glutamate–glutamine cycle, include the EphA5-EphrinA5 pathway (Szepietowska et al. 2012) and the mTORC1/S6K1 pathway, among others (Csibi et al. 2014). Several peripheral metabolites can also affect the glutamate–glutamine cycle, particularly branched-chain amino acids (Kanamori et al. 1998) and acetate (Patel et al. 2010) (Fig. 14.2).

The glutamate–glutamine cycle is highly compartmentalized and tightly regulated to ensure effective extracellular clearance and replenishment of neurotransmitter glutamate, particularly during periods of intense neuronal activity. However, the homeostasis of the cycle can be challenged during conditions of disease, such as brain inflammation (Gras et al. 2006), traumatic brain injury (Bartnik-Olson et al. 2013), hypoxia/ischemia (Haberg et al. 2006), and epilepsy (Eid et al. 2004; Petroff et al. 2002), with abnormal synaptic transmission, seizures, and neuronal loss as some of the deleterious consequences. It is important to understand how these conditions affect the glutamate–glutamine cycle, because such information may lead to novel therapeutic interventions for a wide range of hitherto intractable CNS disorders. The remainder of this chapter will be devoted to the role of the glutamate–glutamine cycle, particularly in the pathophysiology of epilepsy (Fig. 14.2).

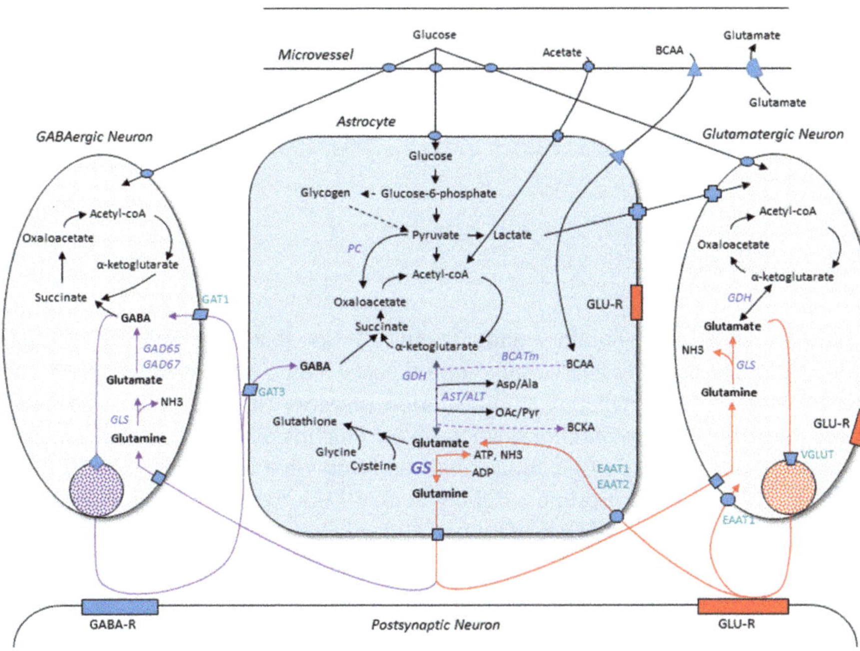

Fig 14.2 The glutamate–glutamine cycle is part of a complex metabolic network. Some of the network components are illustrated here and further discussed in the text. Transmembrane transporter molecules are depicted as blue shapes (triangle, cross, ellipse, etc.). Enzymes are indicated in *blue italics*. *ALT* alanine aminotransferase, *AST* aspartate aminotransferase, *ATP* adenosine triphosphate, *ADP* adenosine diphosphate, *BCAA* branched-chain amino acid, *BCATm* branched-chain aminotransferase mitochondrial isoform, *BCKA* branched chain keto acid, *coA* coenzyme A, *EAAT1-2* excitatory amino acid transporter subtypes 1 and 2, *GABA* gamma-aminobutyric acid, *GABA*-R GABA-receptor, *GAD65/67* glutamic acid decarboxylase isoforms 65 and 67, *GAT1-3* GABA transporters subtypes 1-3, *GDH* glutamate dehydrogenase, *GLS* glutaminase, *GLU-R* glutamate receptor, *GS* glutamine synthetase, *PC* pyruvate carboxylase, *VGLUT* vesicular glutamate receptor

14.2 The Glutamate–Glutamine Cycle in Epilepsy

14.2.1 A Brief Introduction to Epilepsy

Epilepsy is a chronic disorder of the CNS with a prevalence of approximately 1 % of the general population (Hauser et al. 1993). The hallmarks of epilepsy are spontaneous recurrent seizures and an increased incidence of comorbid conditions, particularly anxiety, depression, cognitive impairment, and sudden unexpected death (Tellez-Zenteno et al. 2007). Furthermore, up to 40 % of all people with epilepsy cannot control their disease with medications (Shorvon 1996), which often have significant side effects due to poor pharmacological specificity. A better understanding of the causative mechanisms of epilepsy is important because such understanding will likely facilitate the discovery of more efficacious and specific therapies.

This chapter will focus on the glutamate–glutamine cycle in mesial temporal lobe epilepsy (MTLE), which is one of the most common forms of medically refractory epilepsies. MTLE is of particular interest because perturbations in the glutamate–glutamine cycle are closely associated with the pathophysiology of the disease (Alvestad et al. 2011; During and Spencer 1993; Eid et al. 2004, 2008b; Petroff et al. 2002).

14.2.2 Glutamate

Numerous studies in laboratory animals have shown that increased extracellular brain glutamate or excessive stimulation of postsynaptic glutamate receptors can trigger seizures. Subcutaneous administration of MSG (4 g/kg) to a 1-week-old rhesus monkey (*Macaca mulatta*) lead to severe convulsions and neuronal damage, presumably due to entry of glutamate into the brain via a lesser developed blood–brain barrier in the infant primate (Olney et al. 1972). Systemic administration of relatively high doses of MSG to adult cats (~8 g/kg) (Goodman et al. 1946), adult rats (3.4 g/kg) (Bhagavan et al. 1971), or mice (5–7 g/kg) (Olney et al. 1972) led to severe convulsions and sometimes death. However, it is important to note that the dose of MSG required to induce seizures via the systemic route is quite high due to the effective barrier against glutamate transport from the blood to the brain and the remarkably efficient clearance of extracellular brain glutamate via the EAATs (Danbolt 2001). However, some glutamate receptor agonists such as kainic acid and domoic acid readily cross the blood–brain barrier and can result in seizures and neuronal damage at considerably lower systemic concentrations (Nadler and Cuthbertson 1980; Nadler et al. 1978; Perl et al. 1990).

Increased brain glutamate has also been associated with human epilepsy. Using intracerebral microdialysis in patients with MTLE, During and Spencer discovered in the early 1990s that glutamate increased rapidly in the extracellular compartment of the brain during a spontaneous seizure (During and Spencer 1993). The increase was most pronounced in the hippocampal formation that was the focus of the seizures, and glutamate levels remained high for several minutes after the cessation of the seizure, suggesting impaired extracellular clearance of the transmitter (During and Spencer 1993). In a subsequent study, Cavus et al. demonstrated that extracellular glutamate was chronically elevated in the seizure onset zone in patients with several types of epilepsies, including MTLE and neocortical epilepsies, indicating that the extracellular glutamate excess is a chronic phenomenon that involves several types of focal (i.e., localization-related) epilepsies (Cavus et al. 2005).

Based on our knowledge of the actions of glutamate and glutamate receptor agonists on laboratory animals, it is tempting to speculate that the extracellular glutamate excess is causally related to the triggering of seizures and loss of neurons in human epilepsy. However, the role of the extracellular glutamate excess is not as straightforward as one would think. First, due to the size of the dialysis probe, brain microdialysis is thought to preferentially sample the extrasynaptic extracellular space, rather than the much smaller synaptic extracellular space (van der Zeyden et al. 2008). This is an important distinction because synaptic glutamate may induce a different cellular response (i.e., excitation via ionotropic glutamate receptors) than extrasynaptic glutamate (i.e., inhibition via presynaptic metabotropic glutamate receptors or glial responses via glial glutamate receptors (Scofield et al. 2015)). Second, epileptic seizures are intermittent phenomena while the extracellular glutamate excess is chronic. Third, extracellular glutamate is increased in many CNS conditions, such as stroke (Davalos et al. 2000) and traumatic brain injury (Hinzman et al. 2016); however, not all patients with these conditions experience seizures or develop epilepsy. However, it is reasonable to conclude that increased extracellular brain glutamate definitely causes seizures and neuron loss under certain conditions, and that increased extracellular glutamate likely has significant effects on brain physiology.

14.2.3 Gamma-Aminobutyric Acid

GABA is synthesized from glutamate via the enzyme glutamic acid decarboxylase of which there are two main isoforms in the mammalian brain, GAD65 and GAD67 (Erlander et al. 1991). GABA is the major inhibitory neurotransmitter in the CNS (Capogna and Pearce 2011; DeFelipe 1993); however, the amino acid is also involved in cellular signaling in several organs, including the endocrine pancreas and gut (Erdo and Wolff 1990). In the CNS, GABA binds to ligand-gated $GABA_A$ receptors (Capogna and Pearce 2011) and metabotropic $GABA_B$ receptors (Chen et al. 2005), thereby mediating a variety of effects, such as neuronal inhibition, neuronal excitation, cell differentiation, neurite growth, and synapse formation (Erdo and Wolff 1990). Therapeutic or experimental manipulation of the GABA system is commonly used to modulate seizures. $GABA_A$ receptor agonists such as barbiturates and benzodiazepines are powerful anticonvulsants, and drugs with other actions on the GABA system such as gabapentin and vigabatrin reduce the frequency of many types of seizures (Petroff 2002). The $GABA_A$ antagonist bicuculline (Meldrum and Nilsson 1976) and $GABA_A$ channel blocker picrotoxin (Davidson and Barbeau 1975) both induce seizures in laboratory animals.

While GABA-modulating drugs can have dramatic effects on acute seizures, studies in humans and laboratory animals suggest that perturbations in the GABA system may play a role in the development of epilepsy and manifestation of epileptic (i.e., spontaneous recurrent) seizures. For example, extracellular GABA levels measured by microdialysis increase less in the epileptogenic hippocampal formation than in the nonepileptogenic hippocampal formation during a spontaneous seizure in patients with MTLE (During and Spencer 1993). Moreover, analysis of the GABA response reveals that the glutamate-induced calcium-independent release of GABA in the epileptogenic hippocampal formation is significantly decreased compared to the nonepileptogenic hippocampal formation in these patients (During et al. 1995). Finally, changes in the density, cytoarchitecture, molecular phenotype, and electrophysiological properties of GABAergic neurons have been reported in patients with epilepsy as well as in several animal models of the disease (de Lanerolle et al. 1989; Kobayashi and Buckmaster 2003; Margerison and Corsellis 1966; Mathern et al. 1995). However, it is important to note that the perturbations of the GABA system in epilepsy are complicated by the facts that the transmitter sometimes is excitatory (Ben-Ari et al. 2012) and that free GABA in the brain is in chemical equilibrium with homocarnosine, a dipeptide of GABA and histidine (Pisano et al. 1961), and with pyrrolidinone, the internal lactam of GABA (Seiler 1980). Homocarnosine and pyrrolidinone are particularly abundant in the *human* brain versus the rodent brain (Petroff et al. 1999).

14.2.4 Glutamine

With respect to the glutamate–glutamine cycle, glutamine has been viewed traditionally as a neurologically inert metabolite that merely shuttles and recycles glutamate from astrocytes to neurons (Benjamin and Quastel 1975; van den Berg and Garfinkel 1971). However, the first indication that glutamine was implicated in the pathophysiology of epilepsy, possibly via modulation of the glutamate–glutamine cycle, came from hippocampal slice recordings. Blocking the astroglial synthesis of glutamine by inhibiting glutamine synthetase resulted in slowing of epileptiform activity (Bacci et al. 2002). Laake et al. had previously shown that glutamate became depleted from axon terminals after inhibition of glutamine synthetase in organotypical cultures of the hippocampus, suggesting that decreased release of synaptic glutamate might explain the antiepileptiform properties of glutamine *in vitro* (Laake et al. 1999).

However, the idea that decreased glutamine inhibits epileptiform activity has been challenged by more recent studies of hippocampal glutamine synthetase inhibition *in vivo*. Such inhibition leads to decreased extracellular levels of hippocampal glutamine, recurrent seizures, and preferential loss of neurons in the hippocampal formation (Eid et al. 2008a; Wang et al. 2009). Moreover, humans with mutations in the glutamine synthetase gene exhibit decreased plasma and cerebrospinal fluid glutamine and severe epilepsy (Haberle et al. 2005, 2006, 2011). Finally, Kanamori and Ross showed that spontaneous seizures in the kainic acid rat model of MTLE were associated with a significant *decrease* in the extracellular level of hippocampal glutamine while the level of extracellular glutamate increased (Kanamori and Ross 2011).

Thus, the role of glutamine in epilepsy is complex and the mechanism by which glutamine modulates epilepsy may involve processes other than simply affecting the supply of vesicular glutamate. In fact, glutamine is critical for a variety of biological processes that have linked to neuronal excitability, neurodegeneration, and epilepsy, such as polyglutamine expansion (Banfi et al. 1994; La Spada et al. 1994), ammonia detoxification (Martinez-Hernandez et al. 1977), astrocyte swelling (Albrecht and Norenberg 2006), cell signaling (Marc Rhoads and Wu 2009), and epigenetic regulation (Simpson et al. 2012). All of these processes should be kept in mind when considering possible mechanisms of the glutamine effects in epilepsy.

14.2.5 Enzymes

Several enzymes determine the flux through glutamate–glutamine cycle, and many are altered in epilepsy (Fig. 14.2). In the following, we will focus on three that are altered in human epilepsy, i.e., glutamine synthetase, glutaminase, and glutamate dehydrogenase.

14.2.5.1 Glutamine Synthetase

Glutamate ammonia ligase (GLUL), which encodes for glutamine synthetase (EC 6.3.1.2), is one of the oldest genes known to exist, likely dating more than 1000 million years back in time (Kumada et al. 1993). There are several variants of the gene, which encode for three classes of enzymes. Class I are specific for prokaryotes and consist of 12 identical subunits arranged as two face-to-face hexameric rings, with each subunit having an active catalytic site (Almassy et al. 1986; Liaw and Eisenberg 1994). Class II are specific for eukaryotes and some bacteria and consist of ten identical subunits arranged as two face-to-face pentameric rings (Krajewski et al. 2008). Class III has only been found in a small number of bacteria, e.g., *Bacteroides fragilis and Butyrivibrio fibrisolvens*, and consist of 12 subunits arranged in a similar fashion as the Class I enzyme despite less than 10 % sequence homology (van Rooyen et al. 2006). Glutamine synthetase is present in all extant life forms and catalyzes the ATP-dependent ligation of glutamate and ammonia to glutamine (Fig.14.1). Even though glutamine synthetase is essential for life and therefore widely distributed in nature, the enzyme is restrictively expressed in mammals, being limited to specific cell populations in the liver (Bode et al. 2000; Haussinger 1990; Krebs 1935), kidney (Janicki and Goldstein 1969; Krebs 1935), pancreas (Zhou et al. 2014), adipose tissue (Kuo and Darnell 1989; Palmieri et al. 2014), skin (Danielyan et al. 2009), and CNS (Krebs 1935; Martinez-Hernandez et al. 1977). The astrocyte is the main, if not the only type of cell, expressing glutamine synthetase in the normal adult CNS (Martinez-Hernandez et al. 1977).

The importance of glutamine synthetase for life is reflected by the findings that whole body deletion of the glutamine synthetase gene in mice is embryonically lethal (He et al. 2007) whereas deletion preferentially targeting astrocytes leads to death within a few days after birth (He et al. 2010). Only three cases of whole body GLUL mutations resulting in severe glutamine synthetase deficiency have been described in humans, thus far (Haberle et al. 2005, 2006, 2011). All patients exhibited brain malformations and epilepsy, and two patients developed multiorgan failure, which resulted in death within the first month of life (Haberle et al. 2005, 2006, 2011). Heterozygous deletions in GLUL are compatible with life; however, the susceptibility to experimental febrile seizures is increased (van Gassen et al. 2009). Reduced expression or decreased activity of glutamine synthetase has also been reported to occur in a variety of CNS conditions such as major depression, Alzheimer's disease (Gunnersen and

Haley 1992; Robinson 2001; Tumani et al. 1999), schizophrenia (Bruneau et al. 2005), hepatic encephalopathy (Brusilow et al. 2010; Haussinger et al. 1994, 2000), MTLE (Eid et al. 2004; van der Hel et al. 2005), and neocortical epilepsies (Steffens et al. 2005).

While loss or inhibition of glutamine synthetase leads to a variety of signs and symptoms, depending on the degree of pathology and site of action, seizures and epilepsy often occur, as evident in patients with inherited GLUL mutations (Haberle et al. 2005, 2006, 2011). Moreover, ingestion of domoic acid by California sea lions leads to degradation of GS in astrocytes, followed several weeks later by epileptic seizures and brain pathologies that resemble those of human MTLE (Madl et al. 2014). In the juvenile pilocarpine model of MTLE, glutamine synthetase is lost from astrocytes in the dentate hilus several weeks prior to the onset of epilepsy, whereas the expression of glutamate transporters is unaffected (van der Hel et al. 2014). In the unilateral intracortical kainic acid injection model of epilepsy, astroglial glutamine synthetase is reduced in CA1 of the ipsilateral vs. contralateral hippocampus, 3–7 days after injection, several weeks before the onset of epilepsy (Jordan 2014). If glutamine synthetase is inhibited in the rat hippocampus using the irreversible enzyme blocker methionine sulfoximine, the animals develop spontaneous recurrent seizures and brain damage, similar to human MTLE (Eid et al. 2008a; Wang et al. 2009). Finally, glutamine synthetase inhibition in the amygdala results in epilepsy and depressive-like behaviors in rats, suggesting a possible causal relationship between the glutamate–glutamine cycle in the amygdala and comorbid depression in MTLE (Gruenbaum et al. 2015).

While most studies show *decreases* in glutamine synthetase in epilepsy, glutamine synthetase protein increases temporarily in the kainic acid and kindling models of MTLE (Hammer et al. 2008; Sun et al. 2013), and partial blockade of this increase, using very low doses of methionine sulfoximine, or small interfering RNAs against glutamine synthetase, *slows* the progression of epilepsy (Sun et al. 2013). This seemingly contradictory role of glutamine synthetase in epilepsy may be related to issues such as the dose, duration, and spatial distribution of glutamine synthetase alterations, and the fact that glutamine synthetase changes can modulate both glutamatergic and GABAergic neurotransmission, with potential opposing effects (Liang et al. 2006).

14.2.5.2 Glutaminase

Glutaminase (EC 3.5.1.2) catalyzes the hydrolysis of glutamine to glutamate and ammonia (Fig. 14.2). Four isoforms of glutaminase are known to exist in humans. GLS1 encodes two kidney-type glutaminases and GLS2 encodes two liver-type glutaminases (Campos-Sandoval et al. 2015). GLS1 is the predominant glutaminase transcript expressed in the mammalian brain and initial immunogold electron microscopy studies of perfusion fixed rat brains suggested that the enzyme was preferentially localized to neuronal mitochondria (Laake et al. 1999). However, recent studies by Cardona et al. have shown that astrocytes express GLS1 transcripts and active enzyme in the mitochondria and GLS2 transcripts and active enzyme in the mitochondria and nucleus (Cardona et al. 2015). Moreover, glutaminase is overexpressed in many tumor cells, including malignant gliomas, and strategies aimed at manipulating glutaminase are being explored as a possible treatment for cancer (Cheng et al. 2011).

Using immunohistochemistry and immunogold electron microscopy, we found that the expression of kidney-type glutaminase was increased in surviving neurons in the epileptogenic hippocampal formation in patients with MTLE (Eid et al. 2007). The glutaminase was enriched in mitochondria, and the concentration of the enzyme was correlated with the concentration of mitochondrial glutamate (Eid et al. 2007). We concluded that the increased glutaminase might result in an enhanced capacity for glutamate synthesis in neuronal subsets in the epileptogenic hippocampal formation, thereby possibly contributing to the extracellular glutamate excess in human MTLE (Eid et al. 2007).

14.2.5.3 Glutamate Dehydrogenase

Glutamate dehydrogenase (EC 1.4.1.2) is present in most prokaryotes and in the mitochondria of eukaryotes, and catalyzes the reversible interconversion of glutamate to alpha-ketoglutarate (Dennis et al. 1977). Using enzyme activity measurements of surgically resected brain tissue from patients with medically intractable MTLE we discovered that the glutamate dehydrogenase activity was significantly decreased in the temporal cortex in MTLE patients who had concomitant mesial temporal sclerosis (Gloor 1991; Sommer 1880), i.e., gliosis and patterned loss of neurons in the hippocampal formation, entorhinal cortex, and amygdala (Malthankar-Phatak et al. 2006). There was a nonsignificant trend toward lower glutamate dehydrogenase activity in the hippocampus of these patients. The enzyme activity in the cortex and hippocampus of patients with MTLE and concomitant mesial temporal sclerosis was negatively correlated with the duration since the first intractable seizure (Malthankar-Phatak et al. 2006). The activity of citrate synthase was not changed in the tissue, suggesting that the loss of glutamate dehydrogenase was not due to mitochondrial dysfunction or cell death. We proposed that the loss in glutamate dehydrogenase would lead to decreased oxidation of glutamate with accumulation of mitochondrial glutamate in neurons and astrocytes. The concomitant loss of glutamine synthetase in hippocampal astrocytes in MTLE is expected to further enhance the accumulation of glutamate in these cells, as suggested by quantitative immunogold electron microscopy studies of astrocytes and neurons in the glutamine synthetase-deficient hippocampus in epileptic rats (Eid et al. 2004).

14.2.6 Excitatory Amino Acid Transporters

The mammalian genome contains five genes encoding glutamate (i.e., excitatory amino acid) transporters or EAATs (for reviews, see (Danbolt 2001; Grewer et al. 2014; Vandenberg and Ryan 2013; Zhou and Danbolt 2013)): EAAT1 (GLAST;

slc1a3), EAAT2 (GLT-1; slc1a2), EAAT3 (EAAC1; slc1a1), EAAT4 (slc1a6), and EAAT5 (slc1a7). All the EAATs catalyze coupled transport of $1H^+$, $3Na^+$, and $1K^+$ with 1 substrate molecule. L-glutamate and DL-aspartate are transported with similar affinities while D-glutamate is not. Dihydrokainate selectively blocks EAAT2 (Arriza et al. 1994). EAAT2 was the first glutamate transporter to be isolated in functional form (Danbolt et al. 1990) and localized immunocytochemically (Danbolt et al. 1992; Levy et al. 1993; Danbolt 2016b). It is the most important of the glutamate transporters and accounts for 1 % of total forebrain protein (Lehre and Danbolt 1998) and about 95 % of the total glutamate uptake activity in the normal young adult rat brain (Haugeto et al. 1996). In agreement, deletion of the EAAT2 gene in mice caused an almost complete loss (about 95 %) of glutamate uptake activity and spontaneous epilepsy starting at about 3 weeks of age (Otis and Kavanaugh 2000; Tanaka et al. 1997; Zhou et al. 2014). EAAT2 deficient mice become spontaneously epileptic about 3 weeks after birth and about half of them die suddenly within the subsequent 1–2 weeks (Tanaka et al. 1997; Zhou et al. 2014). The cellular localization of EAAT2 has been much debated as the EAAT2 protein was initially only detected in astrocytes despite presence of EAAT2 mRNA in several neuronal populations (for a review, see (Danbolt et al. 2016a)). It is now clear that there is EAAT2 in nerve terminals in CA1 of the hippocampus (stratum radiatum), at least in the rat where nerve terminal EAAT2 accounts for about 1/10 of the total EAAT2 in young adult rat hippocampus CA1 and all of the nerve terminal glutamate uptake activity (Furness et al. 2008). Several variants of EAAT2 exist due to variable splicing (Grewer and Rauen 2005). The most abundant form is EAAT2a (GLT1a) which is the predominant form both in astrocytes and in axon terminals (Holmseth et al. 2009). The significance of EAAT2 in terminals is unknown; however, selective deletion of EAAT2 in astrocytes is sufficient to create a lethal phenotype (Danbolt et al. 2016a; Petr et al. 2015).

EAAT1 is also highly expressed in the CNS (Lehre and Danbolt 1998), but in contrast to the other glutamate transporters, EAAT1 is selective for astrocytes (Lehre et al. 1995, 1997; Rauen et al. 1996; Regan et al. 2007). Mice lacking EAAT1 (Watase et al. 1998) develop normally but have disturbed functions in the cerebellum, eye, and inner ear (Hakuba et al. 2000; Watase et al. 1998) (for more references see (Zhou and Danbolt 2014)) and display poor nesting behavior, abnormal sociability, reduced alcohol intake, and reward (Karlsson et al. 2009, 2012; Stoffel et al. 2004; Watase et al. 1998). Lack of EAAT1 does not lead to spontaneous seizures like those seen in connection with EAAT2 deficiency, but when seizures are initiated, then lack of EAAT1 increases seizure duration and severity (Watanabe et al. 1999). In humans, mutations in EAAT1 are associated with episodic ataxia (Broer and Palacin 2011; de Vries et al. 2009; Jen et al. 2005).

EAAT3 is selectively expressed in neurons throughout the CNS, but it is only targeted to the cell bodies and dendrites (Holmseth et al. 2012). Within the CNS, EAAT3 is found in the highest concentrations in the hippocampus followed by the neocortex. However, the total tissue content in young adult rat brains is about 100 times lower than that of EAAT2 (Holmseth et al. 2012) explaining why mice lacking EAAT3 (Peghini et al. 1997) do not show any overt neurological phenotype. EAAT3 is highly expressed in the proximal tubules of the kidney, and humans lacking EAAT3 develop dicarboxylic aminoaciduria just like the knockout mice (Bailey et al. 2011). Interestingly, human EAAT3 polymorphisms are associated with obsessive–compulsive disorders (Brandl et al. 2012; Walitza et al. 2010).

Less is known about EAAT4 and EAAT5. These transporters are quite different from those of the other EAATs and may function more like inhibitory glutamate receptors (Dehnes et al. 1998; Schneider et al. 2014; Veruki et al. 2006) due to their very high chloride conductance (Gameiro et al. 2011; Mim et al. 2005). EAAT5 is mostly expressed in the retina (Arriza et al. 1997; Dalet et al. 2012; Eliasof et al. 1998), while EAAT4 is predominantly expressed in the dendrites and cell bodies belonging to the cerebellar Purkinje cells (Dehnes et al. 1998) as well as in a subset of forebrain neurons (de Vivo et al. 2010; Massie et al. 2008). EAAT4 is thereby believed to be neuron selective within the central nervous system, but it does not appear to be of critical importance as EAAT4 knockout mice are viable and appear normal (Huang et al. 2004) albeit with some alteration of receptor activation (Nikkuni et al. 2007).

14.2.7 Glutamine Transporters

The mammalian genome contains at least 14 transporter molecules that can transport glutamine across cell membranes (Bhutia and Ganapathy 2015); however, the mechanism whereby glutamine can enter nerve terminals is not well understood. To date, glutamine transporters have not been identified in nerve terminals in the CNS at sufficiently high expression levels to keep up with glutamate release (Conti and Melone 2006; Mackenzie and Erickson 2004). The system-A glutamine transporters SNAT2 (SAT2; slc38a2) and SNAT1 (SAT1; GlnT; slc38a1) have been reported to be expressed in dendrites and cell bodies of neurons but have not been found in terminals (Conti and Melone 2006; Jenstad et al. 2009; Solbu et al. 2010) and may therefore not be important for glutamatergic activity (Grewal et al. 2009). SNAT3 (slc38a3) is mostly present in the liver whereas the levels in the CNS are very low (Chan et al. 2016; Varoqui and Erickson 2002). SNAT5 is present in the CNS, but is, like SNAT3, predominantly an efflux transporter (Nakanishi et al. 2001; Scalise et al. 2016). Recently, SNAT7 (slc38a7) and SNAT8 (slc38a7) have been reported to be expressed in neurons and may be present in axon terminals (Hagglund et al. 2011, 2015); however, knockout mice have not yet been produced. This means neither the specificity of the antibodies used to localize the transporters nor the functional importance of these transporters has been verified. Slc7a5 may be present at the blood–brain barrier but not in nerve terminals (Dolgodilina et al. 2015). In conclusion, our knowledge about glutamine transporters in the CNS has large gaps that justify further investigation. The reader is referred to Chapter 8 by Leke and Schousboe for a more detailed account of glutamine transporter in the CNS.

14.2.8 GABA Transporters

GABA transporter 1 (GAT1; slc6a1) was the first neurotransmitter transporter to be purified (Radian et al. 1986) and cloned (Conti et al. 2004; Guastella et al. 1990; Zhou and Danbolt 2013). By homology screening it was soon realized that GAT1 was the first member of a whole new family (Eulenburg and Gomeza 2010; Gether et al. 2006; Kanner 2006) which comprises 20 different transporters including another three GABA transporters (GAT2, slc6a13; GAT3, slc6a11; and BGT1, slc6a12) (Liu et al. 1993; Yamauchi et al. 1992). Although there is some GAT1 in astrocytes, particularly in thalamus, the highest levels of GAT1 are found in GABAergic nerve terminals such as the basket cell terminals around the base of the Purkinje cells. Most GABAergic terminals are labeled with the exception of striatonigral and Purkinje cell axon terminals (for reviews, see (Conti et al. 2004; Zhou and Danbolt 2013)). In contrast, GAT3 is considered to be selectively expressed in astrocytes throughout the CNS (for reviews, see (Conti et al. 2004; Zhou and Danbolt 2013)). The highest GAT1 levels are in the cerebral cortex while the highest GAT3 levels are in the brainstem (Conti et al. 2004; Evans et al. 1996).

GAT1 is the major GABA transporter, and deletion of the GAT1 gene leads to reduced aggression (Liu et al. 2007b), hypoalgesia (Xu et al. 2008), reduced anxiety and depression-like behaviors (Liu et al. 2007a), and altered behavioral responses to ethanol (Cai et al. 2006). GAT1 gene deletion also leads to tremor, ataxia, nervousness, and increased GABA-induced tonic conductance in cerebellum (Chiu et al. 2005). The latter signs and symptoms resemble the adverse effects of tiagabine (Vossler et al. 2013), which is a highly selective GAT1 blocker (Quandt et al. 2013). Defects in GAT1 may be associated with panic and anxiety disorders in humans (Thoeringer et al. 2009).

The two last GABA transporters, GAT2 and BGT1 (Zhou et al. 2012a, b), are predominantly expressed in hepatocytes in the liver, but also in the basolateral membranes of the proximal tubules and collecting ducts in the kidney. The transporters are also found in the leptomeninges, and in very low, and probably insignificant concentrations in the brain tissue proper. Deletions of the GAT2 and BGT1 genes in mice do not lead to epilepsy (Lehre et al. 2011; Zhou et al. 2012a).

14.2.9 Glutamate Receptors

14.2.9.1 Introduction

The glutamate receptors are divided into ionotropic and metabotropic types. The ionotropic receptors are cation channels composed of numerous subunits and are named after the structural analogs of glutamate that selectively activate the receptor. The ionotropic glutamate receptors include α-amino-3-hydroxy-5-methyl-4-isoxazolepropionic (AMPA), N-methyl-D-aspartate (NMDA), and 2-carboxy-3-carboxymethyl-4-isopropenylpyrrolidine (kainate) receptors. In contrast to the ionotropic glutamate receptors, the metabotropic glutamate receptors function through G-protein coupling and are subdivided into three groups based both on the second messenger system that they are coupled to, as well as on pharmacological profile. Group I metabotropic glutamate receptors (mGluR1 and mGluR5) are coupled to Gq and thus activate phospholipase C, while Group II glutamate receptors (mGluR2, mGluR3) are coupled to Gi/o and thus inhibit adenylyl cyclase. Group III metabotropic glutamate receptors (mGluR4, mGluR6, mGluR7, mGluR8) are also Gi/o coupled but are categorized based on their pharmacological profile.

14.2.9.2 AMPA Receptors

The AMPA receptor is composed of four subunits (GluR1-4) arranged in a configuration that allows for a channel in the center through which cations can flow. The receptor is permeable to both Na^+ and K^+ ions. If the AMPA receptor lacks a GluR2 subunit, as is the case with AMPA receptors localized on astrocytes and many inhibitory neurons in the amygdala and hippocampus, then the AMPA receptor is also permeable to Ca^{2+} (Geiger et al. 1995; Isa et al. 1996; Mahanty and Sah 1998). Such AMPA receptors are highly expressed in the hippocampal astrocytes of epilepsy patients with Ammon's horn sclerosis (Seifert et al. 2004).

14.2.9.3 NMDA Receptors

Much like the AMPA receptor, the NMDA receptor is also a cation channel permeable to Na^+ and K^+. It can also be Ca^{2+} permeable depending on the subunit composition. The NMDA receptor is composed of seven subunits, NR1, NR2A-D, and NR3A-B. Functional NMDA receptors appear to be made up of at least one NR1 subunit and one NR2 subunit, or one NR1 subunit with both NR2 and NR3 subunits. When expressed with the NR1 and NR2 subunits, the NR3 subunit suppresses Ca^{2+} permeability (Matsuda et al. 2002; Nishi et al. 2001), although expression of the NR3 subunit is mainly limited to the developmental stage, as well as to somatic motor neurons of the brain stem and spinal cord during adulthood (Nishi et al. 2001).

The NMDA receptor requires binding of two agonists: one at the glycine-binding site on the NR1 subunit (Hirai et al. 1996; Kew et al. 2000; Kuryatov et al. 1994; Wafford et al. 1995), and one at the glutamate-binding site on the NR2 subunit (Anson et al. 1998; Laube et al. 1997). Electrophysiological studies demonstrate that NMDA receptor activation requires occupation of two independent glycine sites and two independent glutamate sites (Benveniste and Mayer 1991; Clements and Westbrook 1991), suggesting that a functional NMDA receptor consists of two NR1 subunits and two NR2 subunits. In receptors containing an NR3 subunit, it has been speculated that an NR3 subunit substitutes for one of the NR2 subunits (Kew and Kemp 2005).

14.2.9.4 Kainate Receptors

The kainate receptor is a tetrameric structure that is composed of two subunit families, GluR5-7 and KA1-2. The GluR subunits, but not the KA subunits, can form a homomeric structure. When the KA subunits are included in the heteromeric kainate receptor, they become a high-affinity binding site (Kew and Kemp 2005). The kainate receptor is permeable to Na^+ and K^+, but also can be permeable to Ca^{2+} when expressing the unedited isoforms of the GluR5 or 6 subunits (Kohler et al. 1993). In the hippocampus, GluR6 appears to be an important subunit in both pre- and postsynaptic kainate receptors (Contractor et al. 2000). The GluR5 subunit is abundant in the temporal lobe and, in the amygdala, it is implicated in playing a role in epileptogenesis (Rogawski et al. 2001, 2003).

14.2.9.5 Metabotropic Receptors

The metabotropic glutamate receptors are categorized into three groups based on second messenger system coupling and pharmacological characteristics. The Group I mGluRs, which include mGluR1 and MGluR5 are Gq-coupled receptors; activation of these receptors leads to the stimulation of phospholipase C, which induces the production of diacylglycerol and inositol trisphosphate. Diacylglycerol activates protein kinase C, which leads to a downstream cascade of cellular events, and production of inositol trisphosphate leads to the release of Ca^{2+} from intracellular stores. Group I mGluRs are located primarily postsynaptically on neurons (Manahan-Vaughan 1997), as well as on astrocytes (Porter and McCarthy 1995, 1996); induced intracellular Ca^{2+} release is believed to lead to the release of glutamate from astrocytes, and induce excitatory postsynaptic potentials in neurons. Group II mGluRs include mGluR2 and mGluR3. They are coupled to Gi/o, which when activated, leads to the inhibition of adenylyl cyclase. Group II mGluRs are typically located presynaptically, and activation of these receptors leads to inhibition of neurotransmitter release (Manahan-Vaughan 1997). Group III mGluR receptors include mGluR4, 6, 7, and 8. Much like the Group II receptors, the Group III mGluR receptors are located on neurons, are also coupled to Gi/o, and are also inhibitory in nature (Gereau and Conn 1995). Their categorization is also based on their pharmacological characteristics. As a whole, the metabotropic glutamate receptors have demonstrated a role in epilepsy. For instance, the group I mGluR5 and group II mGluR 2/3 receptors are upregulated in MTLE (Das et al. 2012). Less is known of the role of Group I mGluR1 receptor and group III mGluR receptors in epilepsy.

14.3 Therapeutic Targeting of the Glutamate–Glutamine Cycle

14.3.1 Introduction

As discussed in the preceding sections and in other chapters of this volume, the glutamate–glutamine cycle is perturbed in a number of CNS disorders, justifying the need for therapeutic targeting of the different cycle component. However, such targeting will be challenging for several reasons. First, the glutamate–glutamine cycle is critical for normal brain physiology and for the physiology of several other organs such as the liver, kidney, and pancreas. Second, many of the enzymes, transporters, and metabolites of the glutamate–glutamine cycle are not unique to the cycle but are critically involved in other pathways of the body. Third, the blood–brain barrier will limit many therapeutic options. In the remainder of this chapter, we will discuss several approaches for possible therapeutic targeting. Emphasis will be placed on approaches that is already in clinical use or that can readily be translated to clinical use.

14.3.2 Glutamate Receptor Antagonists

The reader is referred to Table 14.1 for an overview of the most commonly used glutamate receptor blockers with effects on seizures and epilepsy. The full names of the receptor blockers are provided in *Abbreviations.*

14.3.2.1 AMPA Receptor Blockers

Competitive AMPA antagonists have poor selectivity for the AMPA receptor, for they also bind with a similar affinity to the kainate receptor. Somewhat selective competitive antagonists include CNQX, DNQX, NBQX, and PNQX. These antagonists have only modest selectivity for the AMPA-binding site over the glycine-binding site of the NMDA receptor (Kew and Kemp 2005). YM872 is the most selective competitive AMPA antagonist, which fully suppresses amygdala-kindled seizures (Hara et al. 2006). GYKI 52466 is a selective noncompetitive AMPA receptor antagonist that blocks intrahippocampal dendrotoxin K-induced seizures and hippocampal neuronal cell loss. GYKI 53773 is a selective AMPA antagonist that is protective against status epilepticus in a kainic acid-induced neonatal status epilepticus model (Dhir and Chavda 2016).

Table 14.1 Glutamate receptor blockers with effects on seizures

Mechanism of action	Antagonist	Seizure model exhibiting anticonvulsant effect of antagonist	Citation
AMPA receptor antagonist: competitive binding to the glutamate site	YM872	Amygdala kindling	Hara et al. (2006)
AMPA receptor antagonist: noncompetitive binding to the glutamate site	GYKI 52466	Dendrotoxin K	Bagetta et al. (1996)
NMDA receptor antagonist: competitive binding to the glutamate site	D-AP5	Sound	Chapman et al. (1991), Gaoni et al. (1994)
		Strychnine	McAllister (1992)
	CGS 19755	Sound	Chapman et al. (1991)
		Maximal electroshock	Ferkany et al. (1993)
		Hippocampal kindling	Katsumori et al. (1998)
		Lidocaine	McFarlane et al. (1994)
	CGP 37849 & CGP 39551	Sound	Chapman et al. (1991)
		Maximal electroshock	Fagg et al. (1990)
		Light	Chapman et al. (1991)
	NPC 17742	Maximal electroshock	Ferkany et al. (1993)
		Pentylenetetrazole	Ferkany et al. (1993)
	LY 274614 & GYKI 52466	Dendrotoxin K	Bagetta et al. (1996)
	SDZ EAB-515	Maximal electroshock	Urwyler et al. (1996)

NMDA channel blocker	MK-801	Maximal electroshock	Barton et al. (2003), Kleinrok et al. (1995)
		Imipenem	De Sarro et al. (1995)
		Sound	Gill et al. (2002)
		Pentylenetetrazole	Jiang et al. (2004), Lukomskaya et al. (2004)
		Strychnine	McAllister (1992)
		D,L-homocysteine thiolactone	Rasic-Markovic et al. (2011)
		Tetramethylenedisulfotetramine	Shakarjian et al. (2015)
		Dimethoxy-4-ethyl-beta-carboline-3-carboxylate	Tsuda et al. (1997)
		Hippocampal stimulation	Yen et al. (2004)
	Memantine	Pentylenetetrazole	Lukomskaya et al. (2004), Mares and Mikulecka (2009)
		Kainic acid	Serdyuk et al. (2014)
NMDA receptor antagonist: binding to NR2B subunit	Conantokin-R	Sound	White et al. (2000)
		Maximal electroshock	White et al. (2000)
		Pentylenetetrazole	White et al. (2000)
	Traxoprodil	Pentylenetetrazole	Naspolini et al. (2012)
	Ro 63-1908	Sound	Gill et al. (2002)
	Ifenprodil	Pentylenetetrazole	Mares and Mikulecka (2009)
		Strychnine	McAllister (1992)
		Maximal electroshock	Kleinrok et al. (1995), Zarnowski et al. (1994)
		Hippocampal stimulation	Yen et al. (2004)
Kainate receptor antagonist: competitive binding with high affinity for the GluR5 subunit	LY 382884	Maximal electroshock	Barton et al. (2003)
	UBP302	Soman	Apland et al. (2014)

(continued)

Table 14.1 (continued)

Mechanism of action	Antagonist	Seizure model exhibiting anticonvulsant effect of antagonist	Citation
mGluR1 receptor antagonist	AIDA	Acoustic stimulation	Bashkatova et al. (2015), Chapman et al. (1999)
		Pentylenetetrazole	Mares et al. (2010), Thomsen and Dalby (1998)
		Pentetrazole	Watanabe et al. (2010, 2011)
mGluR5 receptor antagonist	MPEP	Sound	Chapman et al. (2000), Bashkatova et al. (2015)
		Maximal electroshock	Barton et al. (2003)
		Pentylenetetrazole	Mares et al. (2010)

14.3.2.2 NMDA Receptor Blockers

NMDA receptor blockers that are competitive antagonists at the glutamate recognition site, only show modest receptor subtype selectivity, but generally show anticonvulsant effects. For example, D-AP5 blocks sound-induced seizures in mice (Chapman et al. 1991; Gaoni et al. 1994) and spinal seizures induced by strychnine (McAllister 1992). CGS 19755, another competitive NMDA antagonist demonstrates anticonvulsant effect against sound-induced seizures in mice (Chapman et al. 1991) and maximum electroshock-induced convulsions in rats (Ferkany et al. 1993). In addition, CGS 19755 reduces the seizures in the partially kindled hippocampal model for epileptic seizures (Katsumori et al. 1998) and increases the dose necessary for lidocaine to induce seizures in rats (McFarlane et al. 1994). CGP 37849 and CGP 39551 block convulsions in the sound-induced (Chapman et al. 1991) and maximal electroshock-induced models of seizures in mice (Fagg et al. 1990). They also reduce photically induced myoclonus in baboons (Chapman et al. 1991). NPC 17742 antagonizes maximal electroshock and pentylenetetrazole-induced seizures (Ferkany et al. 1993). LY 274614 and GYKI 52466 block seizures and protect against hippocampal neuronal loss caused by dendrotoxin K (Bagetta et al. 1996). SDZ EAB-515 blocks maximal electroshock-induced seizures (Urwyler et al. 1996).

NMDA ion channel blockers are among the most commonly used glutamate receptor blockers and include MK-801 (dizocilpine), memantine, and ketamine. MK-801, which is widely used in research, has strong anticonvulsive effects in response to maximal electroshock treatment (Barton et al. 2003; Kleinrok et al. 1995), imipenem (De Sarro et al. 1995), sound (Gill et al. 2002), pentylenetetrazole (Jiang et al. 2004; Lukomskaya et al. 2004), strychnine (McAllister 1992), D,L-homocysteine thiolactone (Rasic-Markovic et al. 2011), tetramethylenedisulfotetramine (Shakarjian et al. 2015), dimethoxy-4-ethyl-beta-carboline-3-carboxylate (Tsuda et al. 1997), and hippocampal stimulation (Yen et al. 2004). Memantine reduces seizures in the pentylenetetrazole (Lukomskaya et al. 2004; Mares and Mikulecka 2009) and kainic acid models (Serdyuk et al. 2014). Finally, ketamine is used in humans to stop refractory status epilepticus (for reviews, see (Fang and Wang 2015; Zeiler 2015)).

With respect to pharmacological agents that show subunit selectivity, NVP-AAM077 shows high selectivity for NR1/NR2A, rather than NR1/NR2B (Chaperon et al. 2003). With respect to the NR2C and NR2D subunits, PPDA is the most potent and selective NR2C/NR2D-preferring antagonist yet reported (Feng et al. 2004). To our knowledge, little data has been collected to determine the effects of these NR1/NR2A, NR2C, and NR2D subunit selective antagonists on seizures. Numerous pharmacological agents that antagonize the NR2B subunit have been tested with respect to their anticonvulsant activity. For example, the NR2B antagonist, conantokin-R (Con-R), prevents sound-induced tonic extension seizures in mice and seizures induced by maximal stimulation, and blocks seizures induced by pentylenetetrazole (White et al. 2000). The highly selective NR2B antagonist, traxoprodil has demonstrated an ability to decrease pentylenetetrazole-induced seizures (Naspolini et al. 2012). The NR2B selective antagonist, Ro 63-1908, is an active inhibitor of sound-induced seizures in mice (Gill et al. 2002). While the earlier NR2B antagonists have demonstrated consistent anticonvulsant effects in the studies tested, mixed results have been obtained with NR2B antagonist, ifenprodil. While ifenprodil has demonstrated anticonvulsant effects (Hrncic et al. 2009; Mares and Mikulecka 2009) (Kleinrok et al. 1995; McAllister 1992; Zarnowski et al. 1994) (Yen et al. 2004), some studies have failed to show such effects. For example, ifenprodil failed to antagonize electrical seizures (Burket et al. 2010) and failed to prevent imipenem-induced convulsions in some studies (De Sarro et al. 1995). In addition, ifenprodil did not block DMCM-induced seizures in mice (Tsuda et al. 1997). In certain instances, ifenprodil had a proconvulsant effect. For example, ifenprodil decreased the latency to the first seizure onset and increased the median number of seizure episodes induced by homocysteine thiolactone (Rasic-Markovic et al. 2011). In addition, at low doses, ifenprodil enhanced amygdala kindling acquisition, and at high doses, blocked amygdala kindling acquisition (Yourick et al. 1999). The results from these studies suggest that the role of the NR2B receptor subunit in anticonvulsant behavior may be mediated in some aspects by the level of activity at the site of ligand binding.

14.3.2.3 Kainate Receptor Blockers

Competitive kainate receptor antagonists show poor selectivity between AMPA receptors and kainate receptors. LY 382884 is a selective kainate receptor antagonist that exhibits high affinity binding at the GluR5 subunit but not at the GluR6 or AMPA-specific subunits. The antagonist protects against maximal electroshock-induced seizures (Barton et al. 2003). UBP-302, another selective GluR5 kainate receptor antagonist reduces the duration of soman-induced status epilepticus and protects against neuronal loss in the hippocampus and amygdala in this model (Apland et al. 2014).

14.3.2.4 Metabotropic Receptor Blockers

AIDA is a group I mGluR antagonist that is selective for mGluR1 versus mGluR5 (Kew and Kemp 2005). AIDA suppresses sound-induced clonic seizures in mice and rats (Bashkatova et al. 2015; Chapman et al. 1999), as well as seizures induced by pentylenetetrazole (Mares et al. 2010; Thomsen and Dalby 1998) SIB-1893 and MPEP are group I mGluR antagonists with selectivity for mGluR5 over mGluR1. MPEP inhibits sound-induced seizures in mice (Chapman et al. 2000), and seizures induced by maximal electroshock treatment (Barton et al. 2003) and by pentylenetetrazole (Mares et al. 2010).

With respect to the group II mGluRs, LY341495 exhibits nanomolar affinity at both mGluR2 and mGluR3 (Kingston et al. 1998). While selective for group II receptors, LY341495 also exhibits some activity at the other group III mGluRs as well as on the group I mGluRs (Kingston et al. 1998). Little is known on the effect of group II mGluR antagonists on seizures. MSOP is a selective group III mGluR antagonist, which is a proconvulsant in mice (Chapman et al. 2001).

14.3.2.5 Use of Glutamate Receptor Blockers in Humans

In recent years, there has been great interest in studying the effects of glutamate receptor antagonists on seizures and brain pathology damage after an acute brain insult (Ikonomidou and Turski 2002). NMDA receptor antagonists have demonstrated neuroprotective effects in animal models of stroke and traumatic brain injury (Lee et al. 1999; McCulloch 1992); however, clinical trials were stopped early because they either had no therapeutic effect in humans or worsened outcomes (Ikonomidou and Turski 2002; Muir 2006). The deleterious effects of NMDA receptor antagonism were thought to be due to interference with normal glutamate signaling, neuronal integrity, and neuronal repair after injury (Ikonomidou and Turski 2002; Zlotnik et al. 2009). In treating epilepsy, however, anticonvulsant effects have been effectively demonstrated with several types of glutamate receptor antagonists. Antagonists of group I mGluRs have demonstrated anticonvulsant and anxiolytic effects in both developing and adult rodents (Mares et al. 2010). Ketamine has been shown to effectively and safely control drug-resistant status epilepticus in both children and adults (Fang and Wang 2015). Furthermore, the anticonvulsant effects of decanoic acid, which increases in the plasma during the ketogenic diet, are though to result from noncompetitive AMPA receptor antagonism (Chang et al. 2016).

14.3.3 Glutamate Scavenging

14.3.3.1 Metabolic Scavenging

Many studies have argued that reducing the extracellular brain glutamate excess in epilepsy may effectively reduce seizure activity and prevent seizure-induced neuronal loss (Eid et al. 2008a, c; Lee et al. 2007). The brain has several mechanisms by which glutamate is removed from the extracellular compartment. In addition to the EAATs located on astrocytes and neurons (see Sect. 14.2.6 and Figs. 14.1 and 14.2), sodium-dependent glutamate transporters present on the antiluminal side of brain capillaries provide another mechanism of reducing excess glutamate from the extracellular space (Fig. 14.2) (O'Kane et al. 1999). In areas of the brain where extracellular glutamate is pathologically increased, glutamate is rapidly transported into the blood, down its concentration gradient via facilitated diffusion (Marliss et al. 1971). Several studies have shown that this brain-to-blood glutamate efflux is a physiological process, which may be an important mechanism by which excess glutamate is removed from pathological areas of the brain (Gottlieb et al. 2003; Hosoya et al. 1999).

Studies have shown that increasing the concentration gradient of glutamate between the brain and the blood can effectively increase the rate of brain-to-blood glutamate efflux (Gottlieb et al. 2003). Thus, decreasing blood glutamate levels is a mechanism by which extracellular brain glutamate may be reduced. The increased gradient has

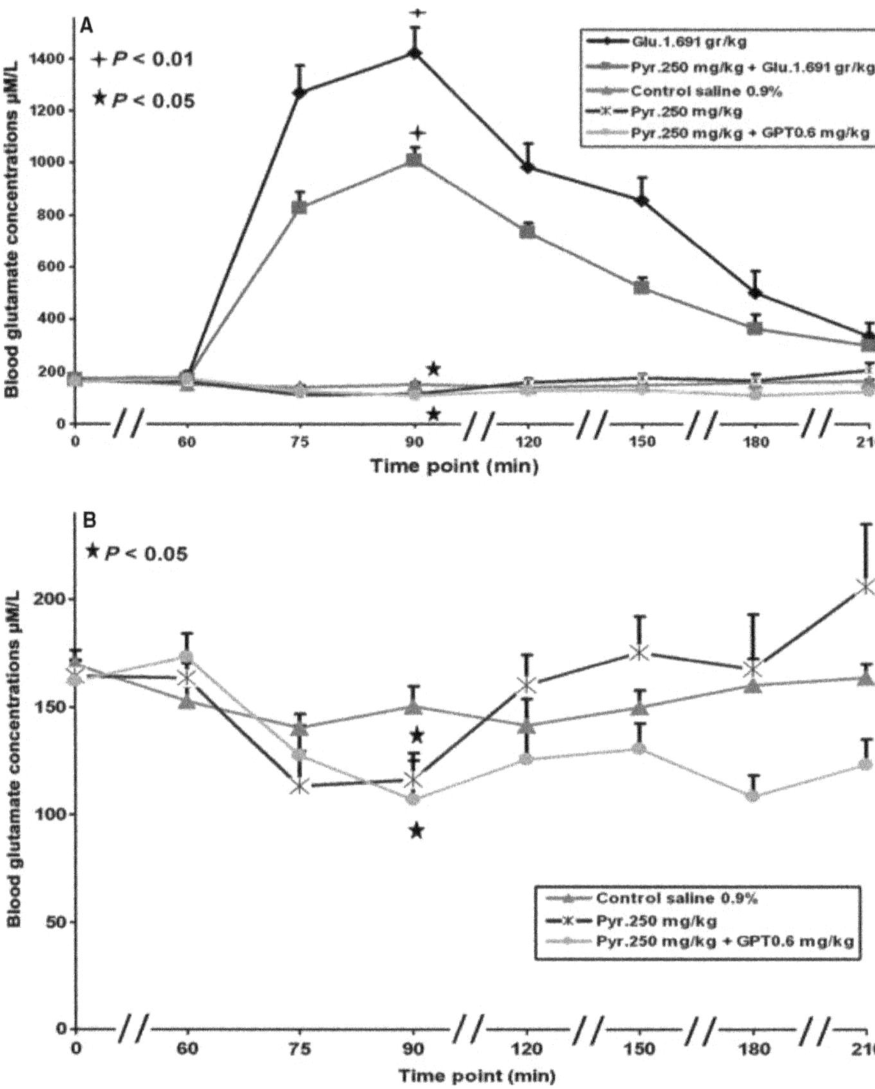

Fig 14.3 Administration of blood glutamate scavengers decrease blood glutamate concentrations over time in a rat model of stroke. Rats administered pyruvate or pyruvate + GPT demonstrate significantly lower blood glutamate concentrations compared with rats administered saline (control) or glutamate. B) Expanded scale of the lower part of A. (* $p<0.05$, + $p<0.01$). Adapted from Boyko et al. Boyko et al. (2013). See text for details

been experimentally demonstrated by peripheral administration of oxaloacetate and pyruvate, which are substrates for aspartate aminotransferase (AST) and alanine aminotransferase (ALT), respectively, converting glutamate to α-ketoglutarate (Fig. 14.3).

Animal studies of various CNS insult paradigms have demonstrated the neuroprotective effects of blood glutamate scavenging, with the strongest evidence in models of traumatic brain injury and stroke (Boyko et al. 2014; Leibowitz et al. 2012). Peripheral administration of blood glutamate scavengers after traumatic brain injury reduced blood glutamate concentrations, improved neurological outcomes, and reduced hippocampal neuronal loss (Zlotnik et al. 2007, 2008, 2009, 2012) (Fig. 14.4). After an ischemic stroke, peripheral administration of blood glutamate scavengers decreased the concentration of brain and blood glutamate, reduced the infarct size and cerebral edema, improved neurological outcomes, and reduced the mortality rate (Boyko et al. 2011; Campos et al. 2011; Nagy et al. 2009). In a rodent model of subarachnoid hemorrhage, administration of blood glutamate scavengers reduced blood and brain glutamate, decreased blood–brain barrier disruption, reduced cerebral edema, and improved the neurological outcome (Boyko et al. 2012). Blood glutamate scavengers may also be neuroprotective in animal models of glioma (Ruban et al. 2012) and in migraine headaches (Campos et al. 2013).

It is possible that blood glutamate scavenging may be an effective therapeutic approach to treat epilepsy. In one study, rats were treated with a single systemic injection of oxaloacetate and pyruvate 30 min after induction of status epilepticus with pilocarpine (Carvalho et al. 2011). Despite showing no reduction in seizures, the study demonstrated prevention of seizure-induced neuronal loss in CA1 of the hippocampus. Furthermore, administration of glutamate scavengers in animal models of organophosphate intoxication resulted in rapid reduction of brain glutamate levels and neuroprotection, but failed to demonstrate a reduction in seizure activity (Ruban et al. 2013, 2015).

Because the rate of brain-to-blood glutamate efflux is dependent on a concentration gradient between the brain and blood, glutamate scavengers are thought to reduce elevated glutamate concentrations preferentially in areas of pathological glutamate excess, with minimal effects on physiological glutamate concentrations (Gottlieb et al. 2003). Furthermore, the rate of efflux from the brain to blood slows as the concentration in the brain approaches that in the blood. Hence, blood glutamate scavengers do not decrease brain concentrations below physiological concentrations, and unlike NMDA receptor antagonists, do not appear to interfere with normal synaptic transmission (Zlotnik et al. 2009).

Despite promising studies in animal models, metabolic glutamate scavenging has limitations. First, the neuroprotective effects of glutamate scavengers are evident only when administered within a short therapeutic window (Boyko et al. 2011; Zlotnik et al. 2007, 2008). The short therapeutic window may limit the utility of scavengers in the treatment of epilepsy, where seizures and brain glutamate surges occur unpredictably (During and Spencer 1993). Second, while the administration of blood glutamate scavengers within 1 h of experimentally-induced traumatic brain injury reduces hippocampal neuronal loss (Zlotnik et al. 2012), it is less clear what effect blood glutamate scavengers would have on hippocampal pathology if administered at later time points. Third, despite reducing brain glutamate levels and protecting against neuronal loss in epilepsy models, blood glutamate scavengers have not demonstrated antiepileptic effects (Carvalho et al. 2011). Therefore, blood glutamate scavengers may be more effective in reducing the secondary brain damage associated with pathologically increased brain glutamate and less effective in treating seizures.

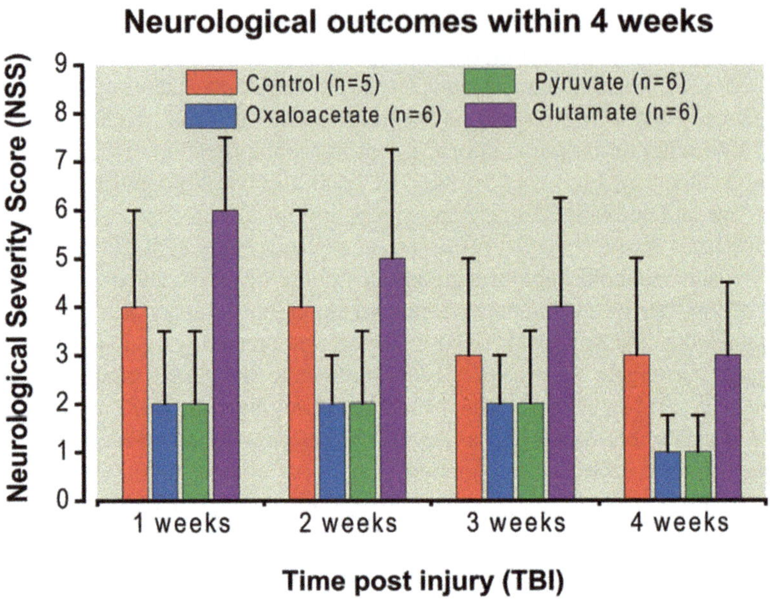

Fig 14.4 Administration of blood glutamate scavengers results in improved neurological outcomes and neuroprotection. (**a**) Administration of the blood glutamate scavengers oxaloacetate and pyruvate results in improved neurological severity score 1–4 weeks after traumatic brain injury (TBI). Adapted from Zlotnik et al. (Zlotnik et al. 2012)

Fourth, whether systemic administration of pyruvate and oxaloacetate at doses needed for brain glutamate scavenging can be safely performed in humans has not yet been established (Boyko et al. 2014).

14.3.3.2 Extracorporeal Methods

The use of extracorporeal methods to scavenge blood glutamate and thereby reduce glutamate-mediated excitotoxicity has gained increased attention in recent years (Zhumadilov et al. 2015). These methods can eliminate pathologically elevated glutamate from the blood, and like glutamate scavengers, can increase the brain-to-blood glutamate efflux. Although the effects of extracorporeal glutamate scavenging has not yet been studied in a setting of epilepsy, the blood glutamate-reducing effects of hemodialysis, peritoneal dialysis, and hemofiltration have been extensively studied in humans and animals (Boyko et al. 2014; Hu et al. 2014; Rogachev et al. 2012, 2013). These methods may be useful in the treatment of refractory epilepsies, particularly refractory status epilepticus.

In conditions where glutamate is pathologically elevated in the brain and blood, studies have shown that dialyzing the blood can eliminate the excess glutamate. A recent study demonstrated that hemodialysis (Rogachev et al. 2012) and peritoneal dialysis (Godino Mdel et al. 2013; Rogachev et al. 2013) can effectively reduce blood glutamate concentrations in patients with chronic kidney disease. Similarly, in a rat model of stroke, the use of peritoneal dialysis was associated with a reduction in blood glutamate levels and a decrease in the cortical infarct area (Godino Mdel et al. 2013).

Despite the effectiveness of lowering blood glutamate concentrations, the hemodialysis procedure requires that patients are anticoagulated. Moreover, hemodialysis is frequently complicated by hypotension. These issues limit the use of hemodialysis in unstable patients and patients with acute brain injury. In contrast, peritoneal dialysis does not require anticoagulation, nor does it cause hemodynamic instability. The insertion of a peritoneal dialysis catheter is safe, minimally invasive, and can be performed at the bedside in approximately 20 min.

Hemofiltration is an attractive alternative extracorporeal method of blood glutamate scavenging. Hemofiltration is unique in that it uses properties of convection to create a positive hydrostatic pressure across a filtering membrane to drive solutes from the blood to the filtrate compartment (Friedrich et al. 2012). Studies have demonstrated that hemofiltration is superior to hemodialysis in clearing medium and larger sized molecules, and may therefore be more effective than hemodialysis in clearing large inflammatory cytokines (Ronco et al. 2003). Hemofiltration requires minimal anticoagulation and results in less hemodynamic instability compared with hemodialysis. Furthermore, hemodialysis typically lasts only 4 h, whereas hemofiltration may be employed continuously for several days. Although there are currently no published reports that examine the effects of hemofiltration on blood glutamate levels and neurological outcomes, two clinical trials are currently underway (Zhumadilov et al. 2015).

In conclusion, extracorporeal methods of blood glutamate scavenging have several favorable properties compared with pharmacological blood glutamate scavengers (Zhumadilov et al. 2015). First, whereas pharmacological agents exploit reversible enzymatic processes, extracorporeal methods definitively remove glutamate from the blood. Second, the timing of pharmacological methods is limited by their pharmacokinetic properties, whereas extracorporeal methods can be employed over several hours or days after an acute brain injury. Third, pharmacological agents may have unknown side effects and actions outside their target area, whereas extracorporeal methods are widely used, reliable, and their peripheral effects are well established and predictable.

14.3.4 Ketogenic Diet

The ketogenic diet has long been known to have antiepileptic properties, and a recent Cochrane review concluded that the diet resulted in seizure control comparable to antiepileptic drugs in children with epilepsy (Levy et al. 2012). Interestingly, although the effects of the ketogenic diet have been known since the 1930s, the underlying mechanisms remain unclear, but may activation of adenosine triphosphate-sensitive potassium channels, inhibition of the mTOR pathway, and inhibition of glutamatergic neurotransmission (Danial et al. 2013).

Studies have demonstrated that acetoacetate and beta-hydroxybutyrate inhibit vesicular glutamate transporters, which function in loading glutamate in vesicles at the glutamatergic neuronal synapse (Juge et al. 2010). The glutamatergic synaptic vesicles require chloride ions outside the vesicle and are dependent on the membrane potential. Both ketones block the transport of glutamate into the synaptic vesicles by altering the dependence of the transporters on chloride, thereby shifting the chloride dependence to higher concentrations. Of the two ketones, acetoacetate is approximately ten times more potent than beta-hydroxybutyrate. Moreover, acetoacetate inhibits glutamate release from cultured hippocampal neurons but not astrocytes and reduces the frequency of excitatory postsynaptic currents in CA1 pyramidal neurons. Acetoacetate reduced seizures in a rodent model of epilepsy and decreased glutamate release during the seizures. Studies have shown that the ketogenic diet does not enhance glutamate reuptake into presynaptic neurons (Bough et al. 2007). Moreover, while acetoacetate had no effect on modulating postsynaptic currents in GABAergic neurons, cerebrospinal fluid levels of GABA were elevated in children 4 months after starting the ketogenic diet (Dahlin et al. 2005).

14.3.5 Branched-Chain Amino Acids

It has been postulated that blood-derived branched-chain amino acids (i.e., leucine, isoleucine, and valine) account for 20 % of de novo synthesis of brain glutamate (Yudkoff 1997). The synthesis occurs via transfer of the branched-chain amino acids' amino group to α-ketoglutarate in the mitochondria of astrocytes, in a reaction catalyzed by branched-chain aminotransferase (Fig. 14.2). Branched-chain amino acids are abundant in the blood, particularly after a meal, and readily enter the brain via the large neutral amino acid transporter, thus providing an important mechanistic link between peripheral metabolism and brain glutamate homeostasis. However, the role of branched-chain amino acids on glutamate metabolism and seizure activity is poorly understood.

There is increasing evidence to suggest that the branched-chain amino acids are critical for several physiological functions in the brain, such as the synthesis of glutamate (and thus glutamine and GABA), intracellular signaling, immune modulation, and mitochondrial health (Blomstrand et al. 2006; De Simone et al. 2013; Lynch and Adams 2014; Rothman et al. 2012). However, there are conflicting data on the role of branched-chain amino acids in the pathophysiology of brain disorders such as epilepsy. While some studies have suggested that branched-chain amino acids may contribute to increased extracellular glutamate and neuronal apoptosis (Alvestad et al. 2011), other studies have demonstrated that branched-chain amino acids and their analogs increase seizure threshold in humans and animal models (Dhaher et al. 2014; Dufour et al. 1999). Importantly, the neuroprotective and anticonvulsant effects of branched-chain amino acids are thought to require glutamate metabolism to glutamine in astrocytes by glutamine synthetase, which would reduce intracellular and extracellular glutamate, increase glutamine clearance to the blood, and increase GABA production in inhibitory neurons (Yudkoff et al. 2004).

The reader is referred to the chapter by Hutson and XXX in this volume for a more comprehensive discussion of the role of branched-chain amino acids in the glutamate–glutamine cycle.

14.3.6 Hypothermia

Therapeutic hypothermia has been used since 1963 to treat a wide range of intracranial pathologies (Bennett et al. 2014) and preclinical studies have demonstrated that hypothermia is effective in improving neurological outcomes in animal models of traumatic brain injury, stroke, and cerebral ischemia (Erecinska et al. 2003; Lampe and Becker 2011). Therapeutic hypothermia reduces the seizure activity in animal models of epilepsy (Motamedi et al. 2006) and is also effective in treating several conditions in humans including refractory status epilepticus (Bennett et al. 2014; Zeiler et al. 2015), hypoxic–ischemic encephalopathy (Azzopardi et al. 2014; Boylan et al. 2015; Orbach et al. 2014), and acute encephalitis and encephalopathy (Imataka et al. 2014).

A recent systematic review identified 13 studies (10 manuscripts and 3 meeting abstracts) in which therapeutic hypothermia was used to treat 40 patients with refractory status epilepticus (Zeiler et al. 2015). Patients were externally cooled to a core temperature of 33 °C, and hypothermia was maintained for a median of 48 h. The studies demonstrated a 15 % reduction in seizures with 62.5 % of patients experiencing complete seizure cessation. About 20 % did not respond to treatment. Therapeutic hypothermia in pediatric epilepsies has resulted in decreased seizure burden during and after refractory status epilepticus and has been suggested to prevent refractory status epilepticus relapse (Guilliams et al. 2013).

Hypoxic–ischemic brain injury is a common cause of seizures in the neonatal period. A recent systematic review examined the evidence from 11 randomized controlled trials, and concluded that therapeutic hypothermia reduces mortality and neurodevelopmental disability in survivors with hypoxic–ischemic encephalopathy (Jacobs et al. 2013). Several recent studies have showed that therapeutic hypothermia may be effective in reducing seizure activity in neonates with hypoxic–ischemic encephalopathy, resulting in a lower seizure burden and shorter seizure duration (Azzopardi et al. 2014). One study observed that cooled neonates with moderate encephalopathy had fewer than half the incidence of seizures compared with neonates that were not cooled, whereas cooling had no effect in neonates with severe encephalopathy (Orbach et al. 2014). Despite a reduction in seizure activity, some studies suggest that therapeutic hypothermia may not reduce the overall need for antiepileptic drugs or duration of treatment (Low et al. 2012).

Several mechanisms have been proposed to explain the antiseizure and neuroprotective effects of therapeutic hypothermia (Motamedi et al. 2013). Hypothermia is known to alter the potassium ion exchange via the sodium–potassium pump, thereby inhibiting the propagation of action potentials (Volgushev et al. 2000). Hypothermia also reduces the cerebral metabolic rate, the intracranial pressure, the release of inflammatory mediators in the brain, and the extent of reperfusion injury (Kentner et al. 2002; Lampe and Becker 2011). Recent studies have suggested that the effects of hyperthermia on seizures and neuronal viability may be glutamate mediated. Hypothermia is thought to reduce glutamate-mediated excitotoxicity by altering the glutamate-binding affinity and synaptic release of glutamate, thereby reducing neurotransmission and epileptic activity (Boucher et al. 2010; Yang et al. 2005). In humans and animal models of stroke and traumatic brain injury, therapeutic hypothermia reduced the glutamate concentration in the extracellular fluid of the brain (Berger et al. 2002, 2004; Winfree et al. 1996).

Boyko demonstrated a strong correlation between body temperature and blood glutamate levels in naïve rats (Boyko et al. 2013). Induction of mild and moderate hypothermia resulted in decreased blood glutamate concentrations (Fig. 14.5) and increased AST concentrations. These findings suggest that the hypothermia-mediated improvement in neurological outcomes might be due to a blood glutamate scavenging mechanism, by creating a more favorable concentration gradient between the brain and blood and increasing the brain to blood glutamate efflux.

Deep hypothermia (20 °C) lasting 30 min in rats terminates refractory status epilepticus within 12 min, reduces seizure activity upon rewarming, and protects against status epilepticus-induced neuronal loss in most animals (Niquet et al. 2015). However, deep hypothermia can also have deleterious effects, as evidenced by increased blood glutamate concentrations, increased extracellular glutamate in cultured astrocytes, overstimulation of glutamate receptors, and worsened neurological outcomes (Bissonnette et al. 1999; Boyko et al. 2013). The worsened neurological outcomes have been attributed to reductions in cerebral blood flow and metabolic rate (Ehrlich et al. 2002). Some authors have postulated that deep hypothermia may cause hepatic and skeletal muscle damage, resulting in increased cellular permeability and increases in blood glutamate concentrations (Blair et al. 1961). This postulate is supported by the observation that rats and rabbits subjected to hypothermia have significantly increased creatinine kinase, lactate dehydrogenase, lipase, and amylase (Altland et al. 1974; Arakawa et al. 1997).

While therapeutic hypothermia is becoming a standard treatment for hypoxia-induced encephalopathy in infants, its use in refractory status epilepticus is increasing. Despite strong preclinical evidence and some clinical evidence, more prospective clinical trials are warranted to examine the effects of therapeutic hypothermia on seizure activity. A major factor that has limited the use of therapeutic hypothermia to treat seizures and other intracranial pathologies has been its risk of complications, including venous thrombosis, infection, cardiac arrhythmias, electrolyte disturbances, and coagulopathy (Zeiler et al. 2015). In neonates with hypoxia-induced encephalopathy, common complications include sinus bradycardia and thrombocytopenia (Jacobs et al. 2013).

14.4 Concluding Remarks

The glutamate–glutamine cycle is perturbed in epilepsy, and manipulations of the different cycle components can stop seizures and prevent seizure-induced neuronal loss. Therapeutic targeting of the cycle in humans is presently limited to a few approaches, such as administration of ketamine, GABA-receptor antagonists (e.g.,

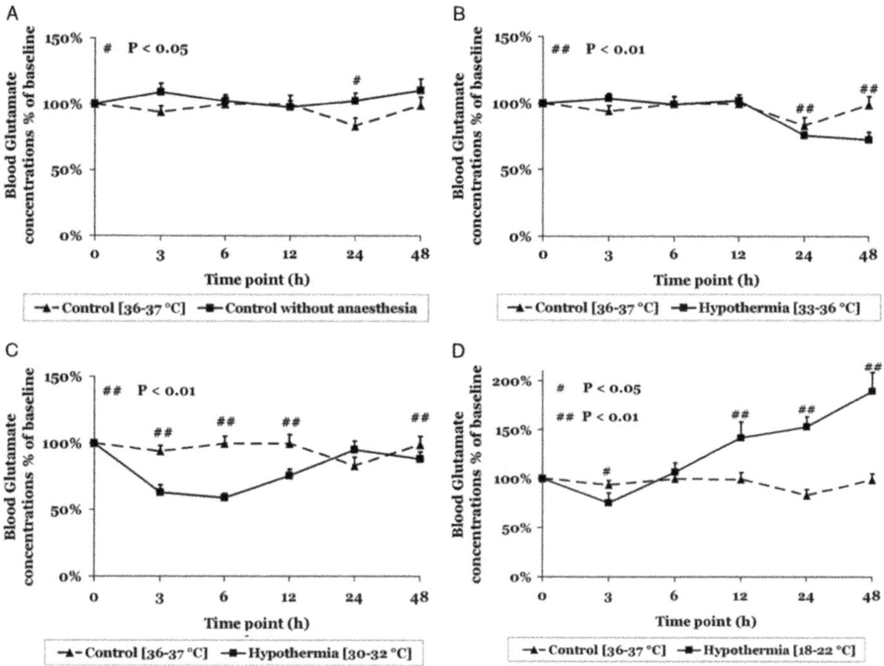

Fig 14.5 Blood glutamate concentration decreases after hypothermia. (**a**) Glutamate is reduced in the control group at 24 h ($p<0.05$). (**b**) Mild hypothermia (33–36 °C) results in decreased glutamate levels at 24 and 48 h after initiation of cooling ($p<0.01$). (**c**) Moderate hypothermia (30–32 °C) results in decreased glutamate levels by 3 h after initiation of cooling ($p <0.01$). The glutamate effect is sustained for 12 h. (**d**) Deep hypothermia results in a sustained increase of blood glutamate from 12 to 48 h ($p<0.01$). Adapted from Boyko et al. (2013)

barbiturates and benzodiazepines), hypothermia, blood glutamate scavengers, and dialysis/hemoperfusion. However, all of these approaches have significant shortcomings such as: (a) markedly affecting normal brain physiology (e.g., ketamine, GABA-receptor antagonists), (b) not being suitable for chronic use or for use outside a health care facility (e.g., blood glutamate scavengers, hemodialysis, hemoperfusion, hypothermia), and (c) exhibiting poor specificity toward the glutamate–glutamine cycle, thereby increasing the risk of adverse effects (e.g., dialysis/hemoperfusion, hypothermia). The lack of effective and specific therapies is largely due to several critical gaps in our understanding of the glutamate–glutamine cycle in health and disease. Areas of markedly deficient knowledge include glutamine transport mechanisms, regulation of key metabolic enzymes, the role of the cycle in cell signaling and differentiation, and the role of peripheral metabolism in the cycle homeostasis. Further studies are required to resolve these issues and to reveal additional therapeutic targets for glutamate–glutamine cycle-related CNS disorders, such as epilepsy, traumatic brain injury, stroke, and hepatic encephalopathy.

Acknowledgements TE and RD are supported by grants from the National Institutes of Health (NIH): NINDS K08 NS058674 and R01 NS070824. SG is supported by a grant from the NIH: T32 GM086287. This work was also made possible by a grant from the National Center for Advancing Translational Sciences (NCATS; UL1 TR000142), a component of the NIH and the NIH roadmap for Medical Research. The authors have no conflicts of interest to declare.

References

Albrecht J, Norenberg MD (2006) Glutamine: a Trojan horse in ammonia neurotoxicity. Hepatology 44:788–794

Almassy RJ, Janson CA, Hamlin R, Xuong NH, Eisenberg D (1986) Novel subunit-subunit interactions in the structure of glutamine synthetase. Nature 323:304–309

Altland PD, Highman B, Sellner RG (1974) Serum enzyme and tissue changes in shaven rabbits exposed to cold. Cryobiology 11:296–304

Alvestad S, Hammer J, Qu H, Haberg A, Ottersen OP, Sonnewald U (2011) Reduced astrocytic contribution to the turnover of glutamate, glutamine, and GABA characterizes the latent phase in the kainate model of temporal lobe epilepsy. J Cereb Blood Flow Metab 31:1675–1686

Anson LC, Chen PE, Wyllie DJ, Colquhoun D, Schoepfer R (1998) Identification of amino acid residues of the NR2A subunit that control glutamate potency in recombinant NR1/NR2A NMDA receptors. J Neurosci 18:581–589

Apland JP, Aroniadou-Anderjaska V, Figueiredo TH, Rossetti F, Miller SL, Braga MF (2014) The limitations of diazepam as a treatment for nerve agent-induced seizures and neuropathology in rats: comparison with UBP302. J Pharmacol Exp Ther 351:359–372

Arakawa H, Kodama H, Matsuoka N, Yamaguchi I (1997) Stress increases plasma enzyme activity in rats: differential effects of adrenergic and cholinergic blockades. J Pharmacol Exp Ther 280:1296–1303

Arriza JL, Fairman WA, Wadiche JI, Murdoch GH, Kavanaugh MP, Amara SG (1994) Functional comparisons of three glutamate transporter subtypes cloned from human motor cortex. J Neurosci 14:5559–5569

Arriza JL, Eliasof S, Kavanaugh MP, Amara SG (1997) Excitatory amino acid transporter 5, a retinal glutamate transporter coupled to a chloride conductance. Proc Natl Acad Sci U S A 94:4155–4160

Azzopardi D, Strohm B, Marlow N, Brocklehurst P, Deierl A, Eddama O, Goodwin J, Halliday HL, Juszczak E, Kapellou O, Levene M, Linsell L, Omar O, Thoresen M, Tusor N, Whitelaw A, Edwards AD, Group, T.S (2014) Effects of hypothermia for perinatal asphyxia on childhood outcomes. N Engl J Med 371:140–149

Bacci A, Sancini G, Verderio C, Armano S, Pravettoni E, Fesce R, Franceschetti S, Matteoli M (2002) Block of glutamate-glutamine cycle between astrocytes and neurons inhibits epileptiform activity in hippocampus. J Neurophysiol 88:2302–2310

Bagetta G, Iannone M, Palma E, Nistico G, Dolly JO (1996) N-methyl-D-aspartate and non-N-methyl-D-aspartate receptors mediate seizures and CA1 hippocampal damage induced by dendrotoxin-K in rats. Neuroscience 71:613–624

Bailey CG, Ryan RM, Thoeng AD, Ng C, King K, Vanslambrouck JM, Auray-Blais C, Vandenberg RJ, Broer S, Rasko JE (2011) Loss-of-function mutations in the glutamate transporter SLC1A1 cause human dicarboxylic aminoaciduria. J Clin Invest 121:446–453

Banfi S, Servadio A, Chung MY, Kwiatkowski TJ Jr, McCall AE, Duvick LA, Shen Y, Roth EJ, Orr HT, Zoghbi HY (1994) Identification and characterization of the gene causing type 1 spinocerebellar ataxia. Nat Genet 7:513–520

Bartnik-Olson BL, Harris NG, Shijo K, Sutton RL (2013) Insights into the metabolic response to traumatic brain injury as revealed by (13)C NMR spectroscopy. Front Neuroenergetics 5:8

Barton ME, Peters SC, Shannon HE (2003) Comparison of the effect of glutamate receptor modulators in the 6 Hz and maximal electroshock seizure models. Epilepsy Res 56:17–26

Bashkatova VG, Sudakov SK, Prast H (2015) Antagonists of Metabotropic Glutamate Receptors Prevent the Development of Audiogenic Seizures. Bull Exp Biol Med 159:1–3
Ben-Ari Y, Khalilov I, Kahle KT, Cherubini E (2012) The GABA excitatory/inhibitory shift in brain maturation and neurological disorders. Neuroscientist 18:467–486
Benjamin AM, Quastel JH (1975) Metabolism of amino acids and ammonia in rat brain cortex slices in vitro: a possible role of ammonia in brain function. J Neurochem 25:197–206
Bennett AE, Hoesch RE, DeWitt LD, Afra P, Ansari SA (2014) Therapeutic hypothermia for status epilepticus: A report, historical perspective, and review. Clin Neurol Neurosurg 126:103–109
Benveniste M, Mayer ML (1991) Kinetic analysis of antagonist action at N-methyl-D-aspartic acid receptors. Two binding sites each for glutamate and glycine. Biophys J 59:560–573
Berger C, Schabitz WR, Georgiadis D, Steiner T, Aschoff A, Schwab S (2002) Effects of hypothermia on excitatory amino acids and metabolism in stroke patients: a microdialysis study. Stroke 33:519–524
Berger C, Schabitz WR, Wolf M, Mueller H, Sommer C, Schwab S (2004) Hypothermia and brain-derived neurotrophic factor reduce glutamate synergistically in acute stroke. Exp Neurol 185:305–312
Bhagavan HN, Coursin DB, Stewart CN (1971) Monosodium glutamate induces convulsive disorders in rats. Nature 232:275–276
Bhutia YD, Ganapathy V (2015) Glutamine transporters in mammalian cells and their functions in physiology and cancer., Biochim Biophys Acta
Bissonnette B, Pellerin L, Ravussin P, Daven VB, Magistretti PJ (1999) Deep hypothermia and rewarming alters glutamate levels and glycogen content in cultured astrocytes. Anesthesiology 91:1763–1769
Blair E, Hook R, Tolley H, Bunce GE (1961) Serum glutamic oxalacetic transaminase content in hypothermia. Science 133:105–106
Blomstrand E, Eliasson J, Karlsson HK, Kohnke R (2006) Branched-chain amino acids activate key enzymes in protein synthesis after physical exercise. J Nutr 136:269S–273S
Bode JG, Peters-Regehr T, Kubitz R, Haussinger D (2000) Expression of glutamine synthetase in macrophages. J Histochem Cytochem 48:415–422
Boucher J, Kroger H, Sik A (2010) Realistic modelling of receptor activation in hippocampal excitatory synapses: analysis of multivesicular release, release location, temperature and synaptic cross-talk. Brain Struct Funct 215:49–65
Bough KJ, Paquet M, Pare JF, Hassel B, Smith Y, Hall RA, Dingledine R (2007) Evidence against enhanced glutamate transport in the anticonvulsant mechanism of the ketogenic diet. Epilepsy Res 74:232–236
Boyko M, Zlotnik A, Gruenbaum BF, Gruenbaum SE, Ohayon S, Kuts R, Melamed I, Regev A, Shapira Y, Teichberg VI (2011) Pyruvate's blood glutamate scavenging activity contributes to the spectrum of its neuroprotective mechanisms in a rat model of stroke. Eur J Neurosci 34:1432–1441
Boyko M, Melamed I, Gruenbaum BF, Gruenbaum SE, Ohayon S, Leibowitz A, Brotfain E, Shapira Y, Zlotnik A (2012) The effect of blood glutamate scavengers oxaloacetate and pyruvate on neurological outcome in a rat model of subarachnoid hemorrhage. Neurotherapeutics 9:649–657
Boyko M, Kuts R, Gruenbaum BF, Melamed I, Gruenbaum SE, Klein M, Shapira Y, Zlotnik A (2013) The role of hypothermia in the regulation of blood glutamate levels in naive rats. J Neurosurg Anesthesiol 25:174–183
Boyko M, Gruenbaum SE, Gruenbaum BF, Shapira Y, Zlotnik A (2014) Brain to blood glutamate scavenging as a novel therapeutic modality: a review. J Neural Transm 121:971–979
Boylan GB, Kharoshankaya L, Wusthoff CJ (2015) Seizures and hypothermia: importance of electroencephalographic monitoring and considerations for treatment. Semin Fetal Neonatal Med 20:103–108
Brandl EJ, Muller DJ, Richter MA (2012) Pharmacogenetics of obsessive-compulsive disorders. Pharmacogenomics 13:71–81

Broer S, Palacin M (2011) The role of amino acid transporters in inherited and acquired diseases. Biochem J 436:193–211

Bruneau EG, McCullumsmith RE, Haroutunian V, Davis KL, Meador-Woodruff JH (2005) Increased expression of glutaminase and glutamine synthetase mRNA in the thalamus in schizophrenia. Schizophr Res 75:27–34

Brusilow SW, Koehler RC, Traystman RJ, Cooper AJ (2010) Astrocyte glutamine synthetase: importance in hyperammonemic syndromes and potential target for therapy. Neurotherapeutics 7:452–470

Burket JA, Mastropaolo J, Rosse RB, Katz EU, Deutsch SI (2010) NMDA NR2B subtype-selective receptor antagonists fail to antagonize electrically-precipitated seizures and elicit popping in mice. Eur Neuropsychopharmacol 20:207–210

Cai YQ, Cai GQ, Liu GX, Cai Q, Shi JH, Shi J, Ma SK, Sun X, Sheng ZJ, Mei ZT, Cui D, Guo L, Wang Z, Fei J (2006) Mice with genetically altered GABA transporter subtype I (GAT1) expression show altered behavioral responses to ethanol. J Neurosci Res 84:255–267

Campos F, Sobrino T, Ramos-Cabrer P, Argibay B, Agulla J, Perez-Mato M, Rodriguez-Gonzalez R, Brea D, Castillo J (2011) Neuroprotection by glutamate oxaloacetate transaminase in ischemic stroke: an experimental study. J Cereb Blood Flow Metab 31:1378–1386

Campos F, Sobrino T, Perez-Mato M, Rodriguez-Osorio X, Leira R, Blanco M, Mirelman D, Castillo J (2013) Glutamate oxaloacetate transaminase: a new key in the dysregulation of glutamate in migraine patients., Cephalalgia

Campos-Sandoval JA, Martin-Rufian M, Cardona C, Lobo C, Penalver A, Marquez J (2015) Glutaminases in brain: multiple isoforms for many purposes. Neurochem Int 88:1–5

Capogna M, Pearce RA (2011) GABA A, slow: causes and consequences. Trends Neurosci 34:101–112

Cardona C, Sanchez-Mejias E, Davila JC, Martin-Rufian M, Campos-Sandoval JA, Vitorica J, Alonso FJ, Mates JM, Segura JA, Norenberg MD, Rama Rao KV, Jayakumar AR, Gutierrez A, Marquez J (2015) Expression of Gls and Gls2 glutaminase isoforms in astrocytes. Glia 63:365–382

Carvalho AS, Torres LB, Persike DS, Fernandes MJ, Amado D, Naffah-Mazzacoratti Mda G, Cavalheiro EA, da Silva AV (2011) Neuroprotective effect of pyruvate and oxaloacetate during pilocarpine induced status epilepticus in rats. Neurochem Int 58:385–390

Cavus I, Kasoff WS, Cassaday MP, Jacob R, Gueorguieva R, Sherwin RS, Krystal JH, Spencer DD, Abi-Saab WM (2005) Extracellular metabolites in the cortex and hippocampus of epileptic patients. Ann Neurol 57:226–235

Ceccarelli C, Grodsky NB, Ariyaratne N, Colman RF, Bahnson BJ (2002) Crystal structure of porcine mitochondrial NADP+-dependent isocitrate dehydrogenase complexed with Mn2+ and isocitrate. Insights into the enzyme mechanism. J Biol Chem 277:43454–43462

Chan K, Busque SM, Sailer M, Stoeger C, Broer S, Daniel H, Rubio-Aliaga I, Wagner CA (2016) Loss of function mutation of the Slc38a3 glutamine transporter reveals its critical role for amino acid metabolism in the liver, brain, and kidney. Pflugers Arch 468:213–227

Chang P, Augustin K, Boddum K, Williams S, Sun M, Terschak JA, Hardege JD, Chen PE, Walker MC, Williams RS (2016) Seizure control by decanoic acid through direct AMPA receptor inhibition. Brain 139:431–443

Chaperon F, Muller W, Auberson YP, Tricklebank MD, Neijt HC (2003) Substitution for PCP, disruption of prepulse inhibition and hyperactivity induced by N-methyl-D-aspartate receptor antagonists: preferential involvement of the NR2B rather than NR2A subunit. Behav Pharmacol 14:477–487

Chapman AG, Graham JL, Patel S, Meldrum BS (1991) Anticonvulsant activity of two orally active competitive N-methyl-D-aspartate antagonists, CGP 37849 and CGP 39551, against sound-induced seizures in DBA/2 mice and photically induced myoclonus in Papio papio. Epilepsia 32:578–587

Chapman AG, Yip PK, Yap JS, Quinn LP, Tang E, Harris JR, Meldrum BS (1999) Anticonvulsant actions of LY 367385 ((+)-2-methyl-4-carboxyphenylglycine) and AIDA ((RS)-1-aminoindan-1,5-dicarboxylic acid). Eur J Pharmacol 368:17–24

Chapman AG, Nanan K, Williams M, Meldrum BS (2000) Anticonvulsant activity of two metabotropic glutamate group I antagonists selective for the mGlu5 receptor: 2-methyl-6-(phenylethynyl)-pyridine (MPEP), and (E)-6-methyl-2-styryl-pyridine (SIB 1893). Neuropharmacology 39:1567–1574

Chapman AG, Talebi A, Yip PK, Meldrum BS (2001) Anticonvulsant activity of a mGlu(4alpha) receptor selective agonist, (1S,3R,4S)-1-aminocyclopentane-1,2,4-tricarboxylic acid. Eur J Pharmacol 424:107–113

Chen K, Li HZ, Ye N, Zhang J, Wang JJ (2005) Role of GABAB receptors in GABA and baclofen-induced inhibition of adult rat cerebellar interpositus nucleus neurons in vitro. Brain Res Bull 67:310–318

Cheng T, Sudderth J, Yang C, Mullen AR, Jin ES, Mates JM, DeBerardinis RJ (2011) Pyruvate carboxylase is required for glutamine-independent growth of tumor cells. Proc Natl Acad Sci U S A 108:8674–8679

Chiu CS, Brickley S, Jensen K, Southwell A, McKinney S, Cull-Candy S, Mody I, Lester HA (2005) GABA transporter deficiency causes tremor, ataxia, nervousness, and increased GABA-induced tonic conductance in cerebellum. J Neurosci 25:3234–3245

Choi DW, Hartley DM (1993) Calcium and glutamate-induced cortical neuronal death. In: Waxman SG (ed) Molecular and cellular approaches to the treatment of neurologic disease. Raven, New York

Clements JD, Westbrook GL (1991) Activation kinetics reveal the number of glutamate and glycine binding sites on the N-methyl-D-aspartate receptor. Neuron 7:605–613

Conti F, Melone M (2006) The glutamine commute: lost in the tube? Neurochem Int 48:459–464

Conti F, Minelli A, Melone M (2004) GABA transporters in the mammalian cerebral cortex: localization, development and pathological implications. Brain Res Brain Res Rev 45:196–212

Contractor A, Swanson GT, Sailer A, O'Gorman S, Heinemann SF (2000) Identification of the kainate receptor subunits underlying modulation of excitatory synaptic transmission in the CA3 region of the hippocampus. J Neurosci 20:8269–8278

Cooper AJ, Kuhara T (2014) alpha-Ketoglutaramate: an overlooked metabolite of glutamine and a biomarker for hepatic encephalopathy and inborn errors of the urea cycle. Metab Brain Dis 29:991–1006

Csibi A, Lee G, Yoon SO, Tong H, Ilter D, Elia I, Fendt SM, Roberts TM, Blenis J (2014) The mTORC1/S6K1 pathway regulates glutamine metabolism through the eIF4B-dependent control of c-Myc translation. Curr Biol 24:2274–2280

Dahlin M, Elfving A, Ungerstedt U, Amark P (2005) The ketogenic diet influences the levels of excitatory and inhibitory amino acids in the CSF in children with refractory epilepsy. Epilepsy Res 64:115–125

Dalet A, Bonsacquet J, Gaboyard-Niay S, Calin-Jageman I, Chidavaenzi RL, Venteo S, Desmadryl G, Goldberg JM, Lysakowski A, Chabbert C (2012) Glutamate transporters EAAT4 and EAAT5 are expressed in vestibular hair cells and calyx endings. PLoS One 7, e46261

Danbolt NC (2001) Glutamate uptake. Prog Neurobiol 65:1–105

Danbolt NC, Pines G, Kanner BI (1990) Purification and reconstitution of the sodium- and potassium-coupled glutamate transport glycoprotein from rat brain. Biochemistry 29:6734–6740

Danbolt NC, Storm-Mathisen J, Kanner BI (1992) An [Na^+ + K^+]coupled L-glutamate transporter purified from rat brain is located in glial cell processes. Neuroscience 51:295–310

Danbolt NC, Furness DN, Zhou Y (2016a) Neuronal vs glial glutamate uptake: resolving the conundrum. Neurochem Int 98:29–45

Danbolt NC, Zhou Y, Furness DN, Holmseth S (2016b) Strategies for immunohistochemical protein localization using antibodies: what did we learn from neurotransmitter transporters in glial cells and neurons. Glia (in press)

Danial NN, Hartman AL, Stafstrom CE, Thio LL (2013) How does the ketogenic diet work? Four potential mechanisms. J Child Neurol 28:1027–1033

Danielyan L, Zellmer S, Sickinger S, Tolstonog GV, Salvetter J, Lourhmati A, Reissig DD, Gleiter CH, Gebhardt R, Buniatian GH (2009) Keratinocytes as depository of ammonium-inducible glutamine synthetase: age- and anatomy-dependent distribution in human and rat skin. PLoS One 4, e4416

Das A, Wallace GCT, Holmes C, McDowell ML, Smith JA, Marshall JD, Bonilha L, Edwards JC, Glazier SS, Ray SK, Banik NL (2012) Hippocampal tissue of patients with refractory temporal lobe epilepsy is associated with astrocyte activation, inflammation, and altered expression of channels and receptors. Neuroscience 220:237–246

Davalos A, Shuaib A, Wahlgren NG (2000) Neurotransmitters and pathophysiology of stroke: evidence for the release of glutamate and other transmitters/mediators in animals and humans. J Stroke Cerebrovasc Dis 9:2–8

Davidson DL, Barbeau A (1975) Locomotor activity and seizures induced by picrotoxin in the fastigeal nucleus of the rat. Exp Neurol 49:622–627

de Lanerolle NC, Kim JH, Robbins RJ, Spencer DD (1989) Hippocampal interneuron loss and plasticity in human temporal lobe epilepsy. Brain Res 495:387–395

De Sarro G, Ammendola D, Nava F, De Sarro A (1995) Effects of some excitatory amino acid antagonists on imipenem-induced seizures in DBA/2 mice. Brain Res 671:131–140

De Simone R, Vissicchio F, Mingarelli C, De Nuccio C, Visentin S, Ajmone-Cat MA, Minghetti L (2013) Branched-chain amino acids influence the immune properties of microglial cells and their responsiveness to pro-inflammatory signals. Biochim Biophys Acta 1832:650–659

de Vivo L, Melone M, Bucci G, Rothstein JD, Conti F (2010) Quantitative analysis of EAAT4 promoter activity in neurons and astrocytes of mouse somatic sensory cortex. Neurosci Lett 474:42–45

de Vries B, Mamsa H, Stam AH, Wan J, Bakker SL, Vanmolkot KR, Haan J, Terwindt GM, Boon EM, Howard BD, Frants RR, Baloh RW, Ferrari MD, Jen JC, van den Maagdenberg AM (2009) Episodic ataxia associated with EAAT1 mutation C186S affecting glutamate reuptake. Arch Neurol 66:97–101

DeFelipe J (1993) Neocortical neuronal diversity: chemical heterogeneity revealed by colocalization studies of classic neurotransmitters, neuropeptides, calcium-binding proteins, and cell surface molecules. Cereb Cortex 3:273–289

Dehnes Y, Chaudhry FA, Ullensvang K, Lehre KP, Storm-Mathisen J, Danbolt NC (1998) The glutamate transporter EAAT4 in rat cerebellar Purkinje cells: a glutamate-gated chloride channel concentrated near the synapse in parts of the dendritic membrane facing astroglia. J Neurosci 18:3606–3619

Dennis SC, Lai JC, Clark JB (1977) Comparative studies on glutamate metabolism in synpatic and non-synaptic rat brain mitochondria. Biochem J 164:727–736

Dhaher R, Damisah EC, Wang H, Gruenbaum SE, Ong C, Zaveri HP, Gruenbaum BF, Eid T (2014) 5-aminovaleric acid suppresses the development of severe seizures in the methionine sulfoximine model of mesial temporal lobe epilepsy. Neurobiol Dis 67:18–23

Dhir A, Chavda V (2016) Pre- and post-exposure talampanel (GYKI 53773) against kainic acid seizures in neonatal rats. Pharmacol Rep 68:190–195

Dolgodilina E, Imobersteg S, Laczko E, Welt T, Verrey F, Makrides V (2015) Brain interstitial fluid glutamine homeostasis is controlled by blood-brain barrier SLC7A5/LAT1 amino acid transporter., J Cereb Blood Flow Metab

Dringen R (2000) Metabolism and functions of glutathione in brain. Prog Neurobiol 62:649–671

Dufour F, Nalecz KA, Nalecz MJ, Nehlig A (1999) Modulation of pentylenetetrazol-induced seizure activity by branched-chain amino acids and alpha-ketoisocaproate. Brain Res 815:400–404

During MJ, Spencer DD (1993) Extracellular hippocampal glutamate and spontaneous seizure in the conscious human brain. Lancet 341:1607–1610

During MJ, Ryder KM, Spencer DD (1995) Hippocampal GABA transporter function in temporal-lobe epilepsy. Nature 376:174–177

Ehrlich MP, McCullough JN, Zhang N, Weisz DJ, Juvonen T, Bodian CA, Griepp RB (2002) Effect of hypothermia on cerebral blood flow and metabolism in the pig. Ann Thorac Surg 73:191–197

Eid T, Thomas MJ, Spencer DD, Runden-Pran E, Lai JC, Malthankar GV, Kim JH, Danbolt NC, Ottersen OP, de Lanerolle NC (2004) Loss of glutamine synthetase in the human epileptogenic hippocampus: possible mechanism for raised extracellular glutamate in mesial temporal lobe epilepsy. Lancet 363:28–37

Eid T, Hammer J, Runden-Pran E, Roberg B, Thomas MJ, Osen K, Davanger S, Laake P, Torgner IA, Lee TS, Kim JH, Spencer DD, Ottersen OP, de Lanerolle NC (2007) Increased expression of phosphate-activated glutaminase in hippocampal neurons in human mesial temporal lobe epilepsy. Acta Neuropathol (Berl) 113:137–152

Eid T, Ghosh A, Wang Y, Beckstrom H, Zaveri HP, Lee TS, Lai JC, Malthankar-Phatak GH, de Lanerolle NC (2008a) Recurrent seizures and brain pathology after inhibition of glutamine synthetase in the hippocampus in rats. Brain 131:2061–2070

Eid T, Williamson A, Lee TS, Petroff OA, de Lanerolle NC (2008b) Glutamate and astrocytes--key players in human mesial temporal lobe epilepsy? Epilepsia 49(Suppl 2):42–52

Eid T, Williamson A, Lee TS, Petroff OA, de Lanerolle NC (2008c) Glutamate and astrocytes-key players in human mesial temporal lobe epilepsy? Epilepsia 49(Suppl 2):42–52

Eliasof S, Arriza JL, Leighton BH, Amara SG, Kavanaugh MP (1998) Localization and function of five glutamate transporters cloned from the salamander retina. Vision Res 38:1443–1454

Erdo SL, Wolff JR (1990) gamma-Aminobutyric acid outside the mammalian brain. J Neurochem 54:363–372

Erecinska M, Thoresen M, Silver IA (2003) Effects of hypothermia on energy metabolism in Mammalian central nervous system. J Cereb Blood Flow Metab 23:513–530

Erlander MG, Tillakaratne NJ, Feldblum S, Patel N, Tobin AJ (1991) Two genes encode distinct glutamate decarboxylases. Neuron 7:91–100

Eulenburg V, Gomeza J (2010) Neurotransmitter transporters expressed in glial cells as regulators of synapse function. Brain Res Rev 63:103–112

Evans JE, Frostholm A, Rotter A (1996) Embryonic and postnatal expression of four gamma-aminobutyric acid transporter mRNAs in the mouse brain and leptomeninges. J Comp Neurol 376:431–446

Fagg GE, Olpe HR, Pozza MF, Baud J, Steinmann M, Schmutz M, Portet C, Baumann P, Thedinga K, Bittiger H et al (1990) CGP 37849 and CGP 39551: novel and potent competitive N-methyl-D-aspartate receptor antagonists with oral activity. Br J Pharmacol 99:791–797

Fang Y, Wang X (2015) Ketamine for the treatment of refractory status epilepticus. Seizure 30:14–20

Feng B, Tse HW, Skifter DA, Morley R, Jane DE, Monaghan DT (2004) Structure-activity analysis of a novel NR2C/NR2D-preferring NMDA receptor antagonist: 1-(phenanthrene-2-carbonyl) piperazine-2,3-dicarboxylic acid. Br J Pharmacol 141:508–516

Ferkany JW, Hamilton GS, Patch RJ, Huang Z, Borosky SA, Bednar DL, Jones BE, Zubrowski R, Willetts J, Karbon EW (1993) Pharmacological profile of NPC 17742 [2R,4R,5S-(2-amino-4,5-(1, 2-cyclohexyl)-7-phosphonoheptanoic acid)], a potent, selective and competitive N-methyl-D-aspartate receptor antagonist. J Pharmacol Exp Ther 264:256–264

Folbergrova J, Passonneau JV, Lowry OH, Schulz DW (1969) Glycogen, ammonia and related metabolities in the brain during seizures evoked by methionine sulphoximine. J Neurochem 16:191–203

Fonnum F (1984) Glutamate: a neurotransmitter in mammalian brain. J Neurochem 42:1–11

Friedrich JO, Wald R, Bagshaw SM, Burns KE, Adhikari NK (2012) Hemofiltration compared to hemodialysis for acute kidney injury: systematic review and meta-analysis. Crit Care 16:R146

Furness DN, Dehnes Y, Akhtar AQ, Rossi DJ, Hamann M, Grutle NJ, Gundersen V, Holmseth S, Lehre KP, Ullensvang K, Wojewodzic M, Zhou Y, Attwell D, Danbolt NC (2008) A quantitative assessment of glutamate uptake into hippocampal synaptic terminals and astrocytes: new insights into a neuronal role for excitatory amino acid transporter 2 (EAAT2). Neuroscience 157:80–94

Gameiro A, Braams S, Rauen T, Grewer C (2011) The discovery of slowness: low-capacity transport and slow anion channel gating by the glutamate transporter EAAT5. Biophys J 100:2623–2632

Gaoni Y, Chapman AG, Parvez N, Pook PC, Jane DE, Watkins JC (1994) Synthesis, NMDA receptor antagonist activity, and anticonvulsant action of 1-aminocyclobutanecarboxylic acid derivatives. J Med Chem 37:4288–4296

Geiger JR, Melcher T, Koh DS, Sakmann B, Seeburg PH, Jonas P, Monyer H (1995) Relative abundance of subunit mRNAs determines gating and Ca2+ permeability of AMPA receptors in principal neurons and interneurons in rat CNS. Neuron 15:193–204
Gereau RWT, Conn PJ (1995) Multiple presynaptic metabotropic glutamate receptors modulate excitatory and inhibitory synaptic transmission in hippocampal area CA1. J Neurosci 15:6879–6889
Gether U, Andersen PH, Larsson OM, Schousboe A (2006) Neurotransmitter transporters: molecular function of important drug targets. Trends Pharmacol Sci 27:375–383
Gill R, Alanine A, Bourson A, Buttelmann B, Fischer G, Heitz MP, Kew JN, Levet-Trafit B, Lorez HP, Malherbe P, Miss MT, Mutel V, Pinard E, Roever S, Schmitt M, Trube G, Wybrecht R, Wyler R, Kemp JA (2002) Pharmacological characterization of Ro 63-1908 (1-[2-(4-hydroxy-phenoxy)-ethyl]-4-(4-methyl-benzyl)-piperidin-4-ol), a novel subtype-selective N-methyl-D-aspartate antagonist. J Pharmacol Exp Ther 302:940–948
Gloor P (1991) Mesial temporal sclerosis: Historical background and an overview from a modern perspective. In: Luders H (ed) Epilepsy Surgery. Raven, New York, pp 689–703
Godino Mdel C, Romera VG, Sanchez-Tomero JA, Pacheco J, Canals S, Lerma J, Vivancos J, Moro MA, Torres M, Lizasoain I, Sanchez-Prieto J (2013) Amelioration of ischemic brain damage by peritoneal dialysis. J Clin Invest 123:4359–4363
Goodman LS, Swinyard EA, Toman JE (1946) Effects of 1 (-) glutamic acid and other agents on experimental seizures. Arch Neurol Psychiatry 56:20–29
Gottlieb M, Wang Y, Teichberg VI (2003) Blood-mediated scavenging of cerebrospinal fluid glutamate. J Neurochem 87:119–126
Gras G, Porcheray F, Samah B, Leone C (2006) The glutamate-glutamine cycle as an inducible, protective face of macrophage activation., J Leukoc Biol
Grewal S, Defamie N, Zhang X, De Gois S, Shawki A, Mackenzie B, Chen C, Varoqui H, Erickson JD (2009) SNAT2 amino acid transporter is regulated by amino acids of the SLC6 gamma-aminobutyric acid transporter subfamily in neocortical neurons and may play no role in delivering glutamine for glutamatergic transmission. J Biol Chem 284:11224–11236
Grewer C, Rauen T (2005) Electrogenic glutamate transporters in the CNS: molecular mechanism, pre-steady-state kinetics, and their impact on synaptic signaling. J Membr Biol 203:1–20
Grewer C, Gameiro A, Rauen T (2014) SLC1 glutamate transporters. Pflugers Arch 466:3–24
Gruenbaum SE, Wang H, Zaveri HP, Tang AB, Lee TS, Eid T, Dhaher R (2015) Inhibition of glutamine synthetase in the central nucleus of the amygdala induces anhedonic behavior and recurrent seizures in a rat model of mesial temporal lobe epilepsy. Epilepsy Behav 51:96–103
Guastella J, Nelson N, Nelson H, Czyzyk L, Keynan S, Miedel MC, Davidson N, Lester HA, Kanner BI (1990) Cloning and expression of a rat brain GABA transporter. Science 249:1303–1306
Guilliams K, Rosen M, Buttram S, Zempel J, Pineda J, Miller B, Shoykhet M (2013) Hypothermia for pediatric refractory status epilepticus. Epilepsia 54:1586–1594
Gunnersen D, Haley B (1992) Detection of glutamine synthetase in the cerebrospinal fluid of Alzheimer diseased patients: a potential diagnostic biochemical marker. Proc Natl Acad Sci U S A 89:11949–11953
Guo C, Pirozzi CJ, Lopez GY, Yan H (2011) Isocitrate dehydrogenase mutations in gliomas: mechanisms, biomarkers and therapeutic target. Curr Opin Neurol 24:648–652
Haberg A, Qu H, Sonnewald U (2006) Glutamate and GABA metabolism in transient and permanent middle cerebral artery occlusion in rat: importance of astrocytes for neuronal survival. Neurochem Int 48:531–540
Haberle J, Gorg B, Rutsch F, Schmidt E, Toutain A, Benoist JF, Gelot A, Suc AL, Hohne W, Schliess F, Haussinger D, Koch HG (2005) Congenital glutamine deficiency with glutamine synthetase mutations. N Engl J Med 353:1926–1933
Haberle J, Gorg B, Toutain A, Rutsch F, Benoist JF, Gelot A, Suc AL, Koch HG, Schliess F, Haussinger D (2006) Inborn error of amino acid synthesis: human glutamine synthetase deficiency. J Inherit Metab Dis 29:352–358
Haberle J, Shahbeck N, Ibrahim K, Hoffmann GF, Ben-Omran T (2011) Natural course of glutamine synthetase deficiency in a 3 year old patient. Mol Genet Metab 103:89–91

Hagglund MG, Sreedharan S, Nilsson VC, Shaik JH, Almkvist IM, Backlin S, Wrange O, Fredriksson R (2011) Identification of SLC38A7 (SNAT7) protein as a glutamine transporter expressed in neurons. J Biol Chem 286:20500–20511

Hagglund MG, Hellsten SV, Bagchi S, Philippot G, Lofqvist E, Nilsson VC, Almkvist I, Karlsson E, Sreedharan S, Tafreshiha A, Fredriksson R (2015) Transport of L-glutamine, L-alanine, L-arginine and L-histidine by the neuron-specific Slc38a8 (SNAT8) in CNS. J Mol Biol 427:1495–1512

Hakuba N, Koga K, Gyo K, Usami SI, Tanaka K (2000) Exacerbation of noise-induced hearing loss in mice lacking the glutamate transporter GLAST. J Neurosci 20:8750–8753

Hammer J, Alvestad S, Osen KK, Skare O, Sonnewald U, Ottersen OP (2008) Expression of glutamine synthetase and glutamate dehydrogenase in the latent phase and chronic phase in the kainate model of temporal lobe epilepsy. Glia 56:856–868

Hara H, Yamada N, Kodama M, Matsumoto Y, Wake Y, Kuroda S (2006) Effect of YM872, a selective and highly water-soluble AMPA receptor antagonist, in the rat kindling and rekindling model of epilepsy. Eur J Pharmacol 531:59–65

Hassel B (2001) Pyruvate carboxylation in neurons. J Neurosci Res 66:755–762

Haugeto Ø, Ullensvang K, Levy LM, Chaudhry FA, Honore T, Nielsen M, Lehre KP, Danbolt NC (1996) Brain glutamate transporter proteins form homomultimers. J Biol Chem 271:27715–27722

Hauser WA, Annegers JF, Kurland LT (1993) Incidence of epilepsy and unprovoked seizures in Rochester, Minnesota: 1935-1984. Epilepsia 34:453–468

Haussinger D (1990) Nitrogen metabolism in liver: structural and functional organization and physiological relevance. Biochem J 267:281–290

Haussinger D, Laubenberger J, Vom Dahl S, Ernst T, Bayer S, Langer M, Gerok W, Hennig J (1994) Proton magnetic resonance spectroscopy studies on human brain myo-inositol in hypo-osmolarity and hepatic encephalopathy. Gastroenterology 107:1475–1480

Haussinger D, Kircheis G, Fischer R, Schliess F, vom Dahl S (2000) Hepatic encephalopathy in chronic liver disease: a clinical manifestation of astrocyte swelling and low-grade cerebral edema? J Hepatol 32:1035–1038

Hayashi T (1954) Effects of sodium glutamate on the nervous system. Keiko J Med 3:183–192

He Y, Hakvoort TB, Vermeulen JL, Lamers WH, Van Roon MA (2007) Glutamine synthetase is essential in early mouse embryogenesis. Dev Dyn 236:1865–1875

He Y, Hakvoort TB, Vermeulen JL, Labruyere WT, De Waart DR, Van Der Hel WS, Ruijter JM, Uylings HB, Lamers WH (2010) Glutamine synthetase deficiency in murine astrocytes results in neonatal death. Glia 58:741–754

Hinzman JM, Wilson JA, Mazzeo AT, Bullock MR, Hartings JA (2016) Excitotoxicity and Metabolic Crisis Are Associated with Spreading Depolarizations in Severe Traumatic Brain Injury Patients., J Neurotrauma

Hirai H, Kirsch J, Laube B, Betz H, Kuhse J (1996) The glycine binding site of the N-methyl-D-aspartate receptor subunit NR1: identification of novel determinants of co-agonist potentiation in the extracellular M3-M4 loop region. Proc Natl Acad Sci U S A 93:6031–6036

Holmseth S, Scott HA, Real K, Lehre KP, Leergaard TB, Bjaalie JG, Danbolt NC (2009) The concentrations and distributions of three C-terminal variants of the GLT1 (EAAT2; slc1a2) glutamate transporter protein in rat brain tissue suggest differential regulation. Neuroscience 162:1055–1071

Holmseth S, Dehnes Y, Huang YH, Follin-Arbelet VV, Grutle NJ, Mylonakou MN, Plachez C, Zhou Y, Furness DN, Bergles DE, Lehre KP, Danbolt NC (2012) The density of EAAC1 (EAAT3) glutamate transporters expressed by neurons in the mammalian CNS. J Neurosci 32:6000–6013

Hosoya K, Sugawara M, Asaba H, Terasaki T (1999) Blood brain barrier produces significant efflux of L-aspartic acid but not D-aspartic acid: in vivo evidence using the brain efflux index method. J Neurochem 73:1206–1211

Hrncic D, Rasic-Markovic A, Susic V, Djuric D, Stanojlovic O (2009) Influence of NR2B-selective NMDA antagonist on lindane-induced seizures in rats. Pharmacology 84:234–239

Hu SL, Wang D, Jiang H, Lei QF, Zhu XH, Cheng JZ (2014) Therapeutic effectiveness of sustained low-efficiency hemodialysis plus hemoperfusion and continuous hemofiltration plus hemoperfusion for acute severe organophosphate poisoning. Artif Organs 38:121–124

Huang YH, Dykes-Hoberg M, Tanaka K, Rothstein JD, Bergles DE (2004) Climbing fiber activation of EAAT4 transporters and kainate receptors in cerebellar Purkinje cells. J Neurosci 24:103–111
Hudson RC, Daniel RM (1993) L-glutamate dehydrogenases: distribution, properties and mechanism. Comp Biochem Physiol B 106:767–792
Hutson SM, Berkich D, Drown P, Xu B, Aschner M, LaNoue KF (1998) Role of branched-chain aminotransferase isoenzymes and gabapentin in neurotransmitter metabolism. J Neurochem 71:863–874
Ikonomidou C, Turski L (2002) Why did NMDA receptor antagonists fail clinical trials for stroke and traumatic brain injury? Lancet Neurol 1:383–386
Imataka G, Wake K, Yamanouchi H, Ono K, Arisaka O (2014) Brain hypothermia therapy for status epilepticus in childhood. Eur Rev Med Pharmacol Sci 18:1883–1888
Isa T, Iino M, Ozawa S (1996) Spermine blocks synaptic transmission mediated by Ca(2+)-permeable AMPA receptors. Neuroreport 7:689–692
Jacobs SE, Berg M, Hunt R, Tarnow-Mordi WO, Inder TE, Davis PG (2013) Cooling for newborns with hypoxic ischaemic encephalopathy. Cochrane Database Syst Rev 1, CD003311
Janicki RH, Goldstein L (1969) Glutamine synthetase and renal ammonia metabolism. Am J Physiol 216:1107–1110
Jen JC, Wan J, Palos TP, Howard BD, Baloh RW (2005) Mutation in the glutamate transporter EAAT1 causes episodic ataxia, hemiplegia, and seizures. Neurology 65:529–534
Jenstad M, Quazi AZ, Zilberter M, Haglerod C, Berghuis P, Saddique N, Goiny M, Buntup D, Davanger S, Hang FM-S, Barnes CA, McNaughton BL, Ottersen OP, Storm-Mathisen J, Harkany T, Chaudhry FA (2009) System A transporter SAT2 mediates replenishment of dendritic glutamate pools controlling retrograde signaling by glutamate. Cereb Cortex 19:1092–1106
Jiang W, Wolfe K, Xiao L, Zhang ZJ, Huang YG, Zhang X (2004) Ionotropic glutamate receptor antagonists inhibit the proliferation of granule cell precursors in the adult brain after seizures induced by pentylenetrazol. Brain Res 1020:154–160
Jordan J (2014) Analysis of the role of dysfunctional astrocytes in temporal lobe epilepsy. Bochum University
Juge N, Gray JA, Omote H, Miyaji T, Inoue T, Hara C, Uneyama H, Edwards RH, Nicoll RA, Moriyama Y (2010) Metabolic control of vesicular glutamate transport and release. Neuron 68:99–112
Kanamori K, Ross BD (2011) Chronic electrographic seizure reduces glutamine and elevates glutamate in the extracellular fluid of rat brain. Brain Res 1371:180–191
Kanamori K, Ross BD, Kondrat RW (1998) Rate of glutamate synthesis from leucine in rat brain measured in vivo by 15N NMR. J Neurochem 70:1304–1315
Kanner BI (2006) Structure and function of sodium-coupled GABA and glutamate transporters. J Membr Biol 213:89–100
Karlsson RM, Tanaka K, Saksida LM, Bussey TJ, Heilig M, Holmes A (2009) Assessment of glutamate transporter GLAST (EAAT1)-deficient mice for phenotypes relevant to the negative and executive/cognitive symptoms of schizophrenia. Neuropsychopharmacology 34:1578–1589
Karlsson RM, Adermark L, Molander A, Perreau-Lenz S, Singley E, Solomon M, Holmes A, Tanaka K, Lovinger DM, Spanagel R, Heilig M (2012) Reduced alcohol intake and reward associated with impaired endocannabinoid signaling in mice with a deletion of the glutamate transporter GLAST. Neuropharmacology 63:181–189
Karmen A, Wroblewski F, Ladue JS (1955) Transaminase activity in human blood. J Clin Invest 34:126–131
Katsumori H, Minabe Y, Osawa M, Ashby CR Jr (1998) Acute effects of various GABA receptor agonists and glutamate antagonists on focal hippocampal seizures in freely moving rats elicited by low-frequency stimulation. Synapse 28:103–109
Kentner R, Rollwagen FM, Prueckner S, Behringer W, Wu X, Stezoski J, Safar P, Tisherman SA (2002) Effects of mild hypothermia on survival and serum cytokines in uncontrolled hemorrhagic shock in rats. Shock 17:521–526
Kew JN, Kemp JA (2005) Ionotropic and metabotropic glutamate receptor structure and pharmacology. Psychopharmacology (Berl) 179:4–29

Kew JN, Koester A, Moreau JL, Jenck F, Ouagazzal AM, Mutel V, Richards JG, Trube G, Fischer G, Montkowski A, Hundt W, Reinscheid RK, Pauly-Evers M, Kemp JA, Bluethmann H (2000) Functional consequences of reduction in NMDA receptor glycine affinity in mice carrying targeted point mutations in the glycine binding site. J Neurosci 20:4037–4049

Kingston AE, Ornstein PL, Wright RA, Johnson BG, Mayne NG, Burnett JP, Belagaje R, Wu S, Schoepp DD (1998) LY341495 is a nanomolar potent and selective antagonist of group II metabotropic glutamate receptors. Neuropharmacology 37:1–12

Kleinrok Z, Turski WA, Czuczwar SJ (1995) Excitatory amino acid antagonists and the anticonvulsive activity of conventional antiepileptic drugs. Pol J Pharmacol 47:247–252

Kobayashi M, Buckmaster PS (2003) Reduced inhibition of dentate granule cells in a model of temporal lobe epilepsy. J Neurosci 23:2440–2452

Kohler M, Burnashev N, Sakmann B, Seeburg PH (1993) Determinants of Ca2+ permeability in both TM1 and TM2 of high affinity kainate receptor channels: diversity by RNA editing. Neuron 10:491–500

Krajewski WW, Collins R, Holmberg-Schiavone L, Jones TA, Karlberg T, Mowbray SL (2008) Crystal structures of mammalian glutamine synthetases illustrate substrate-induced conformational changes and provide opportunities for drug and herbicide design. J Mol Biol 375:217–228

Krebs HA (1935) Metabolism of amino-acids: The synthesis of glutamine from glutamic acid and ammonia, and the enzymic hydrolysis of glutamine in animal tissues. Biochem J 29:1951–1969

Kumada Y, Benson DR, Hillemann D, Hosted TJ, Rochefort DA, Thompson CJ, Wohlleben W, Tateno Y (1993) Evolution of the glutamine synthetase gene, one of the oldest existing and functioning genes. Proc Natl Acad Sci U S A 90:3009–3013

Kuo CF, Darnell JE Jr (1989) Mouse glutamine synthetase is encoded by a single gene that can be expressed in a localized fashion. J Mol Biol 208:45–56

Kuryatov A, Laube B, Betz H, Kuhse J (1994) Mutational analysis of the glycine-binding site of the NMDA receptor: structural similarity with bacterial amino acid-binding proteins. Neuron 12:1291–1300

La Spada AR, Paulson HL, Fischbeck KH (1994) Trinucleotide repeat expansion in neurological disease. Ann Neurol 36:814–822

Laake JH, Takumi Y, Eidet J, Torgner IA, Roberg B, Kvamme E, Ottersen OP (1999) Postembedding immunogold labelling reveals subcellular localization and pathway-specific enrichment of phosphate activated glutaminase in rat cerebellum. Neurosci 88:1137–1151

Lampe JW, Becker LB (2011) State of the art in therapeutic hypothermia. Annu Rev Med 62:79–93

Laube B, Hirai H, Sturgess M, Betz H, Kuhse J (1997) Molecular determinants of agonist discrimination by NMDA receptor subunits: analysis of the glutamate binding site on the NR2B subunit. Neuron 18:493–503

Lee JM, Zipfel GJ, Choi DW (1999) The changing landscape of ischaemic brain injury mechanisms. Nature 399:A7–A14

Lee TS, Mane S, Eid T, Zhao H, Lin A, Guan Z, Kim JH, Schweitzer J, King-Stevens D, Weber P, Spencer SS, Spencer DD, de Lanerolle NC (2007) Gene Expression in Temporal Lobe Epilepsy is Consistent with Increased Release of Glutamate by Astrocytes. Mol Med 13:1–13

Lehre KP, Danbolt NC (1998) The number of glutamate transporter subtype molecules at glutamatergic synapses: chemical and stereological quantification in young adult rat brain. J Neurosci 18:8751–8757

Lehre KP, Levy LM, Ottersen OP, Storm-Mathisen J, Danbolt NC (1995) Differential expression of two glial glutamate transporters in the rat brain: quantitative and immunocytochemical observations. J Neurosci 15:1835–1853

Lehre KP, Davanger S, Danbolt NC (1997) Localization of the glutamate transporter protein GLAST in rat retina. Brain Res 744:129–137

Lehre AC, Rowley NM, Zhou Y, Holmseth S, Guo C, Holen T, Hua R, Laake P, Olofsson AM, Poblete-Naredo I, Rusakov DA, Madsen KK, Clausen RP, Schousboe A, White HS, Danbolt NC (2011) Deletion of the betaine-GABA transporter (BGT1; slc6a12) gene does not affect seizure thresholds of adult mice. Epilepsy Res 95:70–81

Leibowitz A, Boyko M, Shapira Y, Zlotnik A (2012) Blood glutamate scavenging: insight into neuroprotection. Int J Mol Sci 13:10041–10066

Leith E, LaNoue KF, Berkich A, Xu B, Ratz M, Taylor C, Hutson SM (2001) Nitrogen shuttling between neurons and glial cells during glutamate synthesis. J Neurochem 76:1712–1723

Levy LM, Lehre KP, Rolstad B, Danbolt NC (1993) A monoclonal antibody raised against an [Na++K+]coupled l-glutamate transporter purified from rat brain confirms glial cell localization. FEBS Letters 317:79–84

Levy RG, Cooper PN, Giri P (2012) Ketogenic diet and other dietary treatments for epilepsy. Cochrane Database Syst Rev 3, CD001903

Li D, Herault K, Silm K, Evrard A, Wojcik S, Oheim M, Herzog E, Ropert N (2013) Lack of evidence for vesicular glutamate transporter expression in mouse astrocytes. J Neurosci 33:4434–4455

Liang SL, Carlson GC, Coulter DA (2006) Dynamic regulation of synaptic GABA release by the glutamate-glutamine cycle in hippocampal area CA1. J Neurosci 26:8537–8548

Liaw SH, Eisenberg D (1994) Structural model for the reaction mechanism of glutamine synthetase, based on five crystal structures of enzyme-substrate complexes. Biochemistry 33:675–681

Liu QR, Lopez-Corcuera B, Mandiyan S, Nelson H, Nelson N (1993) Molecular characterization of four pharmacologically distinct gamma-aminobutyric acid transporters in mouse brain [corrected]. J Biol Chem 268:2106–2112

Liu GX, Cai GQ, Cai YQ, Sheng ZJ, Jiang J, Mei Z, Wang ZG, Guo L, Fei J (2007a) Reduced anxiety and depression-like behaviors in mice lacking GABA transporter subtype 1. Neuropsychopharmacology 32:1531–1539

Liu GX, Liu S, Cai GQ, Sheng ZJ, Cai YQ, Jiang J, Sun X, Ma SK, Wang L, Wang ZG, Fei J (2007b) Reduced aggression in mice lacking GABA transporter subtype 1. J Neurosci Res 85:649–655

Low E, Boylan GB, Mathieson SR, Murray DM, Korotchikova I, Stevenson NJ, Livingstone V, Rennie JM (2012) Cooling and seizure burden in term neonates: an observational study. Arch Dis Child Fetal Neonatal Ed 97:F267–F272

Lukomskaya NY, Rukoyatkina NI, Gorbunova LV, Gmiro VE, Magazanik LG (2004) Studies of the roles of NMDA and AMPA glutamate receptors in the mechanism of corasole convulsions in mice. Neurosci Behav Physiol 34:783–789

Lynch CJ, Adams SH (2014) Branched-chain amino acids in metabolic signalling and insulin resistance. Nat Rev Endocrinol 10:723–736

Mackenzie B, Erickson JD (2004) Sodium-coupled neutral amino acid (System N/A) transporters of the SLC38 gene family. Pflugers Arch 447:784–795

Madl JE, Duncan CG, Stanhill JE, Tai PY, Spraker TR, Gulland FM (2014) Oxidative stress and redistribution of glutamine synthetase in California sea lions (Zalophus californianus) with domoic acid toxicosis. J Comp Pathol 150:306–315

Mahanty NK, Sah P (1998) Calcium-permeable AMPA receptors mediate long-term potentiation in interneurons in the amygdala. Nature 394:683–687

Malthankar-Phatak GH, de Lanerolle N, Eid T, Spencer DD, Behar KL, Spencer SS, Kim JH, Lai JC (2006) Differential glutamate dehydrogenase (GDH) activity profile in patients with temporal lobe epilepsy. Epilepsia 47:1292–1299

Manahan-Vaughan D (1997) Group 1 and 2 metabotropic glutamate receptors play differential roles in hippocampal long-term depression and long-term potentiation in freely moving rats. J Neurosci 17:3303–3311

Marc Rhoads J, Wu G (2009) Glutamine, arginine, and leucine signaling in the intestine. Amino Acids 37:111–122

Mares P, Mikulecka A (2009) Different effects of two N-methyl-D-aspartate receptor antagonists on seizures, spontaneous behavior, and motor performance in immature rats. Epilepsy Behav 14:32–39

Mares P, Mikulecka A, Ticha K, Lojkova-Janeckova D, Kubova H (2010) Metabotropic glutamate receptors as a target for anticonvulsant and anxiolytic action in immature rats. Epilepsia 51(Suppl 3):24–26

Margerison JH, Corsellis JAN (1966) Epilepsy and the temporal lobes. Brain 89:499–530
Marliss EB, Aoki TT, Pozefsky T, Most AS, Cahill GF Jr (1971) Muscle and splanchnic glutmine and glutamate metabolism in postabsorptive andstarved man. J Clin Invest 50:814–817
Martinez-Hernandez A, Bell KP, Norenberg MD (1977) Glutamine synthetase: glial localization in brain. Science 195:1356–1358
Massie A, Cnops L, Smolders I, McCullumsmith R, Kooijman R, Kwak S, Arckens L, Michotte Y (2008) High-affinity Na+/K+-dependent glutamate transporter EAAT4 is expressed throughout the rat fore- and midbrain. J Comp Neurol 511:155–172
Mathern GW, Babb TL, Pretorius JK, Leite JP (1995) Reactive synaptogenesis and neuron densities for neuropeptide Y, somatostatin, and glutamate decarboxylase immunoreactivity in the epileptogenic human fascia dentata. J Neurosci 15:3990–4004
Matsuda K, Kamiya Y, Matsuda S, Yuzaki M (2002) Cloning and characterization of a novel NMDA receptor subunit NR3B: a dominant subunit that reduces calcium permeability. Brain Res Mol Brain Res 100:43–52
McAllister KH (1992) N-methyl-D-aspartate receptor antagonists and channel blockers have different effects upon a spinal seizure model in mice. Eur J Pharmacol 211:105–108
McCulloch J (1992) Excitatory amino acid antagonists and their potential for the treatment of ischaemic brain damage in man. Br J Clin Pharmacol 34:106–114
McFarlane C, Warner DS, Dexter F, Todd MM (1994) Glutamatergic antagonism: effects on lidocaine-induced seizures in the rat. Anesth Analg 79:701–705
McKenna MC (2007) The glutamate-glutamine cycle is not stoichiometric: fates of glutamate in brain. J Neurosci Res 85:3347–3358
Meldrum BS, Nilsson B (1976) Cerebral blood flow and metabolic rate early and late in prolonged epileptic seizures induced in rats by bicuculline. Brain 99:523–542
Mim C, Balani P, Rauen T, Grewer C (2005) The glutamate transporter subtypes EAAT4 and EAATs 1-3 transport glutamate with dramatically different kinetics and voltage dependence but share a common uptake mechanism. J Gen Physiol 126:571–589
Motamedi GK, Salazar P, Smith EL, Lesser RP, Webber WR, Ortinski PI, Vicini S, Rogawski MA (2006) Termination of epileptiform activity by cooling in rat hippocampal slice epilepsy models. Epilepsy Res 70:200–210
Motamedi GK, Lesser RP, Vicini S (2013) Therapeutic brain hypothermia, its mechanisms of action, and its prospects as a treatment for epilepsy. Epilepsia 54:959–970
Muir KW (2006) Glutamate-based therapeutic approaches: clinical trials with NMDA antagonists. Curr Opin Pharmacol 6:53–60
Nadler J, Cuthbertson G (1980) Kainic acid neurotoxicity toward the hippocampal formation: dependence on specific excitatory pathways. Brain Res 195:47–56
Nadler JV, Perry BW, Cotman CW (1978) Intraventricular kainic acid preferentially destroys hippocampal pyramidal cells. Nature 271:676–677
Nagy D, Marosi M, Kis Z, Farkas T, Rakos G, Vecsei L, Teichberg VI, Toldi J (2009) Oxaloacetate decreases the infarct size and attenuates the reduction in evoked responses after photothrombotic focal ischemia in the rat cortex. Cell Mol Neurobiol 29:827–835
Nakanishi T, Sugawara M, Huang W, Martindale RG, Leibach FH, Ganapathy ME, Prasad PD, Ganapathy V (2001) Structure, function, and tissue expression pattern of human SN2, a subtype of the amino acid transport system N. Biochem Biophys Res Commun 281:1343–1348
Naspolini AP, Cocco AR, Villa Martignoni F, Oliveira MS, Furian AF, Rambo LM, Rubin MA, Barron S, Mello CF (2012) Traxoprodil decreases pentylenetetrazol-induced seizures. Epilepsy Res 100:12–19
Nedergaard M, Verkhratsky A (2012) Artifact versus reality--how astrocytes contribute to synaptic events. Glia 60:1013–1023
Nikkuni O, Takayasu Y, Iino M, Tanaka K, Ozawa S (2007) Facilitated activation of metabotropic glutamate receptors in cerebellar Purkinje cells in glutamate transporter EAAT4-deficient mice. Neurosci Res 59:296–303

Niquet J, Gezalian M, Baldwin R, Wasterlain CG (2015) Neuroprotective effects of deep hypothermia in refractory status epilepticus. Ann Clin Transl Neurol 2:1105–1115

Nishi M, Hinds H, Lu HP, Kawata M, Hayashi Y (2001) Motoneuron-specific expression of NR3B, a novel NMDA-type glutamate receptor subunit that works in a dominant-negative manner. J Neurosci 21:RC185

O'Kane RL, Martinez-Lopez I, DeJoseph MR, Vina JR, Hawkins RA (1999) Na(+)-dependent glutamate transporters (EAAT1, EAAT2, and EAAT3) of the blood-brain barrier. A mechanism for glutamate removal. J Biol Chem 274:31891–31895

Olney JW, Sharpe LG, Feigin RD (1972) Glutamate-induced brain damage in infant primates. J Neuropathol Exp Neurol 31:464–488

Olney JW, Collins RC, Sloviter RS (1986) Excitotoxic mechanims of epileptic brain damage. In: Delgado-Escueta AV, Ward AA Jr, Woodbury DM, Porter RJ (eds) Advances in neurology. Raven, New York, pp 857–877

Orbach SA, Bonifacio SL, Kuzniewicz MW, Glass HC (2014) Lower incidence of seizure among neonates treated with therapeutic hypothermia. J Child Neurol 29:1502–1507

Otis TS, Kavanaugh MP (2000) Isolation of current components and partial reaction cycles in the glial glutamate transporter EAAT2. J Neurosci 20:2749–2757

Palmieri EM, Spera I, Menga A, Infantino V, Iacobazzi V, Castegna A (2014) Glutamine synthetase desensitizes differentiated adipocytes to proinflammatory stimuli by raising intracellular glutamine levels. FEBS Lett 588:4807–4814

Patel AB, de Graaf RA, Rothman DL, Behar KL, Mason GF (2010) Evaluation of cerebral acetate transport and metabolic rates in the rat brain in vivo using 1H-[13C]-NMR. J Cereb Blood Flow Metab 30:1200–1213

Peghini P, Janzen J, Stoffel W (1997) Glutamate transporter EAAC-1-deficient mice develop dicarboxylic aminoaciduria and behavioral abnormalities but no neurodegeneration. Embo J 16:3822–3832

Perl TM, Bedard L, Kosatsky T, Hockin JC, Todd EC, Remis RS (1990) An outbreak of toxic encephalopathy caused by eating mussels contaminated with domoic acid. N Engl J Med 322:1775–1780

Petr GT, Sun Y, Frederick NM, Zhou Y, Dhamne SC, Hameed MQ, Miranda C, Bedoya EA, Fischer KD, Armsen W, Wang J, Danbolt NC, Rotenberg A, Aoki CJ, Rosenberg PA (2015) Conditional deletion of the glutamate transporter GLT-1 reveals that astrocytic GLT-1 protects against fatal epilepsy while neuronal GLT-1 contributes significantly to glutamate uptake into synaptosomes. J Neurosci 35:5187–5201

Petroff OA (2002) GABA and glutamate in the human brain. Neuroscientist 8:562–573

Petroff OA, Hyder F, Collins T, Mattson RH, Rothman DL (1999) Acute effects of vigabatrin on brain GABA and homocarnosine in patients with complex partial seizures. Epilepsia 40:958–964

Petroff OA, Errante LD, Rothman DL, Kim JH, Spencer DD (2002) Glutamate-glutamine cycling in the epileptic human hippocampus. Epilepsia 43:703–710

Pisano JJ, Wilson JD, Cohen L, Abraham D, Udenfriend S (1961) Isolation of gamma-aminobutyrylhistidine (homocarnosine) from brain. J Biol Chem 236:499–502

Porter JT, McCarthy KD (1995) GFAP-positive hippocampal astrocytes in situ respond to glutamatergic neuroligands with increases in [Ca2+]i. Glia 13:101–112

Porter JT, McCarthy KD (1996) Hippocampal astrocytes in situ respond to glutamate released from synaptic terminals. J Neurosci 16:5073–5081

Quandt G, Hofner G, Wanner KT (2013) Synthesis and evaluation of N-substituted nipecotic acid derivatives with an unsymmetrical bis-aromatic residue attached to a vinyl ether spacer as potential GABA uptake inhibitors. Bioorg Med Chem 21:3363–3378

Radian R, Bendahan A, Kanner BI (1986) Purification and identification of the functional sodium- and chloride-coupled gamma-aminobutyric acid transport glycoprotein from rat brain. J Biol Chem 261:15437–15441

Rasic-Markovic A, Hrncic D, Djuric D, Macut D, Loncar-Stevanovic H, Stanojlovic O (2011) The effect of N-methyl-D-aspartate receptor antagonists on D, L-homocysteine thiolactone induced seizures in adult rats. Acta Physiol Hung 98:17–26

Rauen T, Rothstein JD, Wassle H (1996) Differential expression of three glutamate transporter subtypes in the rat retina. Cell Tissue Res 286:325–336
Regan MR, Huang YH, Kim YS, Dykes-Hoberg MI, Jin L, Watkins AM, Bergles DE, Rothstein JD (2007) Variations in promoter activity reveal a differential expression and physiology of glutamate transporters by glia in the developing and mature CNS. J Neurosci 27:6607–6619
Ritthausen KH (1866) Über die Glutaminsäure. Journal fur praktische Chemie 42:454–462
Robinson SR (2001) Changes in the cellular distribution of glutamine synthetase in Alzheimer's disease. J Neurosci Res 66:972–980
Rogachev B, Ohayon S, Saad A, Vorobiovsky V, Gruenbaum BF, Leibowitz A, Boyko M, Shapira Y, Shnaider A, Zlotnik M, Azab AN, Zlotnik A (2012) The effects of hemodialysis on blood glutamate levels in chronic renal failure: implementation for neuroprotection. J Crit Care 27(743):741–747
Rogachev B, Tsesis S, Gruenbaum BF, Gruenbaum SE, Boyko M, Klein M, Shapira Y, Vorobiev M, Zlotnik A (2013) The effects of peritoneal dialysis on blood glutamate levels: implementation for neuroprotection. J Neurosurg Anesthesiol 25:262–266
Rogawski MA, Kurzman PS, Yamaguchi SI, Li H (2001) Role of AMPA and GluR5 kainate receptors in the development and expression of amygdala kindling in the mouse. Neuropharmacology 40:28–35
Rogawski MA, Gryder D, Castaneda D, Yonekawa W, Banks MK, Lia H (2003) GluR5 kainate receptors, seizures, and the amygdala. Ann N Y Acad Sci 985:150–162
Ronco C, Tetta C, Mariano F, Wratten ML, Bonello M, Bordoni V, Cardona X, Inguaggiato P, Pilotto L, d'Intini V, Bellomo R (2003) Interpreting the mechanisms of continuous renal replacement therapy in sepsis: the peak concentration hypothesis. Artif Organs 27:792–801
Rothman DL, De Feyter HM, Maciejewski PK, Behar KL (2012) Is there in vivo evidence for amino acid shuttles carrying ammonia from neurons to astrocytes? Neurochem Res 37:2597–2612
Ruban A, Berkutzki T, Cooper I, Mohar B, Teichberg VI (2012) Blood glutamate scavengers prolong the survival of rats and mice with brain-implanted gliomas. Invest New Drugs 30:2226–2235
Ruban A, Mohar B, Jona G, Teichberg VI (2013) Blood glutamate scavenging as a novel neuroprotective treatment for paraoxon intoxication., J Cereb Blood Flow Metab
Ruban A, Biton IE, Markovich A, Mirelman D (2015) MRS of brain metabolite levels demonstrates the ability of scavenging of excess brain glutamate to protect against nerve agent induced seizures. Int J Mol Sci 16:3226–3236
Sano C (2009) History of glutamate production. International Glutamate Information Service, pp 728S–732S
Scalise M, Pochini L, Galluccio M, Indiveri C (2016) Glutamine transport. From energy supply to sensing and beyond, Biochim Biophys Acta
Schneider N, Cordeiro S, Machtens JP, Braams S, Rauen T, Fahlke C (2014) Functional properties of the retinal glutamate transporters GLT-1c and EAAT5. J Biol Chem 289:1815–1824
Schousboe A (1981) Transport and metabolism of glutamate and GABA in neurons are glial cells. Int Rev Neurobiol 22:1–45
Scofield MD, Boger HA, Smith RJ, Li H, Haydon PG, Kalivas PW (2015) Gq-DREADD selectively initiates glial glutamate release and inhibits cue-induced cocaine seeking. Biol Psychiatry 78:441–451
Seifert G, Huttmann K, Schramm J, Steinhauser C (2004) Enhanced relative expression of glutamate receptor 1 flip AMPA receptor subunits in hippocampal astrocytes of epilepsy patients with Ammon's horn sclerosis. J Neurosci 24:1996–2003
Seiler N (1980) On the role of GABA in vertebrate polyamine metabolism. Physiol Chem Phys 12:411–429
Serdyuk SE, Gmiro VE, Veselkina OS (2014) Combined blockade of NMDA and AMPA receptors prevents acute kainate seizures and chronic kainate lethality in rats. Bull Exp Biol Med 157:15–17
Shakarjian MP, Ali MS, Veliskova J, Stanton PK, Heck DE, Velisek L (2015) Combined diazepam and MK-801 therapy provides synergistic protection from tetramethylenedisulfotetramine-induced tonic-clonic seizures and lethality in mice. Neurotoxicology 48:100–108

Shank RP, Bennett GS, Freytag SO, Campbell GL (1985) Pyruvate carboxylase: an astrocyte-specific enzyme implicated in the replenishment of amino acid neurotransmitter pools. Brain Res 329:364–367
Shorvon SD (1996) The epidemiology and treatment of chronic and refractory epilepsy. Epilepsia 37(Suppl 2):S1–S3
Simpson NE, Tryndyak VP, Pogribna M, Beland FA, Pogribny IP (2012) Modifying metabolically sensitive histone marks by inhibiting glutamine metabolism affects gene expression and alters cancer cell phenotype. Epigenetics 7:1413–1420
Solbu TT, Bjorkmo M, Berghuis P, Harkany T, Chaudhry FA (2010) SAT1, a glutamine transporter, is preferentially expressed in GABAergic neurons. Front Neuroanat 4:1
Sommer W (1880) Erkrankung des Ammonshorns als aetiologisches Moment der Epilepsie. Arch Psychiatr Nervenkr 10:631–675
Steffens M, Huppertz HJ, Zentner J, Chauzit E, Feuerstein TJ (2005) Unchanged glutamine synthetase activity and increased NMDA receptor density in epileptic human neocortex: implications for the pathophysiology of epilepsy. Neurochem Int 47:379–384
Stern JR, Eggleston LV, Hems R, Krebs HA (1949) Accumulation of glutamic acid in isolated brain tissue. Biochem J 44:410–418
Stoffel W, Korner R, Wachtmann D, Keller BU (2004) Functional analysis of glutamate transporters in excitatory synaptic transmission of GLAST1 and GLAST1/EAAC1 deficient mice. Brain Res Mol Brain Res 128:170–181
Sun HL, Zhang SH, Zhong K, Xu ZH, Feng B, Yu J, Fang Q, Wang S, Wu DC, Zhang JM, Chen Z (2013) A transient upregulation of glutamine synthetase in the dentate gyrus is involved in epileptogenesis induced by amygdala kindling in the rat. PLoS One 8, e66885
Szepietowska B, Zhu W, Czyzyk J, Eid T, Sherwin RS (2012) EphA5-EphrinA5 interactions within the ventromedial hypothalamus influence counterregulatory hormone release and local glutamine/glutamate balance during hypoglycemia., Diabetes
Tanaka K, Watase K, Manabe T, Yamada K, Watanabe M, Takahashi K, Iwama H, Nishikawa T, Ichihara N, Kikuchi T, Okuyama S, Kawashima N, Hori S, Takimoto M, Wada K (1997) Epilepsy and exacerbation of brain injury in mice lacking the glutamate transporter GLT-1. Science 276:1699–1702
Tellez-Zenteno JF, Patten SB, Jette N, Williams J, Wiebe S (2007) Psychiatric comorbidity in epilepsy: a population-based analysis. Epilepsia 48:2336–2344
Thoeringer CK, Ripke S, Unschuld PG, Lucae S, Ising M, Bettecken T, Uhr M, Keck ME, Mueller-Myhsok B, Holsboer F, Binder EB, Erhardt A (2009) The GABA transporter 1 (SLC6A1): a novel candidate gene for anxiety disorders. J Neural Transm (Vienna) 116:649–657
Thomsen C, Dalby NO (1998) Roles of metabotropic glutamate receptor subtypes in modulation of pentylenetetrazole-induced seizure activity in mice. Neuropharmacology 37:1465–1473
Tsuda M, Suzuki T, Misawa M (1997) Role of the NMDA receptor complex in DMCM-induced seizure in mice. Neuroreport 8:603–606
Tumani H, Shen G, Peter JB, Bruck W (1999) Glutamine synthetase in cerebrospinal fluid, serum, and brain: a diagnostic marker for Alzheimer disease? Arch Neurol 56:1241–1246
Urwyler S, Campbell E, Fricker G, Jenner P, Lemaire M, McAllister KH, Neijt HC, Park CK, Perkins M, Rudin M, Sauter A, Smith L, Wiederhold KH, Muller W (1996) Biphenyl-derivatives of 2-amino-7-phosphono-heptanoic acid, a novel class of potent competitive N-methyl-D-aspartate receptor antagonists--II. Pharmacological characterization in vivoA mechanism for glutamate removal. Neuropharmacology 35:655–669
van den Berg CJ, Garfinkel D (1971) A stimulation study of brain compartments. Metabolism of glutamate and related substances in mouse brain. Biochem J 123:211–218
van der Hel WS, Notenboom RG, Bos IW, van Rijen PC, van Veelen CW, de Graan PN (2005) Reduced glutamine synthetase in hippocampal areas with neuron loss in temporal lobe epilepsy. Neurology 64:326–333
van der Hel WS, Hessel EV, Bos IW, Mulder S, Verlinde SA, van Eijsden P, de Graan PN (2014) Persistent reduction of hippocampal glutamine synthetase expression after status epilepticus in immature rats., Eur J Neurosci

van der Zeyden M, Oldenziel WH, Rea K, Cremers TI, Westerink BH (2008) Microdialysis of GABA and glutamate: analysis, interpretation and comparison with microsensors. Pharmacol Biochem Behav 90:135–147

van Gassen KL, van der Hel WS, Hakvoort TB, Lamers WH, de Graan PN (2009) Haploinsufficiency of glutamine synthetase increases susceptibility to experimental febrile seizures. Genes Brain Behav 8:290–295

van Rooyen JM, Abratt VR, Sewell BT (2006) Three-dimensional structure of a type III glutamine synthetase by single-particle reconstruction. J Mol Biol 361:796–810

Vandenberg RJ, Ryan RM (2013) Mechanisms of glutamate transport. Physiol Rev 93:1621–1657

Varoqui H, Erickson JD (2002) Selective up-regulation of system a transporter mRNA in diabetic liver. Biochem Biophys Res Commun 290:903–908

Veruki ML, Morkve SH, Hartveit E (2006) Activation of a presynaptic glutamate transporter regulates synaptic transmission through electrical signaling. Nat Neurosci 9:1388–1396

Volgushev M, Vidyasagar TR, Chistiakova M, Yousef T, Eysel UT (2000) Membrane properties and spike generation in rat visual cortical cells during reversible cooling. J Physiol 522(Pt 1):59–76

Vossler DG, Morris GL 3rd, Harden CL, Montouris G, Faught E, Kanner AM, Fix A, French JA, Postmarketing Antiepileptic Drug Survey Group study (2013) Tiagabine in clinical practice: effects on seizure control and behavior. Epilepsy Behav 28:211–216

Wafford KA, Kathoria M, Bain CJ, Marshall G, Le Bourdelles B, Kemp JA, Whiting PJ (1995) Identification of amino acids in the N-methyl-D-aspartate receptor NR1 subunit that contribute to the glycine binding site. Mol Pharmacol 47:374–380

Walitza S, Wendland JR, Gruenblatt E, Warnke A, Sontag TA, Tucha O, Lange KW (2010) Genetics of early-onset obsessive-compulsive disorder. Eur Child Adolesc Psychiatry 19:227–235

Wang Y, Zaveri HP, Lee TS, Eid T (2009) The development of recurrent seizures after continuous intrahippocampal infusion of methionine sulfoximine in rats: a video-intracranial electroencephalographic study. Exp Neurol 220:293–302

Watanabe T, Morimoto K, Hirao T, Suwaki H, Watase K, Tanaka K (1999) Amygdala-kindled and pentylenetetrazole-induced seizures in glutamate transporter GLAST-deficient mice. Brain Res 845:92–96

Watanabe Y, Kaida Y, Takechi K, Kamei C (2010) Anticonvulsant effect of (RS)-1-aminoindan-1,5-dicarboxylic acid on pentetrazol-induced kindled seizures in mice. Biol Pharm Bull 33:647–652

Watanabe Y, Kaida Y, Fukuhara S, Takechi K, Uehara T, Kamei C (2011) Participation of metabotropic glutamate receptors in pentetrazol-induced kindled seizure. Epilepsia 52:140–150

Watase K, Hashimoto K, Kano M, Yamada K, Watanabe M, Inoue Y, Okuyama S, Sakagawa T, Ogawa S, Kawashima N, Hori S, Takimoto M, Wada K, Tanaka K (1998) Motor discoordination and increased susceptibility to cerebellar injury in GLAST mutant mice. Eur J Neurosci 10:976–988

White HS, McCabe RT, Armstrong H, Donevan SD, Cruz LJ, Abogadie FC, Torres J, Rivier JE, Paarmann I, Hollmann M, Olivera BM (2000) In vitro and in vivo characterization of conantokin-R, a selective NMDA receptor antagonist isolated from the venom of the fish-hunting snail Conus radiatus. J Pharmacol Exp Ther 292:425–432

Winfree CJ, Baker CJ, Connolly ES Jr, Fiore AJ, Solomon RA (1996) Mild hypothermia reduces penumbral glutamate levels in the rat permanent focal cerebral ischemia model. Neurosurgery 38:1216–1222

Xu YF, Cai YQ, Cai GQ, Jiang J, Sheng ZJ, Wang ZG, Fei J (2008) Hypoalgesia in mice lacking GABA transporter subtype 1. J Neurosci Res 86:465–470

Yamauchi A, Uchida S, Kwon HM, Preston AS, Robey RB, Garcia-Perez A, Burg MB, Handler JS (1992) Cloning of a Na(+)- and Cl(-)-dependent betaine transporter that is regulated by hypertonicity. J Biol Chem 267:649–652

Yang XF, Ouyang Y, Kennedy BR, Rothman SM (2005) Cooling blocks rat hippocampal neurotransmission by a presynaptic mechanism: observations using 2-photon microscopy. J Physiol 567:215–224

Yen W, Williamson J, Bertram EH, Kapur J (2004) A comparison of three NMDA receptor antagonists in the treatment of prolonged status epilepticus. Epilepsy Res 59:43–50

Yourick DL, Repasi RT, Rittase WB, Staten LD, Meyerhoff JL (1999) Ifenprodil and arcaine alter amygdala-kindling development. Eur J Pharmacol 371:147–152
Yudkoff M (1997) Brain metabolism of branched-chain amino acids. Glia 21:92–98
Yudkoff M, Daikhin Y, Nissim I, Lazarow A, Nissim I (2004) Ketogenic diet, brain glutamate metabolism and seizure control. Prostaglandins Leukot Essent Fatty Acids 70:277–285
Zarnowski T, Kleinrok Z, Turski WA, Czuczwar SJ (1994) The NMDA antagonist procyclidine, but not ifenprodil, enhances the protective efficacy of common antiepileptics against maximal electroshock-induced seizures in mice. J Neural Transm Gen Sect 97:1–12
Zeiler FA (2015) Early use of the NMDA receptor antagonist ketamine in refractory and superrefractory status epilepticus. Crit Care Res Pract 2015:831260
Zeiler FA, Zeiler KJ, Teitelbaum J, Gillman LM, West M (2015) Therapeutic hypothermia for refractory status epilepticus. Can J Neurol Sci 42:221–229
Zhou Y, Danbolt NC (2013) GABA and glutamate transporters in brain. Front Endocrinol (Lausanne) 4:165
Zhou Y, Danbolt NC (2014) Glutamate as a neurotransmitter in the healthy brain. J Neural Transm (Vienna) 121:799–817
Zhou Y, Holmseth S, Guo C, Hassel B, Hofner G, Huitfeldt HS, Wanner KT, Danbolt NC (2012a) Deletion of the gamma-aminobutyric acid transporter 2 (GAT2 and SLC6A13) gene in mice leads to changes in liver and brain taurine contents. J Biol Chem 287:35733–35746
Zhou Y, Holmseth S, Hua R, Lehre AC, Olofsson AM, Poblete-Naredo I, Kempson SA, Danbolt NC (2012b) The betaine-GABA transporter (BGT1, slc6a12) is predominantly expressed in the liver and at lower levels in the kidneys and at the brain surface. Am J Physiol Renal Physiol 302:F316–F328
Zhou Y, Waanders LF, Holmseth S, Guo C, Berger UV, Li Y, Lehre AC, Lehre KP, Danbolt NC (2014) Proteome analysis and conditional deletion of the EAAT2 glutamate transporter provide evidence against a role of EAAT2 in pancreatic insulin secretion in mice. J Biol Chem 289:1329–1344
Zhumadilov A, Boyko M, Gruenbaum SE, Brotfain E, Bilotta F, Zlotnik A (2015) Extracorporeal methods of blood glutamate scavenging: a novel therapeutic modality. Expert Rev Neurother 15:501–508
Zlotnik A, Gurevich B, Tkachov S, Maoz I, Shapira Y, Teichberg VI (2007) Brain neuroprotection by scavenging blood glutamate. Exp Neurol 203:213–220
Zlotnik A, Gurevich B, Cherniavsky E, Tkachov S, Matuzani-Ruban A, Leon A, Shapira Y, Teichberg VI (2008) The contribution of the blood glutamate scavenging activity of pyruvate to its neuroprotective properties in a rat model of closed head injury. Neurochem Res 33:1044–1050
Zlotnik A, Gruenbaum SE, Artru AA, Rozet I, Dubilet M, Tkachov S, Brotfain E, Klin Y, Shapira Y, Teichberg VI (2009) The neuroprotective effects of oxaloacetate in closed head injury in rats is mediated by its blood glutamate scavenging activity: evidence from the use of maleate. J Neurosurg Anesthesiol 21:235–241
Zlotnik A, Sinelnikov I, Gruenbaum BF, Gruenbaum SE, Dubilet M, Dubilet E, Leibowitz A, Ohayon S, Regev A, Boyko M, Shapira Y, Teichberg VI (2012) Effect of glutamate and blood glutamate scavengers oxaloacetate and pyruvate on neurological outcome and pathohistology of the hippocampus after traumatic brain injury in rats. Anesthesiology 116:73–83

About the Editors

Arne Schousboe obtained his M.Sc. in biochemistry from the University of Copenhagen in 1968 at a time where an M.Sc. was considered equivalent to a Ph.D. He subsequently (1978) earned his Doctor of Science (D.Sc.) degree also from the University of Copenhagen. After a postdoc period with Dr. Eugene Roberts at the Department of Neuroscience, City of Hope National Medical Center in Los Angeles 1972–1973, he came back to the University of Copenhagen to resume a tenured position as Associate Professor at the Medical Faculty. In 1990 he moved to the Royal Danish School of Pharmacy, Copenhagen, as a Full Professor of Biochemistry and has remained in this position also when this Institution changed to become an independent University and subsequently a Faculty of Pharmaceutical Sciences, University of Copenhagen, which lately was incorporated in the Faculty of Health and Medical Sciences still at the University of Copenhagen. He has served as the Department Chair from 2005 to 2010 and is now part-time Professor of Neuropharmacology in the Department of Drug Design and Pharmacology at the University of Copenhagen. He has been working on astrocyte function focusing on amino acid neurotransmission during the past more than 40 years and is currently

A. Schousboe, U. Sonnewald (eds.), *The Glutamate/GABA-Glutamine Cycle*, Advances in Neurobiology 13, DOI 10.1007/978-3-319-45096-4

engaged in studies of glutamate and GABA homeostasis and metabolism. He has published over 550 papers on these and related topics. He has served on the Editorial Board on numerous neuroscience journals over the years and is currently the Editor-in-Chief of Neurochemical Research and Editor of the book series Advances in Neurobiology.

Ursula Sonnewald received her Ph.D. in Chemistry at the University of Saskatchewan, Saskatoon (Canada), and subsequent to studies in Hamburg (Germany) and Canada; postdoctoral periods were spent in Salt Lake City, University of Utah, USA, in Copenhagen at the Carlsberg Research Center, Denmark, and at The Brookhaven National Laboratory in Upton, New York, USA. For 5 years she worked in the CNS Department of Novo Nordisk developing antiepileptic drugs. During this period she synthesized the GABA uptake inhibitor Tiagabine, which is on the market as an antiepileptic drug and as a tool for basic research on GABA function and metabolism. In 1990 she moved to Norway and was amongst the first in the area of metabolic studies of brain cell metabolism using magnetic resonance spectroscopy (MRS) in collaboration with the experts (Drs. Arne Schousboe and Niels Westergaard) on nervous cell cultures. Since 1996 she has been a full Professor in the Department of Neuroscience at the Norwegian University of Science and Technology. She applied MRS to animal models of epilepsy, stroke, schizophrenia, Alzheimer's, and other neurological disorders. Exciting new results have been obtained pertinent to the area of neurodegeneration. She has developed tools to analyze mitochondrial function and has applied those to study dysfunction in the aforementioned diseases.

Index

A. Schousboe, U. Sonnewald (eds.), *The Glutamate/GABA-Glutamine Cycle*,
Advances in Neurobiology 13, DOI 10.1007/978-3-319-45096-4

H

I

N

W

www.ingramcontent.com/pod-product-compliance
Ingram Content Group UK Ltd.
Pitfield, Milton Keynes, MK11 3LW, UK
UKHW022216051026
15624UKWH00001B/188
9783319450940